Lecture Notes in Physics

Editorial Board

R. Beig, Wien, Austria
B.-G. Englert, Ismaning, Germany
U. Frisch, Nice, France
P. Hänggi, Augsburg, Germany
K. Hepp, Zürich, Switzerland
W. Hillebrandt, Garching, Germany
D. Imboden, Zürich, Switzerland
R. L. Jaffe, Cambridge, MA, USA
R. Lipowsky, Golm, Germany
H. v. Löhneysen, Karlsruhe, Germany
I. Ojima, Kyoto, Japan
D. Sornette, Nice, France, and Los Angeles, CA, USA
S. Theisen, Golm, Germany
W. Weise, Trento, Italy, and Garching, Germany
J. Wess, München, Germany
J. Zittartz, Köln, Germany

Springer-Verlag Berlin Heidelberg GmbH

Physics and Astronomy ONLINE LIBRARY

http://www.springer.de/phys/

Editorial Policy

The series *Lecture Notes in Physics* (LNP), founded in 1969, reports new developments in physics research and teaching -- quickly, informally but with a high quality. Manuscripts to be considered for publication are topical volumes consisting of a limited number of contributions, carefully edited and closely related to each other. Each contribution should contain at least partly original and previously unpublished material, be written in a clear, pedagogical style and aimed at a broader readership, especially graduate students and nonspecialist researchers wishing to familiarize themselves with the topic concerned. For this reason, traditional proceedings cannot be considered for this series though volumes to appear in this series are often based on material presented at conferences, workshops and schools (in exceptional cases the original papers and/or those not included in the printed book may be added on an accompanying CD ROM, together with the abstracts of posters and other material suitable for publication, e.g. large tables, colour pictures, program codes, etc.).

Acceptance

A project can only be accepted tentatively for publication, by both the editorial board and the publisher, following thorough examination of the material submitted. The book proposal sent to the publisher should consist at least of a preliminary table of contents outlining the structure of the book together with abstracts of all contributions to be included.
Final acceptance is issued by the series editor in charge, in consultation with the publisher, only after receiving the complete manuscript. Final acceptance, possibly requiring minor corrections, usually follows the tentative acceptance unless the final manuscript differs significantly from expectations (project outline). In particular, the series editors are entitled to reject individual contributions if they do not meet the high quality standards of this series. The final manuscript must be camera-ready, and should include both an informative introduction and a sufficiently detailed subject index.

Contractual Aspects

Publication in LNP is free of charge. There is no formal contract, no royalties are paid, and no bulk orders are required, although special discounts are offered in this case. The volume editors receive jointly 30 free copies for their personal use and are entitled, as are the contributing authors, to purchase Springer books at a reduced rate. The publisher secures the copyright for each volume. As a rule, no reprints of individual contributions can be supplied.

Manuscript Submission

The manuscript in its final and approved version must be submitted in camera-ready form. The corresponding electronic source files are also required for the production process, in particular the online version. Technical assistance in compiling the final manuscript can be provided by the publisher's production editor(s), especially with regard to the publisher's own Latex macro package which has been specially designed for this series.

Online Version/ LNP Homepage

LNP homepage (list of available titles, aims and scope, editorial contacts etc.):
http://www.springer.de/phys/books/lnpp/
LNP online (abstracts, full-texts, subscriptions etc.):
http://link.springer.de/series/lnpp/

F. Scheck H. Upmeier W. Werner (Eds.)

Noncommutative Geometry and the Standard Model of Elementary Particle Physics

Springer

Editors

Florian Scheck
Johannes Gutenberg-Universität Mainz
Institut für Physik, Theoretische
Elementarteilchenphysik
Staudinger Weg 7
55099 Mainz, Germany

Harald Upmeier
Philipps-Universität Marburg
Fachbereich Mathematik
und Informatik
Hans-Merwein Strasse
35032 Marburg, Germany

Wend Werner
Westfälische Wilhelms-Universität
Münster
Mathematisches Institut
Einsteinstrasse 62
48149, Germany

Cover Picture: (Determination of c_A and c_V from $e^+e^- \to F^+F^-$ at the Z pole.
From: F. Scheck, Electroweak and Strong Interactions (Springer 1996))

Cataloging-in-Publication Data applied for

A catalog record for this book is available from the Library of Congress.

Bibliographic information published by Die Deutsche Bibliothek

Die Deutsche Bibliothek lists this publication in the Deutsche Nationalbibliografie; detailed bibliographic data is available in the Internet at http://dnb.ddb.de

CR Subject Classification (1998): D.4.5, E.3, C.2.0, H.2.0, K.6.5, K.4.4

ISSN 0075-8450
ISBN 978-3-642-07897-2   ISBN 978-3-540-46082-4 (eBook)
DOI 10.1007/978-3-540-46082-4

This work is subject to copyright. All rights are reserved, whether the whole or part of the material is concerned, specifically the rights of translation, reprinting, reuse of illustrations, recitation, broadcasting, reproduction on microfilm or in any other way, and storage in data banks. Duplication of this publication or parts thereof is permitted only under the provisions of the German Copyright Law of September 9, 1965, in its current version, and permission for use must always be obtained from
Springer-Verlag Berlin Heidelberg GmbH .
Violations are liable for prosecution under the German Copyright Law.
http://www.springer.de

© Springer-Verlag Berlin Heidelberg 2002
Originally published by Springer-Verlag Berlin Heidelberg New York in 2002
Softcover reprint of the hardcover 1st edition 2002

The use of general descriptive names, registered names, trademarks, etc. in this publication does not imply, even in the absence of a specific statement, that such names are exempt from the relevant protective laws and regulations and therefore free for general use.

Typesetting: Camera-ready by the authors/editor
Camera-data conversion by Steingraeber Satztechnik GmbH Heidelberg
Cover design: *design & production*, Heidelberg

Printed on acid-free paper
SPIN: 10887894   54/3141/du - 5 4 3 2 1 0

Preface

A conference on "Noncommutative Geometry and the Standard Model of Elementary Particle Physics" was held at the Hesselberg Academy (in northern Bavaria, Germany) during the week of March 14–19, 1999. The aim of the conference was to give a systematic exposition of the mathematical foundations and physical applications of noncommutative geometry, along the lines developed by Alain Connes. The conference was actually part of a continuing series of conferences at the Hesselberg Academy held every three years and devoted to important developments in mathematical fields, such as geometric analysis, operator algebras, index theory, and related topics together with their applications to mathematical physics.

The participants of the conference included mathematicians from functional analysis, differential geometry and operator algebras, as well as experts from mathematical physics interested in A. Connes' approach towards the standard model and other physical applications. Thus a large range of topics, from mathematical foundations to recent physical applications, could be covered in a substantial way. The proceedings of this conference, organized in a coherent and systematic way, are presented here. Its three chapters correspond to the main areas discussed during the conference:

Chapter 1. Foundations of Noncommutative Geometry and Basic Model Building

Chapter 2. The Lagrangian of the Standard Model Derived from Noncommutative Geometry

Chapter 3. New Directions in Noncommutative Geometry and Mathematical Physics

During the conference the close interaction between mathematicians and mathematical physicists turned out to be quite fruitful and enlightening for both sides. Similarly, it is hoped that the proceedings presented here will be useful for mathematicians interested in basic physical questions and for physicists aiming at a more conceptual understanding of classical and quantum field theory from a novel mathematical point of view. Of course, the whole field is under active development and important aspects, such as field theory of extended objects or advances in renormalization theory, could only be touched upon here; these may be covered in subsequent meetings of the Hesselberg conference series.

From the beginning, the Hesselberg conference series was made possible through financial support from the Volkswagenstiftung. This support is gratefully acknowledged. Thanks are due Martin Briegel, Thomas Eckert and Monika Teubner of the University of Marburg for typing large parts of the manuscript and preparing the final version of the proceedings, and to the participants for their careful work with the individual contributions.

Mainz, Paderborn, Marburg,
2002

Florian Scheck
Wend Werner
Harald Upmeier

List of Contributors

Peter M. Alberti
Wissenschaftspark
Leipzig/Permoserstrasse e.V.
Permoserstraße 15
04318 Leipzig
Germany
alberti@wp.ufz.de

Bernd Ammann
Universität Hamburg
Fachbereich Mathematik
Bundesstrasse 55
20146 Hamburg
Germany

Christian Bär
Universität Hamburg
Fachbereich Mathematik
Bundesstrasse 55
20146 Hamburg
Germany
baer@math.uni-hamburg.de

Karen Elsner
Max-Planck-Institut für Physik
Foehringer Ring 6
80805 München
Germany
elsner@mppmu.mpg.de

Michael Frank
Media and Communication Systems
Software (MCS) GmbH
Altenburger Str. 5
04275 Leipzig
Germany
Michael.Frank@mcs-systemhaus.de

Klaus Fredenhagen
Universität Hamburg
II. Institut für Theoretische Physik
Luruper Chaussee 149
22761 Hamburg
Germany
klaus.fredenhagen@desy.de

Rainer Häußling
Johannes Gutenberg-Universität
Institut für Physik, Theoretische
Elementarteilchenphysik
Staudinger Weg 7
55099 Mainz
Germany

Ralf Holtkamp
Ruhr-Universität Bochum
Fakultät für Mathematik NA 2/32
Universitätsstrasse 150
44780 Bochum
Germany
ralf.holtkamp@ruhr-uni-bochum.de

Bruno Iochum
CNRS, Centre de Physique Theorique
Luminy, case 907
13288 Marseille Cedex 9
France
Bruno.Iochum@cpt.univ-mrs.fr

Giovanni Landi
Universitá di Trieste
Dipartimento di
Scienze Matematiche
Via Valerio 12/b
34127 Trieste
Italy
landi@univ.trieste.it

List of Contributors

Edwin Langmann
Royal Institute of Technology
Mathematical Physics, Department of Physics
SCFAB
10691 Stockholm
Sweden
langmann@theophys.kth.se

Fedele Lizzi
Universitá di Napoli Federico II
C.U. Monte S. Angelo
Via Cintia
80126 Napoli
Italy
fedele.lizzi@na.infn.it

Rainer Matthes
Universität Leipzig
Institut für Theoretische Physik
Augustusplatz 10-11
04109 Leipzig
Germany

Ralf Meyer
Westfälische Wilhelms-Universität Münster
Mathematisches Institut
Einsteinstrasse 62
48149 Münster
Germany
rameyer@math.uni-muenster.de

Ryszard Nest
Institute for Mathematical Sciences
Universitetsparken 5
2100 Kopenhagen Ø
Danemark

Holger Neumann
Philipps-Universität Marburg
Fachbereich Physik
Renthof 5
35032 Marburg
Germany
Holger.Neumann@physik.uni-marburg.de

Florian Scheck
Johannes Gutenberg-Universität
Institut für Physik, Theoretische Elementarteilchenphysik
Street No.
55099 Mainz
Germany
Scheck@thep.physik.uni-mainz.de

Harald Upmeier
Universität Marburg
Fachbereich Mathematik und Informatik
Hans-Meerwein-Strasse, Lahnberge
35032 Marburg
Germany
upmeier@mathematik.uni-marburg.de

Elmar Vogt
Freie Universität Berlin
Fachbereich Mathematik und Informatik
Arnimallee 3
14195 Berlin

Wend Werner
Universität Münster
Fachbereich Mathematik und Informatik
Hittorfstrasse 27
48149 Münster

Raimar Wulkenhaar
Max-Planck-Institut für Mathematik in den Naturwissenschaften
Inselstrasse 22-26
04103 Leipzig
Germany
Raimar.Wulkenhaar@mis.mpg.de

Contents

Part I. Foundations of Noncommutative Geometry and Basic Model Building

Ralf Holtkamp
1 Spectral Triples and Abstract Yang-Mills Functional 4
   1.1 Spectral Triples 4
   1.2 Universal Differential Graded Algebra 5
   1.3 Vector Potentials, Universal Connections. 5
   1.4 Quotient Differential Graded Algebra. 6
   1.5 Inner Product 7
   1.6 Curvature and Yang-Mills Functional. 8

Ralf Meyer
2 Real Spectral Triples and Charge Conjugation 11
   2.1 Real Structures on Even Spectral Triples 11
   2.2 Spin$^{\mathbb{C}}$ Manifolds and Charge Conjugation 13
   2.3 Real Structures via Clifford Algebras 15
   2.4 Real Structures of Odd Dimension 17
   2.5 Relations to Real K-Homology 18
   2.6 Real Structures on the NC Torus 20

Michael Frank
3 The Commutative Case:
   Spinors, Dirac Operator and de Rham Algebra 21
   3.1 The Theorems by Gel'fand and Serre-Swan 21
   3.2 Hermitean Structures and Frames for Sets of Sections 26
   3.3 Clifford and Spinor Bundles, Spin Manifolds 28
   3.4 Spin Connection and Dirac Operator 31
   3.5 The Universal Differential Algebra $\Omega C^{\infty}(M)$
       and Connes' Differential Algebra $\Omega_{\mathcal{D}} C^{\infty}(M)$. 33
   3.6 The Exterior Algebra Bundle $\Lambda(M)$ and the de Rham Complex 35
   3.7 $\Omega_{\mathcal{D}} C^{\infty}(M)$ Versus $\Lambda(M)$ 36

Peter M. Alberti and Reiner Matthes
4 Connes' Trace Formula and Dirac Realization
   of Maxwell and Yang-Mills Action 40
   4.1 Generalities on Traces on C*- and W*-algebras 40
   4.2 Examples of Traces 43
   4.3 Examples of Singular Traces on $B(\mathcal{H})$ 49
   4.4 Calculating the Dixmier Trace 56
   4.5 The Connes' Trace Theorem
       and its Application, Preliminaries 60
   4.6 Connes' Trace Theorem 64
   4.7 Classical Yang-Mills Actions 72

Bernd Ammann and Christian Bär
5 The Einstein-Hilbert Action as a Spectral Action 75
   5.1 Generalized Laplacians and the Heat Equation 75
   5.2 The Formal Heat Kernel 80
   5.3 Dirac Operators and Weitzenböck Formulas 88
   5.4 Integration and Dixmier Trace 91
   5.5 Variational Formulas and the Einstein-Hilbert Action 93
   5.6 Einstein-Hilbert Action and Wodzicki Residue 101

Ryszard Nest, Elmar Vogt, and Wend Werner
6 Spectral Action and the Connes-Chamsedinne Model 109
   6.1 The Spectral Action Principle 109
   6.2 Example: Gravity Coupled to One Gauge Field 111
   6.3 Asymptotic Expansion 113
   6.4 First Example, Final Calculation 117
   6.5 Gravity Coupled to the Standard Model 127

Part II. The Lagrangian of the Standard Model
Derived from Noncommutative Geometry

Herald Upmeier
7 Dirac Operator and Real Structure
   on Euclidean and Minkowski Spacetime 136
   7.1 γ-Matrices on Flat and Curved Spacetime 136
   7.2 Levi-Civita Connection and Dirac Operator 144
   7.3 Real Structure on Spacetime 147
   7.4 Trace Formulas and Inner Products 150

Karen Elsner, Holger Neumann, and Harald Upmeier
8 The Electro-weak Model 152
   8.1 Noncommutative Matter Fields 152
   8.2 Noncommutative Gauge Fields 155
   8.3 Noncommutative Gauge Action Functional 165
   8.4 Noncommutative Matter Action Functional 170

Karen Elsner, Holger Neumann, and Harald Upmeier
9 The Full Standard Model 172
   9.1 Noncommutative Matter Fields 172
   9.2 Noncommutative Gauge Fields 179
   9.3 Noncommutative Gauge Action Functional 206
   9.4 Noncommutative Matter Action Functional 211

Holger Neumann and Harald Upmeier
10 Standard Model Coupled with Gravity 216
    10.1 Generalized Dirac Operators 216
    10.2 Spectral Action and Heat Kernel Invariants 224

Florian Scheck
11 The Higgs Mechanism and Spontaneous Symmetry Breaking 230
    11.1 Historical Note 230
    11.2 Spontaneous Symmetry Breaking and Goldstone Theorem . 232
    11.3 Spontaneous Symmetry Breaking in Yang-Mills Theory .. 234
    11.4 The Case of the Electroweak Model: Bosonic Sector ... 235
    11.5 Electroweak Model: Adding Quarks and Leptons 238
    11.6 Remarks About Fermionic Mass Generation 240

Part III. New Directions in Noncommutative Geometry and Mathematical Physics

Bruno Iochum
12 The Impact of NC Geometry in Particle Physics 244
    12.1 Why Noncommutative Geometry? 244
    12.2 Spectral Triples 245
    12.3 Technical Points 247
    12.4 The Noncommutative Highway 248
    12.5 Computation of Higgs and W Masses 252
    12.6 Parameter Counting 253
    12.7 The Renormalization Machinery 255
    12.8 Noncommutative Relativity 257
    12.9 Conclusions 258

Rainer Häussling
13 The $su(2|1)$ Model of Electroweak Interactions
    and Its Connection to NC Geometry 260
    13.1 Introduction and Motivation 260
    13.2 The Bosonic Part of the Model 260
    13.3 The Fermionic Part of the Model 267
    13.4 The Connection to the Connes-Lott Model 269
    13.5 Conclusions 270

Klaus Fredenhagen
14 Quantum Fields and Noncommutative Spacetime 271
    14.1 Noncommutative Spacetime and Uncertainty Relations 271
    14.2 Noncommutative Spacetime and Quantum Field Theory 273
    14.3 Interactions and Noncommutative Geometry 274
    14.4 Gauge Theories on Noncommutative Spacetime 276

Edwin Langmann
15 NC Geometry and Quantum Fields: Simple Examples 278
    15.1 Introduction 278
    15.2 Preliminaries 279
    15.3 Story I: Chern-Simons Terms from Effective Actions 284
    15.4 Story II: Regularization: Elementary Examples 286
    15.5 Story III: Regularized Traces of Operators 288
    15.6 Story IV: Yang-Mills Actions from Dirac Operators 294
    15.7 Final Remarks 297

Giovanny Landi
16 Dirac Eigenvalues as Dynamical Variables 299
    16.1 Introduction 299
    16.2 Noncommutative Geometry and Gravity 300
    16.3 From the Metric to the Eigenvalues 303
    16.4 Action and Field Equations 307
    16.5 Poisson Brackets for the Eigenvalues 309
    16.6 Final Remarks 311

Raimar Wulkenhaar
17 Hopf Algebras in Renormalization and NC Geometry 313
    17.1 Introductory Remarks 313
    17.2 The Hopf Algebra of Connes–Moscovici 313
    17.3 Rooted Trees 317
    17.4 Feynman Graphs and Rooted Trees 319
    17.5 A Toy Model: Iterated Integrals 321

Fedele Lizzi
18 NC Geometry of Strings and Duality Symmetry 325
    18.1 String Theory and T-duality 325
    18.2 Interacting Strings and Spectral Triples 328
    18.3 Compactification and Noncommutative Torus 333
    18.4 Noncommutative Configuration Space
         and Spectral Geometry 334
    18.5 Conclusions 337

References 338

Part I

Foundations of Noncommutative Geometry and Basic Model Building

This part is devoted to a systematic exposition of the mathematical background necessary for noncommutative geometry and its applications. As a characteristic feature, noncommutative geometry involves algebra, functional analysis and differential geometry in an essential way. Our exposition reflects these various fields. Among the main topics discussed in the following are

- the concept of spectral triple in operator algebras and Fredholm theory,
- analysis of traces (Dixmier trace) in pseudo-differential calculus,
- the heat kernel expansion from a differential-geometric point of view,
- analytic treatment of cut-off functions and their Laplace transforms.

In line with the emphasis towards physics, these mathematical concepts are then used to express the Lagrangian densities of gauge theory (Maxwell and Yang-Mills equations) and general relativity (Einstein-Hilbert action) using the framework of noncommutative geometry. In this way, the basic methods are introduced and applied to the fundamental concepts of physics, without the full algebraic details of the physically realistic models involving fermions.

1 Spectral Triples and Abstract Yang-Mills Functional

Ralf Holtkamp

University of Bochum

The aim of the following note is to give a short introduction to the notions of curvature and Yang-Mills functional for vector potentials ("case α") and for connections on Hermitian finitely generated projective modules ("case β"). At first we recall some foundations of NCG which are needed. We will closely follow [38].

1.1 Spectral Triples

Definition 1 ([38, IV.2.11]). Let $\mathcal{A}$ be a (not necessarily commutative) involutive algebra, $*$-algebra for short. We assume that we have given

(i) a $*$-representation $\mathcal{A} \to \mathcal{L}(\mathcal{H})$ on (separable) Hilbert space $\mathcal{H}$, and
(ii) an unbounded selfadjoint operator D such that $[D, a]$ is bounded (for all $a \in \mathcal{A}$) and such that for all $a \in \mathcal{A}$, $\lambda \notin \mathbb{R}$, $a(D - \lambda)^{-1} \in \mathcal{K}(\mathcal{H})$, i.e. D has compact resolvent.

Then we call $(\mathcal{H}, D)$ a K-cycle over $\mathcal{A}$ and $(\mathcal{A}, \mathcal{H}, D)$ a spectral triple.

Spectral triples are sometimes also called unbounded Fredholm modules.
We will always assume that $\mathcal{A}$ is unitary and 1 acts as identity, $\mathcal{A}$ is a subalgebra of $\mathcal{L}(\mathcal{H})$. Let $U := \{u \in \mathcal{A} : u^*u = uu^* = 1\}$ be the unitary group.
We furthermore assume that $(\mathcal{H}, D)$ is (p, ∞)-summable, i.e. λ_n, the eigenvalues (with multiplicity) of $|D| = \sqrt{D^*D}$, are of order $O(n^{1/p})$. Equivalently, the eigenvalues μ_n of $|D|^{-1}$ are of order $O(n^{-1/p})$.
Then p is the dimension of the spectral triple.

Example 1. For the commutative case, where $\mathcal{A} = C^\infty(M)$, M a compact Riemannian spin manifold, we regard $D = \partial_M$ the Dirac operator, and $\mathcal{H}$ the space of L^2-spinors. This is discussed in section 3.

Remark 1. We will later on define the Yang-Mills functional for connections on Hermitian vector bundles.

Thus instead of (the trivial bundle) $\mathcal{A}$, we are also going to study the more general case, where $\mathcal{E}$ is a finitely generated projective module over a $*$-algebra $\mathcal{A}$, i.e. $\mathcal{E} = e\mathcal{A}^N$ is a direct summand of free module $\mathcal{A}^N$, $e^2 = e = e^* \in M_N(\mathcal{A})$, with an Hermitian structure induced by

$$\langle \xi, \eta \rangle = \sum \xi_i^* \eta_i \quad (\text{all } \xi = (\xi_i), \eta = (\eta_i) \in \mathcal{A}^N).$$

Here a Hermitian structure is defined as follows:

Definition 2 ([38, VI.1.7]). A Hermitian structure on $\mathcal{E}$ is given by a sesquilinear mapping $\langle\,,\,\rangle : \mathcal{E} \times \mathcal{E} \to \mathcal{A}$ such that

(i) $\langle \xi a, \eta b \rangle = a^* \langle \xi, \eta \rangle b$ (all $\xi, \eta \in \mathcal{E}, a, b \in \mathcal{A}$)
(ii) $\langle \xi, \xi \rangle \geq 0$ (all $\xi \in \mathcal{E}$)
(iii) $\mathcal{E}$ is selfdual for $\langle\,,\,\rangle$.

We can replace $\mathcal{H}$ by $\mathcal{E} \otimes_\mathcal{A} \mathcal{H}$ and we can consider the following inner product: $\langle \xi \otimes h, \xi' \otimes h' \rangle = \langle \langle \xi, \xi' \rangle h, h' \rangle$ (and replace D by D_∇, see below).

Now $\mathrm{End}_\mathcal{A}(\mathcal{E}) \cong e\mathcal{A}^N \otimes_\mathcal{A} \mathrm{Hom}_\mathcal{A}(e\mathcal{A}^N, \mathcal{A})$ is $*$-algebra isomorphic to $eM_N(\mathcal{A})e$ and is equipped with the inner product $\langle \xi \otimes \eta^* \xi' \otimes \eta'^* \rangle = \langle\!\langle \xi, \xi' \rangle \eta, \eta' \rangle_\mathcal{A}$. Let
$$U(\mathcal{E}) := \{u \in \mathrm{End}_\mathcal{A}(\mathcal{E}) : u^*u = uu^* = 1\}.$$

In the following, we will refer to this more general case by the name "case β", while "case α" stands for the trivial case $\mathcal{E} = \mathcal{A}$.

1.2 Universal Differential Graded Algebra

Definition 3 ([38, III.1.α.3]). The universal differential graded algebra over $\mathcal{A}$, denoted by $\Omega^*(\mathcal{A})$, with $\Omega^0(\mathcal{A}) = \mathcal{A}$, is the graded $\mathcal{A}$-algebra generated by $\{da : a \in \mathcal{A}\}$, all da of degree 1, subject to the relations

$$d(ab) = (da)b + a(db) \quad (\text{all } a, b \in \mathcal{A})$$
$$d1 = 0.$$

The differential is defined by $d : \Omega^*(\mathcal{A}) \to \Omega^*(\mathcal{A})$, $d(a^0 da^1 \cdots da^n) = da^0 da^1 \cdots da^n$ (all $a^j \in \mathcal{A}$).

We have the following rules:
$$d^2\omega = 0 \quad (\text{all } \omega \in \Omega^*(\mathcal{A}))$$
$$d(\omega_1 \omega_2) = (d\omega_1)\omega_2 + (-1)^{\deg \omega_1} \omega_1 (d\omega_2) \quad (\text{all } \omega_j \in \Omega^*(\mathcal{A})).$$

$\Omega^*(\mathcal{A})$ is a $*$-algebra:
$$(da)^* = -da^*.$$

It is well known that the $\mathcal{A}$-bimodule $\Omega^1 \mathcal{A}$ is isomorphic to $\ker(m : \mathcal{A} \otimes \mathcal{A} \to \mathcal{A})$ via the map $\sum a_i \otimes b_i \mapsto \sum a_i(db_i)$ (e.g. $1 \otimes f - f \otimes 1 \mapsto df$).

1.3 Vector Potentials, Universal Connections

Proposition 1 ([38, VI.1.4(1)]). *We can extend the $*$-representation of $\mathcal{A}$ to a $*$-representation π of $\Omega^*(\mathcal{A})$ on $\mathcal{H}$ (via D) by*

$$\pi(a^0 da^1 \ldots da^n) = a^0 [D, a^1] \ldots [D, a^n] \quad (\text{all } a^j \in \mathcal{A})$$

Note that $\pi\bigl(-(da^{1*})a^{0*}\bigr) = [a^{1*}, D]a^{0*}$.

We have $\pi(\Omega^1) = \{V = \sum a_j^0[D, a_j^1] : a_j^0, a_j^1 \in \mathcal{A}\}$, and a selfadjoint element of $\pi(\Omega^1)$ is called vector potential.

For $u \in U$, let $\omega_u := u[D, u^*]$, $uDu^* = D + \omega_u \in \mathcal{L}(\mathcal{H})$.

Then U acts on vector potentials by

$$\gamma_u(V) = u[D, u^*] + uVu^* \qquad (D + \gamma_u(V) = u(D + V)u^*).$$

We are going to need two definitions for the "case β" ($\nabla : \mathcal{E} \to \mathcal{E} \otimes_\mathcal{A} \Omega^1$ instead of $(d + V)a \in \Omega^1 = \mathcal{A} \otimes_\mathcal{A} \Omega^1$):

Let $\mathcal{E}$ be a finitely generated projective module over a $*$-algebra $\mathcal{A}$ as above.

Definition 4 ([38, VI.1.10]). A universal compatible connection on $\mathcal{E}$ is a linear map $\nabla : \mathcal{E} \to \mathcal{E} \otimes_\mathcal{A} \Omega^1$ such that

a) $\nabla(\xi a) = (\nabla \xi)a + \xi \otimes da$ (all $\xi \in \mathcal{E}, a \in \mathcal{A}$)
b) $\langle \xi, \nabla \eta \rangle - \langle \nabla \xi, \eta \rangle = d\langle \xi, \eta \rangle$ (all $\xi, \eta \in \mathcal{E}$)

where for $\nabla \xi = \sum \xi_i \omega_i$: $\langle \nabla \xi, \eta \rangle = \sum \omega_i^* \langle \xi_i, \eta \rangle$.

We denote by $CC(\mathcal{E})$ the space of universal compatible connections on $\mathcal{E}$.

Example 2. The Grassmannian ∇_0, $\nabla_0((\xi_j)) = e(d\xi_j)$, is an element of $CC(\mathcal{E})$: $\nabla_0((\xi_j a)) = e(d(\xi_j a)) = e(d\xi_j)a + (\xi_j) \, da$ and $d\langle \xi, \eta \rangle = \sum(d\xi_j^*)\eta_j + \sum(d\eta_j)^*\xi_j = -\langle e(d\xi_j), \eta \rangle + \langle \xi, e(d\eta_j) \rangle$.

Definition 5 ([38, VI.1.8]). A compatible connection on $\mathcal{E}$ (not universal) is a linear map $\nabla : \mathcal{E} \to \mathcal{E} \otimes_\mathcal{A} \pi(\Omega^1)$ with a),b) as above.

We denote by $C(\mathcal{E})$ the space of compatible connections on $\mathcal{E}$.

Example 3. $\pi \nabla_0$ is an element of $C(\mathcal{E})$.

Note that $\pi : CC(\mathcal{E}) \to C(\mathcal{E})$ is surjective.

1.4 Quotient Differential Graded Algebra

The definition of dV by $\sum [D, a_j^0][D, a_j^1] \in \pi(\Omega^2)$ would be ambiguous as $\ker \pi$ is not a differential ideal. The following construction helps to define dV unambiguously.

Let $J_0 := \ker \pi$ (graded two-sided ideal), $J_0^{(k)} := J_0 \cap \Omega^k$.

Proposition 2 (Junk ideal, [38, VI.1.4(2)]). $J = J_0 + dJ_0$ is a graded differential two-sided ideal of $\Omega^*(\mathcal{A})$.

Proof. J is a differential ideal (as $d^2 = 0$). It is two-sided: For homogeneous $\omega = \omega_1 + d\omega_2$, $\omega_1 \in J_0^{(k)}, \omega_2 \in J_0^{(k-1)}$ and $\omega' \in \Omega^{k'}$, we have

$$\begin{aligned}\omega \omega' &= \omega_1 \omega' + (d\omega_2)\omega' \\ &= \omega_1 \omega' + \left(-(-1)^{k-1}\omega_2 d\omega' + d(\omega_2 \omega')\right) \in J_0^{(k+k')} + \in dJ_0^{(k'+k-1)}.\end{aligned}$$

Definition 6. We define the quotient differential graded algebra by $\Omega_D^*(\mathcal{A}) := \Omega^*(\mathcal{A})/J$.

1.5 Inner Product

We have
$$\Omega_D^0 = \mathcal{A}$$
(as $\Omega^0 = \mathcal{A} \subset \mathcal{L}(\mathcal{A})$, $J^{(0)} = J_0 \cap \Omega^0 = \{0\}$),
$$\Omega_D^1 = \Omega^1 / \ker \pi \cong \pi(\Omega^1)$$
(as $J \cap \Omega^1 = J_0 \cap \Omega^1 + d(J_0 \cap \Omega^0) = J_0 \cap \Omega_1$),
$$\Omega_D^2 \cong \pi(\Omega^2)/\pi\big(d(J_0 \cap \Omega^1)\big)$$
and similarly for $\Omega_D^3, \Omega_D^4, \ldots$. Every element of Ω_D^2 is of the form

$$\rho = \sum a_j^0 [D, a_j^1][D, a_j^2]$$
$$\text{mod } \{\rho_0 = \sum [D, b_j^0][D, b_j^1] : b_j^k \in \mathcal{A}, \sum b_j^0 [D, b_j^1] = 0\} \quad (1.1)$$

with $a_j^k \in \mathcal{A}$.

We recall the definition of the Dixmier trace $Tr_\omega(T)$. For $T \in \mathcal{K}(\mathcal{H})$ we have $Tr_\omega(T) < \infty$ if the eigenvalues $\mu_0(T) \geq \mu_1(T) \geq \ldots$ of $|T|$ satisfy $\sum_{n=0}^{N-1} \mu_n(T) = O(logN)$ (i.e. T has order 1).

Definition 7 ([38, IV.2.β]).

$$Tr_\omega(T) = \text{Lim}_\omega \frac{1}{logN} \sum_{n=0}^{N-1} \mu_n(T)$$

(for $T \geq 0$, then we can extend the definition).

The fixed lim-procedure Lim_ω is defined via a linear form L on $C_b(\mathbb{R}_+^*)$ with

α) $L(\alpha_n) \geq 0$ for $\alpha_n \geq 0$
β) $L(\alpha_n) = \lim_{n \to \infty} \alpha_n$ for converging α_n
γ) scale invariance: $L(\alpha_1, \alpha_1, \alpha_2, \alpha_2, \ldots) = L(\alpha_1, \alpha_2, \ldots)$

Here γ can be achieved by letting $\text{Lim}_\omega(\alpha_n) := L(M(f_\alpha))$, where $f_\alpha \in C_b(\mathbb{R}_+^*)$ is given by $f_\alpha(\lambda) = \alpha_N, \lambda \in]N-1, N]$ and $M(f)(\lambda) = \frac{1}{log\lambda} \int_1^\lambda f(u) \frac{du}{u}$ is the Cesaro mean. Then

$$Tr_\omega(ST) = Tr_\omega(TS) \text{ (all } T \text{ of order 1, } S \text{ bounded)}$$
$$Tr_\omega(UTU^*) = Tr_\omega(T) \text{ (} U \text{ unitary)}.$$

Now we can define an inner product for $T_j \in \pi(\Omega^2)$ of dimension p by

$$\langle T_1, T_2 \rangle = Tr_\omega\big(T_2^* T_1 |D|^{-p}\big).$$

Let $\mathcal{H}_2$ be the Hilbert space completion of $\pi(\Omega^2)$ and $P\mathcal{H}_2 := \big(\pi\big(d(J_0 \cap \Omega^1)\big)\big)^\perp$.

$P\mathcal{H}_2$ is the Hilbert space completion of $\pi(\Omega_D^2)$ with respect to $\langle P(), P()\rangle$.

1.6 Curvature and Yang-Mills Functional

For vector potentials V, the curvature $\theta(V)$ is defined by $dV + V^2 \in P\mathcal{H}_2$ (as an operator).

Proposition 3 ([38, VI.1.5(& 11)]).

(i) $YM(V) = \langle \theta(V), \theta(V)\rangle$ is a positive functional, quartic and invariant under gauge transformations $\gamma_u(V) = u[D, u^*] + uVu^*$.
(ii) On $\{\alpha \in \Omega^1 : \alpha = \alpha^*\}$, the same holds for
$I(\alpha) = Tr_\omega\big((\pi(d\alpha + \alpha^2))^2 |D|^{-p}\big)$.
(iii) $YM(V) = \inf\limits_{\pi(\alpha)=V} I(\alpha)$.

YM is called Yang-Mills functional.

For a universal compatible connection ("case β"), the curvature $\theta(\nabla) \in \mathrm{Hom}_\mathcal{A}(\mathcal{E}, \mathcal{E} \otimes_\mathcal{A} \Omega^2) \cong \mathcal{E} \otimes_\mathcal{A} \Omega^2 \otimes_\mathcal{A} \mathrm{Hom}_\mathcal{A}(\mathcal{E}, \mathcal{A})$ is given by $\tilde{\nabla}^2\big|_\mathcal{E}$ where $\tilde{\nabla}$ is the (uniquely defined) right Ω^*-module morphism $\mathcal{E} \otimes_\mathcal{A} \Omega^* \to \mathcal{E} \otimes_\mathcal{A} \Omega^*$ with $\tilde{\nabla}\big|_{\mathcal{E} \otimes 1} = \nabla$ and $\tilde{\nabla}(\eta\omega) = (\tilde{\nabla}\eta)\omega + (-1)^{\deg(\eta)}\eta(d\omega)$.

The curvature makes also sense as an operator $\pi(\theta) \in \mathcal{L}(\mathcal{E} \otimes_\mathcal{A} \mathcal{H})$.

Let us furthermore define $D_\nabla \in \mathcal{L}(\mathcal{E} \otimes_\mathcal{A} \mathcal{H})$ by

$$D_\nabla(\xi \otimes h) = \xi \otimes Dh + \big((1 \otimes \pi)\nabla\xi\big)h.$$

For $\nabla \in C(\mathcal{E})$, similarly $\theta = \tilde{\nabla}^2 \in \mathrm{Hom}_\mathcal{A}(\mathcal{E}, \mathcal{E} \otimes_\mathcal{A} \Omega_D^2)$. We can check that

$$\langle \xi \otimes \omega \otimes \eta^*, \xi' \otimes \omega' \otimes \eta'^*\rangle = \langle\langle \xi, \xi'\rangle\omega, \omega'\langle\eta, \eta'\rangle^*\rangle$$

defines an inner product on $\mathrm{Hom}_\mathcal{A}(\mathcal{E}, \mathcal{E} \otimes_\mathcal{A} \Omega_D^2)$.

Proposition' 1 ([38, VI.1.5(& 11)]).

(i') On $C(\mathcal{E})$: $YM(\nabla) = \langle\theta, \theta\rangle$ is a positive, quartic action, invariant under gauge transformations $\gamma_u(\nabla) = u\nabla u^*$.
(ii') On $CC(\mathcal{E})$, the same holds for $I(\nabla) = Tr_\omega\big((\pi(\theta))^2|D_\nabla|^{-p}\big)$.
(iii') $YM(\nabla) = \inf\limits_{\pi(\nabla_1)=\nabla} I(\nabla_1)$.

Proof (Proof (The trivial case)). Ad (iii): The ambiguity in $\pi(d\alpha)$ for α with $\pi(d\alpha) = V$ is $\pi\big(d(J_0 \cap \Omega^1)\big)$ and is deleted by the projection on $P\mathcal{H}_2 = \big(\pi(d(J_0 \cap \Omega^1))\big)^\perp$, while $\pi(d\alpha + \alpha^2) \mapsto dV + V^2$.

Ad (i),(ii): Assertions (i) and (ii) hold, as the following computations show that θ is invariant under gauge transformations:

$$[D, a_j^1 u^*] - a_j^1[D, u^*] = Da_j^1 u^* + a_j^1(-u^*D - Du^* + u^*D) = [D, a_j^1]u^*.$$

Thus

$$\gamma_u(V) = \gamma_u(\sum a_j^0[D, a_j^1]) = u[D, u^*] + \sum ua_j^0[D, a_j^1 u^*] - \sum ua_j^0 a_j^1[D, u^*],$$
$$d\gamma_u(V) = [D, u][D, u^*] + \sum [D, ua_j^0][D, a_j^1 u^*] - \sum [D, ua_j^0 a_j^1][D, u^*]$$

and

$$d\gamma_u(V) + \gamma_u(V)^2 = d\gamma_u(V) + (u[D, u^*] + uVu^*)^2$$
$$= d\gamma_u(V) + u(Du^* - u^*D)u(Du^* - u^*D)$$
$$\quad + u[D, u^*]uVu^* + uVu^*u[D, u^*] + uV^2u^*$$
$$= d\gamma_u(V) - [D, u][D, u^*] - [D, u]Vu^* + uV[D, u^*] + uV^2u^*$$
$$= -\sum [D, ua_j^0 a_j^1][D, u^*] + \sum [D, ua_j^0][D, a_j^1 u^*] - \sum [D, u]a_j^0[D, a_j^1 u^*]$$
$$\quad + \sum [D, u]a_j^0 a_j^1[D, u^*] + \sum ua_j^0[D, a_j^1][D, u^*] + uV^2u^*$$
$$= \sum u[D, a_j^0][D, a_j^1 u^*]$$
$$\quad - \sum ([D, ua_j^0 a_j^1] - [D, u]a_j^0 a_j^1 - ua_j^0[D, a_j^1])[D, u^*] + uV^2u^*$$
$$= \sum u[D, a_j^0]([D, a_j^1 u^*] - a_j^1[D, u^*]) + uV^2u^* = u(dV + V^2)u^*.$$

Example 4. Let us briefly look at an example that will be used to define Connes' Standard Model (c.f. 9). Let $X = \{a, b\}$, $\mathcal{E} = \mathcal{A} = \mathbb{C} \oplus \mathbb{C}$, $e_1 := (1, 0)$, $\mathcal{H} = \mathcal{H}_a \oplus \mathcal{H}_b$, we consider $f \in \mathcal{A}$ on $\mathcal{H}$ as $\begin{pmatrix} f_a & 0 \\ 0 & f_b \end{pmatrix}$. Let $D := \begin{pmatrix} 0 & M^* \\ M & 0 \end{pmatrix}$.

$$\text{Then } [D, f] = (f_b - f_a) \begin{pmatrix} 0 & M^* \\ -M & 0 \end{pmatrix}.$$

$$\Omega^1 = \{\lambda e_1 de_1 + \mu(1 - e_1)de_1\}, \quad \pi\Omega^1 = \left\{ \begin{pmatrix} 0 & -\lambda M^* \\ \mu M & 0 \end{pmatrix} \right\}.$$

$$\Omega^2 = \{\alpha e_1 de_1 de_1 + \beta(1 - e_1)de_1 de_1\}, \quad \pi\Omega^2 = \left\{ \begin{pmatrix} -\alpha M^* M & 0 \\ 0 & -\beta MM^* \end{pmatrix} \right\}.$$

A vector potential is a selfadjoint element: $V = -\bar{\phi} e_1 de_1 + \phi(1 - e_1)de_1$, thus given by $\phi \in \mathbb{C}$, $\pi(V) = \begin{pmatrix} 0 & \phi M^* \\ \phi M & 0 \end{pmatrix}$.

Now

$$\theta = dV + V^2 = -\bar{\phi} de_1 de_1 - \phi de_1 de_1 + (\bar{\phi} e_1 de_1 - \phi(1 - e_1)de_1)^2$$
$$= -(\phi + \bar{\phi} + \phi\bar{\phi})de_1 de_1 = -(|1 + \phi|^2 - 1)de_1 de_1$$

$$\pi(\theta) = -(|1+\phi|^2 - 1)\begin{pmatrix} M^*M & 0 \\ 0 & MM^* \end{pmatrix}$$

and $YM = \langle \theta, \theta \rangle = 2(|1+\phi|^2 - 1)^2 \operatorname{Trace}((M^*M)^2)$,

where Trace denotes the ordinary trace, which is a 0-dimensional analogue of Tr_ω.

2 Real Spectral Triples and Charge Conjugation

Ralf Meyer

University of Münster

You may think of a real structure on a spectral triple as a generalization of the charge conjugation operator acting on the spinor bundle over a spin manifold. The charge conjugation operator is, in fact, an important example and will be considered in detail below. Almost everything in this section is due to Alain Connes [39].

2.1 Real Structures on Even Spectral Triples

Definition 1 ([39]). Let $(A, \mathcal{H}, D)$ be an even spectral triple. A *real structure of dimension* $2p \mod 8$ on $(A, \mathcal{H}, D)$ is a conjugate linear isometry $J \colon \mathcal{H} \to \mathcal{H}$ satisfying:

a) $JD = DJ$, $J^2 = \epsilon$, $J\gamma = \epsilon' \gamma J$;
b) for any $a \in A$, the operators a and $[D, a]$ commute with JAJ^*.

$\epsilon, \epsilon' \in \{+1, -1\}$ depend on $d = 2p \mod 8$ according to the following table:

$d =$	0	2	4	6
$\epsilon =$	1	-1	-1	1
$\epsilon' =$	1	-1	1	-1

$(A, \mathcal{H}, D, J)$ is called a *real spectral triple of dimension* $2p \mod 8$.

Notice that $(\lambda J)^2 = |\lambda|^2 J$ for all $\lambda \in \mathbb{C}$. Hence if J is a real structure of dimension $2p$, then so is λJ for all $\lambda \in \mathbb{C}$, $|\lambda| = 1$.

The crucial part of Def. 1 is condition b). Since A commutes with JAJ^*, we can turn $\mathcal{H}$ into a bimodule over A by putting

$$a\xi b := aJb^*J^*(\xi) \qquad \forall a, b \in A, \ \xi \in \mathcal{H}.$$

Using that J is an isometry (that is, $J^*J = 1$) and that JAJ^* commutes with A, one verifies easily the conditions for a bimodule. In the application to the standard model, this bimodule structure makes sense of $u\xi u^*$ for u in the gauge group $U(A)$ and thus allows us to define the "adjoint" representation of the gauge group $U(A)$ on $\mathcal{H}$.

Condition a) is related to the notion of a "real" algebra. Let us first not worry about the signs ϵ, ϵ'. Ignoring the dimension, we may replace the conditions $J^2 = \epsilon$ and $J\gamma = \epsilon' \gamma J$ by $J^2 = \pm 1$ and $J\gamma = \pm \gamma J$ because any pair $(\epsilon, \epsilon') \in \{\pm 1\} \times \{\pm 1\}$ occurs for a unique $d \in \{0, 2, 4, 6\}$.

Definition 2. Let B be a graded $*$-algebra over $\mathbb{C}$ with grading $x \mapsto x^\gamma$ and involution $x \mapsto x^*$. A *real structure on* B is a conjugate linear homomorphism $B \to B$, $x \mapsto \overline{x}$, satisfying $\overline{\overline{x}} = x$, $\overline{x^*} = (\overline{x})^*$, and $\overline{x^\gamma} = (\overline{x})^\gamma$ for all $x \in B$.

An element $x \in B$ is called *real* iff $\overline{x} = x$.
A "real" graded $*$-algebra is a graded $*$-algebra with real structure.

If B is a "real" graded $*$-algebra, then the set of real elements

$$B_{\mathbb{R}} := \{x \in B \mid \bar{x} = x\}$$

is a graded $*$-algebra over $\mathbb{R}$ such that $B \cong B_{\mathbb{R}} \otimes_{\mathbb{R}} \mathbb{C}$ and the conjugation on B is the standard conjugation on a complexification, $\overline{x \otimes \lambda} := x \otimes \bar{\lambda}$. Hence we could alternatively define a "real" structure as an isomorphism $B \cong B_{\mathbb{R}} \otimes_{\mathbb{R}} \mathbb{C}$ for some $\mathbb{R}$-algebra $B_{\mathbb{R}}$.

Lemma 1. *Let $\mathcal{H}$ be a graded Hilbert space with grading $\gamma \colon \mathcal{H} \to \mathcal{H}$. Let $J \colon \mathcal{H} \to \mathcal{H}$ be a conjugate linear isometry satisfying $J^2 = \pm 1$ and $J\gamma = \pm\gamma J$. Then $\bar{x} := JxJ^*$ defines a real structure on $\mathcal{L}(\mathcal{H})$.*

Conversely, if $x \mapsto \bar{x}$ is a real structure on $\mathcal{L}(\mathcal{H})$, then there is a conjugate linear isometry $J \colon \mathcal{H} \to \mathcal{H}$ satisfying $J^2 = \pm 1$ and $J\gamma = \pm\gamma J$ such that $\bar{x} = JxJ^*$.

In addition, J is unique up to multiplication with scalars of modulus 1.

Proof. Let J be a conjugate linear isometry satisfying $J^2 = \pm 1$ and $J\gamma = \pm\gamma J$. Then $\bar{x} := JxJ^*$ is conjugate linear and $*$-preserving. Since J is an isometry, $\bar{x} \cdot \bar{y} = JxJ^*JyJ^* = JxyJ^* = \overline{x \cdot y}$, that is, conjugation is multiplicative. Since $J^2 = \pm 1$, we have $\bar{\bar{x}} = J^2 x (J^*)^2 = (\pm 1)^2 x = x$ for all $x \in \mathcal{L}(\mathcal{H})$, that is, conjugation is an involution. Finally, $J\gamma = \pm\gamma J$ implies $\gamma J^* = \pm J^* \gamma$ and thus

$$\overline{x^\gamma} = \overline{\gamma x \gamma} = J\gamma x \gamma J^* = (\pm 1)^2 \gamma J x J^* \gamma = \gamma \bar{x} \gamma = (\bar{x})^\gamma$$

for all $x \in \mathcal{L}(\mathcal{H})$. Thus $x \mapsto \bar{x}$ is a real structure on the graded $*$-algebra $\mathcal{L}(\mathcal{H})$ in the sense of Def. 2.

Conversely, let $x \mapsto \bar{x}$ be a real structure on $\mathcal{L}(\mathcal{H})$. This "conjugate automorphism" is necessarily inner in the sense that there is a conjugate linear isometry $J \colon \mathcal{H} \to \mathcal{H}$ such that $\bar{x} = JxJ^*$. To see this, pick any conjugate linear isometry $J' \colon \mathcal{H} \to \mathcal{H}$ and consider $x \mapsto J'\bar{x}(J')^*$. This is a $*$-automorphism of $\mathcal{L}(\mathcal{H})$ and therefore inner.

Since $\bar{\bar{x}} = x$, we have $J^2 x (J^2)^* = x$ for all $x \in \mathcal{L}(\mathcal{H})$. Thus J^2 is in the center of $\mathcal{L}(\mathcal{H})$. Therefore, $J^2 = \lambda$ for some $\lambda \in \mathbb{C}$, $|\lambda| = 1$. Since $\lambda \cdot \mathrm{id} = J^2 = J(J^2)J^* = \overline{\lambda \cdot \mathrm{id}} = \bar{\lambda} \cdot \mathrm{id}$, we automatically have $\lambda \in \{-1, +1\}$. Thus $J^2 = \pm 1$.

Since $\overline{x^\gamma} = (\bar{x})^\gamma$, we have $J\gamma x \gamma J^* = \gamma J x J^* \gamma$ for all $x \in \mathcal{L}(\mathcal{H})$. Equivalently, $(J^*\gamma J\gamma)x(J^*\gamma J\gamma)^*$ for all $x \in \mathcal{L}(\mathcal{H})$, that is, $J^*\gamma J\gamma$ is in the center of $\mathcal{L}(\mathcal{H})$. Thus $J\gamma = \lambda \gamma J$ for some $\lambda \in \mathbb{C}$, $|\lambda| = 1$. Since $J = J\gamma^2 = \lambda\gamma J\gamma = \lambda^2 \gamma^2 J = \lambda^2 J$, it follows that $\lambda \in \{-1, +1\}$, that is, $J\gamma = \pm\gamma J$ as desired.

Finally, conjugate linear isometries J' and J give rise to the same real structure on $\mathcal{L}(\mathcal{H})$ iff $J'J^*$ is in the center of $\mathcal{L}(\mathcal{H})$, that is, $J' = \lambda J$ for some $\lambda \in \mathbb{C}$, $|\lambda| = 1$. Thus J is determined uniquely up to multiplication by scalars of modulus 1.

Thus a real structure on a spectral triple $(A, \mathcal{H}, D)$ is nothing but a real structure on the graded *-algebra $\mathcal{L}(\mathcal{H})$ with the following additional properties:

a)' $\overline{D} = D$;
b)' $\overline{A} := \{\overline{a} \mid a \in A\}$ commutes with A and $[D, A] := \{[D, a] \mid a \in A\}$.

2.2 Spin$^{\mathbb{C}}$ Manifolds and Charge Conjugation

We now come to a crucial example, namely the standard spectral triple $(A, \mathcal{H}, D)$ associated to a Riemannian Spin$^{\mathbb{C}}$ manifold M of dimension $2p$. This spectral triple will be constructed in detail in Section 1.3. We are going to find a "canonical" real structure $J \colon \mathcal{H} \to \mathcal{H}$ on $(A, \mathcal{H}, D)$. The operator J is called *charge conjugation*. Of course, we expect to get a real spectral triple $(A, \mathcal{H}, D, J)$ of dimension $2p$ mod 8. This is indeed the case when the signs ϵ, ϵ' are chosen as in the table in Def. 1.

To construct charge conjugation and the right values ϵ, ϵ', we have to be very careful about sign conventions in the definition of the spectral triple $(A, \mathcal{H}, D)$. Let $F(M) \to M$ be the principal Spin$^{\mathbb{C}}$-bundle describing the Spin$^{\mathbb{C}}$-structure of M. The *spinor bundle* S is associated to this principal bundle. That is, $S = F(M) \times_{\text{Spin}^{\mathbb{C}}} \Sigma$ for a certain graded $\mathbb{C}$-vector space Σ with a grading preserving, unitary representation of Spin$^{\mathbb{C}}$. The algebra $A := C^{\infty}(M)$ acts on the Hilbert space $\mathcal{H} := L^2(M, S)$ of L^2-sections of the spinor bundle by pointwise multiplication. To define Σ, we need the real Clifford algebras. We follow Kasparov [123].

Let $\mathbb{R}^{p,q}$ be the $\mathbb{R}$-vector space with basis $(e_1, \ldots, e_{p+q})$. Endow $\mathbb{R}^{p,q}$ with the symmetric bilinear form Q defined by $Q(e_j, e_k) = 0$ for $j \neq k$, $Q(e_j, e_j) = 1$ for $j = 1, \ldots, p$, and $Q(e_j, e_j) = -1$ for $j = p+1, \ldots, p+q$.

Definition 3. The real *Clifford algebra* $\text{Cl}_{p,q}$ is the universal unital $\mathbb{R}$-algebra generated by the vector space $\mathbb{R}^{p,q}$ with relations $x \cdot x = Q(x, x) \cdot 1$ for all $x \in \mathbb{R}^{p,q}$. That is, $\text{Cl}_{p,q}$ is the universal unital $\mathbb{R}$-algebra with anti-commuting generators $e_1, \ldots, e_{p+q}$ such that $e_j^2 = 1$ for $j = 1, \ldots, p$ and $e_j^2 = -1$ for $j = p+1, \ldots, p+q$.

$\text{Cl}_{p,q}$ is naturally a $\mathbb{Z}_2$-graded *-algebra. The grading is defined by the requirement that $\deg e_j = 1$ for $j = 1, \ldots, p+q$. The involution is defined by $e_j^* = e_j$ for $j = 1, \ldots, p$ and $e_j^* = -e_j$ for $j = p+1, \ldots, p+q$.

We consider $\mathbb{R}^{p,q}$ as a sub-vector space of $\text{Cl}_{p,q}$ in the obvious way.

Fix a basis $e_1, \ldots, e_p$ of the vector space $\mathbb{R}^p$. Endow the exterior algebra $\Lambda \mathbb{R}^p$ with the usual grading and the inner product that is defined by the condition that the vectors $e_{i_1} \wedge \cdots \wedge e_{i_k}$, $i_1 < \cdots < i_k$, form an orthonormal basis. Let $\lambda_x \colon \Lambda \mathbb{R}^p \to \Lambda \mathbb{R}^p$ be exterior multiplication with $x \in \mathbb{R}^p$. Let λ_x^* be the adjoint of λ_x. Then $e_j \mapsto \lambda_{e_j} + \lambda_{e_j}^*$ for $j = 1, \ldots, p$ and $e_{p+j} \mapsto \lambda_{e_j} - \lambda_{e_j}^*$ for $j = 1, \ldots, p$ defines a grading preserving *-homomorphism $\text{Cl}_{p,p} \to$

$\mathcal{L}_\mathbb{R}(\varLambda\mathbb{R}^p) \cong \mathrm{M}_{2^p}(\mathbb{R})$. This representation yields an isomorphism of graded $*$-algebras $\mathrm{Cl}_{p,p} \cong \mathcal{L}_\mathbb{R}(\varLambda\mathbb{R}^p)$.

Let $\mathrm{Cl}_{p,q}^{\mathrm{even}}$ be the even part of $\mathrm{Cl}_{p,q}$. By definition, $\mathrm{Spin}^\mathbb{C}(2p)$ is the group of all unitary elements $\omega \in \mathrm{Cl}_{0,2p}^{\mathrm{even}} \otimes_\mathbb{R} \mathbb{C}$ such that $\omega \mathbb{R}^{0,2p} \omega^{-1} \subset \mathbb{R}^{0,2p}$. Thus $\rho(\omega)(x) := \omega x \omega^{-1}$ defines a representation of $\mathrm{Spin}^\mathbb{C}(2p)$ on $\mathbb{R}^{0,2p}$. One can check that ρ has range $SO(2p)$ and kernel $U(1) = \{\lambda \mid \lambda \in \mathbb{C}, |\lambda| = 1\}$. Thus $\mathrm{Spin}^\mathbb{C}(2p) \cong \mathrm{Spin}(2p) \times_{\mathbb{Z}_2} U(1)$, where $\mathrm{Spin}(2p)$ is the universal covering of $SO(2p)$.

We make $\varSigma := \varLambda\mathbb{C}^p$ a module over $\mathrm{Cl}_{0,2p} \otimes_\mathbb{R} \mathbb{C}$ via the isomorphisms $\mathrm{Cl}_{0,2p} \otimes_\mathbb{R} \mathbb{C} \cong \mathrm{Cl}_{p,p} \otimes_\mathbb{R} \mathbb{C} \cong \mathcal{L}_\mathbb{C}(\varLambda\mathbb{C}^p)$ defined by

$$(e_1,\ldots,e_{2p}) \mapsto (ie_1,\ldots,ie_p,e_{p+1},\ldots,e_{2p})$$
$$\mapsto (i\lambda_{e_1} + i\lambda_{e_1}^*,\ldots,i\lambda_{e_p} + i\lambda_{e_p}^*, \lambda_{e_1} - \lambda_{e_1}^*,\ldots,\lambda_{e_p} - \lambda_{e_p}^*). \quad (2.1)$$

Since $\mathrm{Spin}^\mathbb{C}(2p) \subset \mathrm{Cl}_{0,2p} \otimes_\mathbb{R} \mathbb{C}$, we get a natural action of $\mathrm{Spin}^\mathbb{C}(2p)$ on $\varSigma$. The group $\mathrm{Spin}^\mathbb{C}$ also acts on $\mathrm{Cl}_{0,2p} \otimes_\mathbb{R} \mathbb{C}$ and suitable subspaces by conjugation. For this action, the *Clifford multiplication* $\mathrm{Cl}_{0,2p} \otimes_\mathbb{R} \mathbb{C} \otimes \varSigma \to \varSigma$ is $\mathrm{Spin}^\mathbb{C}$-equivariant. Hence we get induced bundle maps $\mathrm{Cl}(M) \otimes S \to S$, where $\mathrm{Cl}(M)$ is the Clifford algebra bundle of M whose fiber at $x \in M$ is the Clifford algebra of $T_x M$, with symmetric bilinear form $-g$ (where g is the Riemannian metric). Our sign conventions are such that Clifford multiplication by a vector field $X \colon M \to TM$ is *skew*-adjoint. This convention is used by many authors, but unfortunately not by Connes.

Let ∇ be a connection on the principal $\mathrm{Spin}^\mathbb{C}$-bundle $F(M) \to M$. The *Dirac operator* D is defined as the composition

$$C^\infty(M,S) \xrightarrow{\nabla} C^\infty(M,T^*M \otimes S) \xrightarrow{g} C^\infty(M,TM \otimes S) \xrightarrow{\mu} C^\infty(M,S),$$

where g is induced by the metric and μ is Clifford multiplication. Our sign conventions are such that $[D, M_f]$ is Clifford multiplication by the gradient of f.

Let $CC \colon \varSigma \to \varSigma$ be a conjugate linear isometry commuting with the $\mathrm{Spin}^\mathbb{C}$-action. One verifies easily that the induced map $C^\infty(M,S) \to C^\infty(M,S)$ commutes with D iff CC commutes with Clifford multiplication by elements of $\mathbb{R}^{0,2p} \subset \mathrm{Cl}_{0,2p}$. Since $\mathbb{R}^{0,2p}$ generates $\mathrm{Cl}_{0,2p}$, this means that CC commutes with $\mathrm{Cl}_{0,2p}$. The isomorphism $\mathcal{L}(\varSigma) \cong \mathrm{Cl}_{0,2p} \otimes_\mathbb{R} \mathbb{C}$ yields a real structure $x \mapsto \bar{x}$ on the graded $*$-algebra $\mathcal{L}(\varSigma)$. A conjugate linear isometry CC commutes with $\mathrm{Cl}_{0,2p}$ if and only if $CC x CC^* = \bar{x}$ for all $x \in \mathcal{L}(\varSigma)$. By Lemma 1 such an operator CC exists and is unique up to multiplication by scalars of modulus 1. Furthermore, $CC^2 = \pm 1$ and $CC\gamma = \pm \gamma CC$, where γ is the usual grading on $\varLambda\mathbb{C}^p$.

The associated map $J \colon \mathcal{H} \to \mathcal{H}$ satisfies $JD = DJ$ by construction. In fact, it is the only "constant" bundle map with this property up to multiplication by a constant of modulus 1. Furthermore, $J^2 = \epsilon$ and $J\gamma = \epsilon'\gamma J$ for suitable $\epsilon, \epsilon' \in \{-1, +1\}$ because these equations hold for CC. If $f \in A$, then $JM_f J^* = M_{\bar{f}}$ is again a pointwise multiplication operator. Hence JAJ^* commutes with all maps $\mathcal{H} \to \mathcal{H}$ that come from bundle maps $S \to S$. In particular, it commutes with A and $[D, A]$ because $[D, f]$ is Clifford multiplication by the gradient of f. Thus J is a real structure.

It remains to compute the signs ϵ, ϵ'. For that purpose, we derive an explicit formula for $CC \colon \Lambda \mathbb{C}^p \to \Lambda \mathbb{C}^p$. Since $\lambda_x - \lambda_x^*$ and $i\lambda_x + i\lambda_x^*$ are real, an easy computation shows that $\overline{\lambda_x} = -\lambda_x^*$, $\overline{\lambda_x^*} = -\lambda_x$ for all $x \in \mathbb{R}^p$. Thus CC maps

$$\mathbb{C} \cdot 1 = \{\omega \in \Lambda \mathbb{C}^p \mid \lambda_x^*(\omega) = 0 \text{ for all } x \in \mathbb{R}^p\}$$

to

$$\mathbb{C} \cdot (e_1 \wedge \cdots \wedge e_p) = \{\omega \in \Lambda \mathbb{C}^p \mid \lambda_x(\omega) = 0 \text{ for all } x \in \mathbb{R}^p\}.$$

Thus $CC(1) = \lambda e_1 \wedge \cdots \wedge e_p$ for some $\lambda \in \mathbb{C}$. Since CC is isometric, $|\lambda| = 1$. Since CC is only unique up to multiplication by a scalar of modulus 1, anyway, we may assume $CC(1) = e_1 \wedge \cdots \wedge e_p$. Then

$$CC(e_1 \wedge \cdots \wedge e_k) = CC(\lambda_{e_1} \cdots \lambda_{e_k}) CC^* CC(1)$$
$$= (-1)^k \lambda_{e_1}^* \cdots \lambda_{e_k}^* (e_1 \wedge \cdots \wedge e_p) = (-1)^{k(k+1)/2} e_{k+1} \wedge \cdots \wedge e_p.$$

This holds for any oriented orthonormal basis $e_1, \ldots, e_p$ of $\mathbb{R}^p$. Hence CC coincides with the (conjugate linear) Hodge-$*$ operator up to the signs $(-1)^{k(k+1)/2}$ on $\Lambda^k \mathbb{R}^p$.

A straightforward computation shows $CC\gamma = \epsilon' \gamma CC$ for $\epsilon' = (-1)^p$. Since $CC^2 = \epsilon$ is a scalar and $CC^2(1) = CC(e_1 \wedge \cdots \wedge e_p) = (-1)^{p(p+1)/2}$, we have $\epsilon = (-1)^{p(p+1)/2}$. Comparing with the table in Def. 1, we see that J is a real structure of dimension $2p$ as desired.

2.3 Real Structures via Clifford Algebras

The example of the charge conjugation operator shows that the signs ϵ, ϵ' in Def. 1 are related to representations of Clifford algebras. If we decorate spectral triples appropriately with Clifford algebras, we can indeed get rid of the signs as follows:

Definition 4. Let $(A, \mathcal{H}_c, D_c)$ be an even spectral triple with grading automorphism $\gamma_c \in \mathcal{L}(\mathcal{H}_c)$. Let $p, q \in \mathbb{Z}_+$. A $\mathrm{Cl}_{p,q}$-*real structure on* $(A, \mathcal{H}_c, D_c)$ consists of a conjugate linear isometry $J_c \colon \mathcal{H}_c \to \mathcal{H}_c$ and a grading preserving $*$-homomorphism $\phi \colon \mathrm{Cl}_{p,q} \to \mathcal{L}(\mathcal{H}_c)$ such that

Cl. a) $J_c D = D J_c$, $J_c^2 = 1$, $J_c \gamma_c = \gamma_c J_c$; $\phi(\mathrm{Cl}_{p,q})$ commutes with A, D_c, and J_c;

Cl. b) $J_c A J_c^*$ commutes with A and $[D_c, A]$.

We call $(A, \mathcal{H}_c, D_c, J_c, \phi)$ a $\mathrm{Cl}_{p,q}$-*real spectral triple*.

Proposition 1. *Let $p \in \mathbb{Z}_+$. There is a natural bijection between real spectral triples of dimension $2p$ mod 8 and $\mathrm{Cl}_{0,2p}$-real spectral triples. More generally, if $p - q$ is even, there is a natural bijection between $\mathrm{Cl}_{q,p}$-real spectral triples and real spectral triples of dimension $p - q$ mod 8.*

Proof. Let $(A, \mathcal{H}, D, J)$ be a real spectral triple of dimension $2p$ mod 8. Define $\mathcal{H}_c := \mathcal{H} \otimes \Lambda\mathbb{C}^p$, $D_c := D \otimes \mathrm{id}$, $J_c := J \otimes CC$. These tensor products of operators are meant in the *ungraded* sense. Thus, for example, $(J \otimes CC)(\xi \otimes \eta) := J(\xi) \otimes CC(\eta)$. Endow $\mathcal{H}_c$ with the product grading $\gamma_c = \gamma_{\mathcal{H}} \otimes \gamma_{\Lambda\mathbb{C}^p}$. Consider $A \subset \mathcal{L}(\mathcal{H}_c)$ via $a \mapsto a \otimes 1$. Define the representation $\phi \colon \mathrm{Cl}_{0,2p} \to \mathcal{L}(\mathcal{H}_c)$ by $\phi = 1 \otimes \phi_0$ with ϕ_0 as in (2.1). This is a graded $*$-representation.

It is easy to check that $(A, \mathcal{H}_c, D_c, J_c, \phi)$ is a $\mathrm{Cl}_{0,2p}$-real spectral triple in the sense of Def. 4. As an illustration, let us compute $J_c^2 = J^2 \otimes CC^2 = \epsilon^2 \mathrm{id} = \mathrm{id}$ and

$$J_c \gamma_c = (J \cdot \gamma_{\mathcal{H}}) \otimes (CC \cdot \gamma_{\Lambda\mathbb{C}^p}) = (\epsilon')^2 (\gamma_{\mathcal{H}} \cdot J) \otimes (\gamma_{\Lambda\mathbb{C}^p} \cdot CC) = \gamma_c J_c.$$

Conversely, let us start with a $\mathrm{Cl}_{0,2p}$-real spectral triple $(A, \mathcal{H}_c, D_c, J_c, \phi)$. Since $\mathrm{Cl}_{0,2p} \otimes_\mathbb{R} \mathbb{C} \cong \mathcal{L}(\Lambda\mathbb{C}^p)$, the complexified representation $\phi_\mathbb{C} \colon \mathrm{Cl}_{0,2p} \otimes_\mathbb{R} \mathbb{C} \to \mathcal{L}(\mathcal{H}_c)$ is a multiple of the standard representation on $\Lambda\mathbb{C}^p$. In other words, $\mathcal{H}_c \cong \mathcal{H} \otimes \Lambda\mathbb{C}^p$ and $\phi = \mathrm{id} \otimes \phi_0$. The commutant of $\mathrm{Cl}_{0,2p}$ in $\mathcal{L}(\mathcal{H}_c)$ is equal to $\mathcal{L}(\mathcal{H}) \otimes \mathrm{id}$.

Thus the representation $A \to \mathcal{L}(\mathcal{H}_c)$ is of the form $f \otimes \mathrm{id}$ for some $*$-representation $f \colon A \to \mathcal{L}(\mathcal{H})$ and $D_c = D \otimes \mathrm{id}$ for some unbounded self-adjoint operator D on $\mathcal{H}$. Let γ_c and $\gamma_{\Lambda\mathbb{C}^p}$ be the gradings on $\mathcal{H}_c$ and $\Lambda\mathbb{C}^p$, respectively. Then $\gamma_c^{-1}(\mathrm{id} \otimes \gamma_{\Lambda\mathbb{C}^p})$ commutes with $\phi(\mathrm{Cl}_{0,2p})$. Thus $\gamma_c = \gamma_{\mathcal{H}} \otimes \gamma_{\Lambda\mathbb{C}^p}$ for some $\gamma_{\mathcal{H}} \in \mathcal{L}(\mathcal{H})$ that makes $\mathcal{H}$ a graded Hilbert space. Finally, we have $J_c = J \otimes CC$ for some conjugate linear isometry $J \colon \mathcal{H} \to \mathcal{H}$. To see this, choose any conjugate linear isometry $J' \colon \mathcal{H} \to \mathcal{H}$ and argue that $J_c^{-1}(J' \otimes CC)$ commutes with $\phi(\mathrm{Cl}_{0,2p})$. It is straightforward to show that $(A, \mathcal{H}, D, J)$ is a real spectral triple of dimension $2p$ mod 8.

Clearly, the above constructions are inverses of each other. This completes the proof of the special case.

To deal with the general case, choose an isomorphism $\mathrm{Cl}_{q,p} \otimes_\mathbb{R} \mathbb{C} \cong \mathcal{L}(\Lambda\mathbb{C}^{(q+p)/2})$ as in (2.1). The corresponding real structure on $\mathcal{L}(\Lambda\mathbb{C}^{(q+p)/2})$ is implemented by a conjugate linear isometry $CC_{q,p} \colon \Lambda\mathbb{C}^{(q+p)/2} \to \Lambda\mathbb{C}^{(q+p)/2}$ by Lemma 1. An explicit formula for $CC_{q,p}$ yields $\epsilon, \epsilon' \in \{\pm 1\}$ such that $CC_{q,p}^2 = \epsilon$ and $CC_{q,p}\gamma = \epsilon'\gamma CC_{q,p}$. It turns out that ϵ, ϵ' are the same as in the table in Def. 1 for $d = p - q$. We can replace CC by $CC_{q,p}$ in the above construction and conclude that there exists a bijection between real spectral triples of dimension $p - q$ mod 8 and $\mathrm{Cl}_{q,p}$-real spectral triples.

Alternatively, one can use the periodicity of the real Clifford algebras: The graded $*$-algebras $\mathrm{Cl}_{p,q}$ and $\mathrm{Cl}_{p',q'}$ are Morita equivalent iff $p - q \equiv p' - q'$ mod 8. More precisely, there are isomorphisms $\mathrm{Cl}_{p+4,q} \cong \mathrm{Cl}_{p,q+4}$ for all $p, q \geq 0$ (see [123]). Hence $\mathrm{Cl}_{p,q}$ and $\mathrm{Cl}_{p',q'}$ are isomorphic if $p + q = p' + q'$ and $p - q \equiv p' - q'$ mod 8. By definition, $\mathrm{Cl}_{p_1+p_2,q_1+q_2} \cong \mathrm{Cl}_{p_1,q_1} \otimes_{\mathrm{gr}} \mathrm{Cl}_{p_2,q_2}$, where $\otimes_{\mathrm{gr}}$ denotes the graded tensor product. If $p_1 = q_1$, we can replace the graded tensor product by the ordinary tensor product, so that $\mathrm{Cl}_{p_1+p_2,p+q_2} \cong \mathcal{L}(\Lambda\mathbb{C}^p) \otimes \mathrm{Cl}_{p_2,q_2}$.

2.4 Real Structures of Odd Dimension

Def. 4 makes perfectly good sense if $p-q$ is odd. We want to derive a definition for real structures of odd dimension analogous to Def. 1. As in the even case, we may restrict attention to $\mathrm{Cl}_{0,2p-1}$-real spectral triples $(A, \mathcal{H}_c, D_c, J_c, \phi)$. Let $e_1, \ldots, e_{2p-1}$ be the generators of $\mathrm{Cl}_{0,2p-1}$ and let γ_c be the grading operator on $\mathcal{H}_c$. By assumption, γ_c anti-commutes with $\phi(e_j)$ for all j and $\gamma_c^2 = 1$. Thus $\phi(\mathrm{Cl}_{0,2p-1})$ and γ_c together generate a copy of $\mathrm{Cl}_{1,2p-1}$.

The complexification of $\mathrm{Cl}_{1,2p-1}$ is isomorphic to $\mathcal{L}(\Lambda\mathbb{C}^p)$. Thus we can argue as in the proof of Prop. 1: $\mathcal{H}_c \cong \mathcal{H} \otimes \Lambda\mathbb{C}^p$ with $\mathrm{Cl}_{1,2p-1}$ acting only on the second copy. Since A commutes with $\phi(\mathrm{Cl}_{0,2p-1})$ and γ_c, we have $A \subset \mathcal{L}(\mathcal{H}) \otimes \mathrm{id}$. The operator D_c commutes with $\phi(\mathrm{Cl}_{0,2p-1})$ and anti-commutes with γ_c. To make D_c commute with γ_c, multiply it by $\omega := i^p e_1 \cdots e_{2p-1} \in \mathrm{Cl}_{0,2p-1} \otimes_{\mathbb{R}} \mathbb{C}$. A straightforward computation shows that $\omega = \omega^*$, $\omega^2 = 1$, and $e_j\omega = \omega e_j$ for all $j = 0, \ldots, 2p-1$. Since ω is odd, $\phi(\omega)$ commutes with $\mathrm{Cl}_{0,2p-1}$ and anti-commutes with γ_c. Therefore, $\omega D_c = D_c\omega$ commutes with $\mathrm{Cl}_{1,2p-1}$. Thus D_c is of the form $D \otimes \omega$ for an unbounded self-adjoint operator $D: \mathcal{H} \to \mathcal{H}$.

Finally, J_c commutes with $\mathrm{Cl}_{0,2p-1}$ and γ_c and therefore with $\mathrm{Cl}_{1,2p-1}$. Let $CC_{1,2p-1}$ be the conjugate linear isometry : $\Lambda\mathbb{C}^p \to \Lambda\mathbb{C}^p$ that occurred already in the proof of Prop. 1. Thus $CC_{1,2p-1} x CC_{1,2p-1}^* = x$ for all $x \in \mathrm{Cl}_{1,2p-1} \subset \mathcal{L}(\Lambda\mathbb{C}^p)$. Therefore, $J_c = J \otimes CC_{1,2p-1}$ for a suitable conjugate linear isometry $J: \mathcal{H} \to \mathcal{H}$. To see this, let $J': \mathcal{H} \to \mathcal{H}$ be any conjugate linear isometry and observe that conjugating with $J' \otimes CC_{1,2p-1}$ and J_c has the same effect on $\mathrm{Cl}_{1,2p-1}$. It follows that $J^2 = CC_{1,2p-1}^2 = (-1)^{p(p-1)/2}$. Furthermore, $JD \otimes CC_{1,2p-1}\omega = J_c D_c = D_c J_c = DJ \otimes \omega CC_{1,2p-1}$ implies $JD = (-1)^p DJ$ because $CC_{1,2p-1}\omega CC_{1,2p-1}^* = \bar{\omega} = (-1)^p \omega$. Hence we arrive at the following definition:

Definition 5. Let $(A, \mathcal{H}, D)$ be an odd spectral triple. A *real structure of dimension $2p-1$ mod 8* on $(A, \mathcal{H}, D)$ is a conjugate linear isometry $J: \mathcal{H} \to \mathcal{H}$ satisfying

odd. a) $JD = \epsilon' DJ$, $J^2 = \epsilon$;
odd. b) for any $a \in A$, the operators a and $[D, a]$ commute with JAJ^*.

$\epsilon, \epsilon' \in \{-1, +1\}$ depend on $d = 2p-1$ according to the following table

$d =$	1	3	5	7
$\epsilon =$	1	−1	−1	1
$\epsilon' =$	−1	1	−1	1

$(A, \mathcal{H}, D, J)$ is called a *real spectral triple of dimension $2p-1$ mod 8*.

It is left as an exercise for the interested reader to construct the analogue of the charge conjugation operator on a manifold of odd dimension.

The definition of $\mathrm{Cl}_{p,q}$-real spectral triples is quite convenient to treat tensor products. Let $(A_j, \mathcal{H}_j, D_j, J_j, \phi_j)$ be Cl_{p_j,q_j}-real spectral triples for $j = 1, 2$. The tensor product is the $\mathrm{Cl}_{p_1+p_2,q_1+q_2}$-real spectral triple

$$(A_1 \otimes A_2, \mathcal{H}_1 \otimes \mathcal{H}_2, D_1 \otimes 1 + \gamma_1 \otimes D_2, J_1 \otimes J_2, \phi_1 \otimes \phi_2)$$

The above tensor products of operators are all *graded* tensor products. One easily checks the axioms for a $\mathrm{Cl}_{p_1+p_2,q_1+q_2}$-real spectral triple.

For real spectral triples in the sense of Def. 1, tensor products are awkward to define if both dimensions are arbitrary. For the standard model, we only need tensor products in which the first triple has dimension 0 mod 8. Then, the obvious formula

$$(A_1 \otimes A_2, \mathcal{H}_1 \otimes \mathcal{H}_2, D_1 \otimes 1 + \gamma_1 \otimes D_2, J_1 \otimes J_2)$$

works (with all tensor products graded). However, $J_1 \otimes J_2$ is not a real structure if J_1 is odd!

2.5 Relations to Real K-Homology

Let me now indicate the relation of $\mathrm{Cl}_{p,q}$-real spectral triples to Kasparov's real K-homology [123]. An important application is the formulation of Poincaré duality. Let $(A, \mathcal{H}, D, J, \phi)$ be a $\mathrm{Cl}_{p,q}$-real spectral triple.

Definition 6. The *opposite algebra* B^{opp} of a (graded) $*$-algebra B is equal to B as a (graded) $*$-vector space. However, the order of multiplication is reversed: $x^{\mathrm{opp}} \cdot y^{\mathrm{opp}} = (yx)^{\mathrm{opp}}$ for all $x, y \in B$.

Opposite algebras are a device for turning right modules into left modules and vice versa. In our case, the A-bimodule structure on $\mathcal{H}$ is translated to a representation of $A \otimes A^{\mathrm{opp}}$. In addition, we have an action of $\mathrm{Cl}_{p,q}$ that commutes with $A \otimes A^{\mathrm{opp}}$. Therefore, we get a $*$-homomorphism

$$\phi \colon A \otimes A^{\mathrm{opp}} \otimes_\mathbb{R} \mathrm{Cl}_{p,q} \to \mathcal{L}(\mathcal{H}), a \otimes b^{\mathrm{opp}} \otimes x \mapsto aJb^*J^*\phi(x).$$

By construction, $J\phi(a \otimes b^{\mathrm{opp}} \otimes x)J^* = b^*JaJ^*\phi(x) = \phi(b^* \otimes (a^*)^{\mathrm{opp}} \otimes x)$. Thus ϕ becomes a homomorphism of "real" algebras if we endow $A \otimes A^{\mathrm{opp}} \otimes \mathrm{Cl}_{p,q}$ with the natural real structure defined by $\overline{a \otimes b^{\mathrm{opp}} \otimes x} := b^* \otimes (a^*)^{\mathrm{opp}} \otimes x$ for all $a, b \in A$, $x \in \mathrm{Cl}_{p,q}$.

Let $F := \mathrm{sign}\, D$, so that $F = F^*$, $F^2 = 1$. If $[D, x]$ is bounded, then $[F, x]$ is compact. Thus $[F, \phi(x)]$ is a compact operator for all $x \in A \otimes A^{\mathrm{opp}} \otimes \mathrm{Cl}_{p,q}$. Since also $JFJ^* = F$, the data $(\phi, \mathcal{H}, F, J)$ yield an element of Kasparov's real K-homology $KR^0(A \otimes A^{\mathrm{opp}} \otimes \mathrm{Cl}_{p,q}, \mathbb{C}) \cong KR^{q-p}(A \otimes A^{\mathrm{opp}}, \mathbb{C})$. Thus a real spectral triple of dimension d yields an element of $KR^d(A \otimes A^{\mathrm{opp}}, \mathbb{C})$.

Definition 7 ([39]). Let $(A, \mathcal{H}, D, J)$ be a real spectral triple of some dimension. An S^0-*real structure* on this triple is a self-adjoint unitary $\varepsilon \in \mathcal{L}(\mathcal{H})$ (that is, $\varepsilon = \varepsilon^*$, $\varepsilon^2 = 1$) that commutes with A, D, and the grading γ, and anti-commutes with J. We call $(A, \mathcal{H}, D, J, \varepsilon)$ an S^0-*real spectral triple*.

Let $(A, \mathcal{H}, D, J, \varepsilon)$ be an S^0-real spectral triple. Let $\mathcal{H}_{\pm i}$ be the range of the orthogonal projection $\frac{1}{2}(1 \pm \varepsilon)$. Thus $\mathcal{H} \cong \mathcal{H}_i \oplus \mathcal{H}_{-i}$. Since J anti-commutes with ε and A commutes with ε, JAJ^* commutes with ε. Therefore, $A \otimes A^{\mathrm{opp}}$, D, and γ map $\mathcal{H}_i$ to $\mathcal{H}_i$ and $\mathcal{H}_{-i}$ to $\mathcal{H}_{-i}$. Restricting to, say, $\mathcal{H}_i$, we therefore get a spectral triple $(A \otimes A^{\mathrm{opp}}, \mathcal{H}_i, D_i)$ with the additional property that $[D, A]$ commutes with A^{opp}.

Since J anti-commutes with ε, it maps $\mathcal{H}_i$ to $\mathcal{H}_{-i}$ and $\mathcal{H}_{-i}$ to $\mathcal{H}_i$. The restriction of J to $\mathcal{H}_i$ therefore yields an isomorphism $\mathcal{H}_i \cong \overline{\mathcal{H}_{-i}}$ onto the conjugate of $\mathcal{H}_{-i}$. Thus we can reconstruct $\mathcal{H}$ from $\mathcal{H}_{-i}$. Indeed, the S^0-real spectral triple $(A, \mathcal{H}, D, J, \varepsilon)$ is determined uniquely by its restriction $(A \otimes A^{\mathrm{opp}}, \mathcal{H}_i, D_i)$ to $\mathcal{H}_i$ and the dimension $2p$.

Given the data $(A \otimes A^{\mathrm{opp}}, \mathcal{H}_i, D_i)$, we put $\mathcal{H} := \mathcal{H}_i \oplus \overline{\mathcal{H}_i}$ and define ε by $\varepsilon|_{\mathcal{H}_i} = 1$ and $\varepsilon|_{\overline{\mathcal{H}_i}} = -1$. Let $J|_{\mathcal{H}_i}: \mathcal{H}_i \to \overline{\mathcal{H}_i}$ be the identity map, considered as an anti-isomorphism $\mathcal{H}_i \to \overline{\mathcal{H}_i}$. Define $J|_{\overline{\mathcal{H}_i}}: \overline{\mathcal{H}_i} \to \mathcal{H}_i$ such that $J^2 = \pm 1$ gives the sign appropriate for dimension $2p$. The operation of D on $\mathcal{H}_i$ is already prescribed by D_i. Since $JD = DJ$, we have to put $D|_{\mathcal{H}_{-i}} := JD_iJ^*|_{\mathcal{H}_{-i}}$. The operation of A on $\mathcal{H}_i$ is prescribed by the action of $A \otimes A^{\mathrm{opp}}$. Since $(a \otimes b^{\mathrm{opp}})(\xi) = aJb^*J^*(\xi)$, we have $b^*(J^*\xi) = J^*b^{\mathrm{opp}}(\xi)$ for all $\xi \in \mathcal{H}_i$. This determines the action of A on $\mathcal{H}_{-i}$. Finally, the grading γ on $\mathcal{H} \cong \mathcal{H}_i \oplus \mathcal{H}_{-i}$ is $(\gamma_i, (-1)^p J\gamma_i J^*)$. Thus $J\gamma = (-1)^p \gamma J$.

Summing up, there is a natural bijection between S^0-real spectral triples of dimension $2p$ and spectral triples $(A \otimes A^{\mathrm{opp}}, \mathcal{H}_i, D_i)$ with the additional property that $[D_i, A]$ commutes with A^{opp}.

The notion of S^0-reality was introduced by Atiyah in the context of "real" K-theory for topological spaces. A "real" topological space X is a topological space together with an involution $\tau\colon X \to X$. The "real" K-theory of such "real" spaces is concerned with "real" vector bundles over X. A "real" vector bundle is a complex vector bundle with a conjugate linear involution lifting the involution τ on the base space.

Let S^0 be the two-point space $\{\pm i\}$ endowed with the involution that exchanges the two points. The S^0-real K-theory of X is, by definition, the "real" K-theory of $X \times S^0$ with the product involution. Atiyah observes that S^0-real K-theory is nothing but the ordinary complex K-theory of X (regardless of the involution on X). Indeed, $X \times S^0$ is a disjoint union of two copies of X and the involution exchanges these two copies. A "real" vector bundle over X is determined uniquely by the restriction to the first copy of X, which is just a complex vector bundle. Furthermore, any complex vector bundle over X can be extended uniquely to a "real" vector bundle over $X \times S^0$. Thus the "real" K-theory of $X \times S^0$ is the same as the ordinary complex K-theory of X.

2.6 Real Structures on the NC Torus

Finally, let us consider another example of a real structure that illustrates the relation of condition b) of Def. 1 to the modular theory of von Neumann algebras. Let A_θ be a two-dimensional smooth noncommutative torus. That is, A_θ is generated by two unitaries u_1, u_2 subject to the relation $u_2 u_1 = \lambda u_1 u_2$ with $\lambda = \exp(2\pi i \theta)$. Elements of A_θ are described by "Fourier series" $\sum_{n_1, n_2 \in \mathbb{Z}} a_{n_1, n_2} u_1^{n_1} u_2^{n_2}$. The sum of $\sum a_{n_1,n_2} u_1^{n_1} u_2^{n_2}$ belongs to A_θ iff the sequence (a_{n_1,n_2}) is rapidly decreasing. Hence for $\theta = 0$ we get the algebra of smooth functions on the ordinary two-torus. The Lebesgue measure on the torus corresponds in the Fourier picture to the linear functional $\tau(\sum a_{n_1,n_2} u_1^{n_1} u_2^{n_2}) := a_{0,0}$. This continues to be a faithful, positive trace for $\theta \neq 0$. Thus we can construct a Hilbert space $L^2(A_\theta, \tau)$ as the completion of A_θ with respect to the inner product $(a,b) := \tau(ab^*)$. A_θ acts on $L^2(A_\theta, \tau)$ by left multiplication.

We now carry over the Dirac triple on the 2-torus to the case $\theta \neq 0$. For that purpose, we just have to write it down in terms of Fourier series and check that everything makes sense for $\theta \neq 0$. Let $\mathcal{H}^\pm := L^2(A_\theta, \tau)$, put $\mathcal{H} := \mathcal{H}^+ \oplus \mathcal{H}^-$. The $\mathbb{Z}_2$-grading on $\mathcal{H}$ is defined by $\gamma|_{\mathcal{H}^+} = 1$, $\gamma|_{\mathcal{H}^-} = -1$. The continuous linear map $\partial \colon A_\theta \to A_\theta$ defined by

$$\partial(u_1^{n_1} u_2^{n_2}) := \sqrt{2\pi}(in_1 + n_2) u_1^{n_1} u_2^{n_2}$$

is a derivation. View it as an unbounded operator $\mathcal{H}^- \to \mathcal{H}^+$. The adjoint of ∂ is the derivation

$$\partial^*(u_1^{n_1} u_2^{n_2}) := \sqrt{2\pi}(-in_1 + n_2) u_1^{n_1} u_2^{n_2}.$$

Let $D := \begin{pmatrix} 0 & \partial \\ \partial^* & 0 \end{pmatrix}$. The operators ∂, ∂^* are already diagonal. Using this, it is easy to find a basis of eigenvectors for D and to show that D is $(2, \infty)$-summable. One verifies that $(A_\theta, \mathcal{H}, D)$ is a spectral triple.

We now construct a real structure of dimension 2 mod 8 on this spectral triple as follows. Let $J_0 \colon L^2(A_\theta, \tau) \to L^2(A_\theta, \tau)$ be the conjugate linear map defined by $J_0(x) = x^*$ for all $x \in A_\theta$. Since τ is a trace, J_0 is an isometry. It satisfies $J_0^2 = 1$. For $x \in A_\theta$, let L_x and R_x be the operation of left and right multiplication by x on $L^2(A_\theta, \tau)$. The rule $(ab)^* = b^* a^*$ implies that $J_0 L_x J_0^* = R_{x^*}$. Thus $J_0 A_\theta J_0^*$ commutes with A_θ. Furthermore, since ∂ is a derivation, $[\partial, L_x] = L_{\partial(x)}$, so that $J_0 A_\theta J_0^*$ commutes with $[\partial, A_\theta]$. The same holds for ∂^*. Finally, one computes that $J_0 \partial = -\partial^* J_0$ and $J_0 \partial^* = -\partial J_0$. The isometry J_0 is a special case of the Tomita isometry. The above remarks also apply in the more general situation.

We now define our real structure $J \colon \mathcal{H} \to \mathcal{H}$ by $J := \begin{pmatrix} 0 & J_0 \\ -J_0 & 0 \end{pmatrix}$. The computations with J_0 show that J is indeed a real structure on $(A, \mathcal{H}, D)$ of dimension 2 mod 8 as desired.

3 The Commutative Case: Spinors, Dirac Operator and de Rham Algebra

Michael Frank

University of Leipzig

The present chapter is a short survey on the mathematical basics of Classical Field Theory including the Serre-Swan' theorem, Clifford algebra bundles and spinor bundles over smooth Riemannian manifolds, Spin$^{\mathbb{C}}$-structures, Dirac operators, exterior algebra bundles and Connes' differential algebras in the commutative case, among other elements. We avoid the use of principal bundles and put the emphasis on a module-based approach using Serre-Swan's theorem, Hermitian structures and module frames. A detailed proof of the differential algebra isomorphism between the set of smooth sections of the exterior algebra bundle and Connes' differential algebra is presented.

In the first two sections we explain the Gel'fand and the Serre-Swan theorems to explain the background of ideas leading to noncommutative geometry. In section three Hermitean structures on vector bundles and generalized module bases called frames are introduced to have some more structural elements for proving. Furthermore, we give a short introduction to the theory of Clifford and spinor bundles over compact smooth Riemannian manifolds M. Following J. C. Várilly [216] we use the duality between vector bundles and projective finitely generated $C^\infty(M)$-modules as described by the Serre-Swan theorem to give a comprehensive account to the commutative theory. The spectral triple is derived and the crucial properties of the Dirac operator are listed without proof.

3.1 The Theorems by Gel'fand and Serre-Swan

One of the corner stones of the beginning of noncommutative geometry was I. M. Gel'fand's theorem published in 1940. He established an equivalence principle between some topological objects and algebraic-axiomatic structures that can be expressed in the following way (cf. [22,162]):

Theorem 1 (I. M. Gel'fand). *Let A be a commutative C^*-algebra and X the set of its characters. The topology on X should be that one induced by the weak* topology on the dual space A^*. Then X is a locally compact Hausdorff space, and X is compact iff A is unital. The C^*-algebra A is $*$-isomorphic to the commutative C^*-algebra $C_0(X)$ of all continuous functions on X vanishing at infinity.*

In a more contemporary language this bijection can be expressed as a categorical equivalence. We have to add a set of suitable morphisms to the sets of objects 'commutative C*-algebras' and 'locally compact Hausdorff spaces'. They are called proper morphisms: for C*-algebras we have to take $*$-homomorphisms that map approximate identities to approximate identities, and for locally trivial Hausdorff spaces we have to select those continuous maps for which the pre-image of a compact set is always compact. Then we can summarize the categorical equivalence:

commutative C*-algebras $C(X)$	locally compact Hausdorff spaces X
proper $*$-homomorphisms	proper continuous homomorphisms

The noncommutative viewpoint enters the picture removing the commutativity condition on the multiplication in C*-algebras. Algebraically the left side is still a proper category, and many theorems for commutative C*-algebras can be generalized to the noncommutative situation. (But, there are also pure noncommutative structures like those described by Tomita-Takesaki theory.) However, the right side possesses no obvious candidate for a counterpart of the left side generalization to preserve the categorical equivalence. One reason is that the notion of a point that is crucial for any geometry becomes a vacuous notion under such an extension of the theory. Consequently, what we are left with is the algebraic noncommutative picture on the left side.

Looking for further topological and geometrical structures that can be categorically replaced by appropriate algebraic structures J.-P. Serre [201] (1957/58) and R. G. Swan [207](1962) independently established a categorical equivalence between projective finitely generated $C(X)$-modules and locally trivial vector bundles over X for compact Hausdorff spaces X. To describe it in greater detail some preparation is necessary.

To introduce both the notions, first, define a (left) unital A-module $\mathcal{H}$ over a unital algebra to be *projective finitely generated* if it is a direct summand (in an A-module sense) of a free A-module A^n for $n \in \mathbb{N}$, where A^n consists of all n-tuples of elements of A equipped with coordinate-wise addition and an action of A on A^n given as (left) multiplication of any n-tuple entry by fixed elements of A. The set of projective finitely generated A-modules can be equipped with the structure of direct sums $\oplus$ of A-modules. To introduce the structure of a module tensor product we have to consider them as A-bimodules defining another (right) action of A on A^n as a (right) multiplication of any n-tuple entry by fixed elements of A. The module tensor product $\mathcal{H}_1 \otimes_A \mathcal{H}_2$ is the algebraic tensor product of the linear spaces $\mathcal{H}_1$, $\mathcal{H}_2$ factored by the module ideal generated as the linear hull

$$\text{Lin}\{h_1 \otimes ah_2 - h_1 a \otimes h_2 : h_1 \in \mathcal{H}_1, h_2 \in \mathcal{H}_2, a \in A\}.$$

We have associative and distributive laws for the addition and the tensor products and a commutative law for addition. The neutral element of addition is the A-module consisting only of the zero element, and the neutral element of the module tensor product is the bimodule $A^1 = A$.

As the set of *homomorphisms* we consider all A-(bi-)module homomorphisms of projective finitely generated A-(bi-)-modules.

The second structure involved in the stressed for categorical equivalence consists of locally trivial vector bundles over compact Hausdorff spaces X.

The Commutative Case: Spinors, Dirac Operator and de Rham Algebra

Definition 1. Given a topological space E, a compact Hausdorff space X and a continuous mapping $p : E \to X$. Then E is a *locally trivial vector bundle* (E, p, X) *over* X if for every $x \in X$ there exists a finite-dimensional vector space E_x (equipped with the Euclidean topology) and a neighborhood $U_x \subseteq X$ such that a homeomorphism $\phi : U_x \times E_x \to p^{-1}(U_x)$ exists and $p \circ \phi(x, e) = x$ for any $x \in X$. In case $U_x \equiv X$ the vector bundle is *(globally) trivial*.

A map $\phi : (E, p, X) \to (F, q, X)$ is a *vector bundle isomorphism* in case ϕ is bijective, ϕ and ϕ^{-1} are continuous and $\phi(E_x) = F_x$ is linear for any $x \in X$. The map ϕ is a *vector bundle homomorphism* if ϕ is continuous and $\phi(E_x) \subseteq F_x$ is a linear embedding as a subspace.

We call X the *base space*, E the *total space*, $E_x = p^{-1}(\{x\})$ the *fibre over* x and p the projection map.

Note, that the compactness of X implies $\sup(\dim(E_x)) < \infty$. As one of the alternative descriptions of vector bundles in geometry we can describe them in local terms: a vector bundle (E, p, X) is given by an atlas $\{U_\alpha\} \subset X$ of (open) charts and of coordinate homeomorphisms $\{f_\alpha : U_\alpha \times E_x \to p^{-1}(U_\alpha)\}$ ($x \in U_\alpha$) such that the transition functions

$$f_{\alpha\beta} := f_\beta^{-1} f_\alpha : (U_\alpha \cap U_\beta) \times E_x \to (U_\alpha \cap U_\beta) \times E_x$$

are described by $f_{\alpha\beta}(x, e) = (x, \overline{f_{\alpha\beta}}(x)e)$ with continuous functions $\overline{f_{\alpha\beta}} \in GL(n, \mathbb{C})$ fulfilling the law

$$\overline{f_{\alpha\alpha}} = \mathrm{id}_{U_\alpha}, \quad \overline{f_{\alpha\gamma}} \overline{f_{\gamma\beta}} \overline{f_{\beta\alpha}} = \mathrm{id}_{U_\alpha \cap U_\beta \cap U_\gamma}.$$

We can show that the condition $\overline{f_{\alpha\beta}} \in GL(n, \mathbb{C})$ can be always reduced to $\overline{f_{\alpha\beta}} \in U(n)$ (or, for real vector spaces, $\overline{f_{\alpha\beta}} \in O(n)$) changing the coordinate functions in a suitable way, cf. [153,153]. The group $U(n)$ (or $O(n)$) is said to be the *structural group* of the vector bundle.

For further use we introduce the notion of an orientation on vector bundles over orientable compact manifolds.

Definition 2. Let M be an orientable compact manifold. The vector bundle (E, p, M) *is orientable* if there exists an atlas $\{U_\alpha\}$ describing E with transition functions $\{\overline{f_{\alpha\beta}}\} \in GL^+(n, \mathbb{C})$. The corresponding atlas is said to be *an orientation of the vector bundle* (E, p, M).

For a fixed compact Hausdorff space X the set of vector bundles with base space X can be equipped with some algebraic structure. The *Whitney sum of two vector bundles* (E, p, X) and (F, q, X) is the vector bundle $(E \oplus F, p \oplus q, X)$, where

$$E \oplus F := \{(e, f) \in E \times F : p(e) = q(f) \in X\},$$
$$(p \oplus q)(e, f) := p(e) = q(f) \in X.$$

Local triviality is preserved under Whitney sums. The fibres are the vector spaces $E_x \oplus F_x$. The *tensor product of two vector bundles* (E, p, X) and (F, q, X) is the vector bundle $(E \otimes F, p \otimes q, X)$ with the fibres $E_x \otimes F_x$ for $x \in X$ and the transition functions $\overline{f_{\alpha\beta}}(x) := \overline{f_{\alpha\beta,E}}(x) \otimes \overline{f_{\alpha\beta,F}}(x)$ coming from a common atlas $\{U_\alpha\} \subset X$ of the vector bundles (E, p, X) and (F, p, X). We observe that for trivial vector bundles $X \times \mathbb{C}^n =: \overline{n}$ the two operations are related by the isomorphism $E \otimes \overline{n} = \oplus_{i=1}^n E_{(i)}$, where $n \in \mathbb{N}$ is arbitrary. Concerning the algebraic properties of the two operations both they are associative and fulfil the obvious distributivity laws, and the Whitney addition is commutative in the sense of an appropriate isomorphism of vector bundles. The neutral elements are $\overline{0}$ and $\overline{1}$, respectively. One of the central observations is Swan's theorem:

Theorem 2 (R. G. Swan, 1962). *Let (E, p, X) be a locally trivial vector bundle over a compact Hausdorff base space X. There exists a locally trivial vector bundle (F, q, X) over X such that $(E \oplus F, p \oplus q, X)$ is trivial (with finite-dimensional fibre).*

The proof is elaborated, and we refer to R. G. Swan's paper [207] or to [72,153] for different versions of proofs.

Definition 3. A *section* in a vector bundle (E, p, X) is a continuous map $s : X \to E$ such that $(p \circ s)(x) = x$ for every $x \in X$. The set of sections of (E, p, X) is denoted by $\Gamma(E)$.

Proposition 1. *Let X be a compact Hausdorff space. Every locally trivial vector bundle admits non-trivial sections. For every vector bundle (E, p, X) the set $\Gamma(E)$ has the algebraic structure of a $C(X)$-module.*

Any isomorphism of vector bundles induces an isomorphism of the corresponding modules of sections. Whitney sums of vector bundles correspond to direct $C(X)$-module sums of the related modules of sections, tensor products of vector bundles correspond to bimodule tensor products.

For compact X the $C(X)$-module $\Gamma(E)$ is projective and finitely generated, in particular, $\Gamma(X \times \mathbb{C}^n) \cong C(X)^n$ for every $n \in \mathbb{N}$.

Proof. The existence of continuous sections can be proved applying Uryson's Lemma to constant sections in the (trivial) part of the vector bundle over one chart U, getting continuous sections of the whole vector bundle supported in one chart U over which the vector bundle is trivial.

Any bundle homomorphism $\phi : (E, p, X) \to (F, q, X)$ maps sections in E to sections in F. If ϕ is a bundle isomorphism, then $\phi_* : \Gamma(E) \to \Gamma(F)$ is a $C(X)$-module isomorphism.

We observe that $\Gamma(X \times \mathbb{C}^n) \cong C(X)^n$. These $C(X)$-modules are free and finitely generated. Since $C(X)^n \cong \Gamma(E \oplus F) \cong \Gamma(E) \oplus \Gamma(F)$ for a given vector bundle (E, p, X), some vector bundle (F, p, X) and $n < \infty$ by Swan's theorem, $\Gamma(E)$ is projective and finitely generated.

The Commutative Case: Spinors, Dirac Operator and de Rham Algebra

Theorem 3 (J.-P. Serre, 1957/58, R. G. Swan, 1962). *Let X be a compact Hausdorff space and $\mathcal{E}$ be a finitely generated projective $C(X)$-module. If $\mathcal{E} \oplus \mathcal{G} \cong C(X)^n$ for some $n < \infty$, then let P be the projection of $C(X)^n$ onto $\mathcal{E}$ along $\mathcal{G}$. Interpreting P as an element of $M_n(C(X)) \cong C(X, M_n(\mathbb{C}))$ define*

$$\Xi(\mathcal{E}) := \{(x,e) \in X \times \mathbb{C}^n : e \in \mathrm{ran}(P)\}.$$

Then $\Xi(\mathcal{E})$ is a locally trivial vector bundle over X, $\Gamma(\Xi(\mathcal{E})) \cong \mathcal{E}$. Moreover, if $\mathcal{E} = \Gamma(E)$ for some vector bundle E, then $\Xi(\Gamma(E)) \cong E$.

Proof. $\Xi(\Gamma(E)) \cong E$: Assume $E \oplus F \cong X \times \mathbb{C}^n$ by Swan's theorem. Let $\pi_x : \mathbb{C}^n \to E_x$ be the fibrewise projection, $x \in X$. Define $\pi : X \times \mathbb{C}^n \to E$ by $\pi(x,e) = (x, \pi_x(e))$ for $x \in X$, $e \in \mathbb{C}^n$. Then π is a correctly defined surjective bundle homomorphism.

Let $P = \pi_* : \Gamma(X \times \mathbb{C}^n) \to \Gamma(E) \oplus \Gamma(F)$ be the induced $C(X)$-module map, a projection onto $\Gamma(E)$. Note, that $P(x) = \pi_x$ for every $x \in X$. Therefore, $E = \Xi(\Gamma(E))$ by construction.

$\Gamma(\Xi(\mathcal{E})) = \mathcal{E}$: Note, that $(\Xi(\mathcal{E}))_x = \{x\} \times \{e \in \mathbb{C}^n : e \in \mathrm{ran}(P(x))\}$ are the fibres of $\Xi(\mathcal{E})$. The family of projections $\{P(x)\}$ is continuous, and $\Xi(\mathcal{E})$ becomes a locally trivial vector bundle. Thus, $\Gamma(\Xi(\mathcal{E})) = \{f \in C(X, \mathbb{C}^n) : f \in \mathrm{ran}(P) = \mathcal{E}\} \cong \mathcal{E}$.

Formulating the result in a categorical language we obtain a categorical equivalence between an algebraic and a geometric category if suitable sets of $C(X)$-module and bundle homomorphisms are chosen:

projective, finitely generated $C(X)$-modules	locally trivial vector bundles (E, p, X)
proper $C(X)$-module maps	proper bundle homomorphisms

We would like to point out that this categorical equivalence can be extended to the situation of infinite-dimensional fibres, however we will lose local triviality of the Banach bundles if we try to preserve a suitable category of $C(X)$-modules like Banach or Hilbert $C(X)$-modules on the left side. Moreover, most locally trivial bundles over compact Hausdorff spaces X with fibre l_2 turn out to be automatically globally trivial.

Now, we specify the compact Hausdorff space X to be a compact smooth manifold M. The observation to be made is that every locally trivial vector bundle over M with continuous transition functions in some atlas is in fact equipped with an atlas containing smooth transition functions, i.e. there is no reason to distinguish between 'continuous' and 'smooth' vector bundles over smooth compact manifolds M, cf. [153] for a proof.

Lemma 1. *For every vector bundle (E, p, M) there exists an atlas on M such that E is trivial over every chart U_α and the transition functions $\overline{f_{\alpha\beta}} : U_\alpha \cap U_\beta \to GL(n, \mathbb{C})$ are smooth functions.*

The Frèchet algebra $C^\infty(M)$ and the C*-algebra $C(M)$ have the same set of characters: every character on $C^\infty(M)$ is automatically continuous and a measure on M and hence, a character of $C(M)$. Consequently, $C^\infty(M)^n \cong \Gamma^\infty(M \times \mathbb{C}^n)$, and the categorical equivalence between projective $C^\infty(M)$-modules and vector bundles over M is a reduction of Serre-Swan's categorical equivalence. For the Whitney sum and the bundle tensor product we get the following corresponding module operations on the $C^\infty(M)$-modules of smooth sections:

$$\Gamma^\infty(E \oplus F) = \Gamma^\infty(E) \oplus_{C^\infty(M)} \Gamma^\infty(F),$$
$$\Gamma^\infty(E \otimes F) = \Gamma^\infty(E) \otimes_{C^\infty(M)} \Gamma^\infty(F).$$

3.2 Hermitean Structures and Frames for Sets of Sections

As an essential tool we need the existence and the properties of a continuous field of scalar products on the fibres of vector bundles. This structure is not needed to prove the Serre-Swan' theorem, it arises additionally.

Definition 4. Let X be a compact Hausdorff space and (E, p, X) be a vector bundle with base space X. A *$C(X)$-valued inner product on (E, p, X)* is a bilinear mapping $\langle .,. \rangle : \Gamma(E) \times \Gamma(E) \to C(X)$ that is continuous in both the arguments, acts fibrewise (i.e. is $C(X)$-linear in the first argument) and its restriction to any fibre E_x generates a scalar product on it. (Some authors refer to this structure as to a *Hermitean structure* on the vector bundle.)

Theorem 4. *Let X be a compact Hausdorff space and (E, p, X) be a vector bundle with base space X. Then (E, p, X) admits $C(X)$-valued inner products $\langle .,. \rangle$ on the $C(X)$-module $\Gamma(E)$ such that $\Gamma(E)$ is complete with respect to the resulting norm $\|.\| := \langle .,. \rangle^{1/2}$.*
Any two $C(X)$-valued inner products $\langle .,. \rangle_1$, $\langle .,. \rangle_2$ are related by a positive invertible $C(X)$-linear operator S on $\Gamma(E)$ via the formula $\langle .,. \rangle_1 \equiv \langle S(.), . \rangle_2$. If X is a smooth manifold then $\langle .,. \rangle$ restricted to $\Gamma^\infty(E) \times \Gamma^\infty(E)$ takes values in $C^\infty(X)$.

Proof. Because of the categorical equivalence obtained by J.-P. Serre and R. G. Swan we have only to indicate the existence and the properties of $C(X)$-valued inner products on finitely generated projective $C(X)$- or $C^\infty(X)$-modules. For the free C*-module $C(X)^n$ the $C(X)$-valued inner product is defined as $\langle (f_1, ..., f_n), (g_1, ..., g_n) \rangle = \sum_{i=1}^n f_i \overline{g_i}$. For direct summands $P(C(X)^n)$ of $C(X)^n$ we reduce this $C(X)$-valued inner product to elements of them.

The relation between two $C(X)$-valued inner products follows from an analogue of Riesz' representation theorem for $C(X)$-linear bounded module maps from $C(X)^n$ into $C(X)$. (Attention: This may fail for more general $C(X)$-modules with $C(X)$-valued inner products.)

If X is a smooth manifold, then the $C(X)$-valued inner product defined above maps elements with smooth entries to smooth functions on X. A perturbation of the $C(X)$-valued inner product by a positive invertible operator

S that preserves the range $C^\infty(X)$ of it or the restriction to a direct summand of $C^\infty(X)^n$ do not change this fact.

We would like to remark that for more general $*$-algebras $\mathcal{A} \subset C^\infty(X)$ that are closed under holomorphic calculus and contain the identity the property of $\mathcal{A}$-valued inner products on the correspondingly reduced set of sections $\Gamma^\mathcal{A}(E) \subset \Gamma^\infty(E)$ to possess an analogue of the Riesz' property has to be axiomatically supposed, in general.

Now, we indicate the existence of finite sets of generators of $\Gamma^\infty(E)$ as a $C^\infty(M)$-module for vector bundles (E, p, M) over smooth manifolds M. Consider the free $C^\infty(M)$-module $\Gamma^\infty(M \times \mathbb{C}^n) = C^\infty(M)^n$ for $n \in \mathbb{N}$ and a $C^\infty(M)$-valued inner product $\langle.,.\rangle_0$ on it. Then there exists an orthonormal with respect to $\langle.,.\rangle_0$ basis consisting of n elements of this module. Indeed, on free $C(M)$-modules $C(M)^n$ every $C(M)$-valued inner product is related to the canonical $C(M)$-valued inner product by a bounded invertible positive module operator S that fulfills the identity $\langle.,.\rangle_{can.} \equiv \langle S(.),.\rangle$. The restriction of $\langle.,.\rangle_{can.}$ to $C^\infty(M)^n$ is $C^\infty(M)$-valued, and $\langle.,.\rangle_0$ can be extended to $C(M)^n$. So the linking operator S exists on $C(M)^n$, and its restriction to $C^\infty(M)^n$ maps smooth elements to smooth elements. However, the canonical $C^\infty(M)$-valued inner product on $C^\infty(M)^n$ admits an orthonormal basis consisting of smooth elements:

$$\{e_1, ..., e_n \; : \; e_i = (0, ..., 0, 1_{(i)}, 0, ..., 0)\}.$$

Consequently, $\{S^{-1/2}(e_i) \; : \; i = 1, ..., n\}$ is an orthonormal basis of $C^\infty(M)^n$ with respect to the given $C^\infty(M)$-valued inner product $\langle.,.\rangle_0$.

Let $\mathcal{E}$ be a projective finitely generated $C^\infty(M)$-module, i.e. $\mathcal{E} \oplus \mathcal{F} = C^\infty(M)^n$ for a finite integer n. Denote by P the $C^\infty(M)$-linear projection onto $\mathcal{E}$ along $\mathcal{F}$. Then the set $\{P(e_i) \; : \; i = 1, ..., n\}$ of elements of $\mathcal{E}$ has the remarkable property that

$$\xi = \sum_{i=1}^{n} \langle \xi, P(e_i)\rangle_0 P(e_i)$$

for every $\xi \in \mathcal{E}$. The engeneering literature on wavelets calls such sets of generators of Hilbert spaces *(normalized tight) frames*, whereas the literature on conditional expectations calls them *quasi-bases* or *(module) bases*. The notion 'basis' is, however, misleading since the elements of the generator sequence $\{P(e_i) \; : \; i = 1, ..., n\}$ may allow a non-trivial $C^\infty(M)$-linear decomposition of the zero element of $\mathcal{E}$. To see that let $\mathcal{E}$ be simply the subset of all elements admitting only allover equal entries in their n-tuple representation. For more details we refer the reader to [83,84]. To summarize the arguments we formulate

Theorem 5. *Let M be a smooth compact manifold and (E, p, M) be a vector bundle with base space M. Let $\langle.,.\rangle$ be a Hermitean structure on it. Then the*

projective finitely generated $C^\infty(M)$-module $\Gamma^\infty(E)$ possesses a finite subset $\{\eta_i : i \in \mathbb{N}\}$ such that $\Gamma^\infty(E)$ is generated as a $C^\infty(M)$-module by this set and the equality

$$\xi = \sum_{i=1}^{n} \langle \xi, \eta_i \rangle \eta_i$$

is satisfied for every $\xi \in \Gamma^\infty(E)$.

3.3 Clifford and Spinor Bundles, Spin Manifolds

Let (M, g) be a smooth Riemannian manifold, where the Riemannian metric g_x induces a scalar product on T_xM for any $x \in M$. Note, that the tangent space T_xM and the cotangent space T_x^*M are isomorphic via the scalar product on T_xM for any $x \in M$. If (T_xM, g_x) denotes the Hilbert tangent space then let (T_x^*M, g_x^{-1}) denote the resulting Hilbert cotangent space.

Let $Cl(T_xM, g_x)$ be the real Clifford algebra of the tangent space T_xM with respect to the scalar product induced by the Riemannian metric g_x, $x \in X$ arbitrarily fixed. This algebra is defined to be a quotient of the tensor algebra $\mathcal{T}(T_xM)$ generated by the linear space T_xM, i.e. of

$$\mathcal{T}(T_xM) = \mathbb{C} \oplus T_xM \oplus (T_xM \otimes T_xM) \oplus ... \oplus (T_xM \otimes ... \otimes T_xM) \oplus ....$$

More precisely,

$$Cl(T_xM, g_x) := \mathcal{T}(T_xM)/\text{Ideal}(e \otimes e - g_x(e, e) : e \in T_xM).$$

The real Clifford algebra $Cl(T_xM, g_x)$ possesses a $\mathbb{Z}_2$-grading induced by the map $\chi_x : (x, e) \in T_xM \to (x, -e) \in T_xM$, i.e. by the linear operator χ on $Cl(T_xM, g_x)$ with the property $\chi^2 = \text{id}$, with eigen-values $\{1, -1\}$ and isomorphic eigenspaces $Cl^{even}(T_xM, g_x)$, $Cl^{odd}(T_xM, g_x)$ summing up to the algebra itself. The words 'even' and 'odd' refer to the highest degree of the element under consideration and its property to be an even or odd number. If $n = 2m + 1$ then χ is realized as a multiplication by a central element. Extending the isomorphism between tangent space and cotangent space via the scalar product g_x on the first space we obtain a canonical algebraic isomorphism of the Clifford algebras $Cl(T_xM, g_x)$ and $Cl(T_x^*M, g_x)$ for any fixed $x \in X$.

The *Clifford algebra bundle* $\mathbb{Cl}(M)$ is defined fibrewise using the atlas on M induced by the tangent bundle atlas of TM (or the cotangent bundle atlas of T^*M):

$$\mathbb{Cl}_x(M) := Cl(T_xM, g_x) \otimes_{\mathbb{R}} \mathbb{C} \stackrel{\tau}{\cong} \begin{cases} M_{2^m}(\mathbb{C}) & : n = 2m \\ M_{2^m}(\mathbb{C}) \oplus M_{2^m}(\mathbb{C}) & : n = 2m + 1 \end{cases}.$$

Note that the isomorphism τ is quite complicated, and in case $n = 2m + 1$ it maps both the even and the odd part of the Clifford algebra to both

The Commutative Case: Spinors, Dirac Operator and de Rham Algebra   29

the blocks of the matrix sum at the right (see [86, p. 15] for details). The Clifford bundle possesses a $\mathbb{Z}_2$-grading induced from that one on its fibres. The C*-algebra structure of $\Gamma(\mathbb{C}l(M))$ comes from the algebra structure of the Clifford algebra fibres and from the involution induced by $\otimes_{\mathbb{R}} \mathbb{C}$ from $\mathbb{C}$. The C*-norm exists and is uniquely defined since the multiplication and the involution are given and every fibre is finite-dimensional.

Some authors (cf. [216]) prefer to restrict the Clifford algebra bundle to the even part in case the dimension of the manifold is $n = 2m+1$. The loss of that alternative definition is the $\mathbb{Z}_2$-grading. The advantage of that approach is the structure of $\mathbb{C}l(M)$ as a continuous field of *simple* C*-algebras allowing the attempt to interpret this bundle as a homomorphism bundle derived from some other vector bundle with base space M. We prefer to postpone this reduction until the spinor bundle has to be built up.

Consider either the Clifford algebra bundle $\mathbb{C}l(M)$ over M for $n = 2m$ or the first matrix block part $\mathbb{C}l(M)^\dagger$ of the Clifford algebra bundle $\mathbb{C}l(M)$ over M for $n = 2m+1$ (in its matrix representation) locally: for every $x \in X$ we find vector spaces S_x such that the (first part of the) Clifford algebra bundle is locally isomorphic to the trivial homomorphism bundle $\text{Hom}(S_x)$ of the trivial bundle $(U \times S_x, p_U, U)$. The C*-algebra structure on $\Gamma(\mathbb{C}l_x(M))$ (resp., $\Gamma(\mathbb{C}l_x(M)^\dagger)$) induces a unique scalar product on S_x compatible with it. The dimension of the linear spaces S_x is constant and equals $\dim(S_x) = 2^m$ for any manifold dimensions n, $m := [n/2]$.

Whether we can glue these trivial pieces together to obtain a vector bundle S over the compact Riemannian manifold M carrying an irreducible left action of the Clifford bundle (resp., the first part of it) that acts locally in the manner described, or not? Unfortunately, not always. If $n = 2m$ the Clifford bundle $\mathbb{C}l(M)$ serves as a homomorphism bundle for some other vector bundle with the same compact base space M if and only if the Dixmier-Douady class $\delta(\mathbb{C}l(M)) \in H^3(M, \mathbb{Z})$ equals zero, where $\delta(\mathbb{C}l(M))$ also equals the third integral Stiefel-Whitney class $w_3(TM) \in H^3(M, \mathbb{Z})$. If $n = 2m+1$ the first part $\mathbb{C}l(M)^\dagger$ of the Clifford bundle is a homomorphism bundle of some other vector bundle if and only if the second part of it does so, if and only if the Dixmier-Douady class $\delta(\mathbb{C}l(M)^\dagger) \in H^3(M, \mathbb{Z})$ equals zero, where $\delta(\mathbb{C}l(M)^\dagger)$ also equals the third integral Stiefel-Whitney class $w_3(TM) \in H^3(M, \mathbb{Z})$. The first fact was observed by J. Dixmier in [64, Th. 10.9.3] and again investigated in connection with spinor bundles by R. J. Plymen [178].

To formulate the definition of a spinor bundle on a given compact smooth Riemannian manifold M or, equivalently, the definition of the property of M to be a Spin$^{\mathbb{C}}$-manifold we have to introduce the notion of Morita equivalence of certain unital *-algebras. We will do that only for the two *-algebras of interest, for more general cases we refer to [180,178]. Let us fix the unital *-algebra

$$B = \begin{cases} C^\infty(M, \mathbb{C}l(M)) = \Gamma^\infty(\mathbb{C}l(M)) & : n = 2m \\ C^\infty(M, \mathbb{C}l(M)^\dagger) = \Gamma^\infty(\mathbb{C}l(M)^\dagger) & : n = 2m+1 \end{cases}.$$

Definition 5. Let M be a compact smooth Riemannian manifold. Consider the unital $*$-algebras $A = C^\infty(M)$ and B. They are *Morita-equivalent as algebras* if there exists a B-A bimodule $\mathcal{E}$ and an A-B bimodule $\mathcal{F}$ such that $\mathcal{E} \otimes_A \mathcal{F} \cong B$ and $\mathcal{F} \otimes_B \mathcal{E} \cong A$ as B- and A-bimodules, respectively.

In our case $\mathcal{F}$ can be chosen to be a projective and finitely generated (left) module over the unital $*$-algebra $A = C^\infty(M)$ denoted by $\tilde{S}$. As a projective finitely generated $C^\infty(M)$-module $\tilde{S}$ admits a $C^\infty(M)$-valued inner product $\langle .,.\rangle_{C^\infty(M)}$. Then B can be realized as the $*$-algebra of bounded module operators over $\tilde{S}$ generated as $\mathrm{Lin}\{\langle \xi, \eta \rangle_{C^\infty(M,\mathbb{C}l(M))} : \xi, \eta \in \tilde{S}\}$, where

$$\langle \xi, \eta \rangle_{C^\infty(M,\mathbb{C}l(M))}(\nu) := \langle \nu, \xi \rangle_{C^\infty(M)} \eta \quad \text{for} \quad \nu \in \tilde{S}.$$

So the counterpart $\mathcal{E}$ of $\mathcal{F}$ can be described as the set $\{\bar{\xi} : \xi \in \mathcal{F}, \bar{\xi}a := \overline{(a^*\xi)}, a \in A\}$. Obviously, the right action of B on $\mathcal{F}$ is simultaneously swept to a left B-action on $\mathcal{E}$. The C^∞-module $\mathcal{F}$ together with the $C^\infty(M)$-valued inner product $\langle .,.\rangle_{C^\infty(M)}$ is said to be a B-A *imprimitivity bimodule*.

Definition 6 (R. J. Plymen, 1982). Let (M, g) be a compact smooth Riemannian manifold, let $A = C^\infty(M)$ and B as defined above in dependency on the dimension of M. Both A and B are unital $*$-algebras of smooth mappings.

We say that *the tangent bundle TM of M admits a* $\mathrm{Spin}^\mathbb{C}$-*structure* if TM is orientable as a vector bundle and the Dixmier-Douady class $\delta(\mathbb{C}l(M))$ equals zero for $n = 2m$ or, respectively, $\delta(\mathbb{C}l(M)^\dagger) = 0$ for $n = 2m+1$.

If this condition is fulfilled then *the* $\mathrm{Spin}^\mathbb{C}$-*structure on TM* is a pair $(\epsilon, \tilde{S})$ consisting of an orientation ϵ of TM and a B-A imprimitivity bimodule $\tilde{S}$.

The compact smooth Riemannian manifold M *is a* $\mathrm{Spin}^\mathbb{C}$-*manifold* if the tangent bundle TM of M admits a $\mathrm{Spin}^\mathbb{C}$-structure.

Questions like existence or uniqueness of $\mathrm{Spin}^\mathbb{C}$-structures are complicated and depend on several properties of the manifold M. For an accessible and detailed geometrical account see [86].

By the Serre-Swan' theorem the B-A imprimitivity bimodule $\tilde{S}$ can be realized as the $C^\infty(M)$-module of smooth sections $\Gamma^\infty(S)$ of a uniquely determined vector bundle (S, p_S, M) with base space M. The vector bundle (S, p_S, M) is called *the spinor bundle*.

If $n = 2m$ then the spinor bundle admits a non-trivial $\mathbb{Z}_2$-grading arising from the grading of the Clifford bundle $\mathbb{C}l(M)$: $S = S^+ \oplus S^-$, where $\dim(S_x^+) = \dim(S_x^-) = 2^{m-1}$.

Furthermore, the set of smooth sections of S always admits a $C^\infty(M)$-valued inner product $\langle .,.\rangle_{C^\infty(M)}$. The smooth sections of the spinor bundles are called *spinors*, or *chiral vector fields* in physics.

Definition 7. Let H be the Hilbert space

$$H := \overline{\left\{\xi \Gamma^\infty(S) : \int_M \langle \xi, \xi \rangle_{C^\infty(M)} \, dg < +\infty\right\}},$$

Sometimes H is referred to as *the spinor Hilbert space*. The Hilbert space H consists of all square-integrable sections of the spinor bundle S, i.e. $H = L_2(M, S)$.

The spinor Hilbert space H inherits the non-trivial $\mathbb{Z}_2$-grading arising from the grading of the spinor bundle in case $n = 2m$: $H = H^+ \oplus H^-$, where $H^\pm := L_2(M, S^\pm)$.

The sections of the Clifford bundle $\mathbb{C}l(M)$ act naturally on H. To look for details recall that $\Gamma(\mathbb{C}l(M)) = C(M) + \Gamma(T^*M) + ...$ Then the elements of $C(M)$, i.e. of the zeroth component of $\Gamma(\mathbb{C}l(M))$, act as multiplication operators on $\Gamma(S)$ and, hence, on H by continuity. Identifying $\Gamma^\infty(T^*M)$ by the $C^\infty(M)$-module $A^1(M)$ of 1-forms on M, the images of 1-forms under γ fulfill the rule

$$\gamma(\alpha)\gamma(\beta) + \gamma(\beta)\gamma(\alpha) = 2g^{ij}\alpha_i\beta_j \quad \text{for} \quad \alpha, \beta \in A^1(M).$$

Consequently, $\gamma(dx^k)^2 > 0$ and non-trivial 1-forms are faithfully represented. The representation $\gamma : \Gamma(\mathbb{C}l(M)) \to B(H)$ is called *the spin representation*. We will use it again in the last part of the present chapter.

3.4 Spin Connection and Dirac Operator

Let (M, g) be a smooth Riemannian manifold, where the Riemannian metric g_x induces a scalar product in the cotangent spaces T_x^*M for any $x \in M$. The Riemannian metric g on M gives rise to a unique *Levi-Civita connection* ∇^g (or *Riemannian connection*). It is defined on (contra-/covariant) C^∞-tensor fields over M of arbitrary order, ∇^g is symmetric, $\nabla^g(g^{ij}) = 0$ and the torsion of ∇^g vanishes.

In particular, $\nabla^g : A^1(M) \to A^1(M) \otimes_A A^1(M)$ obeying a Leibniz rule:

$$\nabla^g(\omega a) = \nabla^g(\omega)a + \omega \otimes da$$

for $a \in C^\infty(M)$ and arbitrary tensor fields ω on M. Lifting this Levi-Civita connection to the spinor bundle S (where $\Gamma(S)$ is equipped with the $C(M)$-valued inner product arising from $\tilde{S}$) we obtain the spin connection.

Definition 8. The *spin connection* is an operator $\nabla^S : \Gamma^\infty(S) \to \Gamma^\infty(S) \otimes_A A^1(M)$ that is linear and satisfies the two Leibniz rules

$$\nabla^S(\psi a) = \nabla^S(\psi)a + \psi \otimes da,$$

$$\nabla^S(\gamma(\omega)\psi) = \gamma(\nabla^g(\omega))\psi + \gamma(\omega)\nabla^S(\psi)$$

for $a \in A = C^\infty(M)$, $\omega \in A^1(M) = \Gamma^\infty(T^*M)$, $\psi \in \Gamma^\infty(S)$.

The spin connection on the spinor bundle (S, p_S, M) gives rise to the Dirac operator acting on the spinor Hilbert space H.

Definition 9. Let $m : \Gamma^\infty(S) \otimes_A A^1(M) \to \Gamma^\infty(S)$ be the mapping defined by the rule $m(\psi \otimes \omega) = \gamma(\omega)(\psi)$ for $\omega \in A^1(M) = \Gamma^\infty(T^*M) \subset \Gamma^\infty(\mathbb{Cl}(M))$, $\psi \in \Gamma^\infty(S)$. The *Dirac operator on S* is the mapping $\slashed{D} := m \circ \nabla^S$ that acts on the domain $\Gamma^\infty(S) \subset H$ of the spinor Hilbert space H as an unbounded operator.

The Dirac operator $\slashed{D}$ has a number of remarkable properties. We list them without proof for the commutative case. For detailed references see [38,86,153,215]:

- If $n = 2m$ then $\slashed{D} : \Gamma^\infty(S^\pm) \to \Gamma^\infty(S^\mp)$. Moreover, with respect to this decomposition of $\Gamma^\infty(S)$ the Dirac operator can be represented as

$$\slashed{D} = \begin{pmatrix} 0 & \slashed{D}^+ \\ \slashed{D}^- & 0 \end{pmatrix}, \quad \langle \slashed{D}^+(h^+), h^- \rangle = \langle h^+, \slashed{D}^-(h^-) \rangle$$

for $h^\pm \in \Gamma^\infty(S^\pm)$.
- If $n = 2m$ and $\chi : (h^+, h^-) \in H^+ \oplus H^- \to (h^+, -h^-) \in H^+ \oplus H^-$ is the grading operator on the spinor Hilbert space H then $\chi \slashed{D} + \slashed{D} \chi = 0$.
- $\slashed{D}$ is symmetric and extends to an unbounded self-adjoint operator on H. (Same denotation.)
- $[\slashed{D}, a]$ is compact and $[[\slashed{D}, a], b] = 0$ for every $a, b \in C^\infty(M)$.
- If the smooth Riemannian manifold M is compact, then $\slashed{D}$ is a Fredholm operator, i.e. $\ker(\slashed{D})$ is finite-dimensional.
- The operator $\slashed{D}^{-1}$ defined on the orthogonal complement of $\ker(\slashed{D})$ is compact. The eigenvalues $\{\lambda_k\}$ of $\slashed{D}^{-1}$ counted with multiplicity fulfil the relation $\lambda_k \leq C \cdot k^{-1/n}$ for some constant C and $n = \dim(M)$.
- The spectrum of $\slashed{D}$ is discrete and consists of eigenvalues of finite multiplicity.
- $\slashed{D}$ is an elliptic first order differential operator.
- The algebra $A = C^\infty(M)$ is represented on the spinor Hilbert space H by multiplication operators (via γ). We obtain

$$[\slashed{D}, a] = \slashed{D}(a\psi) - a\slashed{D}(\psi) = \gamma(da)\psi$$

for $a \in A = C^\infty(M)$, $\psi \in H$. In particular, since a is smooth and M is compact, the operator $[\slashed{D}, a]$ is bounded with the sup-norm $\|\gamma(da)\|_\infty$ of the multiplication operator by $\gamma(da)$.
- For the geodesic distance of two points $p, q \in M$ we have

$$d(p,q) = \sup\{|\hat{p}(a) - \hat{q}(a)| \: : \: a \in C^\infty(M), \|[\slashed{D}, a]\| \leq 1\},$$

where $\hat{p}$ is the character on $C^\infty(M)$ induced by evaluation in $p \in M$ and $\|\gamma(da)\|_\infty = \|a\|_{Lip} = \|[\slashed{D}, a]\|$.
- The Lichnérowicz formula is valid:

$$\slashed{D}^2 = \Delta^S + \frac{1}{4}R,$$

The Commutative Case: Spinors, Dirac Operator and de Rham Algebra   33

where R is the scalar curvature of the metric and Δ^S is the Laplacian operator lifted to the spinor bundle that can be described in local coordinates by $\Delta^S = -g^{ij}(\nabla_i^S \nabla_j^S - \Gamma_{ij}^k \nabla_k^S)$ with Γ_{ij}^k the Christoffel symbols of the connection.

- For any $f \in C^\infty(M)$ one has the formula

$$\int_M f\, dg = \frac{(-1)^n n \Gamma(n/2)}{2^{[n/2]+1-n} \pi^{-n/2}} \cdot \mathrm{Tr}_\omega(f|\slashed{D}|^{-n}),$$

where Tr_ω denotes the Dixmier trace.

3.5 The Universal Differential Algebra $\Omega C^\infty(M)$ and Connes' Differential Algebra $\Omega_\slashed{D} C^\infty(M)$

As a good source for the commutative approach to differential algebras we can refer to the monograph of G. Landi [136]. Complementary information can be found in [160].

Definition 10. Let M be a compact smooth manifold. Identify a suitable completion of the algebraic tensor product $C^\infty(M) \odot ... \odot C^\infty(M)$ with $C^\infty(M \times ... \times M)$, the same number of $\odot/\times$ operations supposed.

The *universal differential algebra* $\Omega C^\infty(M) = \oplus_p \Omega^p C^\infty(M)$ is defined by the linear spaces:

$$\Omega^0 C^\infty(M) := C^\infty(M)$$
$$\Omega^p C^\infty(M) := \{f \in \overline{\odot_1^{p+1} C^\infty(M)} :$$
$$f(x_1, ..., x_{k-1}, x, x, x_{k+2}, ..., x_{p+1}) = 0\ ,\ \forall k\}$$

The *exterior differential* $\delta : \Omega^p \to \Omega^{p+1}$ is defined by

$$(\delta f)(x_1, x_2) := f(x_2) - f(x_1)$$
$$(\delta f)(x_1, ..., x_{p+1}) := \sum_{k=1}^{p+1} (-1)^{k-1} f(x_1, ..., x_{k-1}, x_{k+1}, ..., x_{p+1})$$

The $C^\infty(M)$-bimodule structure on $\Omega C^\infty(M) := \oplus_p \Omega^p C^\infty(M)$ is given by:

$$(gf)(x_1, ..., x_{p+1}) := g(x_1) f(x_1, ..., x_{p+1})$$
$$(fg)(x_1, ..., x_{p+1}) := f(x_1, ..., x_{p+1}) g(x_{p+1})$$

It extends to a general multiplication by the formula

$$(fh)(x_1, ..., x_{(p+q)+1}) := f(x_1, ..., x_{p+1}) h(x_{p+1}, ..., x_{(p+q)+1})$$

for $f \in \Omega^p C^\infty(M)$, $h \in \Omega^q C^\infty(M)$.

Key properties of the exterior differential are linearity, the Leibniz rule and the vanishing of its square:

$$\delta(ab) = (\delta a)b + (-1)^p a(\delta b) \quad , \quad \delta^2 = 0 ,$$
$$\delta(\alpha a + \beta b) = \alpha(\delta a) + \beta(\delta b)$$

for $a \in \Omega^p C^\infty(M)$, $b \in \Omega C^\infty(M)$, $\alpha, \beta \in \mathbb{C}$. These three properties give rise to another representation of the differential algebra as a linear hull of standard elements as it is used in the noncommutative case:

$$\Omega^p C^\infty(M) = \mathrm{Lin}\{a_0 \delta a_1 ... \delta a_p \ : \ a_i \in C^\infty(M)\} ,$$

$$\delta(a_0 \delta a_1 ... \delta a_p) = \delta a_0 \delta a_1 ... \delta a_p .$$

We take the parity of the degree p as a grading for the differential algebra $\Omega C^\infty(M) = \bigoplus_p \Omega^p C^\infty(M)$.

To go further and to construct Connes' differential algebra we need another property of our compact smooth manifold M – it has to be Riemannian. Then we have a spectral triple $(C^\infty(M), H = L_2(M, S), \slashed{D})$ by construction, and we consider an algebraic representation of $\Omega C^\infty(M)$ on $B(H)$:

$$\pi : \Omega C^\infty \to B(H) \quad , \quad \pi(a_0 \delta a_1 ... \delta a_p) := a_0 [\slashed{D}, a_1]...[\slashed{D}, a_p]$$

where $a_i \in C^\infty(M)$, and $C^\infty(M)$ acts on $\tilde{S} \subseteq H$ by the usual module action. (If one introduces an involution on the differential algebra then π becomes a $*$-representation, however we do not need this additional structure for our purposes.) If we want π to be a representation commuting with the action of the differential in some way we run into difficulty since $\pi(\omega) = 0$ does not imply $\pi(\delta\omega) = 0$, in general. Fortunately, there exists a differential ideal of $\Omega C^\infty(M)$, the 'junk ideal' J.

Lemma 2. *Let $J_0 := \oplus_p J_0^p$ be the graded two-sided ideal of $\Omega C^\infty(M)$ given by*

$$J_0^p := \{\omega \in \Omega^p C^\infty(M) \ : \ \pi(\omega) = 0\} .$$

Then $J := J_0 + \delta J_0$ is a graded differential two-sided ideal of $\Omega C^\infty(M)$.

Proof. Consider an element $\omega = \omega_1 + \delta \omega_2 \in J^p$, $\omega_1 \in J_0^p$, $\omega_2 \in J_0^{p-1}$. For every $\nu \in \Omega^q C^\infty(M)$ we have the equality

$$\omega\nu = \omega_1 \nu + (\delta\omega_2)\nu = \omega_1 \nu \delta(\omega_2 \nu)$$
$$= \omega_1 \nu + \delta(\omega_2 \nu) - (-1)^{p-1} \omega_2 \delta\nu$$
$$= (\omega_1 \nu - (-1)^{p-1} \omega_2 \delta\nu) + \delta(\omega_2 \nu) \in J^{p+q} .$$

Similarly, we obtain $\nu\omega \in J^{p+q}$. Since $\delta^2 = 0$ the ideal J is a differential ideal.

Definition 11 (A. Connes). The *graded differential algebra of Connes' forms* over the algebra $C^\infty(M)$ is defined by

$$\Omega_D C^\infty(M) := \Omega C^\infty(M)/J \cong \pi(\Omega C^\infty(M))/\pi(\delta J_0).$$

The space of Connes' p-forms is $\Omega_D^p C^\infty(M) = \Omega^p C^\infty(M)/J^p$. On $\Omega_D C^\infty(M)$ there exists a differential induced by δ with the usual properties:

$$d : \Omega_D^p C^\infty(M) \to \Omega_D^{p+1} C^\infty(M) \quad , \quad d([\omega]) := [\delta\omega] \cong \pi([\delta\omega]).$$

3.6 The Exterior Algebra Bundle $\Lambda(M)$ and the de Rham Complex

Let M be a compact smooth manifold equipped with an atlas inherited from the cotangent bundle T^*M. Denote by $\Lambda(T_x^*M)$ the real exterior algebra of the cotangent space T_x^*M, $x \in X$. Recall that

$$\Lambda(T_x^*M) := \mathcal{T}(T_x^*M)/\text{Ideal}(e \otimes e : e \in T_x^*M).$$

The real exterior algebra $\Lambda(T_x^*M)$ possesses a $\mathbb{Z}_2$-grading, i.e. a linear operator χ on it with $\chi^2 = \text{id}$, eigenvalues $\{1, -1\}$ and isomorphic eigen-spaces $\Lambda^+(T_x^*M)$, $\Lambda^-(T_x^*M)$ summing up to the algebra itself. The signs $\pm$ stand for the parity of the degree p of the exterior form. An exterior p-form on M is locally given by

$$\omega = \sum_{i_1,\ldots,i_p} a_{i_1,\ldots,i_p} \, dx^{i_1} \wedge \ldots \wedge dx^{i_p}$$

with smooth functions $a_{i_1,\ldots,i_p}(x)$ defined on a chart U.

Definition 12. The *exterior algebra bundle* $\Lambda(M)$ is fibrewise defined using the atlas on M induced by the cotangent bundle atlas of T^*M:

$$\Lambda_x(M) := \Lambda(T_x^*M) \otimes_\mathbb{R} \mathbb{C} , \; x \in X.$$

Consequently, $\Lambda^0(M)$ is a trivial line bundle over M and $\Lambda^1(M) = T^*M$ and $\Lambda^k(M) = 0$ for $k > n = \dim(M)$. The set $\Lambda^p(M)$ is said to be the set of all p-forms, and $\Lambda(M) := \oplus_p \Lambda^p(M)$ is a linear space by definition. The multiplication is fibrewise defined by the $\wedge$-multiplication of $\Lambda_x(M)$, i.e. $\omega_1 \wedge \omega_2 = (-1)^{pq} \omega_2 \wedge \omega_1$ for $\omega_1 \in \Lambda^p$, $\omega_2 \in \Lambda^q$.

The exterior differential $d : \Gamma^\infty(\Lambda^p(M)) \to \Gamma^\infty(\Lambda^{p+1}(M))$ induced by the local differential d_x on $\Lambda(T_x^*M)$ is linear and obeys the rules

$$d(\omega_1 \wedge \omega_2) = d\omega_1 \wedge \omega_2 + (-1)^p \cdot \omega_1 \wedge d\omega_2,$$
$$d(d\omega) \equiv 0$$

for $\omega_1 \in \Lambda^p$, $\omega_2 \in \Lambda^q$. Moreover, in local coordinates we have

$$df = \sum_i \frac{\partial f}{\partial x_i} dx^i \quad \text{for } f \in C^\infty(M),$$

$$d\omega = \sum_{i_1,\ldots,i_p} da_{i_1,\ldots,i_p} \wedge dx^{i_1} \wedge \ldots \wedge dx^{i_p} , \; \omega \in \Lambda^p.$$

As a result we obtain a complex, the de Rham' complex

$$0 \to \Gamma^\infty(\Lambda^0(M)) \xrightarrow{d} \Gamma^\infty(\Lambda^1(M)) \xrightarrow{d} \ldots \xrightarrow{d} \Gamma^\infty(\Lambda^n(M)) \to 0$$

that gives rise to cohomology groups that are isomorphic to the cohomology groups $H^*(M, \mathbb{R})$. The name of the complex comes from the application of de Rham's theorem to this particular situation, cf. [153].

3.7 $\Omega_{\not{D}} C^\infty(M)$ Versus $\Lambda(M)$

The final goal of the present chapter is another theorem relating a structure formally depending on the Riemannian metric on the compact smooth manifold M to another structure that does not depend even on its existence or absence.

Theorem 6 (A. Connes). *Comparing the components of Connes' differential algebra $\Omega_{\not{D}} C^\infty(M)$ and the smooth sections of components of the exterior algebra bundle $\Lambda(M)$ we obtain an isomorphism $\Omega_{\not{D}}^p C^\infty(M) \cong \Gamma^\infty(\Lambda^p(M))$ for every $p \geq 0$. Moreover, it extends to the commutative diagrams*

$$\begin{array}{ccc} \Omega_{\not{D}}^p C^\infty(M) & \xrightarrow{d} & \Omega_{\not{D}}^{p+1} C^\infty(M) \\ \downarrow \cong & & \downarrow \cong \\ \Gamma^\infty(\Lambda^p(M)) & \xrightarrow{d} & \Gamma^\infty(\Lambda^{p+1}(M)) \end{array}$$

showing an equivalence of differential algebras.

Proof. For every $p \geq 0$ consider the subbundle $\mathbb{C}l(M)^{p-ev}$ that consists of the intersection of the subbundle of all elements of $\mathbb{C}l(M)$ of degree at most p with either the subbundle $\mathbb{C}l(M)^{even}$ or $\mathbb{C}l(M)^{odd}$ in accordance with the parity of p. In the same manner we define $\Omega^{p-ev} C^\infty(M) = \oplus_{k=p-ev} \Omega^k C^\infty(M)$, where k runs over all indices between 0 and p differing from p by zero or an even number.

Claim 1: $\pi(\Omega^{p-ev} C^\infty(M)) \equiv \gamma(\Gamma^\infty(\mathbb{C}l^{p-ev}(M)))$ for every $p \in \mathbb{N}$.

We prove the claim by induction. For $p = 0$ a comparison of the definitions shows that $\pi(f) = \gamma(f) = f \cdot \text{id}_H$ for every $f \in C^\infty(M)$. In case $p = 1$ we obtain $\pi(df) = [\not{D}, \pi(f)] = [\not{D}, \gamma(f)] = \gamma(df)$ for every $f \in C^\infty(M)$.

To show the general argument recall that the complexified Clifford algebra of a real vector space V and the complexified exterior algebra of V are related by the isomorphisms $Cl_{\mathbb{C}}^{p-ev}(V)/Cl_{\mathbb{C}}^{(p-2)-ev}(V) \cong \Lambda_{\mathbb{C}}^p(V)$ for every $p \in \mathbb{N}$. So there exist induced *symbol maps* between the components of the Clifford bundle and the exterior algebra bundle over M,

$$\sigma^p : \Gamma^\infty(\mathbb{C}l^{p-ev}(M)) \to \Gamma^\infty(\Lambda^p(M)),$$

$p \in \mathbb{N}$, with kernels $\ker(\sigma^p) = \Gamma^\infty(\mathbb{C}l^{(p-2)-ev}(M))$. Consequently, every smooth section of the p-th component $\mathbb{C}l^p(M)$ of the Clifford bundle can be represented as a finite linear combination of elementary elements of the form $\{f_0\, df_1 \cdot ... \cdot df_p : f_i \in C^\infty(M)\}$, where the central dot denotes the Clifford multiplication. (Apply Swan's theorem and take the projection of the canonical orthonormal basis of the trivial bundle housing $\Lambda^p(M)$ as a direct summand. Then pull this system of generators back via σ^p. Cf. Theorem 5.)

To show the inclusion $\pi(\Omega^{p-ev}C^\infty(M)) \subseteq \gamma(\Gamma^\infty(\mathbb{C}l^{p-ev}(M)))$ we have only to check the canonical elements $f_0\, df_1 df_2 ... df_p \in \Omega^p C^\infty(M)$ since the inclusion is supposed to be already established for lower degrees by induction. We have

$$\pi(f_0\, df_1 df_2 ... df_p) = \pi(f_0)[\slashed{D}, \pi(f_1)]...[\slashed{D}, \pi(f_p)]$$
$$= \gamma(f_0)\gamma(df_1)...\gamma(f_p)$$
$$= \gamma(f_0\, df_1 \cdot ... \cdot df_p).$$

Conversely, we have to show that $\pi(\Omega^{p-ev}C^\infty(M)) \supseteq \gamma(\Gamma^\infty(\mathbb{C}l^{p-ev}(M)))$ for every $p \in \mathbb{N}$.

By the results of our considerations on the symbol maps and by induction we have to verify the inclusion for finite sums $c = \sum_{fin.,l} f_{0,l}\, df_{1,l} \cdot ... \cdot df_{p,l} \in \mathbb{C}l^p(M)$ only. We get

$$\gamma(c) = \sum_{fin.,l} \gamma(f_{0,l})[\slashed{D}, f_{1,l}]...[\slashed{D}, f_{p,l}]$$
$$= \pi\left(\sum_{fin.,l} f_{0,l}\, df_{1,l} ... df_{p,l}\right) \in \pi(\Omega^p C^\infty(M)).$$

This establishes the first claim.

Claim 2: $\pi(d \ker(\pi^{p-1})) \equiv \gamma(\ker(\sigma^p))$ for any $p \in \mathbb{N}$ with $p \geq 2$.

Suppose, $\omega = \sum_{fin.,l} f_{0,l}\, df_{1,l}...df_{p-1,l} \in \ker(\pi^{p-1}) \subset \Omega_{\mathcal{D}}^{p-1}$. By the first step

$$\gamma\left(\sum_{fin.,l} f_{0,l}\, df_{1,l} \cdot ... \cdot df_{p-1,l}\right) = \pi(\omega) = 0$$

and $\sum_{fin.,l} f_{0,l}\, df_{1,l} \cdot ... \cdot df_{p-1,l} = 0$ since γ is injective on such elementary elements. Consider $d\omega = \sum_{fin.,l} df_{0,l} \cdot df_{1,l} \cdot ... \cdot df_{p-1,l} \in \mathbb{C}l^p(M)$:

$$\pi(d\omega) = \gamma\left(\sum_{fin.,l} df_{0,l} \cdot df_{1,l} \cdot ... \cdot df_{p-1,l}\right)$$

$$\sigma^p\left(\sum_{fin.,l} df_{0,l} \cdot df_{1,l} \cdot ... \cdot df_{p-1,l}\right) = \sum_{fin.,l} df_{0,l} \wedge df_{1,l} \wedge ... \wedge df_{p-1,l}$$
$$= d\left(\sum_{fin.,l} f_{0,l}\, df_{1,l} \wedge ... \wedge df_{p-1,l}\right) = d\sigma^{p-1}\left(\sum_{fin.,l} f_{0,l}\, df_{1,l} \cdot ... \cdot df_{p-1,l}\right) = 0.$$

Consequently, $\pi(d\ker(\pi^{p-1})) \subseteq \gamma(\ker(\sigma^p))$ for every $p \in \mathbb{N}$ with $p \geq 2$.

To show the reverse inclusion, let $c \in \Gamma^\infty(\mathbb{Cl}^{(p-2)}(M))$. If $\{u_\alpha\}$ is a partition of unity corresponding to the selected atlas then we can assume $\mathrm{supp}(c) \subset U$ since $\gamma(c) = \gamma(\sum_\alpha u_\alpha c) = \sum_\alpha \gamma(u_\alpha c)$. Furthermore, by the first part of this proof

$$c = \sum_{fin.,l} f_{0,l}\, df_{1,l} \cdot \ldots \cdot df_{p-2,l}$$

for some functions $f_i \in C^\infty(M)$.

Let $h \in C^\infty(M)$ with $h(y) \geq \lambda > 0$ for any $y \in M$ and $\langle d_x h, d_x h\rangle_{g^{-1}}(x) \geq \mu > 0$ for every $x \in U$. This forces $h, h^{-1} \in C^\infty(M)$ and

$$\tilde f_{0,l}(x) := \frac{f_{0,l}(x)}{2\langle d_x h, d_x h^{-1}\rangle_{g^{-1}}} = -\frac{h(x)^2}{2\langle d_x h, d_x h\rangle_{g^{-1}}} \cdot f_{0,l}(x) \in C^\infty(M)$$

for every l. Furthermore,

$$\tilde f_{0,l}\,(dh \cdot dh^{-1} + dh^{-1} \cdot dh) = 2\,\tilde f_{0,l}\langle dh, dh^{-1}\rangle_{g^{-1}} = f_{0,l} \qquad (3.1)$$

for every l, and by the first step we obtain

$$\pi(h\,dh^{-1} + h^{-1}\,dh) = \gamma(h\,dh^{-1} + h^{-1}\,dh) = \gamma(d(hh^{-1})) = 0\,.$$

Therefore, for any l

$$\omega_l := (h\,dh^{-1} + h^{-1}\,dh)\,df_{1,l}df_{2,l}...df_{p-2,l} \in \ker(\pi^{p-1})\,,$$

$$\tilde f_{0,l}\,d\omega_l \in (\ker(\pi^p) + d\ker(\pi^{p-1}))\,,$$

since the latter is an ideal in Ω^p. Finally, by (3.1) and the first step:

$$\gamma(c) = \gamma\Big(\sum_{fin.,l} \tilde f_{0,l}\,(dh \cdot dh^{-1} + dh^{-1} \cdot dh) \cdot df_{1,l} \cdot df_{2,l} \cdot \ldots \cdot df_{p-2,l}\Big)$$

$$= \pi\Big(\sum_{fin.,l} \tilde f_{0,l}\,(dhdh^{-1} + dh^{-1}dh)df_{1,l}df_{2,l}...df_{p-2,l}\Big)$$

$$= \pi\Big(\sum_{fin.,l} \tilde f_{0,l}\,d\omega_l\Big) \in \pi(\ker(\pi^p) + d\ker(\pi^{p-1})) = \pi(d\ker(\pi^{p-1}))\,.$$

We arrive at $\pi(d\ker(\pi^{p-1})) \supseteq \gamma(\ker(\sigma^p))$ for every $p \in \mathbb{N}$ with $p \geq 2$, and claim 2 is proved.

As the final step we list the following chain of identifications and isomorphisms:

$$\Omega^p_{\not{D}} C^\infty(M) = \pi(\Omega^p C^\infty(M))/\pi(d\ker(\pi^{p-1}))$$
$$\cong \Gamma^\infty(\Lambda^p(M)/\ker(\sigma^p) \cong \mathrm{im}(\sigma^p) = \Gamma^\infty(\Lambda^p(M))\,.$$

Corollary 1. $\Omega^p_{\not{D}} C^\infty(M) = 0$ for every $p > \dim(M) = n$.

Acknowledgements

I would like to thank P. M. Alberti and R. Matthes for valuable discussions and the exchange of literature in Leipzig during the time of preparation. I am indebted to C. Bär, K. Fredenhagen and H. Upmeier for comments on the main theorem of Section 1.3.

4 Connes' Trace Formula and Dirac Realization of Maxwell and Yang-Mills Action

Peter M. Alberti and Reiner Matthes

University of Leipzig

4.1 Generalities on Traces on C*- and W*-algebras

The essential ingredients of the trace formula [37,38] of A. Connes are operator algebraic constructs over the W*-algebra $\mathsf{B}(\mathcal{H})$ of all bounded linear operators over some infinite dimensional separable Hilbert space $\mathcal{H}$ which are known as *Dixmier traces*. The constructions will be explained in this section. For a general operator-algebraic background the reader is referred e.g. to [64,187,209,117,118].

Basic topological notions, notations

A C*-algebra M is a Banach *-algebra, with *-operation $x \longmapsto x^*$ and norm $\|\cdot\|$ obeying $\|x^*x\| = \|x\|^2$, for each $x \in M$ (*-quadratic property). Let $M_+ = \{x^*x : x \in M\}$ be the cone of positive elements of M. For $x \in M$ let $x \geq 0$ be synonymous with $x \in M_+$. The Banach space of all continuous linear forms on M (dual) will be denoted by M^*, the dual norm (functional norm) be $\|\cdot\|_1$. As usual, a linear form f over M is termed positive, $f \geq 0$, if $f(x^*x) \geq 0$, for each $x \in M$. Remind that positive linear forms are automatically continuous and are generating for M^*. Thus $M^*_+ = \{f \in M^* : f \geq 0\}$ is the set of all these forms. There is a fundamental result of operator theory saying that if M possesses a unit $\mathbf{1}$ (unital C*-algebra), then $f \geq 0$ if, and only if, $f(\mathbf{1}) = \|f\|_1$. It is common use to refer to positive linear forms of norm one as states. Thus, in a unital C*-algebra the set of all states on M, $\mathcal{S}(M)$, is easily seen to be a convex set which according to the Alaoglu-Bourbaki theorem is $\sigma(M^*, M)$-compact. Here, the $\sigma(M^*, M)$-topology (also w^*-topology in the special context at hand) is the weakest locally convex topology generated by the seminorms ρ_x, $x \in M$, with $\rho_x(f) = |f(x)|$, for each $f \in M^*$. The generalization of the notion of positive linear form on M (which refers to positive linear maps into the special C*-algebra of complex numbers $\mathbb{C}$) is the notion of the positive linear map. Say that a linear mapping $T : M \longrightarrow N$ which acts from one C*-algebra M into another one N is positive if $T(x^*x) \geq \mathbf{0}$ within N for each $x \in M$. In the unital case (for M) one then knows that $\|T(\mathbf{1})\| = \|T\|$ holds ($\|T\|$ refers to the operator norm of T as a linear operator acting from the one Banach space M into the other N). On the other hand, each linear map $T : M \longrightarrow N$ which obeys this relation is known to be positive. If both M and N have a unit, a linear map $T : M \longrightarrow N$ is said to be unital if $T(\mathbf{1}) = \mathbf{1}$ is fulfilled (the units being the respective units). Thus, and in particular, each unital linear map

$T : M \longrightarrow N$ of norm one beween unital C*-algebras has to be a positive map.

Remind that a C*-algebra M is a W*-algebra, that is, is *-isomorphic to some vN-algebra on some Hilbert space $\mathcal{H}$, if and only if, there exists a (unique) Banach subspace M_* (the predual space) of M^* such that M is the (continuous) dual of M_*, $M = (M_*)^*$. Note that a non-zero W*-algebra is always unital. Suppose now that M is a W*-algebra. The forms of M_* will be referred to as normal linear forms. One then knows that the normal positive linear forms $M_{*+} = M_* \cap M_+^*$ (resp. the normal states $\mathcal{S}_0(M) = M_* \cap \mathcal{S}(M)$) are generating for M_* in the sense that each normal linear form may be represented as a complex linear combination of at most four normal states. As a consequence of this, for each ascendingly directed (in the sense of $\leq$) bounded net $\{x_\alpha\} \subset M_+$ there exists a lowest upper bound $l.u.b.\, x_\alpha$ within M_+. On the other hand, a positive linear form $\omega \in M_+^*$ is normal if, and only if, for each ascendingly directed bounded net $\{x_\alpha\} \subset M_+$ the relation $\omega(l.u.b.\, x_\alpha) = l.u.b.\, \omega(x_\alpha) = \lim_\alpha \omega(x_\alpha)$ is valid. Note that in the latter characterization it suffices if the mentioned continuity be fulfilled for ascendingly directed nets of orthoprojections of M.

In contrast to the previous, $\nu \in M_+^*$ is called singular if to each orthoprojection $p \in M$ with $\nu(p) > 0$ there is another orthoprojection $q \in M$ with $0 < q < p$ and $\nu(q) = 0$. According to [208] each $\omega \in M_+^* \backslash M_{*+}$ in a unique way can be decomposed as $\omega = \omega_1 + \omega_2$, with normal $\omega_1 \in M_{*+}$ and singular $\omega_2 \in M_+^*$.

The simplest example of a W*-algebra where singular positive linear forms can exist is the commutative W*-algebra $\ell^\infty = \ell^\infty(\mathbb{N})$ of all bounded sequences $x = (x_n) = (x_1, x_2, \ldots)$ of complex numbers, with norm $\|x\|_\infty = \sup_{n \in \mathbb{N}} |x_n|$, and algebra multiplication $x \cdot y = (x_n y_n)$ and *-operation $x^* = (\bar{x}_n)$ defined componentwise. Recall that in this case the predual space ℓ_*^∞ is the Banach space of all absolutely summable sequences $\ell^1(\mathbb{N})$, with norm $\|\omega\|_1 = \sum_{n \in \mathbb{N}} |\omega_n|$ for $\omega = (\omega_n) \in \ell^1(\mathbb{N})(= \ell^1)$. Thereby, each such ω can be identified with an element in the dual Banach space $(\ell^\infty)^*$ via the identification with the linear functional $\omega(\cdot)$ given as $\omega(x) = \sum_{n \in \mathbb{N}} \omega_n x_n$, for each $x \in \ell^\infty$. For simplicity, also this functional $\omega(\cdot)$ will be referred to as ω, $\omega = \omega(\cdot)$.

In generalizing from the setting of a normal positive linear form, call a positive linear map $T : M \longrightarrow N$ from one W*-algebra M into another W*-algebra N normal if $T(l.u.b.\, x_\alpha) = l.u.b.\, T(x_\alpha)$ holds for each ascendingly directed bounded net $\{x_\alpha\} \subset M_+$. Note that in a W*-algebra the Alaoglu-Bourbaki theorem may be applied on M, and then yields that the (closed) unit ball M_1 of M is $\sigma(M, M_*)$-compact. From this it follows that the unit ball within the bounded linear operators which map the W*-algebra M into itself is compact with respect to the topology determined by the system of seminorms $\rho_{x,f}$, labelled by $x \subset M$ and $f \in M_*$, and which are defined at T by $\rho_{x,f}(T) = |f \circ T(x)|$. Refer to this topology as the $\sigma(M, M_*)$-weak operator topology on the Banach algebra of bounded linear operators $\mathsf{B}(M)$ over the Banach space M. Thereby, by convention for $T, S \in \mathsf{B}(M)$ let the product $TS \in \mathsf{B}(M)$ be defined through successive application of maps to the elements of M in accordance with the rule 'apply right factor first', that is $TS(x) = (T \circ S)(x) = T(S(x))$.

Traces on C^*- and W^*-algebras

We recall the very basic facts on traces as found e.g. in [64, 6.1.]. A function $\tau : M_+ \longrightarrow \overline{\mathbb{R}}_+$ from the positive cone into the extended positive reals is said to be a *trace* provided it is (extended) *additive, positive homogeneous* (that is, $x, y \in M_+$ and $\lambda \in \mathbb{R}_+$ imply $\tau(x+y) = \tau(x) + \tau(y)$ and $\tau(\lambda x) = \lambda \tau(x)$, with $0 \cdot \infty = 0$ by convention) and obeys $\tau(x^*x) = \tau(xx^*)$, for each $x \in M$ (*invariance property*). From the first two properties it follows that $M_{\tau+} = \{x \in M_+ : \tau(x) < \infty\}$ is a *hereditary subcone* of M_+, that is, $M_{\tau+}$ is a cone such that $x \in M_{\tau+}$ and $\mathbf{0} \leq y \leq x$ for $y \in M_+$ implies $y \in M_{\tau+}$. From this it is then easily inferred that $\mathcal{L}_\tau = \{x \in M : \tau(x^*x) < \infty\}$ is a *left ideal* in M. Owing to the invariance condition $\tau(x^*x) = \tau(xx^*)$ however, $\mathcal{L}_\tau$ has to be also a right ideal. Hence, $\mathcal{L}_\tau$ is a *two-sided ideal* of M, which is characteristic for the trace τ. It is known that the complex linear span $[M_{\tau+}]$ is a *-subalgebra of M which is even a two-sided ideal and which is generated by $\mathcal{L}_\tau$ as $[M_{\tau+}] = \mathcal{L}_\tau^2$. Also, on the *-subalgebra $[M_{\tau+}]$ there exists a unique (complex) linear form $\tilde{\tau}$ obeying $\tilde{\tau}(x) = \tau(x)$, for each $x \in M_{\tau+}$. Owing to invariance then also the characteristic commutation property $\tilde{\tau}(yx) = \tilde{\tau}(xy)$ holds, for each $x \in [M_{\tau+}]$ and all $y \in M$.

The two-sided ideal $[M_{\tau+}]$ will be referred to as *defining ideal* of the trace τ. Since the linear extension $\tilde{\tau}$ of τ from the cone of all positive elements of $[M_{\tau+}]$ (which according to the previous is $M_{\tau+}$) onto $[M_{\tau+}]$ is unique, by tacit understanding the notation $\tau(x)$ will be also used at non-positive x of the defining ideal of τ if the evaluation $\tilde{\tau}(x)$ of the linear functional $\tilde{\tau}$ at x is meant.

The trace τ is termed *finite trace* if $M_{\tau+} = M_+$, and *semifinite trace* if $\tau(x) = \sup\{\tau(y) : y \leq x, y \in M_{\tau+}\}$, for each $x \in M_+$. If M is a W^*-algebra, with group of unitary elements $\mathcal{U}(M)$, the above condition on invariance usually is replaced with a seemingly weaker requirement upon *unitary invariance*, that is $\tau(u^*xu) = \tau(x)$ be fulfilled, for each $x \in M_+$ and $u \in \mathcal{U}(M)$. However, both conditions are equivalent there (and are so even on unital C^*-algebras). Also, in the W^*-case the trace τ is said to be *normal* provided for each ascendingly directed bounded net $\{x_\alpha\} \subset M_+$ the relation $\tau(l.u.b. \, x_\alpha) = l.u.b. \, \tau(x_\alpha) = \lim_\alpha \tau(x_\alpha)$ is fulfilled.

Now, suppose $I \subset M$ is a proper two-sided ideal of the W^*-algebra M. Then, I is also a *-subalgebra of M, with generating positive cone $I_+ = I \cap M_+$, that is, $I = [I_+]$ is fulfilled (these facts are consequences of the polar decomposition theorem, essentially). Under these premises we have the following extension principle:

Lemma 1. *Suppose I_+ is a hereditary subcone of M_+. Then, each additive, positive homogeneous and invariant map $\tau_0 : I_+ \longrightarrow \overline{\mathbb{R}}_+$ extends to a trace τ on M, with $M_{\tau+} = \{x \in I_+ : \tau_0(x) < \infty\}$.*

Proof. Define $\tau(x) = \tau_0(x)$, for $x \in I_+$, and $\tau(x) = \infty$ for $x \in M_+ \backslash I_+$. Then, since I_+ is a hereditary cone, for $x,y \in M_+$ with $x + y \in I_+$ one has both $x, y \in I_+$, and thus $\tau(x) + \tau(y) = \tau(x+y)$ is evident from $\tau_0(x) + \tau_0(y) = \tau_0(x+y)$. For $x, y \in M_+$ with $x + y \notin I_+$ at least one of x, y must not be in I_+. Hence, $\tau(x) + \tau(y) = \infty$ and $\tau(x+y) = \infty$ by definition of τ. Thus additivity holds in any case. That τ is also positive homogeneous is clear. Finally, remind that, according to polar decomposition $x = u|x|$ for $x \in M$, one has $xx^* = u(x^*x)u^*$ and $x^*x = u^*(xx^*)u$, with the partial isometry $u \in M$. Since I is a two-sided ideal of M, from this one infers $xx^* \in I_+$ if, and only if, $x^*x \in I_+$. From this in view of the definition and since τ_0 is invariant on I_+ also invariance of τ on M_+ follows.

4.2 Examples of Traces

In the following the classical special cases of traces on M with either $M = \mathsf{B}(\mathcal{H})$ or $M = \mathcal{CB}(\mathcal{H})$ are considered in more detail, where $\mathcal{CB}(\mathcal{H})$ is the C*-subalgebra of $\mathsf{B}(\mathcal{H})$ of all *compact* linear operators on the separable infinite dimensional Hilbert space $\mathcal{H}$. For generalities on the theory of compact operators and proofs from there the reader is referred to [168,189] e.g.; in the following recall only those few facts and details which are important in the context of traces. In all that follows, the scalar product $\mathcal{H} \times \mathcal{H} \ni \{\chi, \eta\} \longmapsto \langle \chi, \eta \rangle \in \mathbb{C}$ on $\mathcal{H}$ by convention is supposed to be linear with respect to the first argument χ, and antilinear in the second argument η, and maps into the complex field $\mathbb{C}$.

Traces on compact linear operators

We start with recalling the characterization of positive compact linear operators in terms of a spectral theorem. Let $x \in \mathsf{B}(\mathcal{H})_+$ be a non-trivial positive (=non-negative) bounded linear operator. Then, x is a (positive) compact operator, $x \in \mathcal{CB}(\mathcal{H})_+$ if, and only if, the following two condition are fulfilled. Firstly, there have to exist a non-increasing infinite sequence $(\mu_1(x), \mu_2(x), \ldots)$ of non-negative reals $\mu_k(x)$, which converge to zero as $k \to \infty$, and an infinite *orthonormal system* (o.n.s. for short) $\{\varphi_n\} \subset \mathcal{H}$ of eigenvectors of x obeying $x\varphi_k = \mu_k(x)\varphi_k$, for each $k \in \mathbb{N}$, and with $x\varphi = 0$, for each $\varphi \in [\{\varphi_n\}]^\perp$ (thus the spectrum of x is $\mathrm{spec}(x) = \{\mu_k(x) : k \in \mathbb{N}\} \cup \{0\}$). And secondly, each non-zero eigenvalue of x has only finite multiplicity, that is, $m(\mu) = \#\{k : \mu_k(x) = \mu\}$ obeys $m(\mu) < \infty$, for each $\mu \in \mathbb{R}_+ \backslash \{0\}$.

Recall that $\mathcal{CB}(\mathcal{H})$ is also a closed *-ideal of $\mathsf{B}(\mathcal{H})$. Hence, according to polar decomposition, $x \in \mathsf{B}(\mathcal{H})$ is compact if, and only if, the module $|x| = \sqrt{x^*x}$ of x is compact, $|x| \in \mathcal{CB}(\mathcal{H})_+$. In line with this and following some common use, for $x \in \mathcal{CB}(\mathcal{H})$ and in view of the above define $\mu_k(x) = \mu_k(|x|)$, for each $k \in \mathbb{N}$, and refer to the ordered sequence $\mu_1(x) \geq \mu_2(x) \geq \ldots$ of eigenvalues of $|x|$ (with each of the non-zero eigenvalues repeated according to its multiplicity) as *characteristic sequence* of x. The terms of this sequence can be obtained by minimizing the distance of the given compact operator x to the finite rank linear operators (which are special compact operators) of a fixed rank as follows:

$$\forall\, k \in \mathbb{N}: \ \mu_k(x) = \min\{\|x - y\| : y \in \mathcal{CB}(\mathcal{H}), \dim y\mathcal{H} \leq k\}. \qquad (4.1\mathrm{a})$$

Alternatively, and yet more important, these values can be obtained also from a representation of the sequence $\{\sigma_n(x)\}$ of their partial sums $\sigma_k(x) = \sum_{j \leq k} \mu_k(x)$ which arises from maximizing the following expression over the unitaries $\mathsf{U}(\mathcal{H})$ of $\mathcal{H}$ and finite orthonormal systems $\{\psi_1, \ldots, \psi_k\} \subset \mathcal{H}$ of cardinality $k \in \mathbb{N}$:

$$\sigma_k(x) = \max\left\{\left|\sum_{j \leq k} \langle ux\psi_j, \psi_j \rangle\right| : u \in \mathsf{U}(\mathcal{H}), \{\psi_1, \ldots, \psi_k\} \text{ o.n.s.}\right\}, \qquad (4.1\mathrm{b})$$

which for positive x simplifies into

$$\forall\, x \in \mathcal{CB}(\mathcal{H})_+: \ \sigma_k(x) = \max\left\{\sum_{j \leq k} \langle x\psi_j, \psi_j \rangle : \{\psi_1, \ldots, \psi_k\} \text{ o.n.s.}\right\}. \qquad (4.1\mathrm{c})$$

Now, let us fix an arbitrary *maximal orthonormal system* (m.o.n.s. for short) $\{\varphi_n\} \subset \mathcal{H}$, and let p be an orthoprojection with $\dim p\mathcal{H} = k < \infty$. Then, for each $x \in \mathsf{B}(\mathcal{H})$ the operator xp is of finite rank, and for each o.n.s. $\{\psi_1, \ldots, \psi_k\}$ which linearily spans $p\mathcal{H}$, by elementary Hilbert space calculus one derives the relation

$$\sum_{n=1}^\infty \langle xp\varphi_n, \varphi_n \rangle = \sum_{j \leq k} \langle x\psi_j, \psi_j \rangle. \qquad (\times)$$

Hence, in case of compact x (4.1b) equivalently reads as

$$\sigma_k(x) = \max\left\{\left|\sum_{n=1}^\infty \langle uxp\varphi_n, \varphi_n \rangle\right| : u \in \mathsf{U}(\mathcal{H}), p = p^* = p^2, \dim p\mathcal{H} \leq k\right\}. \qquad (\star)$$

Note that for $x \geq \mathbf{0}$ the expression of $(\times)$ is positive and with the help of similarly elementary calculations as those which led to $(\times)$ one infers that for each orthoprojection p with $\dim p\mathcal{H} = k < \infty$ and any $x \in \mathsf{B}(\mathcal{H})_+$ the following holds:

$$\sum_{j \leq k} \langle x\psi_j, \psi_j \rangle = \sum_{n=1}^\infty \langle p\sqrt{x}\varphi_n, \sqrt{x}\varphi_n \rangle = \sum_{n=1}^\infty \langle xp\varphi_n, \varphi_n \rangle \geq 0. \qquad (\star\star)$$

Especially, since according to (4.1c) for *positive* compact x maximizing over the unitaries becomes redundant and may be omitted, in view of this and

$(\star\star)$ for each such element the relation $(\star)$ then simplifies into the following well-known form:

$$\forall x \in \mathcal{CB}(\mathcal{H})_+ : \sigma_k(x) = \max\left\{\sum_{n=1}^{\infty}\langle xp\varphi_n,\varphi_n\rangle : p = p^* = p^2, \dim p\mathcal{H} \leq k\right\}. \tag{4.1d}$$

From (4.1b)–(4.1d) one now concludes some useful relations and estimates. The first is a rather trivial consequence of the definition of $\sigma_k(x)$ and says that

$$\forall x \in \mathcal{CB}(\mathcal{H}), \lambda \in \mathbb{C}, k \in \mathbb{N}: \sigma_k(\lambda x) = |\lambda|\,\sigma_k(x). \tag{4.2a}$$

It is stated here only for completeness. In the special case of positive compact operators from (4.1c) we get the following often used estimates:

$$\forall x,y \in \mathcal{CB}(\mathcal{H})_+, k \in \mathbb{N}: y \leq x \implies \sigma_k(y) \leq \sigma_k(x). \tag{4.2b}$$

The next estimate is due to [134] and at once gets obvious from (4.1b), and tells us that the following holds:

$$\forall x,y \in \mathcal{CB}(\mathcal{H}), k \in \mathbb{N}: \sigma_k(x+y) \leq \sigma_k(x) + \sigma_k(y). \tag{4.2c}$$

The third estimate deals with an upper bound of $\sigma_k(x) + \sigma_k(y)$ in case of positive operators $x,y \in \mathcal{CB}(\mathcal{H})_+$ and arises from (4.1d). In line with the latter, let orthoprojections p,q of rank k be given such that $\sigma_k(x) = \sum_{n=1}^{\infty}\langle xp\varphi_n,\varphi_n\rangle$ and $\sigma_k(y) = \sum_{n=1}^{\infty}\langle yq\varphi_n,\varphi_n\rangle$ are fulfilled. Then, the least orthoprojection $p\vee q$ majorizing both p and q has rank $2k$ at most. Thus there is an orthoprojection Q of rank $2k$ and obeying $p\vee q \leq Q$. Hence, in view of the choice of p,q and with the help of $(\star\star)$ one infers that $\sum_{n=1}^{\infty}\langle xQ\varphi_n,\varphi_n\rangle = \sigma_k(x) + \sum_{n=1}^{\infty}\langle x(Q-p)\varphi_n,\varphi_n\rangle \geq \sigma_k(x)$ and $\sum_{n=1}^{\infty}\langle yQ\varphi_n,\varphi_n\rangle = \sigma_k(y) + \sum_{n=1}^{\infty}\langle x(Q-q)\varphi_n,\varphi_n\rangle \geq \sigma_k(y)$. In view of (4.1d) from this then $\sigma_k(x) + \sigma_k(y) \leq \sigma_{2k}(x+y)$ follows. For positive compact operators the previous together with (4.2c) may be summarized into the following one:

$$\forall x,y \in \mathcal{CB}(\mathcal{H})_+, k \in \mathbb{N}: \sigma_k(x+y) \leq \sigma_k(x) + \sigma_k(y) \leq \sigma_{2k}(x+y). \tag{4.2d}$$

Note that, since $\mathcal{CB}(\mathcal{H})$ is a two-sided ideal, from (4.1a) for each $y \in \mathcal{CB}(\mathcal{H})$ and $a,b \in \mathsf{B}(\mathcal{H})$ the estimate

$$\forall k \in \mathbb{N}: \mu_k(ayb) \leq \|a\|\,\|b\|\,\mu_k(y) \tag{4.2e}$$

can be obtained. Thus, under these conditions one has

$$\forall a,b \in \mathsf{B}(\mathcal{H}), y \in \mathcal{CB}(\mathcal{H}), k \in \mathbb{N}: \sigma_k(ayb) \leq \|a\|\,\|b\|\,\sigma_k(y). \tag{4.2f}$$

Especially, if $x = u|x|$ is the polar decomposition of $x \in \mathcal{CB}(\mathcal{H})$ within $\mathsf{B}(\mathcal{H})$, then with the partial isometry $u \in \mathsf{B}(\mathcal{H})$ one has both, $xx^* = ux^*xu^*$ and $x^*x = u^*xx^*u$. In the special cases of (4.2f) with $y = xx^*$, $a = u^*$, $b = u$

and $y = x^*x$, $a = u$, $b = u^*$ we arrive at estimates which fit together into the following assertion:

$$\forall x \in \mathcal{CB}(\mathcal{H}), \, k \in \mathbb{N}: \; \sigma_k(x^*x) = \sigma_k(xx^*). \tag{4.2g}$$

In the following, a trace τ is said to be *non-trivial* if there is at least one $x \geq \mathbf{0}$ with $0 < \tau(x) < \infty$. The relations given in eqs. (4.2) are the key facts that the theory of traces on both algebras $\mathcal{CB}(\mathcal{H})$ and $\mathsf{B}(\mathcal{H})$ can be based on.

Lemma 2. *Let* tr : $\mathcal{CB}(\mathcal{H})_+ \longmapsto \overline{\mathbb{R}}_+$ *be defined by* $\operatorname{tr} x = \lim_{n\to\infty} \sigma_n(x)$, *for each* $x \in \mathcal{CB}(\mathcal{H})_+$. *Then,* tr *is a non-trivial semifinite trace on* $\mathcal{CB}(\mathcal{H})$. *Moreover, to each non-trivial trace* τ *which does not vanish identically on the positive operators of finite rank there exists unique* $\lambda \in \mathbb{R}_+\backslash\{0\}$ *such that* $\lambda \cdot \tau(x) \geq \operatorname{tr} x$ *holds for all* $x \in \mathcal{CB}(\mathcal{H})_+$, *and with equality occuring at each* x *of finite rank.*

Proof. The sequence $\{\sigma_n(x)\}$ is increasing, for each $x \in \mathcal{CB}(\mathcal{H})_+$. Thus in the extended sense $\operatorname{tr} x = \lim_{n\to\infty} \sigma_n(x)$ exists. Especially, from (4.2d) in the limit then additivity of tr follows, whereas from (4.2a) and (4.2g) homogeneity and invariance can be seen. Thus, tr is a trace. By construction $0 < \operatorname{tr} x < \infty$ for each compact positive $x \neq \mathbf{0}$ of finite rank. Thus tr is non-trivial. However, since $\mathcal{H}$ is infinite dimensional, $\operatorname{tr} x = \infty$ will occur for some positive compact operators. To see that tr is semifinite requires to prove that for $x \in \mathcal{CB}(\mathcal{H})_+$ with $\operatorname{tr} x = \infty$ there existed a sequence $\{x_n\} \subset \mathcal{CB}(\mathcal{H})_+$ with $x_n \leq x$ and $\operatorname{tr} x_n < \infty$ such that $\lim_{n\to\infty} \operatorname{tr} x_n = \infty$. Note that by definition of tr, $\operatorname{tr} x = \infty$ implies that x cannot be of finite rank. Hence, x can be written as $x = \sum_{k=1}^{\infty} \mu_k(x)\, p_k$, with infinitely many mutually orthogonal one-dimensional orthoprojections p_k and all $\mu_k(x) \neq 0$. Clearly, for each $n \in \mathbb{N}$ the operators $x_n = \sum_{k=1}^{n} \mu_k(x)\, p_k$ are of finite rank and obey $\mathbf{0} \leq x_1 \leq x_2 \leq x_3 \leq \ldots \leq x$. Also, owing to $\sigma_k(x_n) = \sigma_n(x)$ for $k \geq n$, one has $\operatorname{tr} x_n = \sigma_n(x)$, and therefore $\lim_{n\to\infty} \operatorname{tr} x_n = \infty$ follows. Thus tr is semifinite. Suppose τ is a non-trivial trace. Thus $0 < \tau(y) < \infty$, for some positive compact y. Suppose $\tau(x) > 0$ for some $x \geq \mathbf{0}$ of finite rank. According to additivity and homogeneity of τ there has to exist a one-dimensional subprojection p of a spectral orthoprojection of x with $\tau(p) > 0$. The same arguments for y ensure that $\tau(q) < \infty$, for some one-dimensional subprojection q of some spectral projection of y. But since $q = vv^*$ and $p = v^*v$, with $v \in \mathcal{CB}(\mathcal{H})$, by invariance of τ one has $\tau(q) = \tau(p)$. Hence $\infty > \tau(p) > 0$, and $\tau(q) = \tau(p)$ for each one-dimensional orthoprojection q. Put $\lambda = \tau(p)^{-1}$. Then $\lambda \cdot \tau(q) = \operatorname{tr} q$, and thus $\lambda \cdot \tau(x) = \operatorname{tr} x$ for each positive operator x of finite rank. Finally, if $x \in \mathcal{CB}(\mathcal{H})_+$ is not of finite rank, let $\mathbf{0} \leq x_1 \leq x_2 \leq x_3 \leq \ldots \leq x$ be the above approximating sequence of x by finite rank operators x_n. Also in such case $\lim_{n\to\infty} \operatorname{tr} x_n = \lim_{n\to\infty} \sigma_n(x) = \operatorname{tr} x$ follows. Hence, in view of the above relation over the operators of finite rank, and since $\tau(x_n) \leq \tau(x)$ holds, $\operatorname{tr} x = \lim_{n\to\infty} \operatorname{tr} x_n = \lambda \lim_{n\to\infty} \tau(x_n) \leq \lambda \cdot \tau(x)$.

For completeness, we give yet the most famous formula relating tr and which makes that this trace is so extremely useful.

Corollary 1. *For each maximal orthonormal system* $\{\psi_n\} \subset \mathcal{H}$ *and* $x \in \mathcal{CB}(\mathcal{H})_+$ *one has* $\operatorname{tr} x = \sum_{n=1}^{\infty} \langle x\psi_n, \psi_n \rangle$.

Proof. Let $\{\varphi_j\}$ be an o.n.s. with $x\varphi_k = \mu_k(x)\varphi_k$, for all $k \in \mathbb{N}$, and be p_n the orthoprojection with $p_n \mathcal{H} = [\varphi_1, \ldots, \varphi_n]$. Then, by positivity of x one has $\langle x\psi_n, \psi_n \rangle = \langle \sqrt{x}\psi_n, \sqrt{x}\psi_n \rangle \geq \langle p_k \sqrt{x}\psi_n, \sqrt{x}\psi_n \rangle$, and in view of $(\star\star)$ one gets $\sum_{n=1}^{\infty} \langle x\psi_n, \psi_n \rangle \geq \sum_{n=1}^{\infty} \langle p_k \sqrt{x}\psi_n, \sqrt{x}\psi_n \rangle = \sum_{j=1}^{k} \langle x\varphi_j, \varphi_j \rangle = \sigma_k(x)$. According to Lemma 2 then $\sum_{n=1}^{\infty} \langle x\psi_n, \psi_n \rangle \geq \operatorname{tr} x$ follows. On the other hand, if q_k is the orthoprojection onto $[\psi_1, \ldots, \psi_k]$, according to (4.1d) for each $k \in \mathbb{N}$ one certainly has $\sum_{n=1}^{k} \langle x\psi_n, \psi_n \rangle = \sum_{n=1}^{\infty} \langle xq_k \psi_n, \psi_n \rangle \leq \sigma_k(x)$. From this and Lemma 2 it follows $\sum_{n=1}^{\infty} \langle x\psi_n, \psi_n \rangle \leq \operatorname{tr} x$. Taking together this with the above estimate provides that equality has to occur.

A non-zero trace τ on $\mathcal{CB}(\mathcal{H})$ will be said to be *singular* if $\tau(x) = 0$ for each $x \geq \mathbf{0}$ of finite rank. Relating this and non-trivial traces there is the following result.

Corollary 2. *Let τ be a non-trivial trace on $\mathcal{CB}(\mathcal{H})$. Then, either $\tau = \lambda \cdot \operatorname{tr}$ holds, for a unique $\lambda \in \mathbb{R}_+$, or there exist a singular trace τ_s and a unique $\alpha \in \mathbb{R}_+$ such that $\tau = \tau_s + \alpha \cdot \operatorname{tr}$.*

Proof. If $\tau = \lambda \cdot \operatorname{tr}$ is fulfilled, then $\tau(p) = \lambda \operatorname{tr} p$, for each one-dimensional orthoprojection p. Owing to $\operatorname{tr} p = 1$ (see Corollary 1) then $\lambda = \tau(p)$ follows.

Suppose $\tau \notin \mathbb{R}_+ \operatorname{tr}$. Then, $\tau \neq 0$, and if a decomposition $\tau = \tau_s + \alpha \cdot \operatorname{tr}$ with singular τ_s exists, then $\tau(p) = \alpha$, for some (and thus any) one-dimensional orthoprojection p, and the following two alternatives have to be dealt with: firstly, if τ is vanishing on all positive operators of finite rank, τ is singular, and $\tau = \tau_s$ and $\alpha = 0$ have to be chosen (see above). Secondly, if τ does not vanish on all positive operators of finite rank, according to Lemma 2 there exists unique $\lambda > 0$ with $\lambda \cdot \tau(x) \geq \operatorname{tr} x$, for each $x \in \mathcal{CB}(\mathcal{H})_+$, with equality occuring on any operator of finite rank. Hence, in defining $\tau_s(x) = \tau(x) - \lambda^{-1} \operatorname{tr} x$, for each x with $\operatorname{tr} x < \infty$, and $\tau_s(x) = \infty$ else, we get a positive map τ_s which does not vanish identically on the positive compact operators, but which is vanishing on all positive operators of finite rank. From the previous and since both τ and tr are traces, also additivity, positive homogeneity and invariance of τ_s at once follow. Hence, τ_s is a singular trace, which is easily seen to obey $\tau = \tau_s + \alpha \cdot \operatorname{tr}$, with $\alpha = \lambda^{-1}$.

Traces on $\mathsf{B}(\mathcal{H})$

Remind in short the theory of traces on $M = \mathsf{B}(\mathcal{H})$, with separable infinite dimensional Hilbert space $\mathcal{H}$. Let $\mathcal{F}\mathsf{B}(\mathcal{H})$ be the two-sided ideal of all operators of finit rank in $\mathsf{B}(\mathcal{H})$. In the following an ideal $\mathcal{I}$ will be termed *non-trivial* if $\mathcal{I} \neq \{\mathbf{0}\}$ and $\mathcal{I} \neq \mathsf{B}(\mathcal{H})$. Both $\mathcal{F}\mathsf{B}(\mathcal{H})$ and $\mathcal{CB}(\mathcal{H})$ are non-trivial two-sided ideals. Thereby, the compact operators form a closed ideal, with $\mathcal{F}\mathsf{B}(\mathcal{H})$ being dense within $\mathcal{CB}(\mathcal{H})$. Start with a useful criterion on non-compactness for a positive operator.

Lemma 3. *A positive operator $x \geq \mathbf{0}$ is non-compact if, and only if, there exist real $\lambda > 0$ and infinite dimensional orthoprojection p obeying $\lambda p \leq x$.*

Proof. Note that, in contrast to the spectral characterization of positive compact operators, the spectral theorem in case of a *non-compact* $x \geq \mathbf{0}$ with $\#\operatorname{spec}(x) < \infty$ provides that $\lambda p \leq x$ has to be fulfilled, for some non-zero λ and orthoprojection p with $\dim p\mathcal{H} = \infty$ (for one $\lambda \in \operatorname{spec}(x)\backslash\{0\}$ at least the corresponding spectral eigenprojection p has to meet the requirement). But then, due to normclosedness of the compact operators, and since for each positive x one has $x \in \overline{\{y : \mathbf{0} \leq y \leq x, \#\operatorname{spec}(x) < \infty\}}$ (uniform closure), such type of estimate has to exist in each case of a non-compact positive operator x. On the other hand, if $\lambda p \leq x$ is fulfilled, for some non-zero λ and infinite dimensional orthoprojection p, in view of this relation the equivalence of p with the unit operator $\mathbf{1}$ will imply v^*xv to be invertible, for the partial isometry v achieving $p = vv^*$, $\mathbf{1} = v^*v$. Thus, owing to the non-triviality of the ideal $\mathcal{CB}(\mathcal{H})$, $v^*xv \notin \mathcal{CB}(\mathcal{H})$ has to hold. Due to two-sidedness of $\mathcal{CB}(\mathcal{H})$ the latter requires that also x was non-compact.

Corollary 3. *Both $\mathcal{FB}(\mathcal{H})_+$ and $\mathcal{CB}(\mathcal{H})_+$ are hereditary subcones of $\mathsf{B}(\mathcal{H})_+$.*

Proof. For $\mathcal{FB}(\mathcal{H})_+$ the assertion is trivial. For non-zero $x \in \mathcal{CB}(\mathcal{H})_+$ and positive $y \neq \mathbf{0}$ with $y \leq x$ also y must be compact since otherwise the criterion of Lemma 3 were applicable to y with resulting in a contradiction to the assumed compactness of x, by the same criterion.

The following is likely the most remarkable result relating ideals in $\mathsf{B}(\mathcal{H})$ and descends from [24], see also [189, Lemma 11, Theorem 11].

Theorem 1. *$\mathcal{FB}(\mathcal{H}) \subset \mathcal{I} \subset \mathcal{CB}(\mathcal{H})$, for each non-trivial, two-sided ideal $\mathcal{I}$.*

As a consequence of this the defining ideal of a non-trivial trace τ on $\mathsf{B}(\mathcal{H})$ always is a non-zero ideal of compact operators. Thus especially $\tau(x) = \infty$ must be fulfilled, for each $x \in \mathsf{B}(\mathcal{H})_+\backslash\mathcal{CB}(\mathcal{H})_+$. On the other hand, since according to Corollary 3 $\mathcal{CB}(\mathcal{H})_+$ is a hereditary cone, whenever τ_0 is a non-zero trace on $\mathcal{CB}(\mathcal{H})$, then the extension principle of Lemma 1 can be applied and shows that upon defining $\tau(x) = \tau_0(x)$ for $x \in \mathcal{CB}(\mathcal{H})_+$, and $\tau(x) = \infty$ for $x \in \mathsf{B}(\mathcal{H})_+\backslash\mathcal{CB}(\mathcal{H})_+$, a non-zero trace τ on $\mathsf{B}(\mathcal{H})$ is given. Thus, traces (resp. non-trivial traces) on all bounded linear operators are in one-to-one correspondence with traces (resp. non-trivial traces) on the compact operators.

For the unique extension of the trace tr of Lemma 2 from compact operators onto $\mathsf{B}(\mathcal{H})$ the same notation tr will be used. Note that in view of Lemma 3 with the help of 2.1 ($\star$) and ($\star\star$) easily follows that for non-compact $x \geq \mathbf{0}$ and each m.o.n.s. $\{\varphi_n\}$ one has $\sum_{n=1}^{\infty}\langle x\varphi_n, \varphi_n\rangle = \infty$. Hence, the formula given in Corollary 1 extends on all $x \in \mathsf{B}(\mathcal{H})_+$. From this formula it is plain to see that tr is a non-trivial *normal* trace on $\mathsf{B}(\mathcal{H})$. Up to a positive multiple, tr is also unique on $\mathsf{B}(\mathcal{H})$ as non-trivial trace with this property:

Corollary 4. *A non-trivial normal trace τ has the form $\tau = \alpha \cdot \mathrm{tr}$, with $\alpha > 0$.*

Proof. Let $p_1 < p_2 < p_3 < \ldots < \mathbf{1}$ be a sequence of orthoprojections with $\mathrm{rank}(p_n) = n$, for each $n \in \mathbb{N}$. Then, for each $x \geq \mathbf{0}$, $l.u.b.\sqrt{x}\,p_n\sqrt{x} = x$. Note that $x_n = \sqrt{x}\,p_n\sqrt{x} \in \mathcal{F}B(\mathcal{H})_+$ holds. Since also $\tau|_{\mathcal{C}B(\mathcal{H})_+}$ is a non-trivial trace, by Corollary 2 there is unique $\alpha > 0$ with $\tau(x_n) = \alpha \cdot \mathrm{tr}\,x_n$, for each $n \in \mathbb{N}$. Hence, by normality of τ and since tr is normal, $\tau(x) = \alpha \cdot \mathrm{tr}\,x$ follows, for each $x \geq \mathbf{0}$.

Note that in view of the mentioned one-to-one correspondence with traces on the compact operators Corollary 2 extends to non-trivial traces on $\mathsf{B}(\mathcal{H})$ accordingly. In line with this and Corollary 4 the theory of traces on $\mathsf{B}(\mathcal{H})$ with separable infinite dimensional $\mathcal{H}$ essentially is the theory of the one normal trace tr and myriads of singular traces.

4.3 Examples of Singular Traces on $\mathsf{B}(\mathcal{H})$

Examples of singular traces have been invented by J. Dixmier in [65]. Nowadays this class is referred to as Dixmier traces. In the following, only the singular traces of this class will be constructed and considered. Thereby, in constructing these traces we will proceed in two steps.

In a first step we are going to define some non-trivial two-sided ideal in $\mathsf{B}(\mathcal{H})$, with hereditary positive cone, which later will prove to belong to the defining ideal of each of the singular traces to be constructed. As has been already noticed in context of Theorem 1, each such ideal then is an ideal of compact operators. For such ideals one knows that these can be completely described in terms of the classes (Schatten-classes) of the characteristic sequences coming along with the operators of the ideal, see [189, Theorem 12]. In these sequences, which are in $\ell^\infty(\mathbb{N})_+$, the full information on the ideal is encoded.

In a second step, a class of states on $\ell^\infty(\mathbb{N})$ is constructed which, in restriction to the mentioned sequences from the ideal, yields a map which vanishes on those sequences which correspond to operators of finite rank. If taken as functions on the positive operators of the ideal these maps will be shown to be additive, positive homogeneous and invariant. Hence, the extension via the extension principle of Lemma 1 on all of $\mathsf{B}(\mathcal{H})_+$ finally will provide us with a class of singular traces.

Step one: Some ideal of compact operators

For compact x with the help of the characteristic sequence $\{\mu_n(x)\}$ define

$$\forall k \in \mathbb{N}\setminus\{1\} : \gamma_k(x) = \frac{1}{\log k} \sum_{j \leq k} \mu_j(x) = \frac{\sigma_k(x)}{\log k}. \tag{4.3a}$$

Then, $\{\gamma_n(x) : n > 1\}$ is a sequence of non-negative reals which may be bounded or not. The bounded situation deserves our special interest. Let a subset $L^{1,\infty}(\mathcal{H}) \subset \mathcal{CB}(\mathcal{H})$ be defined as follows:

$$L^{1,\infty}(\mathcal{H}) = \left\{ x \in \mathcal{CB}(\mathcal{H}) : \sup_{n \geq 2} \gamma_n(x) < \infty \right\}. \qquad (4.3b)$$

It is plain to see that by $L^{1,\infty}(\mathcal{H})$ an ideal is given in $\mathsf{B}(\mathcal{H})$, for some corresponding terminology see [154,37,38], and e.g. [95].

Proposition 1. *$L^{1,\infty}(\mathcal{H})$ is a non-trivial two-sided ideal in $\mathsf{B}(\mathcal{H})$, and thus is an ideal of compact operators, with hereditary cone $L^{1,\infty}(\mathcal{H})_+$ of positive elements.*

Proof. In view of the definitions (4.3) and since $\mathcal{CB}(\mathcal{H})$ is a two-sided ideal, the validity of the first assertion follows as an immediate consequence of (4.2a), (4.2c) and (4.2f) together with the fact that for each operator x of finite rank $\{\gamma_n(x) : n > 1\}$ is a null-sequence and thus is bounded. Finally, owing to Corollary 3 for $x \in L^{1,\infty}(\mathcal{H})_+$ and $y \in \mathsf{B}(\mathcal{H})$ with $\mathbf{0} \leq y \leq x$ one infers $y \in \mathcal{CB}(\mathcal{H})_+$, and then $y \leq x$ according to (4.2b) implies also $y \in L^{1,\infty}(\mathcal{H})_+$.

For completeness yet another characterization of $L^{1,\infty}(\mathcal{H})$ will be noted (without proof, see e.g. in [38, IV.2.β]), and a class of $L^{1,\infty}$-elements, which can be characterized through the asymptotic behavior of the singular values, will be given.

Let $L^1(\mathcal{H})$ be the ideal of all operators of *trace-class*, that is, the defining ideal which corresponds to the normal trace tr, cf. Lemma 2 and Corollary 1. From (4.3b) and Lemma 2 then especially follows that the inclusion relation $L^1(\mathcal{H}) \subset L^{1,\infty}(\mathcal{H})$ takes place amongst $L^{1,\infty}(\mathcal{H})$ and the ideal of trace-class operators. Moreover, if in line with [154] another Banach space $L^{\infty,1}(\mathcal{H})$ ($= \mathfrak{S}_\omega$ in [154]) is defined through

$$L^{\infty,1}(\mathcal{H}) = \left\{ y \in \mathcal{CB}(\mathcal{H}) : \sum_{n=1}^{\infty} n^{-1} \mu_n(y) < \infty \right\}, \qquad (4.4)$$

then it is essentially due to (4.2c), (4.2e) and by monotonicity of the sequences of the $\sigma_n(y)$'s and $\frac{1}{n}$'s that also $L^{\infty,1}(\mathcal{H})$ is a non-trivial two-sided ideal (Macaev ideal). Note that in analogy to the above also in this case obviously an inclusion with trace-class operators takes place, $L^1(\mathcal{H}) \subset L^{\infty,1}(\mathcal{H})$. The ideals from (4.3b) and (4.4) are related by the duality given through the 2-form $\Omega(x,y) = \operatorname{tr} xy$. Namely, each $x \in \mathcal{CB}(\mathcal{H})$ obeying $xy \in L^1(\mathcal{H})$, for all $y \in L^{\infty,1}(\mathcal{H})$, is in $L^{1,\infty}(\mathcal{H})$.

Proposition 2. *$L^{1,\infty}(\mathcal{H})$ is the dual to the Macaev ideal.*

Also, in this context note that for each $x \in \mathcal{CB}(\mathcal{H})$ obeying $xz \in L^1(\mathcal{H})$ for all $z \in I$, with an ideal I of compact operators, and each $y \in I$ the relation

$$|\Omega(x,y)| \leq \sum_{n=1}^{\infty} \mu_n(x)\mu_n(y) < \infty \qquad (4.5a)$$

must be fulfilled. In fact, since by assumption for $a, b \in \mathsf{B}(\mathcal{H})$ also $axby \in L^1(\mathcal{H})$ is fulfilled, in view of the polar decomposition of x, y the estimate $|\Omega(x,y)| \leq \sup_{u,v} |\operatorname{tr} u|x|v|y||$ can be easily inferred, with u, v extending over the partial isometries in $\mathsf{B}(\mathcal{H})$. Also, with the help of Corollary 1, and (4.1b) e.g., one finds that $\sup_{u,v} |\operatorname{tr} u|x|v|y|| \leq \sup_{\boldsymbol{r}} \sum_{n=1}^{\infty} \mu_n(x) r_n$ must hold, with $\boldsymbol{r} = (r_1, r_2, \ldots)$ obeying $r_1 \geq r_2 \geq r_3 \geq \ldots \geq 0$ and $\sum_{k \leq n} r_k \leq \sigma_n(y) = \sum_{k \leq n} \mu_k(y)$, for each $n \in \mathbb{N}$ (the ordering of $\mu_n(x)$'s is of importance in this context). Since also the sequence of $\mu_n(y)$'s is in decreasing order, it is not hard to see that by successively exploiting the just mentioned conditions on $\boldsymbol{r}$, for $n \leq N$ with $N \in \mathbb{N}$, the validity of $\sum_{k \leq N} \mu_k(x)(\mu_k(y) - r_k) \geq 0$ can be derived, for each $N \in \mathbb{N}$, and any given $\boldsymbol{r}$ which is subject to the above conditions. From this the left-hand side estimate of (4.5a) gets evident. Now, for any two given compact linear operators x, y in view of the polar decomposition theorem and owing to compactness of both operators partial isometries u, w can be chosen such that $u|x|w|y| \geq \mathbf{0}$ holds, with the singular values of the compact operator $u|x|w|y|$ obeying $\mu_n(u|x|w|y|) = \mu_n(x)\mu_n(y)$, for each $n \in \mathbb{N}$. In accordance with Lemma 2 one then has $|\operatorname{tr} u|x|v|y|| = \operatorname{tr} u|x|w|y| = \lim_{n \to \infty} \sigma_n(u|x|w|y|) = \sum \mu_n(u|x|w|y|) = \sum \mu_n(x)\mu_n(y)$. Hence, since in our particular situation of x, y we have $u|x|w|y| \in L^1(\mathcal{H})$ and the above proved left-hand side estimate of (4.5a) has been shown to hold, (4.5a) is completely seen.

Especially, in view of Proposition 2 the estimate (4.5a) can be applied with $x \in L^{1,\infty}(\mathcal{H})$ and $I = L^{\infty,1}(\mathcal{H})$. Relating asymptotic properties of singular values of $x \in L^{1,\infty}(\mathcal{H})$ we thus get the following information:

$$x \in L^{1,\infty}(\mathcal{H}) \implies \forall y \in L^{\infty,1}(\mathcal{H}) : \sum_{n=1}^{\infty} \mu_n(x)\mu_n(y) < \infty. \qquad (4.5b)$$

Viewing (4.4) and (4.5b) together suggests compact x with asymptotic behavior of singular values like $\mu_n(x) = \mathbf{O}(n^{-1})$ as good candidates for elements of $L^{1,\infty}(\mathcal{H})$.[1] In fact, such asymptotic behavior implies that, with some $C > 0$, for all $n \geq 2$

$$\sigma_n(x) - \sum_{1 \leq k \leq n} \mu_k(x) \leq C\left\{1 + \sum_{2 \leq k \leq n} k^{-1}\right\} \leq C\left\{1 + \int_1^n t^{-1} dt\right\} = C(1 + \log n)$$

is fulfilled. In view of (4.3) we therefore arrive at the following result:

---

[1] As usual, for $g : \mathbb{N} \ni n \mapsto g(n) \in \mathbb{R}_+ \setminus \{0\}$ the notation $x_n = \mathbf{O}(g(n))$ is a shorthand notation for $|x_n| < C g(n)$, with some $C > 0$ (and accordingly defined with $\mathbb{R}_+$ instead of $\mathbb{N}$).

Corollary 5. $x \in \mathcal{CB}(\mathcal{H})$, $\mu_n(x) = \mathbf{O}(n^{-1}) \implies x \in L^{1,\infty}(\mathcal{H})$.

Remark 1. (1) It is easy to see that for compact x with bounded multiplicity function, $m(\lambda) \leq N < \infty$ for all λ, the condition imposed by (4.5b) upon x amounts to $\mu_n(x) = \mathbf{O}(n^{-1})$. Unfortunately, in case of unbounded m this can fail to hold.[2] That this can even occur for x within $L^{1,\infty}(\mathcal{H})$ can be seen by the following counterexample:[3]

(2) Let x be positive and compact with $\mu_1(x) = 1$, and with singular values which for $k \geq 2$ with $(m-1)! < k \leq m!$, $m \geq 2$, are given by $\mu_k(x) = \log m / m!$. One then easily proves that $\sigma_k(x)$ obeys $\sigma_{m!}(x) \leq 1 + \log m!$. Since the function $f(t) = \log(1 + t/m!) - t\{\log(m+1)/(m+1)!\}$ is non-negative for $0 \leq t \leq m \cdot m!$, from the previous also $\sigma_k(x) \leq 1 + \log k$ can be followed whenever $m! < k \leq (m+1)!$ is fulfilled. This conclusion applies for each $m \geq 2$, and thus according to (4.3) we finally get $x \in L^{1,\infty}(\mathcal{H})$. On the other hand, $\lim_{m \to \infty} m! \mu_{m!}(x) = \infty$ holds. Thus in particular $\mu_n(x)$ certainly cannot behave asymptotically like $\mathbf{O}(n^{-1})$.

Step two: Scaling invariant states

Let us come back now to the construction of the Dixmier traces. The construction will be based on considering a certain class of states on the commutative W*-algebra $M = \ell^\infty$. Relating special further notations, for each $k \in \mathbb{N}$ let $e_k \in \ell^\infty$ be the k-th atom in ℓ^∞, with j-th component obeying $(e_k)_j = \delta_{kj}$ (Kronecker symbol), and let E_k be the special orthoprojection of rank k given as $E_k = \sum_{j \leq k} e_j$. The ascendingly directed sequence $\{E_n\}$ obeys $l.u.b. E_n = \mathbf{1}$ and the following equivalence is valid:

$$x \in \ell^\infty : \|\cdot\|_\infty - \lim_{n \to \infty} E_n x = x \iff \lim_n x_n = 0. \quad (4.6)$$

Also, for $x \in \ell^\infty_+$, $\{E_n x\} \subset \ell^\infty_+$ is ascendingly directed, with $l.u.b. E_n x = x$.

For the following, let a mapping $\mathbf{s} : \ell^\infty \longrightarrow \ell^\infty$ (scaling) be defined on $x \in \ell^\infty$ through $\mathbf{s}(x)_j = x_{2j}$, for all $j \in \mathbb{N}$. It is obvious that $\mathbf{s}$ is a normal *-homomorphism onto ℓ^∞. Hence, $\mathbf{s}$ is a unital normal positive linear map onto itself, and $\ell^\infty_{\mathbf{s}} = \{x \in \ell^\infty : \mathbf{s}(x) = x\}$ is a W*-subalgebra of ℓ^∞ (the fixpoint algebra of $\mathbf{s}$).

Lemma 4. *There exists a conditional expectation* $\mathcal{E} : \ell^\infty \longrightarrow \ell^\infty_{\mathbf{s}}$ *projecting onto the fixpoint algebra* $\ell^\infty_{\mathbf{s}}$ *such that the following properties hold*:

(1) $\mathcal{E} \circ \mathbf{s} = \mathcal{E}$;
(2) $\mathcal{E}(x) = (\lim_{n \to \infty} x_n) \cdot \mathbf{1}$, *for each* $x \in \ell^\infty$ *with* $\lim_{n \to \infty} x_n$ *existing*.

---

[2] We are grateful to C. Portenier (Marburg) for mentioning this fact to us.
[3] The counterexample has been communicated to us by J. Várilly, see also [96, Lemma 7.37].

Proof. Let us consider the sequence $\{\mathbf{s}\langle n\rangle\}$ of partial averages $\mathbf{s}\langle n\rangle = \frac{1}{n}\sum_{k\leq n}\mathbf{s}^k$, $n \in \mathbb{N}$. Since these all are unital positive linear maps, by $\sigma(\ell^\infty, \ell^1)$-weak compactness of the closed unit ball in $\mathsf{B}(\ell^\infty)$ the sequence of partial averages then must have a $\sigma(\ell^\infty, \ell^1)$-weak cluster point $\mathcal{E}$ which has to be a unital positive linear map, too. Since then $\mathcal{E} = \sigma(\ell^\infty, \ell^1) - \text{weak}\lim_\lambda \mathbf{s}\langle n_\lambda\rangle$ has to be fulfilled for some appropriately chosen subnet $\{\mathbf{s}\langle n_\lambda\rangle\}$, the inclusion $\ell_\mathbf{s}^\infty \subset \{x \in \ell^\infty : \mathcal{E}(x) = x\}$ gets evident. Since $\mathbf{s}\langle n\rangle \circ \mathbf{s} = \mathbf{s} \circ \mathbf{s}\langle n\rangle$ and $\|\mathbf{s}\langle n\rangle \circ \mathbf{s} - \mathbf{s}\langle n\rangle\| \leq \frac{2}{n}$ hold, for each $n \in \mathbb{N}$, and since owing to normality of $\mathbf{s}$ for each $\omega \in \ell^1$ also $\omega \circ \mathbf{s} \in \ell^1$ is fulfilled, by argueing with the mentioned subnet one infers that $\mathcal{E} \circ \mathbf{s} = \mathbf{s} \circ \mathcal{E} = \mathcal{E}$. From this $\{x \in \ell^\infty : \mathcal{E}(x) = x\} \subset \ell_\mathbf{s}^\infty$ and $\mathbf{s}\langle n\rangle \circ \mathcal{E} = \mathcal{E}$ follow, for each n. Thus in view of the above $\mathcal{E}^2 = \mathcal{E} \circ \mathcal{E} = \mathcal{E}$ follows. Hence, $\mathcal{E}$ is a projection of norm one (conditional expectation) projecting onto the fixpoint algebra of $\mathbf{s}$ and which satisfies (1).

To see (2), note first that owing to $\mathbf{s}(e_k) = \mathbf{0}$ for k odd, and $\mathbf{s}(e_k) = e_{k/2}$ for k even, one certainly has $\mathbf{s}^n(E_k) = \mathbf{0}$, for each $n > \log k / \log 2$. Hence, the action of the n-th average $\mathbf{s}\langle n\rangle$ to the orthoprojection E_k can be estimated as $\|\mathbf{s}\langle n\rangle(E_k)\|_\infty \leq [\log k / \log 2]/n$ (here $[\cdot]$ means the integer part), and thus for all $k \in \mathbb{N}$ one has $\|\cdot\|_\infty - \lim_{n\to\infty}\mathbf{s}\langle n\rangle(E_k) = \mathbf{0}$. From this and $\mathcal{E} = \sigma(\ell^\infty, \ell^1) - \text{weak}\lim_\lambda \mathbf{s}\langle n_\lambda\rangle$ then especially $\omega(\mathcal{E}(E_k)) = 0$ follows, for each $\omega \in \ell^1$. Hence $\mathcal{E}(E_k) = \mathbf{0}$, for each k. Since for each $y \in \ell^\infty$ with $\mathbf{0} \leq y \leq \mathbf{1}$ one has $\mathbf{0} \leq E_k y \leq E_k$, from the previous together with positivity of $\mathcal{E}$ also $\mathcal{E}(E_k y) = \mathbf{0}$ follows. By linearity of $\mathcal{E}$ and since ℓ^∞ is the linear span of $\ell_+^\infty \cap (\ell^\infty)_1$ this remains true for each $y \in \ell^\infty$. But then, for $x \in \ell^\infty$ with $\alpha = \lim_{n\to\infty}x_n$ by continuity of $\mathcal{E}$ and in view of (4.6) one infers $\mathcal{E}(x - \alpha \cdot \mathbf{1}) = \|\cdot\|_\infty - \lim_{k\to\infty} \mathcal{E}(E_k(x - \alpha \cdot \mathbf{1})) = \mathbf{0}$, which is equivalent with (2).

Corollary 6. *There is a state $\omega \in \mathcal{S}(\ell^\infty)$ satisfying the following properties:*

(1) $\omega \circ \mathbf{s} = \omega$;
(2) $\omega(x) = \lim_{n\to\infty} x_n$, provided $\lim_{n\to\infty} x_n$ exists.

The set $\Gamma_\mathbf{s}(\ell^\infty)$ of all such states is a w^*-compact convex subset of singular states. [4]

Proof. Let $\mathcal{E}$ be constructed as in Lemma 4. By positivity and unitality of $\mathcal{E}$, for each $\nu \in \mathcal{S}(\ell^\infty)$ also $\omega = \nu \circ \mathcal{E}$ is a state. In view of (1)–(2) this state then obviously satisfies (1)–(2).[5] That $\Gamma_\mathbf{s}(\ell^\infty)$ is w^*-compact and convex is evident from the linear nature of the conditions (1)–(2). Finally, in accordance with

---

[4] Let another map $\mathbf{d}$ (doubling) over ℓ^∞ be defined at x by $\mathbf{d}(x)_j = x_{[(1+j)/2]}$, $j \in \mathbb{N}$ ($[r]$ refers to the integer part of r). Then, scaling is left-inverse to doubling in $\mathsf{B}(\ell^\infty)$, and thus in addition to (1) one also has $\mathbf{d}$-invariance of each $\omega \in \Gamma_\mathbf{s}(\ell^\infty)$ as well.

[5] The usage of Lemma 4 might be avoided in this context; as we learned from [193] a positivity and separation argument of Hahn-Banach type may be used instead as well.

(2) one has $\omega(E_k) = 0$, for each $\omega \in \Gamma_s(\ell^\infty)$ and all $k \in \mathbb{N}$. Now, let $p \in \ell^\infty$ be any orthoprojection with $\omega(p) > 0$. Then, $p \neq \mathbf{0}$, and owing to $l.u.b.\, E_n p = p$ there has to exist $k \in \mathbb{N}$ with $q = E_k p \neq \mathbf{0}$. Thus $\mathbf{0} < q < p$ and $q \leq E_k$. In view of the above from the latter by positivity of ω then $\omega(q) = 0$ follows. Hence, each $\omega \in \Gamma_s(\ell^\infty)$ is singular.

Constructing the dixmier traces

For given $x \in L^{1,\infty}(\mathcal{H})$, let a sequence $\gamma(x)$ be given through $\gamma(x) = (\gamma_2(x), \gamma_3(x), \ldots)$, with $\gamma_n(x)$ in accordance with (4.3a). Then, by definition (4.3b) one has $\gamma(x) \in \ell_+^\infty$. Hence, if for each fixed scaling invariant state $\omega \in \Gamma_s(\ell^\infty)$, see Corollary 6, following [65] we define

$$\forall x \in L^{1,\infty}(\mathcal{H})_+ \;:\; \mathrm{Tr}_\omega(x) = \omega(\gamma(x)), \qquad (4.7)$$

then according to Proposition 1 and since ω is a positive linear form, we are given a *positive* map $\mathrm{Tr}_\omega : L^{1,\infty}(\mathcal{H})_+ \ni x \longmapsto \mathrm{Tr}_\omega(x) \in \mathbb{R}_+$ defined on the positive cone of the ideal $L^{1,\infty}(\mathcal{H})$. The key idea of [65] is that additivity of Tr_ω can be shown.

Lemma 5. Tr_ω *is an additive, positive homogeneous and invariant map from* $L^{1,\infty}(\mathcal{H})_+$ *into* $\mathbb{R}_+$.

Proof. Since $L^{1,\infty}(\mathcal{H})_+$ is the positive cone of a two-sided ideal of compact operators, for $x, y \in L^{1,\infty}(\mathcal{H})_+$ and $\lambda \in \mathbb{R}_+$ we have that $x + y,\, \lambda x,\, x^* x,\, xx^* \in L^{1,\infty}(\mathcal{H})_+$, and these are compact operators again. Hence, in view of (4.3a) from (4.2a) and (4.2g) both $\lambda \cdot \gamma(x) = \gamma(\lambda x)$ and $\gamma(x^* x) = \gamma(xx^*)$ follow, which in line with (4.7) means that Tr_ω is positive homogeneous and invariant. It remains to be shown that Tr_ω is additive. First note that according to the left-hand side estimate of (4.2d) within ℓ_+^∞ one has $\gamma(x+y) \leq \gamma(x) + \gamma(y)$. Hence, by positivity and linearity of ω, (4.7) yields

$$\mathrm{Tr}_\omega(x + y) \leq \mathrm{Tr}_\omega(x) + \mathrm{Tr}_\omega(y). \qquad (\star)$$

Now, to each compact operator z let $\gamma^0(z) = (\gamma_3(z), \gamma_4(z), \ldots)$, that is, $\gamma^0(z)$ arises from $\gamma(x)$ by application of the one-step left-shift. Also, on ℓ^∞ let a linear map $\mathbf{m}$ be defined by $\mathbf{m}(\beta)_n = \frac{\log 2}{\log(n+1)} \cdot \beta_n$, for all $n \in \mathbb{N}$, at $\beta \in \ell^\infty$. One then has

$$\forall \beta \in \ell^\infty \;:\; \lim_{n \to \infty} \mathbf{m}(\beta)_n = 0. \qquad (\star\star)$$

Note that $\gamma^0(z) \in \ell^\infty$ whenever $z \in L^{1,\infty}(\mathcal{H})$. We are going to estimate $\gamma(z) - \gamma^0(z)$ for $z \in L^{1,\infty}(\mathcal{H})$. Since both $\{\sigma_n(z)\}$ and $\{\log n\}$ are monotoneously increasing, in view of the definition (4.3a) for each $z \in L^{1,\infty}(\mathcal{H})$ the follwing estimates at once can be seen to hold, for all $k \in \mathbb{N} \setminus \{1\}$:

$$\gamma_k(z) - \gamma_{k+1}(z) \leq \gamma_{k+1}(z)\left(\frac{\log(k+1)}{\log k} - 1\right) \leq \frac{\log 2}{\log k} \cdot \gamma_{k+1}(z) \leq \frac{\log 2}{\log k} \|\gamma(z)\|_\infty.$$

On the other hand, we also have

$$\gamma_k(z) - \gamma_{k+1}(z) \geq \frac{\sigma_k(z) - \sigma_{k+1}(z)}{\log(k+1)} = -\frac{\mu_{k+1}(z)}{\log(k+1)} \geq -\frac{1}{\log(k+1)} \cdot \|z\|.$$

From these two estimates we infer that $\Delta(z) = \gamma(z) - \gamma^0(z)$ for $z \in L^{1,\infty}(\mathcal{H})$ is a null-sequence in ℓ^∞, that is, $\lim_{n\to\infty} \Delta(z)_n = 0$ is fulfilled. According to the choice of ω and in accordance with Corollary 6 (2) we thus have the following to hold:

$$\forall z \in L^{1,\infty}(\mathcal{H}), \omega \in \Gamma_{\mathsf{s}}(\ell^\infty) : \omega(\gamma(z)) = \omega(\gamma^0(z)). \qquad (\star\star\star)$$

Let us come back to our above $x, y \in L^{1,\infty}(\mathcal{H})_+$. Having in mind the definitions of the positive linear operators $\mathbf{s}$ and $\mathbf{m}$ as well as the meanings of γ and γ^0, it is easily inferred that from the right-hand side estimate in (4.2d) when divided by $\log k$, and considered for all $k \geq 2$, the estimate $\gamma(x) + \gamma(y) \leq (\sigma_{2k}(x+y)/\log k) = \mathbf{s}(\gamma^0(x+y)) + \mathbf{m} \circ \mathbf{s}(\gamma^0(x+y))$ can be followed to hold in ℓ^∞_+. By positivity and linearity of ω from this then

$$\omega(\gamma(x)) + \omega(\gamma(y)) \leq \omega \circ \mathbf{s}(\gamma^0(x+y)) + \omega \circ \mathbf{m}(\mathbf{s}(\gamma^0(x+y))) \qquad (\circ)$$

follows. Now, in view of $(\star\star)$ and Corollary 6 (2) one has $\omega \circ \mathbf{m}(\mathbf{s}(\gamma^0(x+y))) = 0$, whereas from Corollary 6 (1) and $(\star\star\star)$ one concludes that $\omega \circ \mathbf{s}(\gamma^0(x+y)) = \omega(\gamma(x+y))$. These facts together with $(\circ)$ fit together into the estimate $\omega(\gamma(x)) + \omega(\gamma(y)) \leq \omega(\gamma(x+y))$, which in view of (4.7) says that $\mathrm{Tr}_\omega(x) + \mathrm{Tr}_\omega(y) \leq \mathrm{Tr}_\omega(x+y)$ has to be valid. The latter and $(\star)$ then make that the desired additivity $\mathrm{Tr}_\omega(x) + \mathrm{Tr}_\omega(y) = \mathrm{Tr}_\omega(x+y)$ holds.

The Dixmier trace as a singular trace

We recall that according to Proposition 1 the positive cone of the ideal $L^{1,\infty}(\mathcal{H})$ is hereditary. Thus the extension principle of Lemma 1 according to Lemma 5 for each $\omega \in \Gamma_{\mathsf{s}}(\ell^\infty)$ allows to extend the map Tr_ω of (4.7) to a trace on $\mathsf{B}(\mathcal{H})$. Thereby, the extension constructed in accordance with the proof of Lemma 1 will be the unique one with defining ideal $L^{1,\infty}(\mathcal{H})$. For this trace the same notation Tr_ω will be used henceforth. We refer to this trace as *Dixmier trace* (to the particular $\omega \in \Gamma_{\mathsf{s}}(\ell^\infty)$). The essential properties of Dixmier traces are summarized in the following.

Theorem 2. Tr_ω *is a singular trace on* $\mathsf{B}(\mathcal{H})$, *for each* $\omega \in \Gamma_{\mathsf{s}}(\ell^\infty)$. *The following properties are fulfilled:*

(1) $L^{1,\infty}(\mathcal{H})_+ = \{x \in \mathsf{B}(\mathcal{H})_+ : \mathrm{Tr}_\omega(x) < \infty\}$;
(2) $x \in L^{1,\infty}(\mathcal{H})_+, \exists \lim_{n\to\infty} \gamma_n(x) \implies \mathrm{Tr}_\omega(x) = \lim_{n\to\infty} \gamma_n(x)$.

Proof. The validity of (1) follows since the traces in question all are obtained as extensions of the maps given in (4.7), which satisfy Lemma 5 and which have range $\mathbb{R}_+$ (and not merely $\overline{\mathbb{R}}_+$). Since each state $\omega \in \Gamma_{\mathsf{s}}(\ell^\infty)$ obeys Corollary 6 (2), in view of the previous and (4.7) also (2) follows. Finally, for each $x \in \mathcal{F}B(\mathcal{H})_+$ the sequence $\gamma(x)$ is a null-sequence, and therefore especially $x \in L^{1,\infty}(\mathcal{H})_+$, and as a special case of (2) then $\mathrm{Tr}_\omega(x) = 0$ follows. Hence, Tr_ω is a singular trace.

Note the remarkable feature of the Dixmier traces coming along with Theorem 2 (2) and saying that provided certain circumstances are fulfilled for x, e.g. if the sequence $\{\gamma_n(x)\}$ has a limit, then independent of the state-parameter ω all these Dixmier traces may yield the same common value at this x. It is such case of independence one usually is tacitly addressing to when speaking simply of *the* Dixmier trace of x, whereas the operator itself then is referred to as *measurable operator*, cf. [38, IV.2, Definition 7]. Some criteria of measurability, which however all reduce upon showing that the above mentioned special case of existence of $\lim_{n\to\infty} \gamma_n(x)$ would happen, subsequently will be discussed in more detail.

4.4 Calculating the Dixmier Trace

Simple criteria of measurability

We start with discussing conditions which read in terms of spectral theory and which ensure that – for a given operator $x \in L^{1,\infty}(\mathcal{H})$ which is not simply of finite rank – the above-mentioned special case of measurability occurs, that is, the limit $\lim_{n\to\infty} \gamma_n(x)$ exists. As a first result of that kind one has the following one:[6]

Lemma 6. *Suppose $x \in \mathcal{CB}(\mathcal{H})$, with $\mu_n(x) \sim L \cdot n^{-1}$. Then $\lim_{n\to\infty} \gamma_n(x) = L$.*[7]

Proof. For compact operator x suppose $\lim_{n\to\infty} n\,\mu_n(x) = L$ to be fulfilled. Then, in case of $L > 0$, for δ with $L > \delta > 0$, let $M(\delta) \in \mathbb{N}$ be chosen such that
$$\forall n \geq M = M(\delta): \ (L-\delta)\,n^{-1} \leq \mu_n(x) \leq (L+\delta)\,n^{-1}.$$
From this for each $n \geq M$ we get
$$(L-\delta) \sum_{M<k\leq n} k^{-1} \leq \sum_{M<k\leq n} \mu_n(x) \leq (L+\delta) \sum_{M<k\leq n} k^{-1}. \qquad (\times)$$
Since $0 < t \mapsto t^{-1}$ is a strictly monotone decreasing function and the sequence of the singular values is decreasingly ordered, with the help of
$$\int_{M+1}^n dt\, t^{-1} \leq \sum_{M<k\leq n} k^{-1} \leq \int_M^n dt\, t^{-1}$$

---

[6] We are grateful to C. Portenier, Marburg, for suggesting some details around this and related subjects [177].

[7] For $f: \mathbb{N} \to \mathbb{R}_+$ and $g: \mathbb{N} \to \mathbb{R}_+\backslash\{0\}$ the notation $f(n) \sim L \cdot g(n)$ stands for $\lim_{n\to\infty} f(n)/g(n) = L$ (and accordingly defined with $\mathbb{R}_+$ instead of $\mathbb{N}$).

which holds for $n > M$ the above estimate ($\times$) implies

$$\int_{M+1}^{n} dt\,(L-\delta)\,t^{-1} \leq \sigma_n(x) - \sigma_M(x) \leq \int_{M}^{n} dt\,(L+\delta)\,t^{-1}.$$

From this for all $n > M = M(\delta)$

$$(L-\delta)\{1-\log(M+1)/\log n\} \leq \gamma_n(x)-\sigma_M(x)/\log n \leq (L+\delta)\{1-\log M/\log n\}$$

is obtained. Considering these estimates for $n \to \infty$ then yields

$$(L-\delta) \leq \liminf_{n\to\infty} \gamma_n(x) \leq \limsup_{n\to\infty} \gamma_n(x) \leq L+\delta.$$

Note that in case of $L = 0$ by positivity of all $\gamma_n(x)$ instead of the previous one finds $0 \leq \liminf_{n\to\infty} \gamma_n(x) \leq \limsup_{n\to\infty} \gamma_n(x) \leq \delta$, for any $\delta > 0$. Thus, since $\delta > 0$ can be chosen arbitrarily small, in either case $\lim_{n\to\infty} \gamma_n(x) = L$ follows.

Now, let us suppose $x \in \mathcal{CB}(\mathcal{H})$, with $\mu_n(x) = \mathbf{O}(\frac{1}{n})$. According to Corollary 5 we even have $x \in L^{1,\infty}(\mathcal{H})$, and since $\sum_n n^{-(1+\varepsilon)} < \infty$ is fulfilled for each $\varepsilon > 0$, then $|x|^z$ at each $z \in \mathbb{C}$ with $\operatorname{Re} z > 1$ has to be of trace-class and the definition

$$\zeta_x(z) = \sum_{n \geq 1} \mu_n(x)^z = \operatorname{tr} |x|^z \tag{4.8}$$

will provide us with some holomorphic function ζ_x in the half-plane $\operatorname{Re} z > 1$. For this modification of the Riemann ζ-function the following holds.

Lemma 7. *Let $x \in \mathcal{CB}(\mathcal{H})$, with $\mu_n(x) = \mathbf{O}(\frac{1}{n})$. Suppose ζ_x admits an extension onto the half-plane $\operatorname{Re} z \geq 1$ which is continuous there except for a simple pole with residue L at $z = 1$, at worst. Then even $\mu_n(x) \sim L \cdot n^{-1}$ holds.*

Proof. In case of $x \in \mathcal{FB}(\mathcal{H})$ one has $\lim_n n \cdot \mu_n(x) = 0$ as well as $\lim_{\varepsilon \to 0+} \operatorname{tr} |x|^{1+\varepsilon} = \operatorname{tr} |x| = \sum_n \mu_n(x) < \infty$, by triviality. From the latter $\lim_{s \to 1+}(s-1)\zeta_x(s) = 0$ follows. Hence, for each operator x of finite rank the assertion is true, with $L = 0$.

Suppose now that x is not of finite rank, $x \notin \mathcal{FB}(\mathcal{H})$. In view of definitions (4.3a) and (4.8), upon possibly considering instead of x a scaling λx by a suitably chosen real $\lambda > 0$, without loss of generality it suffices if the assertion for $x \geq \mathbf{0}$ with $\mu_1(x) < 1$ can be shown. In line with this assume such $x \in \mathcal{CB}(\mathcal{H})_+ \setminus \mathcal{FB}(\mathcal{H})$.

By the spectral theorem there exists a spectral representation of x as an operator Stieltjes integral $x = \int_0^{1-} \lambda\,E(d\lambda)$, with projection-valued measure $E(d\lambda)$ derived from a left-continuous spectral family $\{E(\lambda) : \lambda \in \mathbb{R}\}$, that is, a family of orthoprojections obeying $E(t) \leq E(\lambda)$, for $t \leq \lambda$, $E(s) = \mathbf{0}$, for $s \leq 0$ and $E(\lambda-) = l.u.b._{t<\lambda} E(t) = E(\lambda)$, for each $\lambda \leq \infty$, with $l.u.b._{t<\infty} E(t) = \mathbf{1}$.

By convention, for $a < b$, then $\int_a^{b-} E(d\lambda) = E(b) - E(a) = E([a,b[)$ and $\int_a^b E(d\lambda) = E(b+) - E(a) = E([a,b])$, and so on accordingly, where e.g. $E(b+)$ stands for the greatest lower bound $E(b+) = g.l.b._{t>b} E(t)$.

By means of some functional calculus and owing to normality of the trace tr it is easily inferred that (4.8) can be represented as an ordinary Stieltjes integral:

$$\forall z \in \mathbb{C}, \operatorname{Re} z > 1 : \zeta_x(z) = \int_{1+}^{\infty} t^{-z}\, d\alpha(t), \tag{4.9a}$$

with the monotone increasing function α given by

$$\alpha(t) = \operatorname{tr} E([1/t, \infty[). \tag{4.9b}$$

But then, if the assumptions on ζ_x are fulfilled with $\lim_{s\to 1+}(s-1)\zeta_x(s) = L$, all conditions for an application of Ikehara's theorem [109], are given (we refer to the formulation in [222, THEOREM 16]). In line with this the conclusion is that asymptotically

$$\alpha(t) \sim L \cdot t \tag{4.9c}$$

has to be fulfilled as t tends to infinity. Since x is a compact operator, in view of the properties of the spectral resolution E together with normality of tr the definition (4.9b) provides a right-continuous, integral-valued step function which is constant between inverses of neighbouring spectral values of x. Especially, in case of $n \in \mathbb{N}$ with $\mu_n(x) > \mu_{n+1}(x)$ one infers that α for all t with $\mu_n(x)^{-1} \leq t < \mu_{n+1}(x)^{-1}$ yields $\alpha(t) = n$. A moments reflection then shows that with respect to each term of the ordered sequence $n_1 < n_2 < n_3 < \ldots$ of all subscripts where the value of μ_n jumps the relation (4.9c) in view of $\lim_{n\to\infty} \mu_n(x) = 0$ and by continuity of the parameter t in particular also implies both $\lim_{k\to\infty} n_{k+1}\, \mu_{n_{k+1}}(x) = L$ and $\lim_{k\to\infty} n_k\, \mu_{n_{k+1}}(x) = L$ to be fulfilled. But then, since $\mu_n(x) = \mu_{n_{k+1}}(x)$ holds for $n_k < n \leq n_{k+1}$, also $\lim_{n\to\infty} n\, \mu_n(x) = L$ can be obtained from these limit relations. Thus under the condition of the hypothesis also $\mu_n(x) \sim L \cdot n^{-1}$ in case of $x \geq \mathbf{0}$ and which is not of finite rank. In accordance with our preliminary remarks the assertion then has to be true, in either case under the mentioned hypothesis.

Remark 2. (1) Relating Lemma 6 remark that there are examples of operators where $\lim_{n\to\infty} \gamma_n(x) = L$ exists but $\mu_n(x) \not\sim L \cdot n^{-1}$, see [218, Beispiel A.27] or [96, Lemma 7.37].
(2) On the one hand, the conditions imposed on $\mu_n(x)$ and ζ_x in Lemma 7 simply reproduce the usual conditions for the standard results of Tauberian type[8] to become applicable. On the other hand, that the behavior of the extension of ζ_x at the whole line $\operatorname{Re} z = 1$ (and not only at $z = 1$) has

---

[8] This especially concerns theorems of Hardy and Littlewood [107] and Ikehara [109], see [106, Chap. VII, 7.5] and [222, see especially on p. 126 and THEOREM 18].

to be of relevance can be seen also by example: there is $x \in \mathcal{CB}(\mathcal{H})$ with $\mu_n(x) = \mathbf{O}(\frac{1}{n})$ and $\lim_{s \to 1+}(s-1)\zeta_x(s) = 1$ but for which $\mu_n(x) \not\sim 1/n$.[9]

A Residue-formula for the Dixmier trace

The most important from practical point of view special case of measurability for an operator x occurs if the limit $\lim_n \gamma_n(x)$ exists. In particular, according to the previous considerations the latter will happen e.g. provided some function-theoretic assumptions on x can be satisfied. In these cases a formula arises which allows us to calculate the (singular) Dixmier trace with the help of the ordinary trace as a limit of some function-theoretic expression of the operator in question. In fact, in view of Theorem 2 (2) and upon combining Lemma 7 and Lemma 6 we get the following result:

Corollary 7. *For each $x \in \mathcal{CB}(\mathcal{H})_+$ with $\mu_n(x) = \mathbf{O}(\frac{1}{n})$ one has $x \in L^{1,\infty}(\mathcal{H})_+$, and then by $\zeta_x(z) = \operatorname{tr} x^z$ a holomorphic function in the half-plane $\operatorname{Re} z > 1$ is given. Suppose ζ_x extends onto the half-plane $\operatorname{Re} z \geq 1$ and is continuous there except for a simple pole at $z = 1$, at worst. Then the Dixmier trace of x is obtained as*

$$\operatorname{Tr}_\omega(x) = \lim_n \gamma_n(x) = \lim_{s \to 1+}(s-1)\operatorname{tr} x^s. \qquad (4.10\mathrm{a})$$

Especially, when ζ_x extends to a meromorphic function on the whole complex plane, with a simple pole at $z = 1$ at worst, this formula turns into

$$\operatorname{Tr}_\omega(x) = \mathfrak{Res}|_{z=1}(\zeta_x), \qquad (4.10\mathrm{b})$$

with the residue $\mathfrak{Res}(\zeta_x)$ of the extended complex function, taken at $z = 1$.

For completeness remark that by our Corollary 7, which is sufficient to cope with our later needs around Connes' trace theorem, in the special cases at hand the implication (1) $\Rightarrow$ (2) of [38, IV, Proposition 4] is reproduced.

Clearly, from both the theoretical and practical point of view, in context of the previous those situations deserve the main interest where formula (4.10b) could be applied. According to the results in [99, THEOREM 7.1, 7.2] this happens e.g. if the context of the classical pseudodifferential operators of order $-n$ acting on the sections $\Gamma(E)$ of a complex vector bundle $E \to M$ of a n-dimensional compact Riemannian manifold M is considered.

In fact, in [99] one proves that as a consequence of the good function-theoretic properties of ζ_x for each such operator the Weyl's formula of the asymptotic distribution of the spectral values [220] can be seen to hold. Thus, in particular the condition $\mu_k(x) = \mathbf{O}(1/k)$ is then fulfilled automatically and does not appear as an independent condition any longer.

---

[9] J. Várilly has informed us about this fact and examples and counterexamples around this question which can be found in [96]. Also we are very indebted to J. Várilly for some clarifying remarks and hints to the literature.

But then, upon combining formula (4.10b) with a method [99, THEOREM 7.4, 7.5] (or see [224]) of expressing the residue in terms of the principal symbol of the classical pseudodifferential operator in question, one finally will arrive at Connes' trace theorem.

4.5 The Connes' Trace Theorem and its Application, Preliminaries

In the following we are going to comment on the way along to Connes' trace theorem in a more detailed manner and will give some indications on applications of this formula as to classical Yang-Mills theory.

Basic facts about pseudodifferential operators

Let Ω be an open set in $\mathbb{R}^n$, and let $C_0^\infty(\Omega)$ be the space of smooth functions with compact support inside Ω.

Definition 1. *Let $p \in C^\infty(\Omega \times \mathbb{R})$. p is called a symbol of order (at most) $m \in \mathbb{R}$, if it satisfies the estimates*

$$|\partial_\xi^\alpha \partial_x^\beta p(x,\xi)| \leq C_{\alpha\beta K}(1+\|\xi\|)^{m-|\alpha|}, \quad x \in K, \ \xi \in \mathbb{R}^n, \tag{4.11}$$

for any choice of multiindices α, β and compact $K \subset \Omega$. The space of the symbols of order m is denoted by $S^m(\Omega \times \mathbb{R}^n)$ or simply S^m.

Note that our definition corresponds to the special case with $\varrho = 1$ and $\delta = 0$ of a more general class of symbols as considered e.g. in [202, Definition 1.1.], to which and to [74,75] the reader might refer also for other details on pseudodifferential operators.[10] It is obvious that $S^m \subset S^k$ for $m \leq k$. For $p \in S^m$, let $p(x, D)$ denote the operator

$$(p(x,D)u)(x) = (2\pi)^{-n/2} \int p(x,\xi) e^{i\langle x,\xi\rangle} \hat{u}(\xi) d\xi. \tag{4.12}$$

$$\hat{u}(\xi) = (2\pi)^{-n/2} \int e^{-i\langle x,\xi\rangle} u(x) dx \tag{4.13}$$

is the Fourier transform of u. Note that different $p, p' \in S^m$ may lead to the same operator, $p(x, D) = p'(x, D)$.

Definition 2. *A pseudodifferential operator (ψDO) of order (at most) m is an operator of the form*

$$P = p(x, D), \tag{4.14}$$

where $p \in S^m$. The class of ψDO's of order m is denoted by L^m.

---

[10] Relating notions, conventions and terminology, we do not follow the usage of [202] into any detail, but instead join some slightly simplified conventions and notations which are suitable for our purposes and which we borrowed from some survey lectures of E. Zeidler [226], and which are the same as in [76] and [98, 10.4., especially § 10.4.7.].

The mapping $S^m \longrightarrow L^m$, $p \mapsto p(x,D)$, is surjective, but not injective. Its kernel is denoted by $S^{-\infty}$, the corresponding ψDO's form the space $L^{-\infty}$ of smoothing operators. The principal symbol $\sigma_m(P)$ of a ψDO P of order m with symbol $p \in S^m$ is the class of p in S^m/S^{m-1}.

Definition 3. $p \in S^m$ is called classical, if it has an "asymptotic expansion"

$$p \sim \sum_{j=0}^{\infty} p_{m-j}, \qquad (4.15)$$

i.e. $p_{m-j} \in S^{m-j}$ and

$$p - \sum_{j=0}^{N-1} p_{m-j} \in S^{m-N}, \quad \forall N, \qquad (4.16)$$

and if p_{m-j} is positive homogeneous in ξ "away from 0", i.e.

$$p_{m-j}(x, t\xi) = t^{m-j} p_{m-j}(x, \xi), \quad \|\xi\| \geq 1, \ t \geq 1. \qquad (4.17)$$

A ψDO is said to be classical if its symbol is classical. The spaces of classical symbols and ψDO's are denoted by S_{cl}^m and L_{cl}^m respectively.

Let $p_{m-j}^0(x,\xi)$ be homogeneous functions in ξ on $\Omega \times (\mathbb{R}^n \setminus \{0\})$ coinciding with p_{m-j} for $\|\xi\| \geq 1$. These functions are uniquely determined, and one writes also

$$p \sim \sum_{j=0}^{\infty} p_{m-j}^0 \qquad (4.18)$$

instead of (4.15). The principal symbol of a classical ψDO can be identified with the leading term p_m^0 in the asymptotic expansion (4.18).

Theorem 3. Let $F : \Omega' \longrightarrow \Omega$ be a diffeomorphism of domains in $\mathbb{R}^n$.
Then to every ψDO P on Ω with symbol $p \in S^m(\Omega \times \mathbb{R}^n)$ corresponds a ψDO P' on Ω' with symbol $p' \in S^m(\Omega' \times \mathbb{R}^n)$ such that:

$$F^*(Pu) = P'(F^*(u)), \quad u \in C_0^{\infty}(\Omega), \quad F^* - \text{pull-back}, \qquad (4.19)$$

$$p'(x,\xi) - p(F(x), (^tF'(x))^{-1}\xi) \in S^{m-1}(\Omega' \times \mathbb{R}^n). \qquad (4.20)$$

If P is a classical ψDO then so is P'.

The theorem makes it possible to define ψDO's on manifolds. Let M be a paracompact smooth manifold, and consider an operator $A : C_0^{\infty}(M) \longrightarrow C^{\infty}(M)$. If Ω is some coordinate neighborhood of M, there are a natural extension map $i_{\Omega} : C_0^{\infty}(\Omega) \longrightarrow C_0^{\infty}(M)$ and a natural restriction map $p_{\Omega} : C^{\infty}(M) \longrightarrow C^{\infty}(\Omega)$. A is called ψDO of order m if all the local restrictions $A_{\Omega} := p_{\Omega} \circ A \circ i_{\Omega} : C_0^{\infty}(\Omega) \longrightarrow C^{\infty}(\Omega)$ are ψDO of order m. By Theorem 3, this is a good definition, and also classical ψDO can be defined in this manner. Moreover, equation (4.20) says that the principal symbol has an invariant meaning as a function on the cotangent bundle T^*M.

On the other hand, ψDO on a manifold can be constructed by gluing: Let $\bigcup_j \Omega_j = M$ be a locally finite covering of M by coordinate neighbourhoods, and let A_j be ψDO's of order m on Ω_j. Furthermore, let $\sum_j \psi_j = 1$ be a partition of unity subordinate to the given covering, and let $\phi_j \in C_0^\infty(\Omega_j)$ with $\phi_j|_{\text{supp }\psi_j} = 1$. Then $A := \sum_j \phi_j \circ A_j \circ \psi_j$ (ϕ_j, ψ_j considered as multiplication operators) is a ψDO of order m on M whose restrictions A_{Ω_j} coincide with A_j.

ψDO's acting on sections of vector bundles are defined with appropriate modifications: They are glued from local ψDO's which are defined using matrices of symbols. The principal symbol is then a function on T^*M with values in the endomorphisms of E, i.e. a section of the bundle $\pi^*(End(E))$, where $\pi : T^*M \longrightarrow M$ is the projection of the cotangent bundle, and $End(E)$ is the bundle of endomorphisms of E.

ψDO's are operators from $C_0^\infty(M)$ to $C^\infty(M)$. ψDO's of order m can be extended to bounded linear operators $H^s(M) \longrightarrow H^{s-m}(M)$, $s \in \mathbb{R}$ (Sobolev spaces). Notice that, by the Sobolev embedding theorems, every ψDO of order ≤ 0, $H^s \longrightarrow H^{s-m}$, can be considered as an operator $H^s \longrightarrow H^s$. In particular, taking the case $s = 0$, every ψDO of order ≤ 0 may be considered as an operator $L^2 \longrightarrow L^2$. For the case of manifolds, a Riemannian metric is used in the definition of the L^2 scalar products, for vector bundles in addition a fibre metric. $L^2(M, E)$ denotes the corresponding space of L^2 sections. We will need the following list of facts (for some terminology and the corresponding generalities see [202, Definition 3.1., 24.3] and [63, 23.26.12.] e.g.):

1. The product (which exists, if at least one of the factors is "properly supported") of two ψDO's of orders m, m' is a ψDO of order $m + m'$.

2. The principal symbol of the product of two ψDO's is the product of the principal symbols of the factors.

3. A ψDO of order ≤ 0 is bounded. For order < 0 it is compact.

4. A ψDO of order less than $-n$ on a manifold of dimension n is trace class.

5. If A is a ψDO on a manifold, and if ϕ_j and ψ_j are as above, then A may be written
$$A = \sum_j \psi_j A \phi_j + A'$$
with $A' \in L^{-\infty}$ (smoothing operator).

Remark 3. Note that the classical ψDO's form an algebra which is an example of a more abstract object which usually is referred to as Weyl algebra. According to [99], it is a Weyl algebra corresponding to the symplectic cone $Y = T^*M \setminus \{0\}$ ($\{0\}$ the zero section), with its standard symplectic form ω and $\mathbb{R}^+$-action $\rho_t(x,\xi) = (x,t\xi)$. That is, Y is an $\mathbb{R}^+$-principal bundle such that $\rho_t^*\omega = t\omega$. The properties listed above, however, are only part of the conditions assumed in [99, 2., A.1.-E.].

Definition of the Wodzicki residue

There are at least two equivalent definitions of the Wodzicki residue: As a residue of a certain ζ-function and as an integral of a certain local density [224], [124]. We take as starting point the second definition which can be used most directly for writing classical gauge field Lagrangians. The first definition will show up in the second proof of Connes' theorem.

Definition 4. *Let M be an n-dimensional compact Riemannian manifold. Let T be a classical pseudodifferential operator of order $-n$ acting on sections of a complex vector bundle $E \longrightarrow M$. The Wodzicki residue of T is defined by*

$$Res_W(T) = \frac{1}{n(2\pi)^n} \int_{S^*M} \mathrm{tr}_E \sigma_{-n}(T)\mu, \qquad (4.21)$$

where $\sigma_{-n}(T)$ is the principal symbol of T, S^*M is the cosphere bundle $\{\xi \in T^*M : \|\xi\|_g = 1\}$ and μ is the volume element defined by a multiple of the canonical contact form on T^*M. tr_E is the natural pointwise trace on $\pi^*(End(E))$.

The form μ is defined as $\mu = \frac{(-1)^{\frac{n(n+1)}{2}}}{(n-1)!} \alpha \wedge (d\alpha)^{\wedge(n-1)}$, where α is the canonical 1-form on T^*M, $\alpha = \sum_i \xi_i dx^i$ in local coordinates.

Also Res_W is defined for classical ψDO of any order, using the same formula with p_{-n} instead of σ_{-n} for integer order $m \geq -n$, and putting $Res_W = 0$ else.

It should be noted that the Wodzicki residue can be defined without using the Riemannian structure [224]: One starts defining for a ψDO T on a chart domain in $\mathbb{R}^n$ a matrix-valued local density

$$res_x(T) = \left(\int_{\|\xi\|=1} p_{-n}(x,\xi)|\bar{d}\xi|\right) |dx|,$$

where $\bar{d}\xi = \sum_i (-1)^i \xi_i d\xi_1 \wedge \cdots \wedge \hat{d\xi_i} \wedge \cdots \wedge \xi_n$ is the normalized volume form on the standard Euclidean sphere $\|\xi\| = 1$ and dx is the standard volume form in the chart coordinates. Note that $\bar{d}\xi\, dx = \mu$. Then one shows that this has good functorial properties, i.e. is indeed a density (an absolute value of an n-form) on M, and defines

$$Res_W(T) = \frac{1}{n(2\pi)^n} \int_M \mathrm{tr}\, res_x(T).$$

Due to the homogeneity property of $p_{-n}(x,\xi)$ (using the Euler formula), $p_{-n}(x,\xi)\bar{d}\xi$ is a closed form, thus $\|\xi\|=1$ can be replaced by any homologous $n-1$-surface, in particular by any sphere $\|\xi\|_g = 1$ with respect to a chosen Riemannian metric on M. This leads to formula (4.21) used above. Thus, $Res_W(T)$ does not depend on the choice of the Riemannian metric defining the cosphere bundle. It may, however, depend on the metric through a metric-dependence of T.

Remark 4. The residue Res_W defined above coincides, up to a universal factor which depends only on $\dim(M)$, with the residue defined in [99, DEFINITION 6.1].

Properties of the Wodzicki residue (see [224,124] and [99, PROPOSITION 6.1]):

1. Res_W is a linear (in general not positive) functional on classical ψDO's.

2. Res_W is a trace on the algebra of classical ψDO's.

3. It is the only trace if M is connected, $\dim(M) > 1$.

4. Res_W vanishes on operators of order $< -n$ or noninteger.

4.6 Connes' Trace Theorem

Formulation of Connes' trace theorem

We are now ready to formulate the famous trace theorem [37].

Theorem 4. *Let M be a compact Riemannian manifold of dimension $\dim(M) = n$, let $E \longrightarrow M$ be a complex vector bundle over M, and let T be a classical pseudodifferential operator of order $-n$ on $\Gamma(E)$. Then*

(i) The extension of T to the Hilbert space $\mathcal{H} = L^2(M,E)$ belongs to the ideal $L^{1,\infty}(\mathcal{H})$.
(ii) The Dixmier trace $\mathrm{Tr}_\omega(T)$ coincides with the Wodzicki residue,

$$\mathrm{Tr}_\omega(T) = Res_W(T) = \frac{1}{n(2\pi)^n}\int_{S^*M} \mathrm{tr}_E \sigma_{-n}(T)\mu. \tag{4.22}$$

As a consequence, $\mathrm{Tr}_\omega(T)$ does not depend on the choice of the functional ω in this case.

The following two parts will be devoted to proofs of this theorem exclusively. Two variants of proving will be presented:

In the first variant we are following roughly the line of the original arguments given in [37], but see also [136] and [95] for some details [11], and the

---

[11] We are very indebted to B. Crell, Leipzig, who kindly put at our disposal his manuscript [57], the reading of which was strongly facilitating our understanding of some of the peculiarities of Connes' approach towards formula (4.22).

special case of scalar operators is dealt with, essentially. Thereby, to keep short, in some parts the proof will be left a bit sketchy. However, in any case it will be at worst detailed enough to convince the reader of the validity of Connes' trace theorem for the example of the scalar operator $(1+\Delta)^{-n/2}$ on special compact manifold like the n-torus $\mathbb{T}^n$ or the n-sphere $\mathbb{S}^n$, respectively (Δ is the Laplacian there).

The second variant of proving will be based on an application of Corollary 7 and formula (4.10b), together with some of the knowledge gained while proving Connes' theorem in one of the above mentioned special cases which were completely treated in course of the first variant of the proof. Thereby, according to the arguments found in [99], we firstly learn that (4.10b) gets applicable, and secondly see that a complete proof only requires to consider this formula explicitely for a non-trivial example (i.e. one with non-vanishing Dixmier trace). We emphasize that it is due to the pecularity of this second line of argumentation that along with a special case then validity of the theorem in its full generality – not only for scalar operators – can be concluded.

On the proof of Connes' trace theorem

The idea is to see first that the theorem is true if it is true on one manifold and then to prove it on a manifold one likes, e. g. $\mathbb{T}^n$ or $\mathbb{S}^n$.

First, the theorem is true on a manifold M globally iff it is true locally. This is due to property 5. of ψDO's given above and the fact that smoothing operators are in the kernels of both Res_W and Tr_ω. Now one can transport the local situation, using a local diffeomorphism, to a local piece of another manifold M'. Both sides of the desired equation do not change under this transport. Using now again the above local-global argument, we can think of this local operator as part of a global operator on M' (gluing by means of a partition of unity). Thus, if the theorem is true on M', it must also be true on M, otherwise we would have a contradiction.

Let us prove point (i) of the theorem for scalar operators on $\mathbb{T}^n$. First we show $T \in L^{1,\infty}$ for any ψDO of order $-n$ on $\mathbb{T}^n$. The Laplacian Δ (with respect to the standard flat metric on $\mathbb{T}^n$) is a differential operator of order 2, therefore $(1+\Delta)^{-n/2}$ is a ψDO of order $-n$, and T can be written in the form $T = S(1+\Delta)^{-n/2}$, where S is a ψDO of order 0, therefore bounded. Since $L^{1,\infty}$ is an ideal, it is sufficient to see $(1+\Delta)^{-n/2} \in L^{1,\infty}$.

For the proof we need yet a little result from the general theory of compact operators. Suppose $x \geq \mathbf{0}$ is compact but not of finite rank. Let $\lambda_1 > \lambda_2 > \ldots > 0$ be the ordered sequence of the non-zero eigenvalues of x, with multiplicity m_k for λ_k. Then, for each integer $t \in [0, m_{k+1}]$, $k > 2$, let us consider

$$\gamma_{\{\sum_{j\leq k} m_j + t\}}(x) = \frac{\sum_{j\leq k} \lambda_j m_j + \lambda_{k+1} t}{\log\{\sum_{j\leq k} m_j + t\}}, \qquad (4.23a)$$

which yields all terms $\gamma_n(x)$ of the sequence (4.3a) with $\sum_{j\leq k} m_j \leq n \leq \sum_{j\leq k+1} m_j$. From (4.23a) with the help of the properties of the logarithm one then easily infers that for the mentioned t's the following estimate holds:

$$c_{k+1}^{-1} \gamma_{\{\sum_{j\leq k} m_j\}}(x) \leq \gamma_{\{\sum_{j\leq k} m_j + t\}}(x) \leq c_{k+1} \gamma_{\{\sum_{j\leq k+1} m_j\}}(x), \quad (4.23b)$$

with $c_{k+1} = 1 + \{\log(1 + (m_{k+1}/\sum_{j\leq k} m_j))/\log \sum_{j\leq k} m_j\}$. In view of the structure of the latter coefficients from (4.23b) then the following and often useful auxiliary criterion can be seen to hold.

Lemma 8. *Let $x \in \mathcal{CB}(\mathcal{H})_+ \backslash \mathcal{FB}(\mathcal{H})$. Suppose $\lim_{k\to\infty} \gamma_{\{\sum_{j\leq k} m_j\}}(x)$ exists. If the sequence $\{m_{k+1}/\sum_{j\leq k} m_j : k \in \mathbb{N}\}$ is bounded, then also $\lim_{n\to\infty} \gamma_n(x)$ exists.*

Now we are ready to start our considerations around $(1+\Delta)^{-n/2}$. The spectrum of the Laplacian $\Delta = -\sum_{i=1}^n \partial_i^2$ on $\mathbb{T}^n = \mathbb{R}^n/2\pi\mathbb{Z}^n$ is pure point, consisting of the values $\sum_i k_i^2$, $k_i \in \mathbb{Z}$. The corresponding eigenfunctions are e^{ikx}, $k \in \mathbb{Z}^n$, $x \in \mathbb{R}^n$. The multiplicity of an eigenvalue λ of Δ is $m(\lambda) = \#\{k \in \mathbb{Z}^n | \sum_i k_i^2 = \lambda\}$. From this follow analogous facts for the operator $(1+\Delta)^{-n/2}$. Let m_k be the multiplicity of the k-th eigenvalue of the latter. Let

$$\tilde{\gamma}_R((1+\Delta)^{-n/2}) = \frac{\sum_{1+\|k\|^2 \leq R^2} (1+\|k\|^2)^{-n/2}}{\log N_R'},$$

where N_R' is the number of lattice points in $\mathbb{Z}^n$ with $1+\|k\|^2 \leq R^2$. By construction it is easily seen that convergence of $\{\tilde{\gamma}_R((1+\Delta)^{-n/2}) : R \in \mathbb{R}_+ \backslash \{0\}\}$ as $R \to \infty$ implies the limit $\lim_{k\to\infty} \gamma_{\{\sum_{j\leq k} m_j\}}((1+\Delta)^{-n/2})$ of the considered subsequence of the sequence (4.3a) to exist (in which case then both limits have the same value). It is not hard to see that for geometrical reasons with the above multiplicities also the other condition in the hypotheses of Lemma 8 is fulfilled; this e.g. can be concluded as a by-result from our estimates given below and relating the asymptotic behavior of the ratio between the surface of an n-sphere to the volume of the n-ball of the same radius R within $\mathbb{R}^n$. Hence, in view of Lemma 8 the conclusion is that if the limit

$$\lim_{R\to\infty} \tilde{\gamma}_R((1+\Delta)^{-n/2}) = \lim_{R\to\infty} \frac{\sum_{1+\|k\|^2\leq R^2}(1+\|k\|^2)^{-n/2}}{\log N_R'}$$

can be shown to exist, then by Theorem 2 (2) it has to equal $\mathrm{Tr}_\omega (1+\Delta)^{-n/2}$ (independent of ω). Also, it is not hard to see that the latter limit exists if

$$\lim_{R\to\infty} \frac{\sum_{\|k\|\leq R} \|k\|^{-n}}{\log N_R}$$

exists, where N_R is the number of lattice points with $\|k\| \leq R$, in which case then both limits yield the same value. We prove that the latter limit exists, computing its value. It is well known [217] that

$$N_R = V_R + O(R^{\frac{n-1}{2}}),$$

where $V_R = \frac{\Omega_n}{n} R^n$ (volume of the ball of radius R in $\mathbb{R}^n$), $\Omega_n = \frac{2\pi^{n/2}}{\Gamma(n/2)}$ (area of the sphere S^{n-1}). Neglecting terms of lower order in R, we have

$$N_R = V_R + \ldots = \frac{\Omega_n}{n} R^n + \ldots.$$

In order to determine $\sum_{\|k\| \leq R} \|k\|^{-n}$ for large R, we first count the number of lattice points in a spherical shell between R and $R + dR$,

$$N_{R+dR} - N_R = V_{R+dR} - V_R + \ldots = \Omega_n R^{n-1} dR + \ldots.$$

Integrating this, we obtain asymptotically for large R

$$\sum_{\|k\| \leq R} \|k\|^{-n} = \Omega_n \int_1^R r^{-n} r^{n-1} dr + \ldots = \Omega_n \log R + \ldots.$$

Together with $\log N_R = n \log R + \log \Omega_n - \log n + \ldots$ this leads to

$$\lim_{R \to \infty} \frac{\sum_{\|k\| \leq R} \|k\|^{-n}}{\log N_R} = \frac{\Omega_n}{n},$$

i.e.

$$\operatorname{Tr}_\omega (1 + \Delta)^{-n/2} = \frac{\Omega_n}{n}. \tag{4.24}$$

It is much easier to determine the Wodzicki residue of $(1 + \Delta)^{-n/2}$: the principal symbol of $(1 + \Delta)^{-n/2}$ is $\sigma_{-n}((1 + \Delta)^{-n/2})(x, \xi) = \|\xi\|^{-n}$, where $\|.\|$ denotes the standard euclidean metric on $\mathbb{R}^n$. Therefore,

$$\operatorname{Res}_W ((1 + \Delta)^{-n/2}) = \frac{1}{n(2\pi)^n} \int_{S^* \mathbb{T}^n} d\xi dx = \frac{1}{n(2\pi)^n} \Omega_n \int_{\mathbb{T}^n} dx =$$

$$= \frac{1}{n(2\pi)^n} \Omega_n (2\pi)^n = \frac{\Omega_n}{n},$$

coinciding with the result for the Dixmier trace. Thus, the theorem is already proved for a special operator on $\mathbb{T}^n$.

To prove point (ii) of the theorem, we start with some general remarks about Tr_ω. It is a positive linear functional on the space L^{-n} of ψDO's of order $-n$ with $L^{-n-1} \subset \ker \operatorname{Tr}_\omega$, because elements of L^{-n-1} are trace class (see property 4. above). Using $L^{-n} \simeq S^{-n}/S^{-\infty}$, it follows that Tr_ω may be considered as a linear functional on S^{-n}/S^{-n-1}, the space of principal symbols of ψDO's of order $-n$. By restriction, it is also a linear functional on the space of principal symbols of classical ψDO's of order $-n$. On the other hand, this latter space and the space $C^\infty(S^*M)$ coincide, since every element of $C^\infty(S^*M)$ by homogeneity defines a classical principal symbol, ([63], 23.29.11.). Thus, we end up with a linear functional on $C^\infty(S^*M)$. It follows from symbol calculus that this functional is positive (see [57]). Thus, we have a positive distribution, which is always given by a positive measure on S^*M ([62], 17.6.2).

Since an isometry of M gives rise to a unitary transformation of $L^2(M, v_g)$, and the spectrum of an operator does not change under unitary transformations, the Dixmier trace is invariant under isometries. Therefore, the corresponding measure on S^*M is invariant under isometries.

Considering now the case $M = \mathbb{S}^n$, the standard n-sphere with the metric induced from the euclidean metric on $\mathbb{R}^{n+1}$, the group of isometries is $SO(n+1)$, and S^*M is a homogeneous space under the induced action of $SO(n+1)$. It is easy to see that the volume form of the induced Riemannian metric on $S^*\mathbb{S}^n$ is invariant under the action of $SO(n+1)$. Uniqueness of the invariant measure on a homogeneous space shows that the positive measure corresponding to the Dixmier trace must be proportional to the measure given by this volume form v_g,

$$\mathrm{Tr}_\omega(T) = const. \int_{S^*M} \sigma_{-n}(T) v_g.$$

It is an easy exercise to show that v_g coincides in this case with the form μ defined above. Moreover, the constant, which neither depends on the operator nor on the Riemannian manifold, is determined by the example of the torus.

Remark 5. Note that it follows from Corollary 7 and 4 that the Wodzicki residue in some cases coincides with a residue of the zeta function $\zeta_T(z) = \mathrm{tr}\, T^z$ (see also below).

An alternative proof of the Connes' trace formula

In this part another derivation of the trace formula (4.22) will be presented. Thereby, the line will be to make the abstract formula (4.10b) directly accessible for Connes' trace formula in the special case of a classical pseudodifferential operator T of order $-n$. Thereby, by some continuity argument, it suffices if the case of positive-definite operators of that kind can be dealt with.

According to the hypotheses in Corollary 7 under which (4.10b) is supposed to hold, for this aim it is sufficient to know that the 'right' asymptotic behavior of the spectral values of T holds, and in which case then the extensions of the ζ_T-function of (4.8) have to be analyzed.

Both exercises can be achieved at once and almost without proof by means of V. Guillemin's methods given in [99]. In demonstrating this way towards formula (4.22) we finally will end up with an alternative proof of Theorem 4.

Before doing this, we recall the special settings corresponding to the assumptions of Theorem 4. In line with these and in accordance with *Remark 3* the Weyl algebra $\mathcal{W}$ of all classical ψDO's corresponding to the symplectic cone $Y = T^*M\backslash\{0\}$ will be considered. We then have the Hilbert space $\mathcal{H} = L^2(M, E)$, which is the completion of the sections $\Gamma(E)$ under a scalar product descending from a symplectic volume element μ on Y and a fibre metric on E. If these are fixed, each classical ψDO T which is at most of order 0 corresponds to a bounded linear operator, and can be identified with its unique bounded linear extension $x = T$ from sections $\Gamma(E)$ onto the whole $L^2(M, E)$. Accordingly, in such case we use the same notation T for both the ψDO and its unique bounded linear extension on all of $L^2(M, E)$.

Now, let x be a bounded linear operator on $\mathcal{H}$. Then, in the special case that x is positive-definite, there exists in a unique way the inverse of x on $\mathcal{H}$, that is, a densely defined, positive-definite self-adjoint linear operator x^{-1} on $\mathcal{H}$, with $x^{-1}x = 1$ and $xx^{-1} \subset 1$. In this case we then define P_x to be the n-th (positive) root of this inverse, $P_x = \sqrt[n]{x^{-1}}$.

Especially, if $T \in L_{cl}^{-n}(= \mathcal{W}^{-n}$ in the terminology of [99]) is supposed to be positive-definite, by compactness and in view of the definition of – and the properties coming along with – the term ellipticity, for $x = T$ the operator x^{-1}, and thus also P_x, is positive-definite, self-adjoint, elliptic and of order one.

Now, from [99, (1.1)] one in particular learns that for the asymptotic growth of the singular values of a positive-definite, self-adjoint elliptic differential operator P of order one the Weyl's formula [220] holds. This equivalently says that the singular values behave like $\mu_k(P) \sim l\sqrt[n]{k}$, with some constant l, see [202, (13.18) and Proposition 13.1]. Hence, in the above-mentioned special case $P = P_x$ this asymptotic law amounts to $\mu_k(T) = \mu_k(x) \sim L \cdot k^{-1}$, with some constant $L \geq 0$, for each positive-definite $T \in L_{cl}^{-n}(= \mathcal{W}^{-n})$. Especially, in line with Lemma 6 we then have $T \in L^{1,\infty}(\mathcal{H})$, for each such operator. Since $L^{1,\infty}(\mathcal{H})$ is an ideal from this especially $L_{cl}^{-n} \subset L^{1,\infty}(\mathcal{H})$ is seen, which demonstrates that Theorem 4 (i) can be equivalently followed also from the main result of [99].

On the other hand, owing to Weyl's asymptotic law, the applicability of formula (4.10b) now will rely on the extension properties of ζ_x from the half-plane $\mathrm{Re}\, z > 1$ onto $\mathbb{C}$ exclusively. This matter we are going to discuss now.

We start with some preliminary considerations about various existing definitions relating to ζ-functions which can be associated to some positive operator.[12]

Firstly, in accordance with the above and Corollary 7 the ζ_x-function as given in (4.8) for compact positive $x \in L_{cl}^{-n} \subset L^{1,\infty}(\mathcal{H})$ is holomorphic in the half-plane $\mathrm{Re}\, z > 1$. Thus for given fixed $n \in \mathbb{N}$ upon defining $\zeta^x(w) = \zeta_x(z)$ at $w = n(1-z)$, $\mathrm{Re}\, z > 1$, one gets another complex function $w \mapsto \zeta^x(w)(= \zeta_n^x(w))$ which is holomorphic in the half-plane $\mathrm{Re}\, w < 0$. Recall that in view of the arguments given in context of (4.8) the operator family $\mathrm{Re}\, z > 1 : z \mapsto x^z$ consists of trace-class operators. But then, by functional calculus this operator family may be equivalently re-defined at each $w \in \mathbb{C}$ with $\mathrm{Re}\, w < 0$ and $z = (1 - w/n)$ as $\mathrm{Re}\, w < 0$, $w \mapsto xP_x^w$. Hence, if another

---

[12] We have to thank H. Upmeier who outlined to us some of the relevant details.

ζ-function
$$\forall w \in \mathbb{C}, \operatorname{Re} w < 0 : \ \zeta(x,P)(w) = \operatorname{tr} x P^w \tag{4.25}$$
is defined for bounded-invertible, positive-definite self-adjoint linear operator P, and bounded $x \geq \mathbf{0}$ such that $xP^w \in L^1(\mathcal{H})$ for all w with $\operatorname{Re} w < 0$, in view of the above in the special case of $P = P_x$ we may summarize as follows.

Lemma 9. *Suppose $x \in L_{cl}^{-n}$ to be positive-definite. Then, in the special case of $P = P_x$ the complex function $w \mapsto \zeta(x,P)(w)$ is holomorphic in the half-plane $\operatorname{Re} w < 0$, and there $\zeta(x,P)(w) = \zeta_x(z)$ is fulfilled, at $z = (1 - w/n)$.*

Now, from [99, THEOREM 7.4] one also knows that the nuclear dimension of $\mathcal{W}$ is n (the conclusion of Theorem 4 (i) being in accordance with this), where n is the dimension of the basic manifold M. Hence and especially, if $x = T \in L^{-n}$ is a classical ψDO ($\in \mathcal{W}^{-n}$ in the terminology of [99]), then for each positive-definite, self-adjoint *elliptic* operator $P \in \mathcal{W}^1$ (that is, P is of order one, and among other facts, has to be bounded-invertible on $\mathcal{H}$, e.g.) the ζ-function $w \mapsto \zeta(T,P)(w)$ of (4.25) may be considered. In fact, as a consequence of positive-definiteness and ellipticity of P and since T is of order $-n$ (which is the negative of the nuclear dimension of $\mathcal{W}$) all the conditions under which (4.25) is to hold are fulfilled, and thus the restriction to the negative half-plane of the trace of the operators $T(w)$, whose operator family is given as $T(w) = TP^w \in \mathcal{W}^{w-n} \subset \mathcal{W}^w$ for $w \in \mathbb{C}$, makes sense. Thereby, the mentioned operator family itself is known to possess a canonical property; it is a so-called *holomorphic* family of operators [13].

The latter especially means that the conditions of the hypothesis of [99, THEOREM 7.1] are fulfilled, and then in line with the conclusion of this result $\zeta(T,P)(w)$ has to be holomorphic in the half-plane $\operatorname{Re} w < 0$ and has a meromorphic extension to the whole complex plane, and at $w = 0$ has, at worst, a simple pole. Moreover, according to [99, THEOREM 7.4] the residue of this meromorphic extension depends only on the symbol $\sigma_{-n}(T)$ of T, and has the form
$$\mathfrak{Res}|_{w=0} \zeta(T,P) = g_0 \operatorname{Res}(\sigma_{-n}(T)), \tag{4.26a}$$
with a non-zero constant, $g_0 \neq 0$, which depends only on the Weyl algebra $\mathcal{W}$ under consideration. Now, remind that we are in the special context described in *Remark 3*. But then, the cosphere bundle $S^*M = \{\xi \in T^*M : \|\xi\| = 1\}$ appears as the (compact) base of the symplectic cone $Y = T^*M \setminus \{0\}$. Denoting by $\pi' : Y \to S^*M$ the projection of Y, one defines for a homogeneous (of degree $-n$) C^∞-section f of the bundle $\pi^* End(E) \to Y$, according to [99, DEFINITION 6.1],
$$\operatorname{Res}(f) = \int_{S^*M} \operatorname{tr}_E f \tilde{\mu} = \int_{S^*M} \mu_f. \tag{4.26b}$$

---
[13] For the precise definition and basic properties around *ellipticity* and *holomorphy* for ψDO's of a given order and operator families, respectively, we refer to [99, DEFINITION 2.1, PROPOSITION 4.1, and eq. (3.18)]

Here $\tilde{\mu} = \alpha \wedge \omega^{\wedge(n-1)}$ is the volume element defined by a salar multiple of the canonical contact form on T^*M (and thus in accordance with *Remark 3* one has $\rho_t^* \tilde{\mu} = t^n \tilde{\mu}$, μ_f is the uniquely determined $(2n-1)$-form defined through $\mathrm{tr}_E f \tilde{\mu} = \pi'^*(\mu_f)$ and tr_E is the natural pointwise trace on $\pi^*(End(E))$. As mentioned above there is a basic fact saying that homogeneous C^∞-sections of $\pi^* End(E) \to Y$ in our case exactly yield all the principal symbols to classical ψDO's of order $-n$.

Thus, the above-mentioned conclusions about the possibility of a meromorphic extension of $\zeta(T,P)$ and its singularity structure at $w = 0$ apply with $P = P_T$. In view of Lemma 9 and as a consequence of the just said, upon changing the complex variable w into z in accordance with $w = n(1-z)$ we will see that analogous facts hold in respect of ζ_T and at $z = 1$, accordingly. That is, ζ_T possesses a meromorphic extension into the whole z-plane, with a simple pole at $z = 1$, at worst. Having in mind this, and taking into account that from $w = n(1-z)$ a geometric factor $1/n$ arises while passing from the residue of the one extension at $w = 0$ to the residue of the transformed extension at $z = 1$, in view of (4.26a)–(4.26b) we then may summarize as follows:

Corollary 8. *Let $T \in L_{cl}^{-n}$ be positive-definite. The holomorphic function $\zeta_T(z)$ has a meromorphic extension from the half-plane $\mathrm{Re}\, z > 1$ into the whole complex plane with, at worst, a simple pole at $z = 1$. The residue of the extension obeys*

$$\mathfrak{Res}|_{z=1} \zeta_T(z) = (g_0/n) \int_{S^*M} \mathrm{tr}_E \sigma_{-n}(T) \mu \,,$$

with a constant $g_0 \neq 0$ which does not depend on the special operator T.

Foremost, according to Corollary 7 and by the above-mentioned asymptotic spectral properties the Corollary guarantees that formula (4.10b) can be applied for positive-definite classical ψDO's of order $-n$, with the result that

$$\mathrm{Tr}_\omega(T) = (g_0/n) \int_{S^*M} \mathrm{tr}_E \sigma_{-n}(T) \mu \qquad (4.26c)$$

has to be fulfilled, for each positive-definite T of order $-n$. Also, in order to fix the constant g_0 it obviously suffices to deal with one particular case of such an operator. Moreover, once more again according to the local-global and the M-to-M' arguments, which we have already mentioned at the beginning of 4.6 while proceeding the first variant of the proof of Theorem 4, we have to conclude that the constant g_0 within (4.26c) has to be the same, in each case of an n-dimensional compact manifold M. Hence, we may content with the known result for $T = (1+\Delta)^{-n/2}$ on the n-torus $M = \mathbb{T}^n$. According to our calculations therefore $g_0 = 1/(2\pi)^n$ has to hold, and then (4.26c) will yield that (4.22) has to be valid, for each positive-definite T of order $-n$ on an arbitrary compact n-dimensional manifold M. From this the validity for all positive T of order $-n$ can be concluded, since for fixed positive-definite T_0 of order $-n$ and each positive T the family $T(\varepsilon) = T + \varepsilon T_0$, $\varepsilon > 0$, consists of positive-definite ψDO's of order $-n$, for which according to the above the assertion of Connes formula holds. In fact, according to Lemma 5 one knows $\text{Tr}_\omega(T(\varepsilon)) = \text{Tr}_\omega(T) + \varepsilon\,\text{Tr}_\omega(T_0)$. On the other hand, the map $A \mapsto \sigma_{-n}(A)$ between ψDO's of order $-n$ and their principal symbols is a homomorphism, and therefore also $\sigma_{-n}(T(\varepsilon)) = \sigma_{-n}(T) + \varepsilon\,\sigma_{-n}(T_0)$. Hence, since the expression on the right-hand side of (4.26c) obviously is a linear form with respect to the $\sigma_{-n}(T)$-variable, the validity of (4.22) in the general case can be obtained simply via the just mentioned linearity and upon taking the difference between a relation of type (4.26c), taken at one particular $T(\varepsilon)$, for some $\varepsilon > 0$, and a multiple of the relation of type (4.26c) at T_0 with ε.

4.7 Classical Yang-Mills Actions

Here we make some remarks about the construction of the bosonic part of classical (pure) gauge field actions in terms of the Dixmier trace and the classical Dirac operator. This is considered in more detail in Section 1.1 and Chapter 2. We will make use of the fact that the de Rham algebra of exterior forms is isomorphic to the differential algebra $\Omega_D(C^\infty(M))$ coming from the classical spectral triple $(\mathcal{A} = C^\infty(M), \mathcal{H} = L^2(M,S), D)$, D the Dirac operator on the compact n-dimensional Riemannian spin manifold M, S the spinor bundle (see Section 1.3). The representation π of $\mathcal{A}$ on $\mathcal{H}$ is given by sending $f \in \mathcal{A}$ to the operator of multiplication with the function f.

The Classical Dirac Operator and Integration on Manifolds

Proposition 3. *Consider $f \in C^\infty(M)$ as left multiplication operator on $L^2(M,S)$. Then*

$$\text{Tr}_\omega\left(f|D|^{-n}\right) = \frac{1}{c(n)} \int_M f\, v_g, \tag{4.27}$$

where v_g denotes the Riemannian volume element, $c(n) = 2^{n-[n/2]-1}\pi^{n/2} n \cdot \Gamma(n/2)$, and Tr_ω is the Dixmier trace with respect to any invariant mean ω.

Proof. (see [136], p. 98) The principal symbol of the Dirac operator is $\gamma(\xi)$ (Clifford multiplication on spinors), thus D is a first order (elliptic) ψDO. Multiplication with f is a zero order operator, therefore $f|D|^{-n}$ is a ψDO of order $-n$. Its principal symbol is $\sigma_{-n}(x,\xi) = f(x)\|\xi\|^{-n}1_{2^{[n/2]}}$, where $1_{2^{[n/2]}}$ is the identity map of the fibre S_x of S. This principal symbol reduces on the cosphere bundle S^*M to $f(x)1_{2^{[n/2]}}$. Thus, Theorem 4 gives

$$\text{Tr}_\omega\left(f|D|^{-n}\right) = \frac{1}{n(2\pi)^n} \int_{S^*M} \text{tr}_S(f(x)1_{2^{[n/2]}}) dx d\bar{\xi}$$

$$= \frac{2^{[n/2]}}{n(2\pi)^n} \int_{S^{n-1}} d\bar{\xi} \int_M f(x) dx.$$

The area $\int_{S^{n-1}} d\bar{\xi} = \frac{2\pi^{n/2}}{\Gamma(n/2)}$ of the unit sphere S^{n-1} leads to the right factor $c(n)$.

Since $|D|^{-n}$ is in $L^{1,\infty}$ we can define the following inner product on $\pi(\Omega^k \mathcal{A})$:

$$\langle T_1, T_2 \rangle_k := \mathrm{Tr}_\omega \left(T_1^* T_2 |D|^{-n} \right). \tag{4.28}$$

In order to really have an inner product, one needs some assumptions about $\mathcal{A}$ which are fulfilled in the classical case, but which also hold in more general situations of spectral triples, see [35,136,216].[14] The orthogonal complement with respect to this inner product of the subspace $\pi(d(J_0 \cap \Omega^{k-1})) \subset \pi(\Omega^k \mathcal{A})$ is isomorphic to $\Omega_D^k \mathcal{A}$ (both are images of surjections with the same kernel). Thus, the inner product (4.28) can be tranported to $\Omega_D^k \mathcal{A}$.

Proposition 4. *Under the isomorphism between $\Omega_D^k \mathcal{A}$ and $\Gamma(\Lambda_\mathbb{C} T^* M)$ the inner product on $\Omega_D^k \mathcal{A}$ is proportional to the usual Riemannian inner product,*

$$\langle \omega_1, \omega_2 \rangle_k = (-1)^k \lambda(n) \int_M \omega_1 \wedge {}^*\omega_2 \tag{4.29}$$

for $\omega_i \in \Omega_D^k \mathcal{A} \simeq \Gamma(\Lambda_\mathbb{C} T^* M)$, where

$$\lambda(n) = \frac{2^{[n/2]+1-n} \pi^{-n/2}}{n \Gamma(n/2)}.$$

Proof. We refer to [136], p. 120.

Classical gauge field actions in terms of Dixmier trace

Now, in usual gauge theory, the gauge field F may be interpreted as a two-form with values in the endomorphisms of a vector bundle E over M (curvature of a connection). (Such vector bundles typically arise as bundles associated to a principal bundle with the group of inner symmetries as structure group). The (pure) gauge field action is then constructed by combining the scalar product on the right-hand side of (4.29) with a (fiberwise) product and trace of the endomorphisms,

$$YM(\nabla) = const. \int_M \mathrm{tr}(F \wedge {}^*F).$$

By (4.29) this can be written equivalently in terms of the differential algebra $\Omega_D^k \mathcal{A}$ and the scalar product there.

---

[14] Thanks to J. Várilly for pointing out this fact to us.

Moreover, it is almost obvious from the definition of the inner product that the classical YM action can be obtained as an infimum over a "universal" YM action defined over universal connections (which are elements of $\Gamma(EndE) \otimes_{C^\infty(M)} \Omega^2(\mathbb{C}^\infty(M))$. More precisely [38], [136], one shows that

$$1 \otimes \pi : \mathcal{E} \otimes_\mathcal{A} \Omega^1 \mathcal{A} \longrightarrow \mathcal{E} \otimes_\mathcal{A} \Omega^1_D \mathcal{A}$$

($\mathcal{E} := \Gamma(E)$) gives rise to a surjection from universal connections to usual connections. If θ is the curvature of a universal connection ∇_{un}, one defines

$$I(\nabla_{un}) = \mathrm{Tr}_\omega \left(\{1 \otimes \pi\}(\theta)^2 |1 \otimes D|^{-n} \right).$$

and finds

$$YM(\nabla) = const. \inf\{I(\nabla_{un})|\pi(\nabla_{un}) = \nabla\}.$$

Thus the classical Yang-Mills action can be entirely written in terms of objects which have a straightforward generalization to the noncommutative situation.

Acknowledgements

We thank the organizers for the invitation. Special thanks go to Harald Upmeier for insisting on the use of Guillemin's results for a proof of Connes' formula. In particular, we are grateful to him for hints relating ζ-functions, and for supplying Portenier's notes.

We are very indebted to Gianni Landi (Trieste, Italy), Claude Portenier (Marburg) and Joseph C. Várilly (San José, Costa Rica) for helpful comments, examples and hints to the literature.

Last but not least, we are grateful to all those collegues who helped us in reconstructing and extracting some of the valuable informations reminiscent of the 'Oberseminar zur Mathematischen Physik' which took place as a joint seminar during the academic course 1994/95 at the Mathematics and Theoretical Physics Departments of Leipzig University. In this respect, special thanks go to Bernd Crell, Konrad Schmüdgen and Eberhard Zeidler (all from Leipzig) whose notices of their contributed lectures to the Oberseminar together with many clarifying discussions around the subject proved particularly useful for us.

5 The Einstein-Hilbert Action as a Spectral Action

Bernd Ammann and Christian Bär

University of Hamburg

5.1 Generalized Laplacians and the Heat Equation

We start by examining the analysis of so-called generalized Laplacians. A detailed exposition can be found in [12]. Throughout this section let M be a compact Riemannian manifold, let $E \to M$ be a Riemannian or Hermitian vector bundle over M. Let ∇ be a metric connection on E, i.e. for smooth sections φ and ψ in E and $X \in TM$ we have

$$\partial_X \langle \varphi, \psi \rangle = \langle \nabla_X \varphi, \psi \rangle + \langle \varphi, \nabla_X \psi \rangle.$$

Here $\langle \cdot, \cdot \rangle$ denotes the Riemannian resp. Hermitian metric on E. If φ is a smooth section in E, then $\nabla \varphi$ is a smooth section in $T^*M \otimes E$. Note that the Riemannian metric and the Levi-Civita connection on M together with the metric and ∇ on E induce a metric and a compatible connection on $T^*M \otimes E$, again denoted $\langle \cdot, \cdot \rangle$ and ∇. Similarly, the kth covariant derivative, $\nabla^k \varphi$, is a section in $\underbrace{T^*M \otimes \ldots \otimes T^*M}_{k \text{ times}} \otimes E$ and this bundle carries a natural metric and connection. For φ a smooth section in E, $\varphi \in C^\infty(E)$, we define the L^2-scalar product

$$(\varphi, \psi)_{L^2} := \int_M \langle \varphi, \psi \rangle \, dV$$

and the associated L^2-norm

$$\|\varphi\|_{L^2}^2 := \int_M \langle \varphi, \varphi \rangle \, dV.$$

More generally, for any k we have the *Sobolev-norms*

$$\|\varphi\|_{H^k}^2 := \|\varphi\|_{L^2}^2 + \|\nabla \varphi\|_{L^2}^2 + \ldots + \|\nabla^k \varphi\|_{L^2}^2.$$

The completions of $C^\infty(E)$ with respect to these norms are denoted $L^2(E)$ and $H^k(E)$, the spaces of square-integrable sections and Sobolev-sections in E.

The C^k-*norm* is defined in a similar manner,

$$\|\varphi\|_{C^0} := \sup_M |\varphi|,$$
$$\|\varphi\|_{C^k} := \max\{\|\varphi\|_{C^0}, \|\nabla \varphi\|_{C^0}, \ldots, \|\nabla^k \varphi\|_{C^0}\}.$$

The two families of norms, $\|\cdot\|_{H^k}$ and $\|\cdot\|_{C^k}$, are equivalent in the following sense: It is trivial to see that $\|\cdot\|_{H^k}$ can be estimated against $\|\cdot\|_{C^k}$,

$$\|\varphi\|_{H^k} \leq \operatorname{vol}(M)^{\frac{1}{2}} \cdot (k+1)^{\frac{1}{2}} \cdot \|\varphi\|_{C^k}.$$

Conversely, we have [146, Thm. III.2.5]

Proposition 1 (Sobolev Embedding Theorem). *For each k there exists a constant $c = c(k, M, \Delta)$ such that*

$$\|\varphi\|_{C^k} \leq c \cdot \|\varphi\|_{H^\ell}$$

whenever $\ell > k + \frac{n}{2}$, $n = \dim(M)$.

Now let ∇^* be the L^2-adjoint of ∇, i.e. $(\nabla\varphi, \psi)_{L^2} = (\varphi, \nabla^*\psi)$ for all $\varphi \in C^\infty(E)$, $\psi \in C^\infty(T^*M \otimes E)$. For any symmetric endomorphism field $\mathcal{K} \in C^\infty(\operatorname{End}(E))$ the operator

$$\Delta := \nabla^*\nabla + \mathcal{K} : C^\infty(E) \to C^\infty(E)$$

is called a *generalized Laplacian*.

Since Δ^k is a differential operator of order $2k$ we have

$$\|\Delta^k \varphi\|_{L^2} \leq C \cdot \|\varphi\|_{H^{2k}}.$$

But Δ is *elliptic* and this implies the following converse [146, Thm. III.5.2]

Proposition 2 (Elliptic Estimates). *For each $k \in \mathbb{N}$ there is a constant $C = C(k, M, \Delta)$ such that*

$$\|\varphi\|_{H^{2k}} \leq C \cdot \left(\|\varphi\|_{L^2} + \|\Delta^k \varphi\|_{L^2} \right).$$

Finally, we need the following fundamental result [146, Thm. III.5.8]

Theorem 1. *There exists a Hilbert space orthonormal basis $\varphi_1, \varphi_2, \ldots$ of $L^2(E)$ and real numbers $\lambda_1, \lambda_2, \ldots$ such that*

$$\Delta \varphi_k = \lambda_k \cdot \varphi_k,$$

$\lambda_1 \leq \lambda_2 \leq \lambda_3 \leq \ldots \nearrow +\infty$, *and each λ_k is repeated only finitely many times. All φ_k are smooth, $\varphi_k \in C^\infty(E)$.*

The theorem says in particular that the eigenvalues tend to $+\infty$. To get started we need some control on how fast they grow. The following proposition will later be improved considerably, c.f. Theorem 2.

Proposition 3. *There exists a positive constant $c = c(M, \Delta)$ such that for all k*

$$\lambda_k \geq c \cdot k^{\frac{4}{n(n+6)}} + \lambda_1 - 1.$$

Proof. Replacing $\mathcal{K}$ by $\mathcal{K} - \lambda_1 \cdot \mathrm{id}$ will shift the spectrum of Δ by λ_1. Hence we can assume w.l.o.g. that $\lambda_1 = 0$. Now let $\epsilon > 0$ and let $\{p_1, \ldots, p_N\}$ be a minimal ϵ-dense subset of M, i.e. $M = \bigcup_{i=1}^{N} B(p_i, \epsilon)$ with N minimal. Here $B(p, \epsilon)$ denotes the ball of radius ε about p. It is not hard to see that there is a constant $c_1 = c_1(M)$ such that for all $\epsilon > 0$

$$N = N(\epsilon) \leq c_1 \cdot \epsilon^{-n}.$$

Let $V \subset L^2(E)$ be the subspace spanned by $\varphi_1, \ldots, \varphi_k$. Consider $\varphi = \sum_{i=1}^{k} \alpha_i \varphi_i \in V$ and assume $\varphi(p_i) = 0$, $i = 1, \ldots, N$. Given $x \in M$ choose p_i such that $\mathrm{dist}(x, p_i) < \epsilon$. Differentiation along a shortest geodesic from p_i to x yields

$$|\varphi(x)| = |\varphi(x)| - |\varphi(p_i)| \leq \epsilon \cdot \|\nabla \varphi\|_{C^0} \leq \epsilon \cdot \|\varphi\|_{C^1}.$$

Integration over M gives

$$\|\varphi\|_{L^2} \leq \epsilon \cdot \|\varphi\|_{C^1} \cdot \mathrm{vol}(M)^{\frac{1}{2}}.$$

Let $\ell := \left[\frac{n}{2}\right] + 2$. By the Sobolev embedding theorem we have

$$\|\varphi\|_{C^1} \leq c_2 \cdot \|\varphi\|_{H^\ell}.$$

By the elliptic estimates

$$\|\varphi\|_{H^\ell} \leq c_3 \cdot \left(\|\varphi\|_{L^2} + \|\Delta^{\left[\frac{\ell+1}{2}\right]}\varphi\|_{L^2}\right) \leq c_3 \cdot \left(1 + \lambda_k^{\left[\frac{\ell+1}{2}\right]}\right) \cdot \|\varphi\|_{L^2}$$
$$\leq c_3 \cdot (1+\lambda_k)^{\left[\frac{\ell+1}{2}\right]} \cdot \|\varphi\|_{L^2} \leq c_3 \cdot (1+\lambda_k)^{\frac{n}{4}+\frac{3}{2}} \cdot \|\varphi\|_{L^2}.$$

Combining these estimates we obtain

$$\|\varphi\|_{L^2} \leq \epsilon \cdot \mathrm{vol}(M)^{\frac{1}{2}} \cdot c_2 \cdot c_3 \cdot (1+\lambda_k)^{\frac{n+6}{4}} \cdot \|\varphi\|_{L^2}$$
$$= \epsilon \cdot c_4 \cdot (1+\lambda_k)^{\frac{n+6}{4}} \cdot \|\varphi\|_{L^2}.$$

For $\epsilon = \frac{1}{2c_4} \cdot (1+\lambda_k)^{-\frac{n+6}{4}}$ we conclude $\|\varphi\|_{L^2} \leq \frac{1}{2}\|\varphi\|_{L^2}$, hence $\varphi = 0$. Thus for this ε the linear mapping

$$V \longrightarrow E_{p_1} \oplus \cdots \oplus E_{p_N}$$
$$\varphi \longmapsto \left(\varphi(p_1), \ldots, \varphi(p_N)\right)$$

is injective. Therefore

$$k = \dim V \leq \dim(E_{p_1} \oplus \ldots \oplus E_{p_N}) = N \cdot \mathrm{rk}(E)$$
$$\leq c_1 \cdot \epsilon^{-n} \cdot \mathrm{rk}(E) = c_5 \cdot (1+\lambda_k)^{\frac{n(n+6)}{4}}.$$

Hence
$$1 + \lambda_k \geq \left(\frac{k}{c_5}\right)^{\frac{4}{n(n+6)}} = c_6 \cdot k^{\frac{4}{n(n+6)}}.$$

The main purpose of this section is to study the *heat equation*

$$\frac{\partial \varphi_t}{\partial t} + \Delta \varphi_t = 0$$

where φ_t is a smooth section in E for each $t \geq 0$ and φ_t depends smoothly on t.

The connection ∇ on E induces a connection, again denoted ∇, on the dual bundle E^*. The endomorphism field $\mathcal{K}$ of E gives the endomorphism field $\mathcal{K}^*$ on E^*. Hence we obtain a generalized Laplacian $\Delta = \nabla^* \nabla + \mathcal{K}^*$ on E^*.

For a section φ in E we define the section φ^* in E^* by

$$\varphi^*(\psi) := \langle \varphi, \psi \rangle \qquad \forall \psi \in E.$$

One easily checks $\nabla_X(\varphi^*) = (\nabla_X \varphi)^*$ for $X \in TM$, $(\mathcal{K}\varphi)^* = \mathcal{K}^* \varphi^*$, and $(\Delta \varphi)^* = \Delta(\varphi^*)$. Hence if $\varphi_1, \varphi_2, \ldots$ is an orthonormal basis of $L^2(E)$ consisting of eigenvectors of Δ, then we get an orthonormal eigenbasis of $L^2(E^*)$ by $\varphi_1^*, \varphi_2^*, \ldots$ for the same eigenvalues.

Now we form the bundle $E \boxtimes E^*$ over $M \times M$ whose fiber over $(x, y) \in M \times M$ is given by

$$(E \boxtimes E^*)_{(x,y)} = E_x \otimes E_y^* = \mathrm{Hom}(E_y, E_x).$$

Again, we get an induced connection $\tilde{\nabla}$ on $E \boxtimes E^*$. We put $\tilde{\mathcal{K}} = \mathcal{K} \otimes \mathrm{id} + \mathrm{id} \otimes \mathcal{K}^*$ and obtain the corresponding generalized Laplacian

$$\tilde{\Delta} = \tilde{\nabla}^* \tilde{\nabla} + \tilde{\mathcal{K}}.$$

If φ and ψ are sections in E we get a section $\varphi \boxtimes \psi^*$ in $E \boxtimes E^*$ by

$$(\varphi \boxtimes \psi^*)(x, y) = \varphi(x) \otimes \psi^*(y).$$

One sees that $\tilde{\Delta}(\varphi_j \boxtimes \varphi_k^*) = (\Delta \varphi_j) \boxtimes \varphi_k^* + \varphi_j \boxtimes (\Delta \varphi_k^*) = (\lambda_j + \lambda_k)(\varphi_j \boxtimes \varphi_k^*)$. Hence $\varphi_j \boxtimes \varphi_k^*$, $j, k \geq 1$, form an orthonormal basis of $L^2(E \boxtimes E^*)$ consisting of eigensections for $\tilde{\Delta}$. The infinite sum

$$k_t(x, y) := \sum_{j=1}^{\infty} e^{-t \lambda_j} \varphi_j(x) \otimes \varphi_j^*(y),$$

$x, y \in M$, $t > 0$, is called the *heat kernel* of Δ on M.

Proposition 4. *Let $t_0 > 0$. Then the heat kernel and all its t-derivatives converge uniformly in $t \geq t_0$ in all H^k-norms and all C^k-norms. In particular, $k_t(x, y)$ is smooth in t, x, and y, and we can differentiate term by term.*

Proof. In view of the Sobolev embedding theorem it is sufficient to prove the proposition for the H^k-norms. All but finitely many λ_j fulfill $\lambda_j \geq 1$. By the elliptic estimates we then have

$$\|e^{-t\lambda_j}\varphi_j \boxtimes \varphi_j^*\|_{H^{2k}} \leq c_1 \cdot e^{-t\lambda_j} \cdot \left(\|\varphi_j \boxtimes \varphi_j^*\|_{L^2} + \|\tilde{\Delta}^k(\varphi_j \boxtimes \varphi_j^*)\|_{L^2}\right)$$
$$= c_1 \cdot e^{-t\lambda_j} \cdot \left(1 + (2\lambda_j)^k\right) \leq c_2 \cdot \lambda_j^k \cdot e^{-t\lambda_j} \leq c_2 \cdot \lambda_j^k \cdot e^{-t_0\lambda_j}.$$

Since for large enough x we have $x^k e^{-t_0 x/2} \leq 1$ we have for almost all j:

$$\|e^{-t\lambda_j}\varphi_j \boxtimes \varphi_j^*\|_{H^{2k}} \leq c_2 \cdot e^{-t_0\lambda_j/2}.$$

By Proposition 3 we have for $\alpha = \frac{4}{n(n+6)}$

$$\lambda_j \geq c_3 \cdot j^\alpha + c_4$$

and therefore

$$\|e^{-t\lambda_j}\varphi_j \boxtimes \varphi_j^*\|_{H^{2k}} \leq c_5 \cdot e^{-c_6 \cdot j^\alpha}.$$

Convergence of the series $\sum_j e^{-c_6 \cdot j^\alpha}$ follows from finiteness of the integral

$$\int_0^\infty e^{-c_6 \cdot t^\alpha} dt = c_7 \cdot \int_0^\infty e^{-s} \cdot s^{\frac{1-\alpha}{\alpha}} ds = c_7 \cdot \Gamma\left(\frac{1}{\alpha}\right).$$

We have shown that

$$\sum_{j=1}^\infty e^{-t\lambda_j}\varphi_j \boxtimes \varphi_j^*$$

converges in each H^k-norm, uniformly in $t \geq t_0$. The same argument applies to the t-derivatives

$$\sum_{j=1}^\infty \left(\frac{d}{dt}\right)^m \left(e^{-t\lambda_j}\varphi_j \boxtimes \varphi_j^*\right) = \sum_{j=1}^\infty (-\lambda_j)^m e^{-t\lambda_j}\varphi_j \boxtimes \varphi_j^*.$$

Since we are allowed to differentiate term by term we compute for y fixed

$$\frac{\partial}{\partial t} k_t(x,y) = \frac{\partial}{\partial t} \sum_j e^{-t\lambda_j}\varphi_j \boxtimes \varphi_j^* = \sum_j \frac{\partial}{\partial t} e^{-t\lambda_j}\varphi_j \boxtimes \varphi_j^*$$
$$- \sum_j (-\lambda_j) e^{-t\lambda_j}\varphi_j \boxtimes \varphi_j^* - - \sum_j e^{-t\lambda_j} (\Delta\varphi_j) \boxtimes \varphi_j^* - -\Delta_x k_t(x,y).$$

For $u_0 \in L^2(E)$ we put $u_t(x) := \int_M k_t(x,y) u_0(y)\, dV(y)$ and see

$$\frac{\partial u_t}{\partial t} + \Delta u_t = 0.$$

Hence u_t solves the heat equation. Moreover,

$$\int_M k_t(x,y)\varphi_k(y)\, dV(y) = \sum_j e^{-t\lambda_j}\varphi_j \cdot (\varphi_j, \varphi_k)_{L^2} = e^{-t\lambda_k}\varphi_k.$$

Thus $k_t(x,y)$ is the integral kernel of the operator $e^{-t\Delta}$. As $t \searrow 0$ the heat kernel becomes singular. Indeed, since $e^{-0\cdot\Delta} = \text{id}$ we expect the heat kernel to concentrate along the diagonal $\{(y,y) \in M \times M \mid y \in M\}$. We next want to examine the asymptotic behavior of $k_t(x,y)$ for $t \searrow 0$.

5.2 The Formal Heat Kernel

We start with the *Euclidean heat kernel*

$$q_t : M \times M \to \mathbb{R}, \quad q_t(x,y) = (4\pi t)^{-\frac{n}{2}} \exp\left(-\frac{\text{dist}(x,y)^2}{4t}\right).$$

A formal series

$$\tilde{k}_t(x,y) = q_t(x,y) \cdot \sum_{j=0}^{\infty} t^j \cdot \Phi_j(x,y),$$

$\Phi_j \in C^{\infty}(E \boxtimes E^*)$, is called a *formal heat kernel* if for each $N \in \mathbb{N}$ there exists m_0 such that for all $m \geq m_0$

$$\left(\frac{\partial}{\partial t} + \Delta_x\right)\left\{q_t \cdot \sum_{j=0}^{m} t^j \cdot \Phi_j\right\} = q_t \cdot O(t^N).$$

Lemma 1. *Let Δ_0 denote the standard Laplace-Beltrami operator acting on functions. Then*

$$\left(\frac{\partial}{\partial t} + \Delta_{0,x}\right) q_t(x,y) = \frac{a(x,y)}{t} \cdot q_t(x,y)$$

where a is smooth on $(M \times M)_{\epsilon_0}$ and a vanishes along the diagonal, $a(x,x) = 0$. In geodesic polar coordinates about y we have

$$a(x,y) = \frac{r}{2}\frac{d}{dr}\left(\ln \det(d\exp_y(rX))\right),$$

$x = \exp_y(rX)$, $X \in T_yM$, $\|X\| = 1$. *Hence a is essentially given by the radial logarithmic derivative of volume distortion of the exponential map.*

Here $\exp_y : T_yM \to M$ denotes the Riemannian exponential map.

Proof. Fix $y \in M$. We express Δ_0 in polar coordinates about y:

$$\Delta_0 = \Delta^{S_r} - \frac{\partial^2}{\partial r^2} + (n-1) \cdot H \cdot \frac{\partial}{\partial r}.$$

Here $S_r = \{x \in M \mid \text{dist}(x, y) = r\}$ denotes the distance sphere of radius r and H is its mean curvature. A direct calculation yields

$$\left(\frac{\partial}{\partial t} + \Delta_{0,x}\right) q_t$$

$$= \left(\frac{\partial}{\partial t} + \Delta^{S_r} - \frac{\partial^2}{\partial r^2} + (n-1) \cdot H \cdot \frac{\partial}{\partial r}\right) \left((4\pi t)^{-\frac{n}{2}} \exp\left(-\frac{r^2}{4t}\right)\right)$$

$$= -(n-1)\frac{1+Hr}{2t} \cdot q_t.$$

Hence $a(x, y) = -\frac{n-1}{2}(1 + Hr)$.

In order to identify this term we fix $X \in T_y M$, $\|X\| = 1$, and let $c(r) = \exp_y(rX)$ be the unit speed geodesic emanating from y in direction X. Let $e_1 = X, e_2, \ldots, e_n$ be an orthonormal basis of $T_y M$. Let V_i be the Jacobi field along c determined by the initial condition $V_i(0) = 0$ and $\frac{\nabla}{dr} V_i(0) = e_i$, $i = 1, \ldots, n$. It is well-known that [121, 1.2.2] the differential of the exponential map at the point rX is given by

$$d\exp_y(rX)(e_i) = \frac{1}{r} V_i(r).$$

Thus $\left(\frac{\nabla}{dr} d\exp_y(rX)\right)(e_i) = -\frac{1}{r^2} V_i(r) + \frac{1}{r} \frac{\nabla}{dr} V_i(r)$. In particular, $V_1(r) = rc'(r)$ and hence $\left(\frac{\nabla}{dr} d\exp_y(rX)\right)(e_1) = 0$. For $i = 2, \ldots, n$ we have $\frac{\nabla}{dr} V_i(r) = -B(V_i(r))$ where B is the Weingarten map (second fundamental form) of S_r [121, 1.2.6]. It follows

$$\left(\frac{\nabla}{dr} d\exp_y(rX)\right)(e_i) = \left(-\frac{1}{r^2}\text{id} - \frac{1}{r}B\right) V_i(r)$$

$$= \left(-\frac{1}{r}\text{id} - B\right) d\exp_y(rX)(e_i)$$

and thus

$$\frac{d}{dr} \det(d\exp_y(rX)) = \det(d\exp_y(rX)) \text{tr}\left(\left(\frac{\nabla}{dr} d\exp_y(rX)\right) \cdot (d\exp_y(rX))^{-1}\right)$$

$$= \det(d\exp_y(rX)) \text{tr}\left(-\frac{1}{r}\text{id}_{X^\perp} - B\right)$$

$$= \det(d\exp_y(rX)) \left(-\frac{n-1}{r} - (n-1)H\right)$$

$$= \frac{2}{r} \cdot \det(d\exp_y(rX)) \cdot a.$$

Hence

$$a = \frac{r}{2} \det(d\exp_y(rX))^{-1} \cdot \frac{d}{dr} \det(d\exp_y(rX)) = \frac{r}{2}\frac{d}{dr} \ln \det(d\exp_y(rX)).$$

Proposition 5. *Let ϵ_0 be the injectivity radius of M. Then there exists a unique formal heat kernel with Φ_j defined and smooth on $(M \times M)_{\epsilon_0} := \{(x, y) \in M \times M \mid \operatorname{dist}(x, y) < \epsilon_0\}$ such that*

$$\Phi_0(x, x) = \operatorname{id}_{E_x} \in \operatorname{Hom}(E_x, E_x) = E_x \otimes E_x^*.$$

Proof. We first show uniqueness of the Φ_j. To do this we differentiate the formal series $\tilde{k}_t(x, y)$ term by term, order the result by powers of t and equate the resulting coefficients to zero. We use the formula

$$\Delta(f \cdot \varphi) = (\Delta_0 f) \cdot \varphi - 2\nabla_{\operatorname{grad} f}\varphi + f\Delta\varphi$$

where f is a function and φ a section in E. Now

$$\left(\frac{\partial}{\partial t} + \Delta_x\right)\tilde{k}_t = \left(\left(\frac{\partial}{\partial t} + \Delta_{0,x}\right)q_t\right) \cdot \sum_j t^j \Phi_j - 2\nabla_{\operatorname{grad}_x q_t}\sum_j t^j \Phi_j$$

$$+ q_t\left(\frac{\partial}{\partial t} + \Delta_x\right)\sum_j t^j \Phi_j$$

$$= \frac{a}{t}q_t \cdot \sum_j t^j \Phi_j + \frac{1}{2t} \cdot q_t \cdot \nabla_{\operatorname{grad}_x(r^2)}\sum_j t^j \Phi_j$$

$$+ q_t \cdot \sum_j t^j \Delta_x \Phi_j + q_t \cdot \sum_j jt^{j-1}\Phi_j$$

$$= q_t \cdot \sum_{j=-1}^{\infty} t^j \cdot \{a \cdot \Phi_{j+1} + r\nabla_{\operatorname{grad}_x r}\Phi_{j+1} + \Delta_x\Phi_j + (j+1)\Phi_{j+1}\}$$

where again $r = \operatorname{dist}(x, y)$, y fixed, and with the convention that $\Phi_{-1} := 0$. Along any unit speed geodesic $c(r) = \exp_y(rX)$ emanating from y we obtain singular ordinary differential equations $(\Phi_j(r) := \Phi_j(\exp_y(rX), y))$:

$$(j + 1 + a(r))\Phi_{j+1}(r) + r\frac{\nabla}{dr}\Phi_{j+1}(r) + (\Delta_x\Phi_j)(r) = 0. \qquad (5.1)$$

To solve this equation we introduce the *integrating factor*

$$R_j(r) = r^{j+1} \cdot \exp\left(\int_0^r \frac{a(\rho)}{\rho}\, d\rho\right).$$

Then we have

$$\frac{r}{R_j(r)} \cdot \frac{\nabla}{dr}(R_j(r)\Phi_{j+1}(r))$$

$$= \frac{r}{R_j(r)} \cdot \left\{\frac{j+1}{r}R_j(r)\Phi_{j+1}(r) + R_j(r) \cdot \frac{a(r)}{r} \cdot \Phi_{j+1}(r) + R_j(r)\frac{\nabla}{dr}\Phi_{j+1}(r)\right\}$$

$$= -(\Delta_x\Phi_j)(r).$$

We denote parallel translation along $c(r)$ from $c(r_1)$ to $c(r_2)$ by π_{r_1,r_2} and we obtain

$$R_j(r)\Phi_{j+1}(r) = -\int_0^r \frac{R_j(\rho)}{\rho}\pi_{\rho,r}(\Delta_x\Phi_j)(\rho)\,d\rho + \pi_{0,r}C_j.$$

Evaluating this equation for $j = -1$ at $r = 0$ yields

$$1\cdot \mathrm{id}_{E_y} = 0 + C_{-1}.$$

Hence $C_{-1} = \mathrm{id}_{E_y}$ and

$$\Phi_0(r) = \frac{1}{R_{-1}(r)}\cdot \pi_{0,r}\cdot C_{-1} = \exp\left(-\int_0^r \frac{a(\rho)}{\rho}\,d\rho\right)\pi_{0,r}\mathrm{id}_{E_y}$$

$$= \det\left(d\exp_y(rX)\right)^{-\frac{1}{2}}\cdot \pi_{0,r}.$$

We have computed Φ_0:

$$\Phi_0(x,y) = \det\left(d(\exp_y^{-1})(x)\right)^{\frac{1}{2}}\cdot \pi_{y,x}$$

where $\pi_{y,x}$ denotes parallel translation from y to x (along the unique shortest geodesic connecting y and x).

For $j \geq 0$ we get at $r = 0$:

$$0\cdot \Phi_{j+1}(0) = 0 + C_j.$$

Hence $C_j = 0$ and

$$\Phi_{j+1}(r) = -\frac{1}{R_j(r)}\int_0^r \frac{R_j(\rho)}{\rho}\pi_{\rho,r}(\Delta_x\Phi_j)(\rho)\,d\rho.$$

This way we can recursively determine the Φ_j and uniqueness is proven. For the existence part simply use the above equations to define the Φ_j recursively.

Remark. By assumption we have

$$\Phi_0(y,y) = \mathrm{id}_{E_y}.$$

Plugging $r = 0$ into (5.1) for $j = 0$ we obtain

$$\Phi_1(0) = -(\Delta_x\Phi_0)(0).$$

Let us compute this term. We use the Taylor expansion of the metric in normal coordinates about $y(\hat{=}0)$:

$$g_{ij}(x) = \delta_{ij} + \frac{1}{3}\sum_{kl}R_{ikjl}(0)x^k x^l + O(\|x\|^3). \tag{5.2}$$

Hence
$$\det(d\exp_y) = \det((g_{ij})_{i,j=1,\ldots,n})^{\frac{1}{2}}$$
$$= \left[1 + \operatorname{tr}\left(\frac{1}{3}\sum_{kl} R_{ikjl}(0)x^k x^l + O(\|x\|^3)\right) + O(\|x\|^4)\right]^{\frac{1}{2}}$$
$$= 1 - \frac{1}{6}\sum_{kl}\operatorname{ric}_{kl}(0)x^k x^l + O(\|x\|^3)$$

Here $\operatorname{ric}_{kl} = \sum_{ij} g^{ij} R_{iklj} = -\sum_{ij} g^{ij} R_{ikjl}$ denotes *Ricci curvature*. Thus $\det(d\exp_y)^{-\frac{1}{2}} = 1 + \frac{1}{12}\sum_{kl}\operatorname{ric}_{kl}(0)x^k x^l + O(\|x\|^3)$ and therefore

$$\Delta_{0,x}\left(\det(d\exp_y)^{-\frac{1}{2}}\right) = -\frac{1}{6}\sum_k \operatorname{ric}_{kk}(0) + O(\|x\|) = -\frac{1}{6}\operatorname{scal}(0) + O(\|x\|).$$

Here $\operatorname{scal} = \sum_k \operatorname{ric}_{kk}$ denotes the *scalar curvature*.

Now $(\Delta_x \Phi_0)(x,y) = \left(\Delta_{0,x}\left(\det(d\exp_y)^{-\frac{1}{2}}\right)\right) \cdot \pi_{y,x} + \det(d\exp_y)^{-\frac{1}{2}} \cdot \mathcal{K}_x \circ \pi_{y,x}$ and therefore $\Delta_x \Phi_0(y,y) = -\frac{1}{6}\operatorname{scal}(y) + \mathcal{K}_y$.

We have shown
$$\Phi_1(y,y) = \frac{1}{6}\operatorname{scal}(y) \cdot \operatorname{id}_{E_y} - \mathcal{K}_y.$$

This is of greatest importance to us because this function will give us the Einstein-Hilbert action.

It remains to see what the formal heat kernel and the true heat kernel have to do with each other. Pick a smooth cut-off function $\chi : \mathbb{R} \to \mathbb{R}$, such that $\chi(r) = 1$ for $r \leq \frac{\epsilon_0}{3}$, $\chi(r) = 0$ for $r \geq \frac{2\epsilon_0}{3}$, and $0 \leq \chi \leq 1$ everywhere. We define
$$\widehat{k}_t(x,y) := \widetilde{k}_t(x,y) \cdot \chi(\operatorname{dist}(x,y)).$$
Hence $\widehat{k}_t$ coincides with the formal heat kernel $\widetilde{k}_t$ on a neighborhood of the diagonal, but $\widehat{k}_t$ is defined and smooth on all of $M \times M$ (or, more precisely, its finite partial sums $\widehat{k}_t^{(m)}(x,y) := \chi(\operatorname{dist}(x,y)) \cdot q_t(x,y) \cdot \sum_{j=0}^m t^j \Phi_j \cdot (x,y)$).

Proposition 6. $\widehat{k}_t$ *is asymptotic to* k_t, *in symbols*
$$k_t \stackrel{t \searrow 0}{\sim} \widehat{k}_t,$$
in the following sense: For each $N \in \mathbb{N}$ *there exists* $m_0 \in \mathbb{N}$ *and* $t_0 > 0$ *such that for all* $m \geq m_0$ *there is a constant* $C_{N,m} > 0$ *with*
$$|k_t(x,y) - \widehat{k}_t^{(m)}(x,y)| \leq C_{N,m} \cdot t^N$$
for all $t \in (0,t_0)$, $x,y \in M$.

Proof. Let $\varphi \in C^0(E)$ such that the support of φ is contained in a ball of radius $\frac{\varepsilon_0}{2}$. Recall that ε_0 is the injectivity radius of M. Since q_t is the Euclidean heat kernel we see

$$\lim_{t\searrow 0}\int_M q_t(x,y)\Phi_0(x,y)\varphi(y)\,dV(y) = \Phi_0(x,x)\varphi(x) = \varphi(x).$$

A partition of unity argument yields for arbitrary $\varphi \in C^0(E)$

$$\lim_{t\searrow 0}\int_M \widehat{k}_t^{(0)}(x,y)\varphi(y)\,dV(y) = \varphi(x).$$

Since higher powers of t do not contribute to the limit for $t \searrow 0$ we have

$$\lim_{t\searrow 0}\int_M \widehat{k}_t^{(m)}(x,y)\varphi(y)\,dV(y) = \varphi(x)$$

for all $m \in \mathbb{N}$ and $\varphi \in C^0(E)$. On the other hand, since $e^{-t\Delta}$ tends to $e^{-0\cdot\Delta} = \mathrm{id}$, we also have

$$\lim_{t\searrow 0}\int_M k_t(x,y)\varphi(y)\,dV(y) = \varphi(x).$$

Thus for $\delta_t^{(m)} := k_t - \widehat{k}_t^{(m)}$ we get

$$\lim_{t\searrow 0}\int_M \delta_t^{(m)}(x,y)\varphi(y)\,dV(y) = 0.$$

Now put $\left(\frac{\partial}{\partial t} + \Delta_x\right)\delta_t^{(m)} =: \eta_t^{(m)}$ and $\widetilde{\delta}_t^{(m)} := \int_0^t e^{-(t-\tau)\Delta_x}\eta_\tau^{(m)}\,d\tau$. We know that

$$\eta_t^{(m)} = -\left(\frac{\partial}{\partial t} + \Delta_x\right)\widehat{k}_t^{(m)} = -\left(\frac{\partial}{\partial t} + \Delta_x\right)\left(\chi\cdot\bar{k}_t^{(m)}\right)$$

where $\chi(x,y) = \chi(\mathrm{dist}(x,y))$. Hence

$$\eta_t^{(m)} = -\chi\cdot\left(\frac{\partial}{\partial t} + \Delta_x\right)\bar{k}_t^{(m)} + \underbrace{(\Delta_{0,x}\chi)\cdot\bar{k}_t^{(m)} - 2\nabla_{\mathrm{grad}_x\chi}\bar{k}_t^{(m)}}_{=:R_t^{(m)}}$$

$$= q_t \cdot O(t^N) + R_t^{(m)}.$$

Now $R_t^{(m)}$ is of the form $q_t \times$ smooth section vanishing for $\mathrm{dist}(x,y) < \frac{\varepsilon_0}{3}$. For $\mathrm{dist}(x,y) > \frac{\varepsilon_0}{3}$ we have

$$q_t(x,y) \le c_1 \cdot \exp\left(-\frac{c_2}{t}\right)\cdot q_{2t}(x,y)$$

for suitable constants $c_1, c_2 > 0$. Therefore
$$\eta_t^{(m)} = q_{2t} \cdot O(t^N).$$

From the definition of $\widetilde{\delta}_t^{(m)}$ we have
$$\frac{\partial}{\partial t}\widetilde{\delta}_t^{(m)} = e^{-(t-t)\Delta_x}\eta_t^{(m)} + \int_0^t -\Delta_x e^{-(t-\tau)\Delta_x}\eta_\tau^{(m)}d\tau$$
$$= \eta_t^{(m)} - \Delta_x \widetilde{\delta}_t^{(m)}.$$

Therefore $\left(\frac{\partial}{\partial t} + \Delta_x\right)\widetilde{\delta}_t^{(m)} = \eta_t^{(m)}$ and $\left(\frac{\partial}{\partial t} + \Delta_x\right)\left(\widetilde{\delta}_t^{(m)} - \delta_t^{(m)}\right) = 0$. Since $\widetilde{\delta}_t^{(m)} - \delta_t^{(m)} \overset{t \searrow 0}{\longrightarrow} 0$ it follows $\widetilde{\delta}_t^{(m)} - \delta_t^{(m)} = e^{-t\Delta}0 = 0$, thus
$$\delta_t^{(m)} = \widetilde{\delta}_t^{(m)} = \int_0^t e^{-(t-\tau)\Delta_x}\eta_\tau^{(m)}d\tau$$

and hence
$$\|\delta_t^{(m)}\|_{H^k} \le t \cdot \sup_{\tau \in [0,t]} \|e^{-(t-\tau)\Delta_x}\|_{H^k, H^k} \cdot \sup_{\tau \in [0,t]} \|\eta_\tau^{(m)}\|_{H^k} = O\left(t^{N+1}\right).$$

The Sobolev embedding theorem implies for $k > \frac{n}{2}$
$$\|k_t - \widehat{k}_t^{(m)}\|_{C^0} = \|\delta_t^{(m)}\|_{C^0} = O\left(t^{N+1}\right).$$

Corollary 1.
$$k_t(x,x) \overset{t \searrow 0}{\sim} \widehat{k}_t(x,x) = \widetilde{k}_t(x,x)$$
$$= (4\pi t)^{-\frac{n}{2}} \cdot \left\{ \text{id}_{E_x} + t \cdot \left(\frac{1}{6}\text{scal}(x) \cdot \text{id}_{E_x} - \mathcal{K}_x\right) + O(t^2)\right\}.$$

Corollary 2.
$$\sum_{i=1}^\infty e^{-t\lambda_i} = \text{Tr}\left(e^{-t\Delta}\right) = \int_M \text{tr}\left(k_t(x,x)\right) dV(x) \overset{t \searrow 0}{\sim}$$
$$(4\pi t)^{-\frac{n}{2}} \cdot \left\{\text{rk}(E) \cdot \text{vol}(M) + t \cdot \left(\frac{\text{rk}(E)}{6}\int_M \text{scal}(x)\,dV(x) - \int_M \text{tr}(\mathcal{K}_x)\,dV(x)\right) + O(t^2)\right\}.$$

Lemma 2 (Karamata). Let $d\mu$ be a positive measure on $(0,\infty)$, let $\alpha > 0$ and $C > 0$. We assume
$$\int_0^\infty e^{-t\lambda}d\mu(\lambda) < \infty$$

for all $t > 0$ and

$$\lim_{t \searrow 0} t^\alpha \int_0^\infty e^{-t\lambda} d\mu(\lambda) = C.$$

Then for all continuous functions f on $[0,1]$ the following holds:

$$\lim_{t \searrow 0} t^\alpha \int_0^\infty f\left(e^{-t\lambda}\right) e^{-t\lambda} d\mu(\lambda) = \frac{C}{\Gamma(\alpha)} \int_0^\infty f\left(e^{-t}\right) t^{\alpha-1} e^{-t} dt.$$

Proof. By Weierstrass' theorem the polynomials lie dense in $C^0([0,1])$ (w.r.t. the C^0-norm). Hence it is sufficient to prove the lemma for f a polynomial. Then we can assume w.l.o.g. that $f(x) = x^k$. For the left hand side we get

$$\lim_{t \searrow 0} t^\alpha \int_0^\infty f\left(e^{-t\lambda}\right) e^{-t\lambda} d\mu(\lambda) = \lim_{t \searrow 0} t^\alpha \int_0^\infty e^{-(k+1)t\lambda} d\mu(\lambda)$$

$$= \lim_{s \searrow 0} \left(\frac{s}{k+1}\right)^\alpha \int_0^\infty e^{-s\lambda} d\mu(\lambda) = \frac{C}{(k+1)^\alpha}.$$

The right hand side turns out to be the same

$$\frac{C}{\Gamma(\alpha)} \int_0^\infty f\left(e^{-t}\right) t^{\alpha-1} e^{-t} dt = \frac{C}{\Gamma(\alpha)} \int_0^\infty t^{\alpha-1} e^{-(k+1)t} dt$$

$$= \frac{C}{\Gamma(\alpha)} \int_0^\infty \left(\frac{s}{k+1}\right)^{\alpha-1} \cdot e^{-s} \cdot \frac{ds}{k+1} = \frac{C}{\Gamma(\alpha)} \cdot \frac{\Gamma(\alpha)}{(k+1)^\alpha}.$$

Theorem 2 (Weyl). *Let $\Delta : C^\infty(E) \to C^\infty(E)$ be a generalized Laplace operator over an n-dimensional compact Riemannian manifold. For each $\lambda \in \mathbb{R}$ let $N(\lambda)$ be the number of eigenvalues of Δ less than λ. Then*

$$\lim_{\lambda \to \infty} \frac{N(\lambda)}{\lambda^{\frac{n}{2}}} = \frac{\mathrm{rk}(E) \cdot \mathrm{vol}(M)}{(4\pi)^{\frac{n}{2}} \cdot \Gamma\left(\frac{n}{2}+1\right)}.$$

Proof. Since a shift of the spectrum by a constant will not alter the limit $\lim_{\lambda \to \infty} \frac{N(\lambda)}{\lambda^{\frac{n}{2}}}$ we may w.l.o.g. assume that all eigenvalues λ_i are positive. We apply Karamata's lemma with $\alpha = \frac{n}{2}$, $C = (4\pi)^{-\frac{n}{2}} \mathrm{rk}(E) \mathrm{vol}(M)$, and the spectral measure $d\mu = \sum_{i=1}^\infty \delta_{\lambda_i}$. Since

$$\int_0^\infty e^{-t\lambda} d\mu(\lambda) = \sum_{i=1}^\infty e^{-t\lambda_i} = \mathrm{Tr}\left(e^{-t\Delta}\right) < \infty$$

and

$$\lim_{t \searrow 0} t^\alpha \cdot \int_0^\infty e^{-t\lambda} d\mu(\lambda) = \lim_{t \searrow 0} t^{\frac{n}{2}} \cdot \mathrm{Tr}\left(e^{-t\Delta}\right) = C$$

by Corollary 2 the assumptions in Karamata's lemma are satisfied.

Let $\epsilon > 0$ and pick a continuous function $f : [0,1] \to \mathbb{R}$ such that $f(x) = 0$ for $x \leq e^{-(1+\epsilon)}$, $f(x) = x^{-1}$ for $x \geq e^{-1}$ and $0 \leq f(x) \leq x^{-1}$ everywhere. For the left hand side in Karamata's lemma we get

$$\lim_{t \searrow 0} t^{\frac{n}{2}} \int_0^\infty f\left(e^{-t\lambda}\right) e^{-t\lambda} d\mu(\lambda) = \lim_{t \searrow 0} t^{\frac{n}{2}} \int_0^{(1+\epsilon)t^{-1}} f\left(e^{-t\lambda}\right) e^{-t\lambda} d\mu(\lambda)$$

$$\geq \limsup_{t \searrow 0} t^{\frac{n}{2}} \int_0^{t^{-1}} d\mu(\lambda) = \limsup_{t \searrow 0} t^{\frac{n}{2}} N(t^{-1}) = \limsup_{\lambda \to \infty} \frac{N(\lambda)}{\lambda^{\frac{n}{2}}}.$$

For the right hand side we obtain

$$\frac{C}{\Gamma(\alpha)} \int_0^\infty f\left(e^{-t}\right) t^{\alpha-1} e^{-t} dt = \frac{C}{\Gamma(\alpha)} \int_0^{1+\epsilon} f\left(e^{-t}\right) t^{\alpha-1} e^{-t} dt$$

$$\leq \frac{C}{\Gamma(\alpha)} \int_0^{1+\epsilon} t^{\alpha-1} dt = \frac{C \cdot (1+\epsilon)^\alpha}{\Gamma(\alpha) \cdot \alpha} = \frac{C \cdot (1+\epsilon)^\alpha}{\Gamma(\alpha+1)}.$$

Thus

$$\limsup_{\lambda \to \infty} \frac{N(\lambda)}{\lambda^{\frac{n}{2}}} \leq \frac{C \cdot (1+\epsilon)^\alpha}{\Gamma(\alpha+1)}$$

and $\epsilon \searrow 0$ yields

$$\limsup_{\lambda \to \infty} \frac{N(\lambda)}{\lambda^{\frac{n}{2}}} \leq \frac{C}{\Gamma(\alpha+1)} = \frac{\mathrm{rk}(E) \cdot \mathrm{vol}(M)}{(4\pi)^{\frac{n}{2}} \Gamma\left(\frac{n}{2}+1\right)}.$$

The proof of $\liminf_{\lambda \to \infty} \frac{N(\lambda)}{\lambda^{\frac{n}{2}}} \geq \frac{C}{\Gamma(\alpha+1)}$ is completely analogous. One uses continuous functions $f : [0,1] \to \mathbb{R}$ satisfying $f(x) = 0$ for $x \leq e^{-1}$, $f(x) = x^{-1}$ for $x \geq e^{-1+\epsilon}$ and $0 \leq f(x) \leq x^{-1}$ everywhere.

5.3 Dirac Operators and Weitzenböck Formulas

Again, let M be a compact Riemannian manifold. Let $\mathrm{Cl}(M)$ denote the *Clifford bundle* of M, i.e. at each point $p \in M$ the fiber $\mathrm{Cl}(M)_p$ is the Clifford algebra of T_pM. There is a canonical vector bundle isomorphism $\mathrm{Cl}(M) \xrightarrow{\cong} \bigoplus_{k=0}^n \Lambda^n TM$ which we use to define the Levi-Civita connection ∇

on Cl(M). For an orthonormal basis $e_1, \ldots, e_n$ of T_pM this isomorphism is given by $e_{i_1} \cdot \ldots \cdot e_{i_k} \mapsto e_{i_1} \wedge \ldots \wedge e_{i_k}$, $i_1 < i_2 < \ldots < i_k$. Note that this is **not** an algebra homomorphism.

Now let $E \to M$ be a Cl(M)-module bundle, i.e. for each $p \in M$ there is an action of Cl(M)$_p$ on E_p. We will assume that this action depends smoothly on p. Suppose furthermore that E carries a Hermitian or Riemannian metric with respect to which the action of vectors $X \in T_pM \subset $ Cl(M)$_p$ is skew-adjoint,

$$\langle X \cdot \varphi, \psi \rangle = -\langle \varphi, X \cdot \psi \rangle, \qquad \varphi, \psi \in E_p,$$

and a metric connection ∇^E which is compatible with the Levi-Civita connection in the following sense:

$$\nabla^E_X(\omega \cdot \varphi) = (\nabla_X \omega) \cdot \varphi + \omega \cdot \nabla^E_X \varphi$$

for all $X \in TM$, $\omega \in C^\infty(\text{Cl}(M))$, $\varphi \in C^\infty(E)$.

Now the *Dirac operator* $D : C^\infty(E) \to C^\infty(E)$ is defined by

$$D\varphi := \sum_{k=1}^n e_k \cdot \nabla^E_{e_k} \varphi.$$

This definition is independent of the choice of local orthonormal frame $e_1,\ldots,e_n$. The Dirac operator is an elliptic differential operator of first order. It is self-adjoint in $L^2(E)$ with domain $H^1(E)$.

Example. If M is a Riemannian **spin** manifold, then we can take $E := \varSigma M$, the *spinor bundle*. The resulting operator D is the *classical Dirac operator*, sometimes also called *Atiyah-Singer operator*.

Example. If E is a Cl(M)-module bundle as above and V is another Hermitian or Riemannian vector bundle over M with a metric connection, then $E \otimes V$ is again a Cl(M)-module bundle. Here the Cl(M)-action is on the first factor,

$$\omega \cdot (\varphi \otimes v) = (\omega \cdot \varphi) \otimes v, \qquad \omega \in \text{Cl}(M)_p, \varphi \in E_p, v \in V_p,$$

and $E \otimes V$ carries the induced metric and connection, $\nabla^{E \otimes V} = \nabla^E \otimes \text{id} + \text{id} \otimes \nabla^V$. The resulting Dirac operator is called a *twisted Dirac operator* with coefficients in V.

For any Dirac operator direct computation yields

$$D(f \cdot \varphi) = \text{grad} f \cdot \varphi + f \cdot D\varphi \tag{5.3}$$

for $\varphi \in C^\infty(E)$ and a smooth function f on M. This can also be expressed by saying that the principal symbol of D is given by Clifford multiplication.

The link to the previous section is now established by

Proposition 7 (Bochner-Weitzenböck formula). *Let E be a Cl(M)-module bundle over M. Then the square of its Dirac operator is a generalized Laplace operator*

$$D^2 = (\nabla^E)^* \nabla^E + \mathcal{K}$$

where $\mathcal{K} = \frac{1}{2} \sum_{i,j=1}^{n} e_i \cdot e_j \cdot R^E(e_i, e_j)$.

Proof. Fix $p \in M$ and choose an orthonormal frame $e_1, \ldots, e_n$ near p *synchronous* at p, i.e. $(\nabla e_k)(p) = 0$ for all k. Then at p

$$D^2\varphi = \sum_{ij} e_i \nabla^E_{e_i} \left(e_j \nabla^E_{e_j} \varphi \right) = \sum_{ij} e_i e_j \nabla^E_{e_i} \nabla^E_{e_j} \varphi$$

$$= \sum_i e_i^2 \nabla^E_{e_i} \nabla^E_{e_i} \varphi + \sum_{i<j} \left(e_i e_j \nabla^E_{e_i} \nabla^E_{e_j} + e_j e_i \nabla^E_{e_j} \nabla^E_{e_i} \right) \varphi$$

$$= -\sum_i \nabla^E_{e_i} \nabla^E_{e_i} \varphi + \sum_{i<j} e_i e_j \left(\nabla^E_{e_i} \nabla^E_{e_j} - \nabla^E_{e_j} \nabla^E_{e_i} \right) \varphi$$

$$= (\nabla^E)^* \nabla^E \varphi + \sum_{i<j} e_i e_j R^E(e_i, e_j) \varphi.$$

Example. In the case of the classical Dirac operator acting on spinors the *curvature endomorphism* $\mathcal{K}$ takes a very simple form [147,195]

$$\mathcal{K}_p = \frac{1}{4} \mathrm{scal}(p) \cdot \mathrm{id}_{\Sigma_p M}.$$

Example. In the case of a twisted Dirac operator we have

$$R^{E \otimes V}(X, Y) = R^E(X, Y) \otimes \mathrm{id} + \mathrm{id} \otimes R^V(X, Y)$$

and hence

$$\mathcal{K}^{E \otimes V}(\varphi \otimes v) = \frac{1}{2} \sum_{ij} e_i e_j R^{E \otimes V}(e_i, e_j)(\varphi \otimes v)$$

$$= \frac{1}{2} \sum_{ij} e_i e_j R^E(e_i, e_j) \varphi \otimes v + \frac{1}{2} \sum_{ij} e_i e_j \varphi \otimes R^V(e_i, e_j) v$$

$$= \mathcal{K}^E \varphi \otimes v + \mathcal{F}^V(\varphi \otimes v),$$

i.e. $\mathcal{K}^{E \otimes V} = \mathcal{K}^E \otimes \mathrm{id} + \mathcal{F}^V$.

Here $\mathcal{F}^V$ is the so-called *twisting curvature*. If E is the spinor bundle as in the previous example, then the twisted classical Dirac operator has

$$\mathcal{K} = \frac{1}{4} \mathrm{scal} \cdot \mathrm{id} + \mathcal{F}^V$$

as its curvature endomorphism. In particular, we can write down the heat asymptotics. By Corollary 1 we have for the heat kernel of D^2

$$k_t(x,x) \overset{t \searrow 0}{\sim} (4\pi t)^{-\frac{n}{2}} \cdot \left\{ \mathrm{id}_{\Sigma_x M} + t \cdot \left(\frac{1}{6} \mathrm{scal}(x) \cdot \mathrm{id}_{\Sigma_x M} - \mathcal{K}_x \right) + O(t^2) \right\}$$

$$= (4\pi t)^{-\frac{n}{2}} \cdot \left\{ \mathrm{id}_{\Sigma_x M} - t \cdot \left(\frac{1}{12} \mathrm{scal}(x) \cdot \mathrm{id}_{\Sigma_x M} + \mathcal{F}^V_x \right) + O(t^2) \right\}.$$

Since the rank of the spinor bundle is $2^{[n/2]}$ integration yields

$$\operatorname{Tr}\left(e^{-tD^2}\right) \overset{t \searrow 0}{\sim}$$

$$2^{[n/2]}(4\pi t)^{-\frac{n}{2}} \cdot \left\{ \operatorname{vol}(M) - t \cdot \int_M \left(\frac{1}{12}\operatorname{scal}(x) + 2^{-[n/2]}\operatorname{tr}(\mathcal{F}_x^V) \right) dV(x) + O(t^2) \right\} \tag{5.4}$$

5.4 Integration and Dixmier Trace

In noncommutative geometry one replaces a "classical" compact Riemannian spin manifold M by the tripel $(\mathcal{A}, \mathcal{H}, D)$ where $\mathcal{A} = C^\infty(M)$ is the pre-C^*-algebra of smooth functions on M (with respect to the C^0-norm), $\mathcal{H} = L^2(\Sigma M)$ is the Hilbert space of square-integrable spinors, and D is the classical Dirac operator. The algebra $\mathcal{A}$ acts on $\mathcal{H}$ by pointwise multiplication. For any $f \in \mathcal{A}$ the commutator of f and D is given by Clifford multiplication with the gradient of f, cf. (5.3). Hence the condition $\|[D,f]\|_{C^0} \leq 1$ means that the gradient of f is bounded by 1. This observation is important since it implies that we can reconstruct the distance function and hence the metric on M from the triple $(\mathcal{A}, \mathcal{H}, D)$:

$$\operatorname{dist}(x,y) = \sup \{|f(x) - f(y)| \mid f \in \mathcal{A}, \ \|[D,f]\|_{C^0} \leq 1\}.$$

In order to get noncommutative generalizations we have to express classical geometric operations in terms of the triple $(\mathcal{A}, \mathcal{H}, D)$. We will do this now for integration of functions over M.

Let $\lambda_1, \lambda_2, \ldots$ be the eigenvalues of a generalized Dirac operator, ordered by increasing absolute values, $|\lambda_1| \leq |\lambda_2| \leq \ldots \nearrow \infty$. We assume that 0 is not an eigenvalue of D. The square D^2 is a generalized Laplacian with eigenvalues $0 < \lambda_1^2 \leq \lambda_2^2 \leq \ldots \nearrow \infty$. By Weyl's theorem $\lim_{k \to \infty} k/|\lambda_k|^n = C$ with $C = (\operatorname{rk}(E) \cdot \operatorname{vol}(M))/((4\pi)^{n/2} \cdot \Gamma(n/2 + 1))$. In particular, there exists a constant $C' > 0$ such that

$$|\lambda_k| \geq C' \cdot k^{1/n}$$

for all but finitely many k. Therefore

$$\operatorname{Tr}_\omega(|D|^{-n}) := \lim_{N \to \infty} \frac{1}{\log(N)} \sum_{k=1}^N |\lambda_k|^{-n} \leq C'' \cdot \lim_{N \to \infty} \frac{1}{\log(N)} \sum_{k=1}^N \frac{1}{k} < \infty.$$

The number $\operatorname{Tr}_\omega(|D|^{-n})$ is called the *Dixmier trace* of $|D|^{-n}$. Let $\Psi \in C^\infty(\operatorname{End}(E))$ be an endomorphism field, for example $\Psi = f \cdot \operatorname{id}$ where $f \in C^\infty(M)$. Connes' trace theorem tells us that the Dixmier trace is a residue, more precisely

$$\operatorname{Tr}_\omega(\Psi \circ |D|^{-n}) = \frac{1}{n} \lim_{p \searrow n} (p - n) \operatorname{Tr}(\Psi \circ |D|^{-p}). \tag{5.5}$$

In order to apply this we have to control the integral kernel of $|D|^{-p}$ with $p > n$. Let k_t be the heat kernel of the generalized Laplacian D^2. We perform the following *Mellin transformation*: After restriction to the λ-eigenspace of D we have

$$\Gamma\left(\frac{p}{2}\right)|D|^{-p} = |\lambda|^{-p}\int_0^\infty e^{-t}t^{p/2-1}dt$$

$$= |\lambda|^{-p}\int_0^\infty e^{-s\lambda^2}\left(s\lambda^2\right)^{p/2-1}\lambda^2 ds = \int_0^\infty e^{-s\lambda^2}s^{p/2-1}ds,$$

hence

$$|D|^{-p} = \frac{1}{\Gamma\left(\frac{p}{2}\right)}\int_0^\infty t^{p/2-1}e^{-tD^2}dt.$$

Therefore $|D|^{-p}$ has the integral kernel

$$k(x,y;|D|^{-p}) = \frac{1}{\Gamma\left(\frac{p}{2}\right)}\int_0^\infty t^{p/2-1}k_t(x,y)dt.$$

Then $\Psi \circ |D|^{-p}$ has integral kernel

$$k(x,y;\Psi \circ |D|^{-p}) = \frac{1}{\Gamma\left(\frac{p}{2}\right)}\int_0^\infty t^{p/2-1}\Psi(x)\circ k_t(x,y)dt.$$

Therefore

$$\mathrm{Tr}(\Psi \circ |D|^{-p}) = \int_M \mathrm{tr}\bigl(k(x,x;\Psi \circ |D|^{-p})\bigr)\,dV(x)$$

$$= \frac{1}{\Gamma\left(\frac{p}{2}\right)}\int_0^\infty t^{p/2-1}\int_M \mathrm{tr}\bigl(\Psi(x)k_t(x,x)\bigr)\,dV(x)dt.$$

For any $t_0 > 0$ the integral

$$\int_{t_0}^\infty t^{p/2-1}\int_M \mathrm{tr}\bigl(\Psi(x)k_t(x,x)\bigr)\,dV(x)dt$$

remains bounded for $p \searrow n$ (remember that $e^{-tD^2} \leq e^{-t\lambda_1^2}$ tends to zero exponentially fast for $t \to \infty$) and hence does not contribute to the residue. For $0 < t < t_0$, t_0 sufficiently small, we have by Corollary 1 that

$$k_t(x,x) = (4\pi t)^{-\frac{n}{2}}\mathrm{id} + O(t^{-\frac{n}{2}+1}).$$

Thus

$$\int_0^{t_0} t^{p/2-1}\int_M \mathrm{tr}\left(\Psi(x)k_t(x,x)\right)\,dV(x)dt$$

$$= (4\pi)^{-\frac{n}{2}}\int_0^{t_0}\left(\int_M t^{\frac{p-n}{2}-1}\mathrm{tr}\left(\Psi(x)\right)\,dV(x) + O(t^{\frac{p-n}{2}})\right)dt$$

$$= (4\pi)^{-\frac{n}{2}}\frac{2}{p-n}t_0^{\frac{p-n}{2}}\int_M \mathrm{tr}\left(\Psi(x)\right)\,dV(x) + O(1)$$

and therefore by (5.5)
$$\operatorname{Tr}_\omega(\Psi \circ |D|^{-n}) = (4\pi)^{-\frac{n}{2}} \frac{2}{n\Gamma(n/2)} \int_M \operatorname{tr}(\Psi(x))\, dV(x).$$

We have shown

Proposition 8. *Let $\Psi \in C^\infty(\operatorname{End}(E))$. Then*
$$\operatorname{Tr}_\omega(\Psi \circ |D|^{-n}) = (4\pi)^{-\frac{n}{2}} \frac{2}{n\Gamma(n/2)} \int_M \operatorname{tr}(\Psi(x))\, dV(x).$$

In particular, for $\Psi = f \cdot \operatorname{id}$
$$\operatorname{Tr}_\omega(f \cdot |D|^{-n}) = (4\pi)^{-\frac{n}{2}} \frac{2\operatorname{rk}(E)}{n\Gamma(n/2)} \int_M f(x)\, dV(x).$$

and for $f = 1$
$$\operatorname{Tr}_\omega(|D|^{-n}) = (4\pi)^{-\frac{n}{2}} \frac{2\operatorname{rk}(E)}{n\Gamma(n/2)} \operatorname{vol}(M).$$

This justifies to call $|D|^{-n}$ the operator theoretic volume element and to interprete Tr_ω as integration, c.f. [38,119].

5.5 Variational Formulas and the Einstein-Hilbert Action

In this section we want to calculate the variation of the *gravity action*
$$\int_M (\operatorname{scal}_g + \lambda)\, dV_g$$
under changes of the Riemannian metric g. Here λ is twice the cosmological constant as we will see at the end of this section. The Euler-Lagrange equations of this functional will turn out to be the Einstein equations of General Relativity. In this section we follow [17, Ch. 1.K].

Let $\mathcal{M}$ be the space of smooth semi-Riemannian metrics on a manifold M. In contrast to all other sections of this article the manifold M need not be compact and g need not be Riemannian. We view the Riemannian curvature tensor R as a functional $\mathcal{M} \to C^\infty(T^{3,1}M)$, $g \mapsto R_g$, where $T^{i,j}M$ denotes the bundle of (i,j)-tensors on M.

The corresponding differential R'_g at g is defined as
$$R'_g h(X,Y)Z = \frac{d}{dt}\bigg|_{t=0} R_{g+th}(X,Y)Z$$
where h is an arbitrary smooth symmetric $(2,0)$-tensor on M. Similarly we consider the Ricci curvature ric, the scalar curvature scal and the Levi-Civita

connection ∇, and we denote their differentials in the direction of h by $\mathrm{ric}'_g h$, $\mathrm{scal}'_g h$ and $\nabla'_g h$.

Connections are not tensorial in the second slot, but differences of two connections are. Therefore $\nabla'_g h$ is a $(2,1)$-tensor.

If v and w are symmetric $(2,0)$-tensors, we define the *composition* $v \circ w$ to be the $(2,0)$-tensor given by

$$(v \circ w)(X, Y) = \sum_{i=1}^{n} \varepsilon_i v(X, e_i) w(e_i, Y) \qquad X, Y \in T_p M,$$

where $e_1, \ldots, e_n$ is an orthonormal basis of $T_p M$, i.e. $g(e_i, e_j) = \varepsilon_i \delta_{ij}$ with $\varepsilon_i = \pm 1$. In the Riemannian case all $\varepsilon_i = +1$.

The Riemannian curvature tensor acts on symmetric $(2,0)$-tensors via

$$\mathring{R}_g h(X, Y) := \sum_{i=1}^{n} \varepsilon_i h(R(e_i, X) Y, e_i).$$

In Corollary 3 it will be proven that $\mathring{R}_g h$ is actually a symmetric $(2,0)$-tensor.

The *Lichnerowicz Laplacian* Δ_L on symmetric $(2,0)$-tensors is defined by

$$\Delta_L h := \nabla^* \nabla h + \mathrm{ric}_g \circ h + h \circ \mathrm{ric}_g - 2\mathring{R}_g h.$$

The semi-Riemannian metric g on M defines a scalar product on the bundle $T^*M \otimes T^*M$ given locally by

$$\langle h_1, h_2 \rangle_g := \sum_{i,j=1}^{n} \varepsilon_i \varepsilon_j h_1(e_i, e_j) h_2(e_i, e_j)$$

where $e_1, \ldots, e_n$ is an orthonormal frame, i.e. $g(e_i, e_j) = \varepsilon_i \delta_{ij}$.

Now we can formulate the variation formulas for the curvature.

Proposition 9. *Let (M, g) be a semi-Riemannian manifold and let h be a symmetric $(2,0)$-tensor on M. The differentials of R, ric, scal and ∇ at g, in the direction of h, are given by the formulas:*

(a) Levi-Civita connection

$$g\left(\nabla'_g h(X, Y), Z\right) = \frac{1}{2} \left\{ (\nabla_X h)(Y, Z) + (\nabla_Y h)(X, Z) - (\nabla_Z h)(X, Y) \right\},$$

(b) Riemannian curvature tensor

$$R'_g h(X, Y) Z = \left(\nabla_X \nabla'_g h\right)(Y, Z) - \left(\nabla_Y \nabla'_g h\right)(X, Z),$$

(c) Ricci tensor

$$\mathrm{ric}'_g h = \frac{1}{2} \Delta_L h - \delta^*_g (\delta_g h) - \frac{1}{2} \nabla_g d(\mathrm{tr}_g h),$$

(d) *scalar curvature*
$$\text{scal}'_g h = \Delta_g(\text{tr}_g h) + \delta_g(\delta_g h) - \langle \text{ric}_g, h \rangle_g.$$

Proof (Proof of (a)). We set $g_t := g + th$. Then
$$T_t(X, Y, Z) := g(\nabla^t_X Y, Z) - g(\nabla^0_X Y, Z)$$
is a $(3,0)$-tensor field on M for any t near 0. We want to compute $\frac{\partial}{\partial t} T_t\big|_{t=0}$. We can assume that X, Y and Z are vectorfields on M that are synchronous for g at a fixed point $p \in M$. That is, $\nabla^0_W X = \nabla^0_W Y = \nabla^0_W Z = 0$ for any $W \in T_p M$. This implies that the commutators of X, Y and Z vanish at p, too.

By the Koszul formula we get at p
$$2g_t(\nabla^t_X Y, Z) = 2g(\nabla^0_X Y, Z) + t\{\partial_X(h(Y,Z)) + \partial_Y(h(X,Z)) - \partial_Z(h(X,Y))\}$$
$$= 2g(\nabla^0_X Y, Z) + t\{(\nabla_X h)(Y, Z) + (\nabla_Y h)(X, Z) - (\nabla_Z h)(X, Y)\}.$$

On the other hand, the left hand side is equal to
$$2g(\nabla^t_X Y, Z) + 2th(\underbrace{\nabla^0_X Y}_{=0 \text{ at } p}, Z) + O(t^2).$$

Therefore
$$\frac{\partial}{\partial t}\bigg|_{t=0} T_t(X,Y,Z) = \frac{1}{2}\{(\nabla_X h)(Y,Z) + (\nabla_Y h)(X,Z) - (\nabla_Z h)(X,Y)\}$$
which proves part (a).

Proof (Proof of (b)). The Riemannian curvature tensor is defined as
$$R_g(X,Y)Z := \nabla_X \nabla_Y Z - \nabla_Y \nabla_X Z - \nabla_{[X,Y]} Z.$$

We will calculate its differential $(R'_g h)$ at the point $p \in M$. For the calculation of $(R'_g h)(X,Y,Z)$ we can assume that X, Y and Z are synchronous vector fields at p.
$$(R'_g h)(X,Y,Z) = (\nabla'_g h)(X, \nabla_Y Z) + \nabla_X((\nabla'_g h)(Y,Z))$$
$$- (\nabla'_g h)(Y, \nabla_X Z) - \nabla_Y((\nabla'_g h)(X,Z)) - (\nabla'_g h)([X,Y], Z)$$
$$= (\nabla_X (\nabla'_g h))(Y,Z) - (\nabla_Y (\nabla'_g h))(X,Z).$$

Before we go on proving the proposition, we will prove a lemma and a corollary.

Lemma 3. *Let h be a symmetric $(2,0)$-tensor, let $e_1, \ldots, e_n$ be a locally defined orthonormal frame, i.e. $g(e_i, e_j) = \varepsilon_i \delta_{ij}$, $\varepsilon_i = \pm 1$.*

Then for any $X \in TM$
$$\sum_{i=1}^n \varepsilon_i h(\nabla_X e_i, e_i) = 0.$$

Proof. We write $\nabla_X e_i = \sum_{j=1}^{n} \alpha_{ji} e_j$. Differentiation of the orthogonality relation yields

$$0 = \partial_X(g(e_i, e_j)) = g(\nabla_X e_i, e_j) + g(e_i, \nabla_X e_j)$$
$$= \sum_{k=1}^{n} (\alpha_{ki}\varepsilon_k \delta_{kj} + \alpha_{kj}\varepsilon_k \delta_{ki}) = \alpha_{ji}\varepsilon_j + \alpha_{ij}\varepsilon_i.$$

Using this we calculate

$$\sum_{i=1}^{n} \varepsilon_i h(\nabla_X e_i, e_i) = \sum_{i,k=1}^{n} \varepsilon_i h(\alpha_{ki} e_k, e_i)$$
$$= \frac{1}{2} \sum_{i,k=1}^{n} (\alpha_{ki}\varepsilon_i + \alpha_{ik}\varepsilon_k) h(e_k, e_i) = \frac{1}{2} \sum_{i,k=1}^{n} \varepsilon_i \varepsilon_k \underbrace{(\alpha_{ki}\varepsilon_k + \alpha_{ik}\varepsilon_i)}_{=0} h(e_k, e_i) = 0.$$

Corollary 3. *Let h be a symmetric $(2,0)$-tensor. Then $\mathring{R}_g h$, defined as above, is a symmetric $(2,0)$-tensor.*

Proof. We have to show that

$$\left(\mathring{R}_g h\right)(X, Y) = \left(\mathring{R}_g h\right)(Y, X).$$

By definition

$$\left(\mathring{R}_g h\right)(X, Y) = \sum_{i=1}^{n} \varepsilon_i h(R(e_i, X)Y, e_i)$$

for any orthonormal frame $e_1, \ldots, e_n$. Using the Bianchi identity this is equal to

$$-\sum_{i=1}^{n} \varepsilon_i h(R(X, Y)e_i, e_i) - \sum_{i=1}^{n} \varepsilon_i h(R(Y, e_i)X, e_i)).$$

The second term is just $(\mathring{R}_g h)(Y, X)$, so we have to show that the first term vanishes.

$$h(R(X,Y)e_i, e_i) = h(\nabla_X \nabla_Y e_i, e_i) - h(\nabla_Y \nabla_X e_i, e_i) - h(\nabla_{[X,Y]} e_i, e_i)$$
$$= \partial_X(h(\nabla_Y e_i, e_i)) - (\nabla_X h)(\nabla_Y e_i, e_i) - h(\nabla_Y e_i, \nabla_X e_i)$$
$$- \partial_Y(h(\nabla_X e_i, e_i)) + (\nabla_Y h)(\nabla_X e_i, e_i) + h(\nabla_X e_i, \nabla_Y e_i)$$
$$- h(\nabla_{[X,Y]} e_i, e_i).$$

If we apply Lemma 3 to the symmetric $(2,0)$-tensors h, $\nabla_X h$ and $\nabla_Y h$ we get

$$\sum_{i=1}^{n} \varepsilon_i h(R(X,Y)e_i, e_i) = 0.$$

We return to the proof of Proposition 9.

Proof (Proof of (c)). The Ricci curvature is defined as
$$\mathrm{ric}_g(X,Y) := \mathrm{tr} R_g(\cdot, X)Y.$$
Since here tr denotes the trace of a linear map it does not depend on the metric. Therefore tr commutes with differentiation in direction h. Using (b) we get
$$\mathrm{ric}'_g h(X,Y) = \sum_{i=1}^n \varepsilon_i \left\{ (\nabla_{e_i}(C_g h))(X,Y,e_i) - (\nabla_X(C_g h))(e_i,Y,e_i) \right\}, \quad (5.6)$$
with $C_g h(X,Y,Z) := g(\nabla'_g h(X,Y), Z)$.

The second term can easily be computed using (a). We will suppose that X and Y and the orthonormal frame $e_1, \ldots, e_n$ are synchronous at p. Then we get at p:
$$\sum_{i=1}^n \varepsilon_i (\nabla_X(C_g h))(e_i, Y, e_i)$$
$$= \frac{1}{2} \sum_{i=1}^n \varepsilon_i \partial_X \left\{ (\nabla_{e_i} h)(Y, e_i) + (\nabla_Y h)(e_i, e_i) - (\nabla_{e_i} h)(e_i, Y) \right\}$$
$$= \frac{1}{2} \partial_X \partial_Y (\mathrm{tr}_g h) = \frac{1}{2} (\nabla d(\mathrm{tr}_g h))(X,Y).$$

Now we turn to the first term of (5.6). Applying (a) shows that the first term is equal to
$$\frac{1}{2} \sum_{i=1}^n \varepsilon_i \left\{ (\nabla^2_{e_i, X} h)(Y, e_i) + (\nabla^2_{e_i, Y} h)(X, e_i) - (\nabla^2_{e_i, e_i} h)(X,Y) \right\}. \quad (5.7)$$

The last term hereof is one half the connection Laplacian
$$\nabla^* \nabla h = -\sum_{i=1}^n \varepsilon_i \nabla^2_{e_i, e_i} h.$$

The first term of (5.7) can be rewritten using the curvature tensor on the bundle of (2,0)-tensors:
$$R_{X,Y} - \nabla^2_{X,Y} - \nabla^2_{Y,X},$$
$$(\nabla^2_{e_i, X} h)(Y, e_i) = (\nabla^2_{X, e_i} h)(Y, e_i) + (R_{e_i, X} h)(Y, e_i). \quad (5.8)$$
The curvature of a (2,0)-tensor can be expressed in terms of the Riemannian curvature tensor:
$$(R_{A,B} h)(V,W) = -h(R(A,B)V, W) - h(V, R(A,B)W).$$

On the other hand note that

$$(h \circ \mathrm{ric}_g)(X,Y) = \sum_{i,j=1}^n \varepsilon_i \varepsilon_j h(X,e_i) g(R_g(e_j,e_i)Y,e_j)$$
$$= \sum_{j=1}^n \varepsilon_j h(X, R_g(Y,e_j)e_j)$$

and similarly

$$(\mathrm{ric}_g \circ h)(X,Y) = \sum_{j=1}^n \varepsilon_j h(R_g(X,e_j)e_j, Y).$$

Altogether we obtain

$$\mathrm{ric}'_g h(X,Y) = -\frac{1}{2} \left(\nabla d(\mathrm{tr}_g h)\right)(X,Y)$$
$$+ \frac{1}{2} \left[\nabla^* \nabla h + \mathrm{ric}_g \circ h + h \circ \mathrm{ric}_g - 2\mathring{R}_g h\right](X,Y)$$
$$+ \frac{1}{2} \sum_{i=1}^n \varepsilon_i \left[\left(\nabla^2_{X,e_i} h\right)(Y,e_i) + \left(\nabla^2_{Y,e_i} h\right)(X,e_i)\right]. \quad (5.9)$$

Now we calculate the divergence of h

$$\delta_g h = -\sum_{i=1}^n \varepsilon_i \left(\nabla_{e_i} h\right)(e_i, \cdot)$$

and applying the formal adjoint, δ_g^*, we get

$$(\delta_g^* \delta_g h)(X,Y) = -\frac{1}{2} \sum_{i=1}^n \varepsilon_i \left\{ \left(\nabla^2_{X,e_i} h\right)(e_i, Y) + \left(\nabla^2_{Y,e_i} h\right)(e_i, X)\right\}$$

which is up to a sign the last term of (5.9). As the second term of (5.9) is one half the Lichnerowicz Laplacian, we have

$$\mathrm{ric}'_g h = \frac{1}{2} \Delta_L h - \delta_g^* \delta_g h - \frac{1}{2} \nabla d(\mathrm{tr}_g h)$$

which proves (c).

In the following we will generalize our previous definition of the composition: If A and B are tensors, then $A \circ B$ means contraction of $A \otimes B$ in the last slot of A with the first slot of B.

For the semi-Riemannian metric g which is a $(2,0)$-tensor there is a unique $(0,2)$-tensor $L(g)$ such that $L(g) \circ g = \mathrm{id}|_{TM}$. If $e_1, \ldots, e_n$ are orthonormal with respect to g, i.e. $g(e_i, e_j) = \varepsilon_i \delta_{ij}$, then $L(g) = \sum_{i=1}^n \varepsilon_i e_i \otimes e_i$.

Now the metric trace $\operatorname{tr}_g(h)$ of a $(2,0)$-tensor h can be expressed as a metric-independent trace via

$$\operatorname{tr}_g(h) = \operatorname{tr}\bigl(L(g) \circ h\bigr)$$

and the metric on symmetric $(2,0)$-tensors h_1, h_2 fulfills

$$\langle h_1, h_2 \rangle_g = \operatorname{tr}\Bigl(L(g) \circ h_1 \circ L(g) \circ h_2\Bigr).$$

Lemma 4. *For symmetric $(2,0)$-tensors h and r we have*

$$\left.\frac{d}{dt}\right|_{t=0} \operatorname{tr}_{g+th}(r) = - \langle h, r \rangle_g.$$

Proof. Because of

$$0 = \left.\frac{d}{dt}\right|_{t=0} \Bigl(L(g+th) \circ (g+th)\Bigr) = L'_g h \circ g + L(g) \circ h$$

we get

$$L'_g h = -L(g) \circ h \circ L(g)$$

and therefore

$$\left.\frac{d}{dt}\right|_{t=0} \operatorname{tr}_{g+th}(r) = \left.\frac{d}{dt}\right|_{t=0} \operatorname{tr}(L(g+th) \circ r)$$
$$= \operatorname{tr}\left(L'_g h \circ r\right) = -\operatorname{tr}\Bigl(L(g) \circ h \circ L(g) \circ r\Bigr) = - \langle h, r \rangle_g.$$

Now we are ready to calculate the variation of the scalar curvature.

Proof (Proof of (d)).

$$\operatorname{scal}'_g h = \left.\frac{d}{dt}\right|_{t=0} \operatorname{tr}_{g+th}(\operatorname{ric}_{g+th})$$
$$= \left(\left.\frac{d}{dt}\right|_{t=0} \operatorname{tr}_{g+th}(\operatorname{ric}_g)\right) + \operatorname{tr}_g\left(\operatorname{ric}'_g h\right)$$
$$= - \langle h, \operatorname{ric}_g \rangle_g + \operatorname{tr}_g \left(\frac{1}{2}\Delta_L h - \delta^*_g \delta_g h - \frac{1}{2}\nabla d(\operatorname{tr}_g h)\right).$$

Note that $\Delta_g f = -\operatorname{tr}_g(\nabla df)$. Furthermore for any 1-form ω we have

$$\delta^*_g \omega(X, Y) = \frac{1}{2}\Bigl((\nabla \omega)(X, Y) - (\nabla \omega)(Y, X)\Bigr)$$

$$\Rightarrow \operatorname{tr}_g(\delta^*_g \omega) = \sum_{i=1}^{n} \varepsilon_i (\nabla \omega)(e_i, e_i) = -\delta_g \omega.$$

Now we want to compute $\operatorname{tr}_g \Delta_L h$. It is straightforward to show that
$$\operatorname{tr}_g(h \circ \operatorname{ric}_g) = \operatorname{tr}_g(\operatorname{ric}_g \circ h) = \operatorname{tr}_g(\mathring{R}_g h).$$
So we have $\operatorname{tr}_g \Delta_L h = \operatorname{tr}_g(\nabla^* \nabla h)$. On the other hand, since $L(g)$ is parallel we get
$$\Delta_g(\operatorname{tr}_g h) = \Delta_g(\operatorname{tr}(L(g) \circ h)) = -\sum_{i=1}^n \varepsilon_i \nabla^2_{e_i, e_i} \operatorname{tr}(L(g) \circ h)$$
$$= -\operatorname{tr}\left(L(g) \circ (\sum_{i=1}^n \varepsilon_i \nabla^2_{e_i, e_i} h)\right) = \operatorname{tr}_g(\nabla^* \nabla h).$$
Hence
$$\Delta_g(\operatorname{tr}_g h) = \operatorname{tr}_g(\Delta_L h).$$
Putting everything together we obtain
$$\operatorname{scal}'_g h = \Delta_g(\operatorname{tr}_g h) + \delta_g \delta_g h - \langle \operatorname{ric}_g, h \rangle_g$$
and therefore the proposition is proven.

As a next step we want to calculate the variation of the volume element.

Proposition 10. *Let (M, g) be a semi-Riemannian manifold. Then the differential of the volume element dV_g is given by*
$$dV'_g h = \frac{1}{2} (\operatorname{tr}_g h) dV_g.$$

Proof. We consider dV as a map from symmetric $(2,0)$-tensors to volume densities, locally given by
$$\sum_{i,j=1}^n g_{ij} dx^i \otimes dx^j \mapsto \sqrt{|\det(g_{ij})|}\, dx^1 dx^2 \cdots dx^n.$$
For $A \in \operatorname{End}(TM)$ we write $g\&A(X,Y) := g(AX, AY)$. Then
$$dV_{g\&A} = dV_g \cdot |\det A|.$$
If h is a symmetric $(2,0)$-tensor, then $H := L(g) \circ h \in \operatorname{End}(TM)$ satisfies
$$h(X, Y) = g(HX, Y) = g(X, HY)$$
and
$$h \circ h(X, Y) = g(HX, HY) = g\&H(X, Y),$$
$$g\&(\operatorname{id} + tH) = g + 2th + t^2 h \circ h,$$
$$dV_{g + 2th + t^2 h \circ h} = dV_g\, |\det(\operatorname{id} + tH)|.$$
We differentiate w.r.t. t at $t = 0$ and get
$$2 dV'_g h = dV_g \operatorname{tr}(H) = (\operatorname{tr}_g h)\, dV_g.$$

Now we calculate the Euler-Lagrange equations for the gravity action $\int(\mathrm{scal}_g + \lambda)\, dV_g$.

In order to have a finite integral, we suppose that the variation h of the metric has compact support contained in an open and relatively compact subset $U \subset M$.

The variation of $S_M := \int_M (\mathrm{scal}_g + \lambda)\, dV_g$ is given by

$$S'_U h = S'_M h = \int_M \mathrm{scal}'_g h\, dV_g + \int_M (\mathrm{scal}_g + \lambda)\, dV'_g h$$

$$= \int_M \left\{ \Delta_g(\mathrm{tr}_g h) + \delta_g(\delta_g h) - \langle \mathrm{ric}_g, h \rangle_g \right\} dV_g$$

$$+ \frac{1}{2} \int_M (\mathrm{scal}_g + \lambda)(\mathrm{tr}_g h)\, dV_g.$$

The first two summands of the first integral vanish since they are divergences. We rewrite $(\mathrm{tr}_g h)$ as $\langle g, h \rangle_g$.

$$S'_M h = -\int_M \left\langle \mathrm{ric}_g - \frac{1}{2}\mathrm{scal}_g \cdot g - \frac{1}{2}\lambda \cdot g,\, h \right\rangle_g dV_g.$$

We have shown

Proposition 11. *Stationarity of the functional S_M at g is equivalent to the Einstein equations*

$$\mathrm{ric}_g - \frac{1}{2}\mathrm{scal}_g \cdot g - \Lambda \cdot g = 0$$

of the vacuum with cosmological constant $\Lambda = \lambda/2$.

5.6 Einstein-Hilbert Action and Wodzicki Residue

In the fourth section we have seen how to characterize integration of functions over a closed Riemannian manifold using the Dirac operator and the Dixmier trace. This was based on the first coefficient Φ_0 in the heat asymptotics. In the previous section we have shown that the total scalar curvature functional gives rise to the field equations of General Relativity. But this is exactly the second term Φ_1 in the heat asymptotics. Therefore the question arises if we can extract the second heat coefficient using some kind of a trace. This is what we do in this section. Here we follow closely the work of Kalau and Walze [119].

Let $P : C^\infty(E) \to C^\infty(E)$ be a classical pseudo-differential operator of order m over the closed Riemannian manifold M. After choosing a system of local coordinates and a trivialization of the bundle E we can look at the total sysmbol σ^P of P and develop it into a formal series

$$\sigma^P(x, \xi) \sim \sum_{k=0}^{\infty} \sigma^P_{m-k}(x, \xi), \qquad (5.10)$$

where each σ_j^P is a matrix valued function homogeneous of degree j in ξ (for $\xi \geq \varepsilon > 0$) and satisfies an estimate

$$|\partial_x^\alpha \partial_\xi^\beta \sigma_j^P(x,\xi)| \leq c_{\alpha\beta}(1+\|\xi\|)^{j-|\beta|}$$

for all multiindices α and β. Conversely, given a formal series as in (5.10) there exists a classical pseudodifferential operator with this development. The pseudodifferential operator is unique up to smoothing operators.

The "Leibniz rule" gives us a multiplication in the space of formal developments of symbols which corresponds to the composition of operators [146, Ch. III]

$$\sigma^{P_1 \circ P_2}(x,\xi) \sim \sigma^{P_1} \circ \sigma^{P_2} = \sum_{|\alpha|=0}^{\infty} (-i)^{|\alpha|} \frac{1}{\alpha!} \partial_\xi^\alpha \sigma^{P_1} \partial_x^\alpha \sigma^{P_2}. \tag{5.11}$$

Except for the leading part σ_m, the *principal symbol*, the total symbol does depend on the choice of local coordinates and trivialization. However, for $p \in M$ the quantity

$$\int_{S_p^{n-1}} \mathrm{tr}\left(\sigma^P_{-n}(p,\xi)\right) d\xi$$

is invariantly defined and independent of the choices [223]. Here integration is over the unit sphere S_p^{n-1} in the cotangent bundle T_p^*M. One further integration over the manifold gives us the *Wodzicki residue*,

$$\mathrm{Res}(P) = \frac{\Gamma(\frac{n}{2})}{2\pi^{n/2}} \int_M \int_{S_p^{n-1}} \mathrm{tr}\left(\sigma^P_{-n}(p,\xi)\right) d\xi \, dV(p).$$

Lemma 5. *In Riemannian normal coordinates x^ρ based at the point p we have for the Christoffel symbols $\Gamma_{\mu\nu}^\kappa$,*

$$\sum_{\mu,\nu=1}^n \delta^{\mu\nu} \partial_{x^\rho} \Gamma_{\mu\nu}^\kappa = \frac{2}{3} \mathrm{ric}_\rho^\kappa$$

where $\delta^{\mu\nu}$ is the Kronecker symbol.

Proof. We use the Einstein summation convention in order to keep notation at a reasonable size. The Koszul formula for the Levi-Civita connection reads in coordinates

$$2\,\Gamma_{\mu\nu}^\kappa = g^{\kappa\lambda}(\partial_{x^\mu} g_{\nu\lambda} + \partial_{x^\nu} g_{\mu\lambda} - \partial_{x^\lambda} g_{\mu\nu})$$

which together with (5.2) implies

$$\Gamma_{\mu\nu}^\kappa = \frac{1}{6} g^{\kappa\lambda} \Big(R_{\nu\mu\lambda\delta} + R_{\nu\delta\lambda\mu} + R_{\mu\nu\lambda\delta}$$
$$+ R_{\mu\delta\lambda\nu} - R_{\mu\lambda\nu\delta} - R_{\mu\delta\nu\lambda} \Big) x^\delta + O(\|x\|^2).$$

So we get

$$\partial_{x^\rho} \Gamma^\kappa_{\mu\nu} = \frac{1}{6} g^{\kappa\lambda} \Big(R_{\nu\mu\lambda\rho} + R_{\nu\rho\lambda\mu} + R_{\mu\nu\lambda\rho}$$
$$+ R_{\mu\rho\lambda\nu} - R_{\mu\lambda\nu\rho} - R_{\mu\rho\nu\lambda} \Big) + O(\|x\|)$$

and therefore at the base point p (corresponding to $x^\mu = 0$)

$$\delta^{\mu\nu} \partial_{x^\rho} \Gamma^\kappa_{\mu\nu} = \frac{2}{3} g^{\kappa\lambda} \mathrm{ric}_{\rho\lambda} = \frac{2}{3} \mathrm{ric}^\kappa_\rho.$$

Now the main result is

Theorem 3 (Kalau-Walze[119], Kastler[125]). *Let M be a compact Riemannian manifold of dimension n, n even, $n \geq 4$. Let*

$$\Delta = \nabla^* \nabla + \mathcal{K}$$

be an invertible generalized Laplacian over M. Then for each $p \in M$

$$\frac{\Gamma(\frac{n}{2})}{2\pi^{n/2}} \int_{S_p^{n-1}} \mathrm{tr}\left(\sigma^{\Delta^{-(n/2)+1}}_{-n}(p, \xi) \right) d\xi = \frac{n-2}{2} \mathrm{tr}\left(\Phi_1(p, p) \right).$$

In particular,

$$\mathrm{Res}\left(\Delta^{-(n/2)+1} \right) = \frac{n-2}{2} \int_M \mathrm{tr}\left(\Phi_1(p, p) \right) dV(p).$$

Proof. (i) With respect to any system of local coordinates and to any local trivialization of the bundle E we write down the total symbol of Δ, $\sigma^\Delta(x, \xi) := \sigma_2 + \sigma_1 + \sigma_0$. In particular, σ_2 is proportional to $\mathrm{id}_{\mathrm{End}(E)} =: \mathbf{1}$. We introduce a new pseudodifferential operator P by inverting the principal symbol of Δ, $\sigma^P(x, \xi) = \sigma^P_{-2} := (\sigma_2)^{-1}$. By (5.11) we have

$$\sigma^{\Delta \circ P^{-1}} \sim \sum_{|\alpha|=0}^\infty (-i)^{|\alpha|} \frac{1}{\alpha!} \partial_\xi^\alpha \sigma^\Delta \partial_x^\alpha \sigma_2^{-1} - \mathbf{1}$$

$$\sim \sum_{k=1}^2 \sum_{|\alpha|=0}^k (-i)^{|\alpha|} \frac{1}{\alpha!} \partial_\xi^\alpha \sigma_{|\alpha|+2-k} \partial_x^\alpha \sigma_2^{-1} =: -r(x, \xi)$$

In other words $\sigma^\Delta \circ (\sigma^P \circ (1 - r)^{-1}) \sim 1$. Using the geometric series in symbol-space (this can be done because r is of order -1) we obtain

$$\sigma^{\Delta^{-1}}(x, \xi) \sim \sigma_2^{-1} \circ \sum_{k=0}^\infty r^{\circ k}.$$

We begin to compute

$$r_{-k}(x,\xi) = -\sum_{|\alpha|=0}^{k}(-i)^{|\alpha|}\frac{1}{\alpha!}\partial_\xi^\alpha \sigma_{|\alpha|+2-k}\,\partial_x^\alpha \sigma_2^{-1}\,,$$

$$r_{-1}(x,\xi) = -\sigma_2^{-1}\sigma_1 - i\,\sigma_2^{-2}\,\partial_{\xi_\mu}\sigma_2\,\partial_{x^\mu}\sigma_2\,,$$

$$r_{-2}(x,\xi) = -\sigma_2^{-1}\sigma_0 - \sigma_2^{-2}(i\,\partial_{\xi_\mu}\sigma_1\,\partial_{x^\mu}\sigma_2 + \tfrac{1}{2}\,\partial_{\xi_\mu}\partial_{\xi_\nu}\sigma_2\,\partial_{x^\mu}\partial_{x^\nu}\sigma_2) \quad (5.12)$$
$$+\sigma_2^{-3}\,\partial_{\xi_\mu}\partial_{\xi_\nu}\sigma_2\,\partial_{x^\mu}\sigma_2\,\partial_{x^\nu}\sigma_2\,,$$

$$r_{-k}(x,\xi) = 0 \qquad \forall\, k > 2\,.$$

Furthermore we write

$$\sum_{k=0}^{\infty} r^{\circ k} = \sum_{j=0}^{\infty} s_{-j} \quad \text{with} \quad s_0 = 1,\ s_{-1} = r_{-1},\ s_{-2} = r_{-1}^2 + r_{-2},\ \ldots$$

From this we can read off the symbol of Δ^{-1}:

$$\sigma^{\Delta^{-1}}(x,\xi) \sim \sum_{l=2}^{\infty} \sigma_{-l}^{\Delta^{-1}} \text{ with } \sigma_{-l}^{\Delta^{-1}}(x,\xi) = \sum_{|\alpha|=0}^{l-2}(-i)^{|\alpha|}\frac{1}{\alpha!}\partial_\xi^\alpha \sigma_2^{-1}\,\partial_x^\alpha s_{|\alpha|+2-l}\,.$$

We will only need the first three non-vanishing terms:

$$\sigma_{-2}^{\Delta^{-1}}(x,\xi) = \sigma_2^{-1}\,, \qquad \sigma_{-3}^{\Delta^{-1}}(x,\xi) = \sigma_2^{-1} r_{-1}\,, \qquad (5.13)$$
$$\sigma_{-4}^{\Delta^{-1}}(x,\xi) = \sigma_2^{-1}(r_{-1}^2 + r_{-2}) + i\,\sigma_2^{-2}\,\partial_{\xi_\mu}\sigma_2\,\partial_{x^\mu}r_{-1}\,.$$

More generally we get

$$\sigma^{\Delta^{-m}}(x,\xi) \sim \sigma^{\Delta^{-m+1}} \circ \sigma^{\Delta^{-1}} \sim \sum_{|\alpha|=0}^{\infty}(-i)^{|\alpha|}\frac{1}{\alpha!}\partial_\xi^\alpha \sigma^{\Delta^{-m+1}}\,\partial_x^\alpha \sigma^{\Delta^{-1}} = \sum_{l=2m}^{\infty}\sigma_{-l}^{\Delta^{-m}}\,,$$

with $\quad \sigma_{-l}^{\Delta^{-m}}(x,\xi) = \displaystyle\sum_{|\alpha|=0}^{l-2m}\sum_{k=2}^{2+l-|\alpha|-2m}(-i)^{|\alpha|}\frac{1}{\alpha!}\partial_\xi^\alpha \sigma_{|\alpha|+k-l}^{\Delta^{-m+1}}\,\partial_x^\alpha \sigma_{-k}^{\Delta^{-1}}\,.$

Using this and $\sigma_{-2m}^{\Delta^{-m}} = \sigma_2^{-m}$ we get the recursion relations

$$\sigma_{3-2k}^{\Delta^{-k+2}}(x,\xi) = \sigma_{5-2k}^{\Delta^{-k+3}}\,\sigma_2^{-1} + \sigma_2^{-k+3}\,\sigma_{-3}^{\Delta^{-1}} - i\,\partial_{\xi_\mu}\sigma_2^{-k+3}\,\partial_{x^\mu}\sigma_2^{-1} \qquad (5.14)$$

and

$$\sigma_{-2k}^{\Delta^{-k+1}}(x,\xi) = \sum_{|\alpha|=0}^{2}\sum_{j=2}^{4-|\alpha|}(-i)^{|\alpha|}\frac{1}{\alpha!}\partial_\xi^\alpha \sigma_{|\alpha|+j-2k}^{\Delta^{-k+2}}\partial_x^\alpha \sigma_{-j}^{\Delta^{-1}}$$

$$= \sigma_{2-2k}^{\Delta^{-k+2}}\sigma_2^{-1} + \sigma_{3-2k}^{\Delta^{-k+2}}\sigma_{-3}^{\Delta^{-1}} + \sigma_2^{-k+2}\sigma_{-4}^{\Delta^{-1}} - i\partial_{\xi_\mu}\sigma_{3-2k}^{\Delta^{-k+2}}\partial_{x^\mu}\sigma_2^{-1}$$

$$- i\partial_{\xi_\mu}\sigma_2^{-k+2}\partial_{x^\mu}\sigma_{-3}^{\Delta^{-1}} - \tfrac{1}{2}\partial_{\xi_\mu}\partial_{\xi_\nu}\sigma_2^{-k+2}\partial_{x^\mu}\partial_{x^\nu}\sigma_2^{-1}.$$
(5.15)

(ii) Since the formula for $Res(\Delta^{-(n/2)+1})$ in an arbitrary coordinate system contains a lot of terms it is more convenient to specialize our formulas to Riemannian normal coordinates x^μ about the base point p for which $x^\mu = 0$. We will also use the Einstein summation convention for all Greek indices. By (5.2) the $(0,2)$-tensor corresponding to the metric has the Taylor expansion

$$g^{\mu\nu} = \delta^{\mu\nu} - \tfrac{1}{3}R^\mu{}_\gamma{}^\nu{}_\delta(p)\,x^\gamma x^\delta + O(\|x\|^3)\,.$$

We also have to trivialize the bundle E. Then the connection is given by $\nabla_{x^\mu} = \partial_{x^\mu} + A_\mu$ with A_μ matrix-valued functions. We choose the trivialization such that it simplifies the calculations. Parallel translation of a basis of E_p along the radial geodesics emanating from p yields a trivialization such that $A_\mu(p) = 0$.

In these coordinates the generalized Laplacian takes the form

$$\Delta = \nabla^*\nabla + \mathcal{K}$$
$$= -g^{\mu\nu}\left\{\nabla_\mu\nabla_\nu - \nabla_{\nabla_\mu \partial_{x^\nu}}\right\} + \mathcal{K}$$
$$= -g^{\mu\nu}\left\{(\partial_{x^\mu} + A_\mu)(\partial_{x^\nu} + A_\nu) - \Gamma_{\mu\nu}^\beta(\partial_{x^\beta} + A_\beta)\right\} + \mathcal{K}$$
$$= g^{\mu\nu}\left\{-\partial_{x^\mu}\partial_{x^\nu} - 2A_\mu\partial_{x^\nu} + \Gamma_{\mu\nu}^\beta\partial_{x^\beta} - (\partial_{x^\mu}A_\nu) - A_\mu A_\nu + \Gamma_{\mu\nu}^\beta A_\beta\right\} + \mathcal{K}.$$

Therefore Δ has the symbols

$$\sigma_2 = g^{\mu\nu}\xi_\mu\xi_\nu$$
$$\sigma_1 = -2ig^{\mu\nu}A_\mu\xi_\nu + ig^{\mu\nu}\Gamma_{\mu\nu}^\beta\xi_\beta$$
$$\sigma_0 = -g^{\mu\nu}(\partial_{x^\mu}A_\nu) - g^{\mu\nu}A_\mu A_\nu + g^{\mu\nu}\Gamma_{\mu\nu}^\beta A_\beta + \mathcal{K}.$$

At p we get

$$\sigma_2(p,\xi) = \delta^{\mu\nu}\xi_\mu\xi_\nu$$
$$\sigma_1(p,\xi) = 0$$
$$\sigma_0(p,\xi) = -\delta^{\mu\nu}\partial_{x^\mu}A_\nu + \mathcal{K} \qquad (5.16)$$
$$\partial_{x^\mu}\sigma_1(p,\xi) = -2i(\partial_{x^\mu}A_\nu)\xi^\nu + i\underbrace{\delta^{\rho\nu}\partial_{x^\mu}\Gamma_{\rho\nu}^\beta}_{=(2/3)\mathrm{ric}_\mu^\beta}\xi_\beta$$

$$\partial_{x^\mu}\sigma_2(p,\xi) = 0$$
$$\partial_{x^\gamma}\partial_{x^\delta}\sigma_2(p,\xi) = -\frac{2}{3} R^\mu{}_\gamma{}^\nu{}_\delta \, \xi_\mu \xi_\nu$$
$$\partial_{\xi_\mu}\sigma_2(p,\xi) = 2\, \xi^\mu.$$

So we obtain

$$\partial_{\xi_\delta}\sigma_2(p,\xi)\, \partial_{x^\gamma}\partial_{x^\delta}\sigma_2(p,\xi) = -\frac{4}{3} R^\mu{}_\gamma{}^{\nu\delta} \xi_\mu \xi_\nu \xi_\delta = 0. \qquad (5.17)$$

With these quantities we can calculate

$$r_{-1}(p,\xi) = 0, \qquad r_{-2}(p,\xi) = -\sigma_2^{-1}\sigma_0 + \tfrac{2}{3}\sigma_2^{-2}\delta^{\rho\sigma} R^\mu{}_\rho{}^\nu{}_\sigma \xi_\mu \xi_\nu,$$

$$\partial_{x^\mu} r_{-1}(p,\xi) = -\sigma_2^{-1}\partial_{x^\mu}\sigma_1 - i\sigma_2^{-2} 2\xi^\nu \partial_{x^\mu}\partial_{x^\nu}\sigma_2$$
$$= -\sigma_2^{-1}\partial_{x^\mu}\sigma_1 + \tfrac{4}{3} i\sigma_2^{-2} R^\gamma{}_\mu{}^\delta{}_\nu \xi^\nu \xi_\gamma \xi_\delta$$
$$\stackrel{(5.17)}{=} -\sigma_2^{-1}\partial_{x^\mu}\sigma_1$$

$$\sigma_{-2}^{\Delta^{-1}}(p,\xi) = \sigma_2^{-1}, \qquad \sigma_{-3}^{\Delta^{-1}}(p,\xi) = 0,$$

$$\sigma_{-4}^{\Delta^{-1}}(p,\xi) = -\sigma_2^{-2}\sigma_0 + \sigma_2^{-3}\left(-2i\, \partial_{x^\mu}\sigma_1\, \xi^\mu + \tfrac{2}{3}\delta^{\rho\sigma} R^\mu{}_\rho{}^\nu{}_\sigma \xi_\mu \xi_\nu\right).$$
$$= -\sigma_2^{-2}\sigma_0 + \sigma_2^{-3}\left(-4(\partial_{x^\mu}A_\nu)\xi^\mu\xi^\nu + 2\cdot\tfrac{2}{3}\mathrm{ric}_{\mu\beta}\xi^\mu\xi^\beta - \tfrac{2}{3}\mathrm{ric}_{\mu\nu}\,\xi^\mu\xi^\nu\right)$$
$$= -\sigma_2^{-2}\sigma_0 + \sigma_2^{-3}\left(-4\partial_{x^\mu}A_\nu + \tfrac{2}{3}\mathrm{ric}_{\mu\nu}\right)\xi^\mu\xi^\nu.$$
$$(5.18)$$

We define
$$a_k := \sigma_{3-2k}^{\Delta^{-k+2}}(p,\xi).$$
It is easy to check that $a_2 = \sigma_{-1}^{\mathrm{id}}(p,\xi) = 0$. The recursion formula (5.14) gives

$$a_k = a_{k-1}\sigma_2^{-1},$$

and therefore
$$a_k = \sigma_2^{-k+2} a_2 = 0.$$

Now we set
$$b_k := \sigma_{-2k}^{\Delta^{-k+1}}(p,\xi)\, \sigma_2^k(p,\xi).$$

Obviously, we have $b_1 = 0$ and the recursion formula (5.15) yields for b_k

$$b_k = b_{k-1} + \sigma_2^2 \sigma_{-4}^{\Delta^{-1}} - i\sigma_2^k \partial_{\xi_\mu}\sigma_2^{-k+2} \partial_{x^\mu}\sigma_{-3}^{\Delta^{-1}}$$
$$-\tfrac{1}{2}\sigma_2^k \partial_{\xi_\mu}\partial_{\xi_\nu}\sigma_2^{-k+2} \partial_{x^\mu}\partial_{x^\nu}\sigma_2^{-1} \qquad (5.19)$$

The Einstein-Hilbert Action as a Spectral Action

The term $\sigma_{-3}^{\Delta^{-1}}$ can be expressed in quantities we know already

$$\sigma_{-3}^{\Delta^{-1}} \stackrel{(5.13)}{=} \sigma_2^{-1} r_{-1} \stackrel{(5.12)}{=} -\sigma_2^{-2}\sigma_1 - i\sigma_2^{-3}\partial_{\xi_\nu}\sigma_2 \partial_{x^\nu}\sigma_2,$$

so its x^μ-derivative at p is

$$\begin{aligned}
\left(\partial_{x^\mu}\sigma_{-3}^{\Delta^{-1}}\right)(p,\xi) &= -\sigma_2^{-2}\partial_{x^\mu}\sigma_1 - i\sigma_2^{-3}\partial_{\xi_\nu}\sigma_2 \partial_{x^\mu}\partial_{x^\nu}\sigma_2 \\
&\stackrel{(5.17)}{=} -\sigma_2^{-2}\partial_{x^\mu}\sigma_1 \\
&= 2i\sigma_2^{-2}\left(\partial_{x^\mu}A_\nu\right)\xi^\nu - \frac{2}{3}i\sigma_2^{-2}\mathrm{ric}_{\mu\beta}^\beta \xi_\beta.
\end{aligned}$$

Now we are ready to calculate the summands of the recursion formula (5.19). We already know the second summand. The third one yields

$$\begin{aligned}
-i\,\sigma_2^k\,\partial_{\xi_\mu}\sigma_2^{-k+2}\,\partial_{x^\mu}\sigma_{-3}^{\Delta^{-1}} &= -2i\,\sigma_2^k\,(-k+2)\,\sigma_2^{-k+1}\,\xi^\mu\left(\partial_{x^\mu}\sigma_{-3}^{\Delta^{-1}}\right) \\
&= 4(-k+2)\,\sigma_2^{-1}\,\xi^\mu\left(\partial_{x^\mu}A_\nu\right)\xi^\nu - 2\cdot\frac{2}{3}(-k+2)\,\sigma_2^{-1}\,\xi^\mu\,\mathrm{ric}_{\mu\beta}\,\xi^\beta.
\end{aligned}$$

It is straightforward to transform the last summand of (5.19).

$$-\frac{1}{2}\sigma_2^k\,\partial_{\xi_\mu}\partial_{\xi_\nu}\sigma_2^{-k+2}\,\partial_{x^\mu}\partial_{x^\nu}\sigma_2^{-1} = \frac{2}{3}(-k+2)\sigma_2^{-1}\mathrm{ric}_{\mu\nu}\,\xi^\mu\xi^\nu.$$

The above formulas yield

$$\begin{aligned}
b_k &= b_{k-1} + \sigma_2^2\,\sigma_{-4}^{\Delta^{-1}} + \sigma_2^{-1}(-k+2)\,\xi^\mu\xi^\nu\left\{4\partial_{x^\mu}A_\nu - \frac{4}{3}\mathrm{ric}_{\mu\nu}\right\} \\
&\quad + \frac{2}{3}(-k+2)\sigma_2^{-1}\mathrm{ric}_{\mu\nu}\,\xi^\mu\xi^\nu \\
&= b_{k-1} + \sigma_2^2\,\sigma_{-4}^{\Delta^{-1}} + \sigma_2^{-1}(-k+2)\,\xi^\mu\xi^\nu\left\{4\partial_{x^\mu}A_\nu - \frac{2}{3}\mathrm{ric}_{\mu\nu}\right\}.
\end{aligned}$$

Using $b_1 = 0$ and an induction over k we get

$$\begin{aligned}
b_k &= (k-1)\sigma_2^2\,\sigma_{-4}^{\Delta^{-1}} - \sigma_2^{-1}(k-1)(k-2)\left\{2\partial_{x^\mu}A_\nu - \frac{1}{3}\mathrm{ric}_{\mu\nu}\right\} \\
&\stackrel{(5.18)}{=} (k-1)\left\{-\sigma_o + \sigma_2^{-1}\left(\left(-4\left(\partial_{x^\mu}A_\nu\right) + \frac{2}{3}\mathrm{ric}_{\mu\nu}\right)\xi^\mu\xi^\nu\right.\right. \\
&\quad \left.\left. + \sigma_2^{-1}\left((-2k+4)\left(\partial_{x^\mu}A_\nu\right) + \frac{k-2}{3}\mathrm{ric}_{\mu\nu}\right)\xi^\mu\xi^\nu\right\} \\
&= (k-1)\left\{-\sigma_o + \sigma_2^{-1}\left(-2k\left(\partial_{x^\mu}A_\nu\right) + \frac{k}{3}\mathrm{ric}_{\mu\nu}\right)\xi^\mu\xi^\nu\right\}.
\end{aligned}$$

Now we want to integrate $\mathrm{End}(E)$-valued $(0,2)$-tensors over the unit sphere in T^*M. For this we have the formula

$$\int_{S^{n-1}} d\xi\, T^{\mu\nu} \xi_\mu \xi_\nu = \frac{2\pi^{n/2}}{n\,\Gamma(n/2)} g_{\mu\nu} T^{\mu\nu}.$$

Note that $\frac{2\pi^{n/2}}{\Gamma(n/2)}$ is the volume of $S^{n-1} \subset \mathbb{R}^n$. We get for every fixed p

$$\int_{S^{n-1}} d\xi\, \sigma_{-n}^{\Delta^{-(n/2)+1}} = \int_{S^{n-1}} d\xi\, b_{n/2}(\xi)$$
$$= \frac{2\pi^{n/2}\left(\frac{n}{2}-1\right)}{n\,\Gamma(n/2)} \left\{ -n\,\sigma_0 - n\,(\delta^{\mu\nu} \partial_{x^\mu} A_\nu) + \frac{n}{6} \mathrm{scal} \right\}.$$

From (5.16) we know that $\mathcal{K} = \sigma_0 + (\delta^{\mu\nu} \partial_\mu A_\nu)$, hence

$$\int_{S^{n-1}} d\xi\, b_{n/2}(\xi) = \frac{2\pi^{n/2}\left(\frac{n}{2}-1\right)}{\Gamma(n/2)} \left\{ -\mathcal{K} + \frac{1}{6} \mathrm{scal} \right\}.$$

Remark. It was noted by Ackermann [1] that Theorem 1.5.22 is a special case of a more general relationship between the Wodzicki residue of certain powers of an elliptic operator and the asymptotic expansion of the trace of the corresponding heat operator.

6 Spectral Action and the Connes-Chamsedinne Model

Ryszard Nest, Elmar Vogt, and Wend Werner

University of Copenhagen, Freier Iniversität Berlin, University of Münster

We have seen so far that the Dirac operator not only comprises the (geodesic) metric of a spin manifold: It also reproduces the Einstein Hilbert action in form of a coefficient of the asymptotic expansion of the heat kernel for the operator D^{-n+2}.

The universality of this operator has led Alain Connes to the conjecture that a great deal of physical information is encoded in properly chosen Dirac operators. If one tries to couple gauge fields to gravity within such a framework, a first idea might be to substitute the pure Dirac operator D in the action

$$\text{Res}(D^{-n+2})$$

by a twisted Dirac operator D_1 defined on a bundle $\mathcal{S} \otimes \mathcal{F}$, where $\mathcal{S}$ is the spin bundle and $\mathcal{F}$ a bundle carrying the coupled gauge field. Unfortunately, the resulting gauge potential is traceless and does not show up in such an action [119], [125].

A more successful attack uses real spectral triples, and is the objective in the following: We will briefly explain the *Spectral Action Principle*, and then quite closely follow the lines of the paper by Connes and Chamseddine [31]: We start with an example illustrating the basic technique in some detail and finally sketch how to obtain a coupling of the standard model to gravity. A more detailed exposition of the standard model coupled to gravity can be found in Chapter 2.

6.1 The Spectral Action Principle

In the following, M denotes a compact spin manifold of dimension 4, equipped with a (positive definite) Riemannian metric g and a fixed spin structure. In all applications of noncommutative geometry to physics, the (underlying classical) geometry is Euclidean, i.e. we change the signature of the Riemannian metric from $---+$ to $----$ and $++++$ (Wick-rotation), hoping that this won't ruin the underlying physics. We furthermore will denote by Int $\mathfrak{A}$, Out $\mathfrak{A}$ and Aut $\mathfrak{A}$ the group of inner, outer and, respectively, all automorphisms of an algebra $\mathfrak{A}$.

The invariance group underlying general relativity is $\text{Diff}(M)$, the group of diffeomorphisms of M. For an additional gauge field F with (global) gauge group U minimal coupling (which restricts the present consideration to low energies) leads to an action functional

$$\mathfrak{J} = \mathfrak{J}_E + \mathfrak{J}_F.$$

The natural invariance group for such a functional is the semidirect product

$$G = U \rtimes \mathrm{Diff}(M).$$

of the (local) gauge group U with $\mathrm{Diff}(M)$. A basic idea behind the following is the similarity of the exact sequence

$$1 \longrightarrow U \longrightarrow G \longrightarrow \mathrm{Diff}(M) \longrightarrow 1$$

with

$$1 \longrightarrow \mathrm{Int}\,\mathfrak{A} \longrightarrow \mathrm{Aut}\,\mathfrak{A} \longrightarrow \mathrm{Out}\,\mathfrak{A} \longrightarrow 1$$

for certain algebras $\mathfrak{A}$. For example, if $\mathfrak{A} = C^\infty(M) \otimes M_n(\mathbb{C})$, then

$$\mathrm{Int}\,\mathfrak{A} = \{f \mapsto u^* f u \mid u \in C^\infty(M, U(n))\},$$

$$\mathrm{Out}\,\mathfrak{A} = \mathrm{Diff}(M),$$

$$\mathrm{Aut}\,\mathfrak{A} = \{f \circ \varphi \mid f \in \mathrm{Int}\,\mathfrak{A} \text{ and } \varphi \in \mathrm{Out}\,\mathfrak{A}\}$$

$$= \mathrm{Int}\,\mathfrak{A} \rtimes \mathrm{Out}\,\mathfrak{A},$$

which is the expected invariance group for gravitation coupled to one $U(n)$-gauge field.

A little care should be taken in the presence of fermions. In such a case we wouldn't like diffeomorphisms to change the spin structure and so it might be appropriate to expect $G = U \rtimes \mathrm{Diff}^0(M)$ as a group of invariance and make changes in the above accordingly.

Motivated by this observation, it is very convincing to regard the group of inner automorphisms of the algebra $\mathfrak{A}$ in a real spectral 'triple' $(\mathcal{H}, \mathfrak{A}, \pi, D_0, J)$ as an underlying gauge group. Physically relevant information must be invariant under the action of this group, and an important object which is invariant under $\mathrm{Int}\,\mathfrak{A}$ (acting on $\mathcal{H}$ via $h \to uJuJ^{-1}h$) is the spectrum of the Dirac operator. The spectral action principle as proposed by Connes has as its basic axiom that the physical information is all contained in the spectrum of the Dirac operator D_0 of a given real spectral triple. The exact recipe that eventually leads to an action functional is the following:

First step. Choose an appropriate real spectral triple $(\mathcal{H}, \mathfrak{A}, \pi, D_0, J)$

Second step. Identifying elements of $\mathfrak{A}$ with operators on $\mathcal{H}$ via π, we denote by $\mathcal{A}$ the operator $a_i[D_0, b_i]$. Suppose then that $\mathcal{A}$ is self-adjoint and calculate the 'gauged' Dirac operator

$$D = D_0 + \mathcal{A} + J\mathcal{A}J^{-1}$$

Third step. Compute the spectral action

$$\mathfrak{I}_{D,\mathcal{A},\psi} = \mathrm{tr}\,\widehat{\chi}\left(\frac{D}{m_0}\right) + \langle \psi, D\psi \rangle,$$

where ψ is an element of $\mathcal{H}$, $\widehat{\chi}$ ideally is a cut-off at 1, like $\chi_{[0,1]}$, the characteristic function of the unit interval, and m_0 is a cut-off parameter.

Let us remark at this point that, for technical reasons (see [78, Section 5]), $\chi_{[0,1]}$ actually is not a good choice for the cut-off function $\widehat{\chi}$. Rather, $\widehat{\chi}$ should be sufficiently fast decreasing and non-negative, equal to 1 at the origin and have higher derivatives at 0 equal to zero. Among other things, $\widehat{\chi}(\frac{D}{m_0})$ then is of trace class. We also point out that a cut-off is necessary here in order to avoid terms which are known to be non-renormalizable — a topic which we won't touch here.

Note also that in order to obtain the time dependence of the system in a forth step, the Dirac-operator itself has become a parameter in the Euler-Lagrange equation. More on how such derivatives can be carried out can be found in [20].

6.2 Example: Gravity Coupled to One Gauge Field

Before we examine a first example, we fix some notational conventions since they vary substantially in the literature. To begin with, tensor components are expressed either in local coordinates with Greek indices, i.e. $\partial_\mu = \frac{\partial}{\partial x^\mu}$, $\mu = 1, \ldots, 4$, or dx^μ, $\mu = 1, \ldots, 4$, or in terms of local orthonormal frames of TM or T^*M with Latin indices and Greek letters, usually γ. Thus γ_a, $a = 1, \ldots, 4$, is a local orthonormal frame of TM and γ^a, $a = 1, \ldots, 4$ is a local orthonormal frame of T^*M.

Lowering and raising of indices is done with the help of the metric tensor $g_{\mu\nu} = g(\partial_\mu, \partial_\nu)$ or $g^{\mu\nu} = g(dx^\mu, dx^\nu)$ via

$$T^{\mu_2\ldots\mu_s}_{\nu_1\ldots\nu_r\lambda} = g_{\lambda\mu}T^{\mu\mu_2\ldots\mu_s}_{\nu_1\ldots\nu_r} \quad \text{and}$$

$$T^{\lambda\mu_1\ldots\mu_s}_{\nu_1\ldots\nu_{r-1}} = g^{\lambda\nu}T^{\mu_1\ldots\mu_s}_{\nu_1\ldots\nu_{r-1}\nu}$$

For Latin indices $g_{ab} = g^{ab} = \delta_{ab}$, so that

$$T^{a_1\ldots a_s}_{b_1\ldots b_r} = T_{b_1\ldots b_r a_1\ldots a_s}.$$

To avoid any ambiguity, for us the Clifford algebra associated to a bilinear form b on the vector space V is by definition the algebra with unit generated by V satisfying

$$x \cdot y + y \cdot x = -2b(x,y) \cdot 1 \quad \text{for} \quad x, y \in V,$$

and for an element $\xi \in T_x^*M$ we denote by $c(\xi)$ the Clifford multiplication by the element $v \in T_xM$ such the $\xi(w) = g(v,w)$ for $w \in T_xM$.

Let us see how the spectral action principle applies to the simplest feasible example, one N-dimensional gauge field (minimally) coupled to gravity: Fix a compact (oriented) spin manifold M, denote the spinor bundle on M by S, write $L^2(M,S)$ for the space of measurable sections of S which are square integrable with respect to the volume form, and denote by ∇^S the spin connection defined on smooth sections of S. If we let D^S be the Dirac operator, then, in a local frame γ^a we have $D^S\psi = c(\gamma^a)\nabla^S_{\gamma^a}\psi$ for any $\psi \in C^\infty(S)$. We also write ψ^c for the (pointwise) charge conjugation, as defined previously. Then a real spectral triple is given by

$$\mathfrak{A} := C^\infty(M) \otimes M_N(\mathbb{C})$$
$$H := L^2(M,S) \otimes M_N(\mathbb{C})$$
$$\text{with } \langle \psi_1 \otimes B_1, \psi_2 \otimes B_2 \rangle_H = \langle \psi_1, \psi_2 \rangle_{L^2(M,S)} \operatorname{tr}(B_1 B_2^*)$$
$$\pi(f \otimes B)(\psi \otimes C) := f\psi \otimes BC$$
$$J(\psi \otimes B) := \psi^c \otimes B^*$$
$$D_0 := D^S \otimes \operatorname{Id}.$$

We first put

$$D = D^S \otimes \operatorname{Id} + a_i[D^S \otimes \operatorname{Id}, b_i] + J\left(a_i[D^S \otimes \operatorname{Id}, b_i]\right) J^{-1}$$

into a more convenient form. Omitting explicit reference to π throughout, we have in local coordinates x^μ and for elements $a_m = (a_{ij}^m)$, $b_m = (b_{ij}^m) \in C^\infty(M) \otimes M_N(\mathbb{C})$ and $\psi = (\psi_{ij}^\mu) \in C^\infty(S) \otimes M_n(\mathbb{C})$ that the operator $\mathcal{A} = a_m[D_0, b_m]$ satisfies

$$\mathcal{A}\psi = c(dx^\mu) a_{ij}^m \frac{db_{jk}^m}{dx^\mu} \psi_{kl} =: c(dx^\mu) G_\mu \psi$$

Since $\mathcal{A}^* = -[D_0, b_m^*]a_m^* = -c(dx^\mu)G_\mu^*$ and $\mathcal{A}$ was supposed to be self-adjoint, we have $G_\mu = -G_\mu^*$, and G turns out to be an element of $C^\infty(M) \otimes \mathfrak{u}(N)$. Invoking the definition of J yields

$$\left(\mathcal{A} + J\mathcal{A}J^{-1}\right)\psi = c(dx^\mu)\left(G_\mu \psi + \psi G_\mu^*\right) = c(dx^\mu)[G_\mu \psi - \psi G_\mu].$$

In this action, however, the trace part of G is irrelevant. So $\mathcal{A} + J\mathcal{A}J^{-1}$, which is of the form $dx^\mu \otimes A_\mu$, might be understood as the action of an $\mathfrak{su}(N)$-valued 1-form (i.e. the potential of an $SU(N)$-gauge field) via the adjoint representation. It also follows that D is a twisted Dirac operator on $S \otimes M_N(\mathbb{C})$, the tensor product of S with the trivial bundle with fiber $M_N(\mathbb{C})$ and (spin-) connection

$$\nabla = \nabla^S \otimes 1 + 1 \otimes A$$

on $S \otimes M_N(\mathbb{C})$. This means that D is the composition of

$$C^\infty(S \otimes M_N(\mathbb{C})) \xrightarrow{\nabla^S \otimes 1 + 1 \otimes A} C^\infty(T^*M \otimes S \otimes M_N(\mathbb{C})) \xrightarrow{c \otimes 1} C^\infty(S \otimes M_N(\mathbb{C}))$$

In order to better understand the action of $\mathfrak{J}_{D_0,\mathcal{A},\psi}$ we will restrict attention to the bosonic action, $\mathfrak{J}^B_{D_0,\mathcal{A}} = \operatorname{tr}\widehat{\chi}\left(\frac{D^2}{m_0}\right)$, since the summand $\langle\psi, D\psi\rangle$ essentially is the usual fermionic action (see e.g. [16, 12.4]). For the bosonic part we will slightly change $\widehat{\chi}$ and actually write

$$\mathfrak{J}^B_{D_0,\mathcal{A}} = \operatorname{tr}\chi\left(\frac{D^2}{m_0^2}\right).$$

This is formally closer to the Lagrangian of gauge field theory and will permit to make use of techniques available for the square of a twisted Dirac operator.

6.3 Asymptotic Expansion

We first collect some features of the Laplace transform which will be used later on. Details are in [198] and [221, II.2, II.5].

Proposition 1. (i) *Suppose $F: H_\sigma = \{z \in \mathbb{C} \mid \operatorname{Re} z > \sigma\} \to \mathbb{C}$ is analytic, and $F(z) = O(|z|^\alpha)$, as $z \to \infty$, with $\alpha > 1$. Then there is a function $\widehat{F}: \mathbb{R}_0^+ \to \mathbb{R}$ so that for $\operatorname{Re} z > \sigma$, $F(z)$ is the Laplace transform of $\widehat{F}$,*

$$F(z) = \int_0^\infty e^{-sz}\widehat{F}(s)\,ds.$$

(ii) *Suppose $F(z)$ is, for $\operatorname{Re} z > \sigma_0$, the Laplace transform of the function $\widehat{F}$, then:*
  (a) *F is analytic in the open half plane H_σ, and*

$$F^{(n)}(z) = (-1)^n \int_0^\infty s^n e^{-sz}\widehat{F}(s)\,ds.$$

  (b) *For any $\sigma > \sigma_0$*

$$\int_0^s \widehat{F}(\tau)\,d\tau = o(e^\sigma s).$$

Recall the following definition: Suppose (f_n) is a sequence of functions and t_0 a point in $\widehat{\mathbb{R}}$ so that $f_n(t) \neq 0$ for $t \neq t_0$ and $f_{n+1}(t) = o(f_n(t))$ as $t \to t_0$. A function f is said to have an asymptotic expansion $f \sim \sum a_n f_n$ as $t \to t_0$, iff for each $n \in \mathbb{N}$

$$f(t) = \sum_{\nu=0}^n a_\nu f_\nu + O(f_{n+1}(t)), \quad \text{as } t \to t_0$$

In this way, every smooth function has, for example, the asymptotic expansion

$$f \sim \sum_n f^{(n)}(t_0)\frac{(t-t_0)^n}{n!}, \quad \text{as } t \to t_0.$$

Our goal will be to find an asymptotic expansion in m_0 for $\mathfrak{I}^B_{D_0,\mathcal{A}}$. To this end, we will use the coefficients of the asymptotic expansion of the heat kernel of D^2,

$$\operatorname{tr} e^{-tD^2} \sim \sum_{n\geq 0} t^{\frac{n-4}{2}} a_n(D^2), \qquad \text{as } t \to 0.$$

They vanish for odd values of n. For every $\sigma \in \mathbb{R}$, denote by H_σ the open half-plane $\{z \in \mathbb{C} \mid \operatorname{Re} z > \sigma\}$.

Lemma (Werner) 1. *Let k be a natural number with $k > 3$. Then there exists a constant $C_k > 0$ so that for all $K \in \mathbb{N}_0$ and $t \geq 0$*

$$\Psi_K(t) \leq C_k t^4 \lambda_K^{\frac{k}{2}} \exp\left(-\lambda_K^{1/2} t^2\right)$$

as well as

$$-C_k t^5 \lambda_K^{\frac{k}{2}} \exp\left(-\lambda_K^{1/2} t^2\right) \leq \Psi'_K(t)$$

Proof. By Weyl's Theorem, we have $\lambda_k = O(k^{1/2})$ as $k \to \infty$, and it follows that for some constant C, independent of K,

$$\sum_{k=K}^\infty \exp\left(-\lambda_k t^2\right) \leq C \int_{\lambda_K^{1/2}}^\infty \sigma \exp\left(-\sigma t^2\right) d\sigma.$$

But, if $\lambda_K \geq 1$,

$$\int_{\lambda_K^{1/2}}^\infty \sigma \exp\left(-\sigma t^2\right) d\sigma \leq \overline{C}_k \frac{\lambda_K^{\frac{k}{2}}}{t^2} \exp\left(-\lambda_K^{1/2} t^2\right),$$

which remains true if we (possibly have to) extend the integration to $[\lambda_K^{1/2}, 1]$. This proves the claim for $t \geq 1$. Since the functions Ψ_K are uniformly bounded on the interval $[0, 1]$ the assertion follows for these. The estimate for Ψ'_K is proven similarly.

Theorem 1. *If $\chi \in C^\infty[0,\infty)$ is the Laplace transform of a function $\widehat{\chi}$ in the function space $\mathcal{S}(0,\infty) = \{\varphi \in \mathcal{S} \mid \varphi(x) = 0, \ x \leq 0\}$ then $\chi(tD^2)$ is an operator of trace class, and*

$$\operatorname{tr} \chi(tD^2) \sim \sum_{n\geq 0} \chi_n t^{n-2} a_{2n}(D^2), \qquad \text{as } t \to 0,$$

where the coefficients χ_n are given by

$$\chi_0 = \int_0^\infty u\chi(u)\,du, \quad \chi_1 = \int_0^\infty \chi(u)\,du, \quad \chi_n = (-1)^n \chi^{(n)}(0) \quad \text{for } n \geq 2.$$

Proof. Suppose that $0 < \lambda_0 \leq \lambda_1 \leq \ldots$ are the eigenvalues of D^2 and define, for $K = 0, 1, 2, \ldots$, functions $\Psi_K : [0, \infty) \to \mathbb{R}$ by

$$\Psi_K(t) := t^4 \sum_{k=K}^{\infty} \exp\left(-\lambda_k t^2\right).$$

We also let $\Psi = \Psi_0$. Since $\exp\left(-t^2 P\right)$ is infinitely smoothing the function

$$4t^3 \operatorname{tr} \exp\left(-t^2 P\right) - 2t^5 \operatorname{tr}\left(P \exp\left(-t^2 P\right)\right)$$

is well defined and is easily seen to coincide for $t > 0$ with the derivative of Ψ. Note that also Ψ_K is differentiable for positive values of t. For $t \to 0_+$ it follows from the central result of [80] that Ψ' has an asymptotic expansion

$$\Psi'(t) \sim \sum_{k=0}^{\infty} t^k (k+1) a_{k+1}(P).$$

and hence, Ψ is differentiable at 0. If

$$\Psi(t) = \sum_{k=0}^{N} t^k a_k(P) + R_{\Psi,N}(t) \quad \text{and} \quad \Psi'(t) = \sum_{k=0}^{N} t^k (k+1) a_{k+1}(P) + R_{\Psi',N}(t)$$

this also means that for $t > 0$

$$R'_{\Psi,N}(t) = R_{\Psi',N-1}(t). \tag{6.1}$$

Any twice differentiable function χ_2 with $\chi_2'' = \chi$ is by Lemma 1 and Proposition 1(i) the Laplace transform of a function $\widehat{\chi}_2$ such that $\int_0^s \widehat{\chi}_2(\sigma) \sigma^n \, d\sigma$ is for all $N \in \mathbb{N}_0$ a bounded function of s. We now observe that for $t > 0$

$$t^4 \operatorname{tr} \chi(t^2 P) = \sum_k \int_0^{\infty} s^{1/2} t^4 \exp\left(-st^2 \lambda_k\right) \widehat{\chi}_2(s) \, ds = \int_0^{\infty} \Psi(s^{1/2} t) \widehat{\chi}_2(s) \, ds. \tag{6.2}$$

In fact,

$$\left| \int_0^S \Psi_K(s^{1/2} t) \widehat{\chi}_2(s) \, ds \right| \leq$$

$$\Psi_K(S^{1/2} t) \left| \int_0^S \widehat{\chi}_2(s) \, ds \right| + \int_0^S \left| t s^{-1/2} \Psi_K'(s^{1/2} t) \right| \left| \int_0^s \widehat{\chi}_2(\sigma) \, d\sigma \right| ds$$

so that by Lemma 1, the assumptions made on χ and Proposition 1(ii) there are positive constants B_2 and α, independent of K, with

$$\left| \int_0^{\infty} \Psi_K(s^{1/2} t) \widehat{\chi}_2(s) \, ds \right| \leq B_2 t^6 \int_0^{\infty} s^{-1/2} \lambda_K^{\alpha} \exp\left(-st^2 \lambda_K^2\right) ds.$$

Since $\lambda_K \to \infty$, this shows (2). Proceeding similarly, we will see in a moment that, whenever $t > 0$,

$$\int_0^\infty \Psi(s^{1/2}t)\widehat{\chi}_2(s)\,ds = \lim_{\varepsilon \to 0_+} \int_0^\infty e^{-\varepsilon s}\Psi(s^{1/2}t)\widehat{\chi}_2(s)\,ds. \qquad (6.3)$$

Namely, in quite the same way as above we have for $A \geq 1$

$$\left|\int_A^\infty \left(1 - e^{-\varepsilon s}\right)\Psi(s^{1/2}t)\widehat{\chi}_2(s)\,ds\right| \leq$$

$$\int_A^\infty \left|(1 - e^{-\varepsilon s})ts^{-1/2}\Psi'(s^{1/2}t) + \varepsilon e^{-\varepsilon s}\Psi(s^{1/2}t)\right| \left|\int_A^s \widehat{\chi}_2(\sigma)\,d\sigma\right| ds \leq$$

$$B_3(1+\varepsilon)\max\{t^6, t^4\} \int_A^\infty \lambda_K^\alpha \exp\left(-st^2\lambda_0^2\right) ds.$$

Choosing A large enough to make the first expression small and then ε such that $\left|\int_0^A \left(1 - e^{-\varepsilon s}\right)\Psi(s^{1/2}t)\widehat{\chi}_2(s)\,ds\right|$ becomes small yields (3). We next fix $N \in \mathbb{N}$ and note that for $R_{\Psi,N}(t)$ with $\Psi(t) = \sum_{n=1}^N t^n a_n(P) + R_{\Psi,N}(t)$ we have by (1)

$$|R_{\Psi,N}(t)| \leq K_N \min\{t^N, t^{N+1}\} \quad \text{and} \quad |R_{\Psi',N}(t)| \leq K_N \min\{t^{N-1}, t^N\}$$

for all $t \in \mathbb{R}_0^+$. Write $e^{-\varepsilon s} R_{\Psi,N}(s^{1/2}t)\widehat{\chi}_2(s) = R_{\Psi,N}(s^{1/2}t)s^{-N-1}e^{-\varepsilon s}s^{N+1}\widehat{\chi}_2(s)$ and note that it follows from Proposition 1 that for fixed N and small $\varepsilon > 0$ the integrals $\int_0^s e^{-\varepsilon \sigma}\sigma^N \widehat{\chi}_2(\sigma)\,d\sigma$ are uniformly bounded in s. Then, if $T \geq 1$,

$$\left|\int_T^\infty e^{-\varepsilon s} R_{\Psi,N}(s^{1/2}t)\widehat{\chi}_2(s)\,ds\right|$$

$$\leq \int_T^\infty \left|\frac{s^{1/2}t}{2}\frac{R_{\Psi',N-1}(s^{1/2}t)}{s^{N+2}} - (N+1)\frac{R_{\Psi,N}(s^{1/2}t)}{s^{N+2}}\right|$$

$$\cdot \left|\int_T^s \exp\left(-\varepsilon\sigma\right)\sigma^{N+1}\widehat{\chi}_2(\sigma)\,d\sigma\right| ds$$

$$\leq \widetilde{K}_{N+1}t^{N+1}\int_T^\infty s^{N/2-N-2}\,ds,$$

and, similarly,

$$\left|\int_0^T e^{-\varepsilon s} R_{\Psi,N}(s^{1/2}t)\widehat{\chi}_2(s)\,ds\right|$$

$$\leq \left|R_{\Psi,N}(T^{1/2}t)\int_0^T e^{-\varepsilon s}\widehat{\chi}_2(s)\,ds\right|$$

$$+ \int_0^T \left|\frac{s^{1/2}t}{2}R_{\Psi',N-1}(s^{1/2}t)\right| \left|\int_0^s \exp\left(-\varepsilon\sigma\right)\widehat{\chi}_2(\sigma)\,d\sigma\right| ds$$

$$\leq K_0 t^{N+1} \left(T^{\frac{N+1}{2}} + \frac{1}{2} \int_0^T s^{\frac{N+1}{2}} \, ds \right).$$

It follows that there are constants M_N with

$$|\rho_N(\varepsilon, t)| \leq M_N t^{N+1}, \tag{6.4}$$

where

$$\rho_N(\varepsilon, t) := \int_0^\infty e^{-\varepsilon s} R_{\Psi, N}(s^{1/2} t) \widehat{\chi}_2(s) \, ds.$$

Combining (2), (3), and (4) we find

$$t^n \operatorname{tr} \chi(t^2 P) = t^n \sum_k \chi(\lambda_k t^2) = \lim_{\varepsilon \to 0_+} \int_0^\infty e^{-\varepsilon s} \Psi(s^{1/2} t) \widehat{\chi}_2(s) \, ds$$

$$= \sum_{k=0}^N a_k(P) t^k \lim_{\varepsilon \to 0_+} \int_0^\infty e^{-\varepsilon s} s^{k/2} \widehat{\chi}_2(s) \, ds + \rho_N(t),$$

where $\rho_N(t) = \lim_{\varepsilon \to 0_+} \rho_N(\varepsilon, t)$ with $|\rho_N(t)| \leq M_N t^{N+1}$ as $t \to 0_+$. But

$$\lim_{\varepsilon \to 0_+} \int_0^\infty e^{-\varepsilon s} s^{k/2} \widehat{\chi}_2(s) \, ds = \lim_{\varepsilon \to 0_+} \chi^{\left(\frac{k-4}{2}\right)}(x) = \chi^{\left(\frac{k-4}{2}\right)}(0),$$

and Theorem 1 has been proven.

We are finally faced with the question of how many functions actually satisfy the hypothesis of Theorem 1. It is answered by the following result from [73].

Theorem 2. *For any sequence of complex numbers* (a_n), $n \in \mathbb{Z}$ *there exists a function* $\widehat{\chi}$ *in* $\mathcal{S}(0, \infty)$ *such that*

$$a_n = \int_0^\infty t^n \widehat{\chi}(t) \, dt \quad \text{for all } n \in \mathbb{Z}.$$

6.4 First Example, Final Calculation

We are finally faced with calculating the coefficients $a_n(D^2)$. Let us fix some further notation: With regard to Christoffel symbols and curvature tensors of the Riemannian manifold M we follow the conventions of Besse [17, pp. 30,31,43]. Thus,

$$\Gamma^\rho_{\mu\nu} = \frac{1}{2} g^{\rho\sigma} \left(\partial_\mu g_{\nu\sigma} + \partial_\nu g_{\mu\sigma} - \partial_\sigma g_{\mu\nu} \right)$$

which agrees with the convention of Connes–Chamseddine. The Riemannian connection on M is denoted by ∇^{LC} for Levi–Civita, so that

$$\nabla^{LC}_{\partial_\mu} \partial_\nu = \Gamma^\rho_{\mu\nu} \partial_\rho.$$

The corresponding curvature 2–form we denote by K^{LC} and the Riemannian tensor $R^\sigma_{\mu\nu\rho}$ is given by

$$K^{LC}(\partial_\mu, \partial_\nu)\partial_\rho = R^\sigma_{\mu\nu\rho}\partial_\sigma,$$
$$K^{LC}(X,Y)Z = [\nabla^{LC}_X, \nabla^{LC}_Y]Z - \nabla^{LC}_{[X,Y]}Z,$$

$X, Y, Z \in C^\infty(TM)$. (This convention differs from Connes–Chamseddine: their $R^\sigma_{\mu\nu\rho}$ is our $R^\sigma_{\rho\nu\mu}$). The Ricci tensor is the 2–tensor on M given by

$$\mathrm{Ric}(X,Y) = \mathrm{tr}(Z \longrightarrow K^{LC}(X,Z)Y).$$

This agrees with [31] as does the definition of the scalar curvature of M which is just the trace of the map $\mathrm{ric} : TM \longrightarrow TM$ given by

$$g(\mathrm{ric}(X), Y) = \mathrm{Ric}(X, Y).$$

Attention: A straightforward calculation, using coordinates around a point a such that $g_{\mu\nu}(a) = \delta_{\mu\nu}$ and $\Gamma^\rho_{\mu\nu}(a) = 0$, shows that the scalar curvature of the unit 2–sphere equals -2.

In terms of the Riemann tensor $R^\sigma_{\mu\nu\rho}$ the Ricci–tensor is given by $R_{\mu\rho} := \mathrm{Ric}(\partial_\mu, \partial_\rho) = R^\nu_{\mu\nu\rho}$, and the scalar curvature is given by

$$R := R^\mu_\mu (= R_{\mu\nu}g^{\nu\mu}) = R^\sigma_{\mu\sigma\nu}g^{\nu\mu}.$$

With respect to the calculation of D^2 (Lichnérowicz formula) some more terminology will be useful. To each connection ∇ on a bundle V over a Riemannian manifold M one has the associated Laplacian $P_\nabla : C^\infty(V) \longrightarrow C^\infty(M)$ defined as the composition

$$C^\infty(V) \xrightarrow{\nabla} C^\infty(T^*M \otimes V)$$
$$\xrightarrow{\nabla^{LC}\otimes 1 + 1\otimes \nabla} C^\infty(T^*M \otimes T^*M \otimes V) \xrightarrow{-g\otimes 1} C^\infty(V)$$

where as before ∇^{LC} is the Levi–Civita connection extended to forms. Thus locally

$$(\nabla^{LC} \otimes 1 + 1 \otimes \nabla) \circ \nabla = (\nabla^{LC} \otimes 1 + 1 \otimes \nabla)(dx^\rho \otimes G_\rho)$$
$$= -dx^\mu \otimes \Gamma^\rho_{\mu\nu} dx^\nu \otimes G_\rho + dx^\rho \otimes dx^\mu G_\mu G_\rho.$$

Therefore, we obtain

$$P_\nabla = -g^{\mu\nu}(G_\mu G_\nu - \Gamma^\lambda_{\mu\nu} G_\lambda). \tag{6.5}$$

This formula will be useful for our description of $a_n(D^2)$. To do this we follow [91, Section 4.8]. There $a_n(P)$ is calculated for any second order operator P on a vector bundle V over the Riemannian manifold (M,g) with the leading symbol of P equal to $g^{\mu\nu}$, i.e. operators of the form

$$-g^{\mu\nu}\partial x_\mu \partial x_\nu + A_\mu \partial x_\mu + B,$$

with A_μ, B endomorphisms of V. For any such operator one finds a connection ∇ and bundle endomorphism E such that $P = P_\nabla - E$. It is not difficult to show that ∇ and E are determined by P. Then $a_n(P)$ is expressed in terms of covariant derivatives of the Riemann tensor R_{abcd}, of the curvature 2–form Ω of ∇, and of the endomorphism E.

In our situation, the connection $\nabla = \nabla^S \otimes 1 + 1 \otimes A$ is the one we need to obtain $D^2 = P_\nabla - E$.

Lemma (Werner) 2. *(Lichnérowicz).* Let $\nabla = \nabla^S \otimes 1 + 1 \otimes A$ and locally $\nabla = dx^\mu \otimes G_\mu$. Let $D = c(dx^\mu)G_\mu$. Then

$$D^2 = -g_{\mu\nu}(G_\mu G_\nu - \Gamma^\lambda_{\mu\nu}G_\lambda) + \frac{1}{2}c(dx^\mu)c(dx^\nu)[G_\mu, G_\nu].$$

Proof:

$$D^2 = c(dx^\mu)G_\mu c(dx^\nu)G_\nu = c(dx^\mu)c(dx^\nu)G_\mu G_\nu + c(dx^\mu)[G_\mu, c(dx^\nu)]G_\nu$$

$$= \frac{1}{2}\left(c(dx^\mu)c(dx^\nu) + c(dx^\nu)c(dx^\mu)\right)G_\mu G_\nu$$

$$+ \frac{1}{2}c(dx^\mu)c(dx^\nu)[G_\mu, G_\nu] + c(dx^\mu)[G_\mu, c(dx^\nu)]G_\nu.$$

But ∇ is a twisted spin connection and thus compatible with Clifford multiplication. Consequently,

$$[G_\mu, c(dx^\nu)] = c(\nabla^{LC}_{\partial_\mu} dx^\nu) = -\Gamma^\nu_{\mu\lambda}c(dx^\lambda).$$

Therefore the last summand equals $-c(dx^\mu)c(dx^\nu)\Gamma^\lambda_{\mu\nu}G_\lambda$. We use the defining relation for the Clifford algebra

$$c(dx^\mu)c(dx^\nu) + c(dx^\nu)c(dx^\mu) = -2g^{\mu\nu}$$

and $\Gamma^\nu_{\mu\lambda} = \Gamma^\nu_{\lambda\mu}$ (since ∇^{LC} is torsion free) to obtain the desired formula. □

Now $[G_\mu, G_\nu] = \Omega(\partial_\mu, \partial_\nu)$ for the curvature 2–form Ω of ∇ because $[\partial_\mu, \partial_\nu] = 0$. Therefore

$$-\frac{1}{2}c(dx^\mu)c(dx^\nu)[G_\mu, G_\nu]$$

is an endomorphism of $A \otimes M_N(\mathbb{C})$ which equals the E in [91]. In our situation $\nabla = \nabla^S \otimes 1 + 1 \otimes A$ and we can proceed further in describing E.

Lemma (Werner) 3. *If R is the scalar curvature of M and K^A the curvature 2–form of the connection A of $M \times M_N(\mathbb{C})$, then*

$$E = \frac{1}{4} R \otimes 1 - \frac{1}{2} c(dx^\mu) c(dx^\nu) \otimes K^A(\partial_\mu, \partial_\nu).$$

Proof. We have $[G_\mu, G_\nu] = \Omega(\partial_\mu, \partial_\nu)$ for the curvature 2–form Ω of ∇. Thus for any local frame of TM and its dual, in particular for the orthonormal frame $\{\gamma_a\}$ and its dual $\{\gamma^a\}$ we obtain

$$E = -\frac{1}{2} c(\gamma^a) c(\gamma^b) \Omega(\gamma_a, \gamma_b).$$

Furthermore,

$$\Omega(\gamma_a, \gamma_b) = K^S(\gamma_a, \gamma_b) \otimes 1 + 1 \otimes K^A(\gamma_a, \gamma_b),$$

where K^S is the curvature 2–form of the spin connection and K^A is the curvature 2–form of A. The 2–form K^S can be calculated from the curvature K^{LC} of the Levi–Civita connection, the metric g, and Clifford multiplication:

$$K^S(\gamma_a, \gamma_b) = \frac{1}{4} g\left(K^{LC}(\gamma_a, \gamma_b)\gamma_c, \gamma_d\right) c(\gamma_c) c(\gamma_d)$$

In terms of the Riemann tensor we obtain

$$c(\gamma^a) c(\gamma^b) K^S(\gamma_a, \gamma_b) = \frac{1}{4} c(\gamma^a) c(\gamma^b) c(\gamma^c) c(\gamma^d) R_{abcd}.$$

It is easy to check that for any a, b, c we have

$$c(\gamma^a) c(\gamma^b) c(\gamma^c) =$$
$$\frac{1}{6} \sum_{\sigma \in S_3} \text{sign}(\sigma) c(\gamma^{\sigma(a)}) c(\gamma^{\sigma(b)}) c(\gamma^{\sigma(c)}) - \delta^{ab} c(\gamma^c) - \delta^{bc} c(\gamma^a) + \delta^{ac} c(\gamma^b),$$

where S_3 is the symmetry group of the three symbols $\{a, b, c\}$.

Also, by the algebraic Bianchi identity ($R_{abcd} + R_{cabd} + R_{bcad} = 0$) the antisymmetrization of R_{abcd} with respect to a, b, c is 0. Thus we obtain

$$c(\gamma^a) c(\gamma^b) c(\gamma^c) c(\gamma^d) R_{abcd} = -c(\gamma^a) c(\gamma^d) R_{abbd} + c(\gamma^b) c(\gamma^d) R_{cbcd}$$
$$= -2 c(\gamma^a) c(\gamma^d) R_{abbd} = 2 R_{abba} = -2 R_{abab}$$

using the antisymmetry of R_{abcd} with respect to a, b. With our conventions R_{abab} is the scalar curvature of M:

$$R := R^\sigma_{\mu\sigma\nu} g^{\nu\mu} = R^{\mu\sigma}_{\mu\sigma} = R^{ab}_{ab} = R_{abab}.$$

This establishes the lemma.

We are now in a position to apply the results of [91, Theorem 4.8] to our situation. Thus $a_{2n+1}(D^2) = 0$ for all $n \geq 0$ and

$$a_0(x, P) = \frac{1}{16\pi^2} \operatorname{tr}(I) = \frac{N}{4\pi^2} \tag{i}$$

$$a_2(x, P) = \frac{1}{6 \cdot 16\pi^2} \operatorname{tr}(-R \cdot I + 6E) = \frac{N}{48\pi^2} R \tag{ii}$$

The last equation holds by Lemma 2 and the fact that $K^A(\partial_\mu, \partial_\nu)$ is trace-free since A is an $SU(N)$–connection. Moreover

$$a_4(x, P) = \frac{1}{16\pi^2} \cdot \frac{1}{360} \cdot \operatorname{tr}(-12 R_{;aa} \cdot I + 5R^2 I \tag{iii}$$

$$- 2\operatorname{Ric}_{ab} \cdot \operatorname{Ric}_{ab} \cdot I + 2 R_{abcd} R_{abcd} \cdot I$$

$$- 60 R \cdot E + 180 E^2 + 60 E_{;aa} + 30 \Omega_{ab} \Omega_{ab})$$

where everything is expressed in an orthonormal frame and $T_{;aa}$ means summation of the second covariant derivative of T in direction γ_a, expressed in the orthonormal frame $\{\gamma_a\}$. Traces of these endomorphisms give divergences which will vanish when integrated over M (after multiplication with the volume form). We use Lemma 2, the equation

$$\Omega_{ab} = K^S(\gamma_a, \gamma_b) \otimes 1 + 1 \otimes K^A(\gamma_a, \gamma_b)$$

and express elements of $\mathfrak{su}(N)$ in terms of a basis $i \cdot T^j$ with $\operatorname{tr}(T^j T^k) = 2\delta^{jk}$. The T^j are sometimes called Gell-Mann matrices and will be very useful when calculating traces of compositions. Writing (conventions differ by a factor $-\frac{1}{2}$ from [31])

$$K^A(\gamma_a, \gamma_b) = i \cdot F_{ab}^j T^j$$

we obtain

$$\operatorname{tr} \Omega_{ab} \Omega_{ab} = \frac{N}{16} \operatorname{tr}\left(R_{abcd} R_{abc'd'} c(\gamma_c) c(\gamma_d) c(\gamma_{c'}) c(\gamma_{d'})\right) - 4 \cdot Tr(F_{ab}^j F_{ab}^k T^j T^k)$$

Now, the calculation

$$\operatorname{tr}\left(c(\gamma_c) c(\gamma_d) c(\gamma_{c'}) c(\gamma_{d'})\right) = \operatorname{tr}\left(c(\gamma_{c'}) c(\gamma_c) c(\gamma_d) c(\gamma_{d'})\right) =$$

$$= \operatorname{tr}\left(c(\gamma_c) c(\gamma_d) c(\gamma_{d'}) c(\gamma_{c'})\right) = -\operatorname{tr}\left(c(\gamma_c) c(\gamma_d) c(\gamma_{c'}) c(\gamma_{d'})\right),$$

if $c' \neq c, d$ and $c' \neq d'$, shows that

$$\operatorname{tr} \Omega_{ab}\Omega_{ab} = -\frac{N}{2}R_{abcd}R_{abcd} - 8 \cdot F^j_{ab}F^j_{ab}$$

$$\operatorname{tr} R \cdot E = N \cdot R^2, \quad \text{since the } T^j \text{ are trace-free,}$$

$$\operatorname{tr} E_{;aa} = N \cdot R_{;aa}$$

$$\operatorname{tr} E^2 = \frac{N}{4}R^2 - \frac{1}{2}\operatorname{tr}\left(c(\gamma_a)c(\gamma_b)c(\gamma_c)c(\gamma_d) \otimes F^j_{ab}F^j_{cd}\right)$$

$$= \frac{N}{4}R^2 + 4F^j_{ab}F^j_{ab}.$$

Altogether we obtain

$$a_4(x, P) = \frac{N}{16\pi^2} \cdot \frac{1}{360}\Bigg[-48R_{;aa} + 60R_{;aa} + 20R^2 - 60R^2 - 8\operatorname{Ric}_{ab}\operatorname{Ric}_{ab}$$

$$+ 8R_{abcd}R_{abcd} + 45R^2 - 15R_{abcd}R_{abcd} + \left(-\frac{240}{N} + \frac{720}{N}\right)F^j_{ab}F^j_{ab}\Bigg]$$

$$= \frac{N}{16\cdot 360\cdot \pi^2}\left(12R_{;aa} + 5R^2 - 8\operatorname{Ric}_{ab}\operatorname{Ric}_{ab} - 7R_{abcd}R_{abcd} + \frac{480}{N}F^j_{ab}F^j_{ab}\right)$$

or in local coordinates (here we use the fact that for an orthonormal frame one can freely raise and lower indices and that for an (n, n)–tensor the coefficients $T^{i_1\cdots i_n}_{i_1\cdots i_n}$ are independent of the frame)

$$a_4(x,P) = \frac{N}{16\cdot 360\cdot \pi^2}(12R^\mu_{;\mu} + 5R^2 - 8\operatorname{Ric}_{\mu\nu}\operatorname{Ric}^{\mu\nu} - 7R_{\mu\nu\rho\sigma}R^{\mu\nu\rho\sigma} + \frac{480}{N}F^j_{\mu\nu}F^{\mu\nu j})$$

(note that our $F^j_{\mu\nu}$ equal $-\frac{1}{2}F^j_{\mu\nu}$ in [31]). We can further simplify the expression above by using

Lemma (Werner) 4. *For a closed oriented Riemannian 4–manifold M the Euler form (with respect to the Levi–Civita connection)*

$$\frac{\operatorname{vol}_M}{128\pi^2}\sum_{\tau,\sigma\in S_4}\operatorname{sign}\sigma\,\operatorname{sign}\tau R_{\sigma_1\sigma_2\tau_1\tau_2}R_{\sigma_3\sigma_4\tau_3\tau_4}$$

equals

$$\frac{\operatorname{vol}_M}{32\pi^2}(R_{abcd}R_{abcd} - 4\operatorname{Ric}_{ab}\operatorname{Ric}_{ab} + R^2)$$

where vol_M is the volume form.

Proof: The proof is easy but a little lengthy and uninspiring. We need to show that

$$4R_{abcd}R_{abcd} - 16\operatorname{Ric}_{ab}\operatorname{Ric}_{ab} + 4R^2 = \sum_{\sigma,\tau\in S_4}\operatorname{sign}\sigma\cdot\operatorname{sign}\tau R_{\sigma_1\sigma_2\tau_1\tau_2}R_{\sigma_3\sigma_4\tau_3\tau_4}\,.$$

For this we split each summand on the left hand side into subsummands which can be brought into the form

$$R_{\sigma_1\sigma_2\tau_1\tau_2} R_{\sigma_3\sigma_4\tau_3\tau_4} \text{ for some } \sigma, \tau \in S_4$$

and the rest. We show that the rest vanishes and that the other summands correspond bijectively (including signs) to the summands on the right hand side.

(i) $R_{abcd}R_{abcd} = R_{abcd}R_{cdab}$, so we obtain the desired form if $(abcd) = (\sigma_1\sigma_2\sigma_3\sigma_4)$ is a permutation. Then

$$4R_{abcd}R_{abcd} = R_{abcd}R_{cdab} - R_{abdc}R_{cdab} + R_{abdc}R_{cdba} - R_{abcd}R_{cdba}.$$

In this way we get four summands $R_{\sigma_1\sigma_2\tau_1\tau_2}R_{\sigma_3\sigma_4\tau_3\tau_4}$ with $(\sigma_1\sigma_2\sigma_3\sigma_4) = (abcd)$ in all four cases and four different τ's. Notice that sign σ sign τ corresponds to the sign in the equation above.

(ii) $\text{Ric}_{ab}\text{Ric}_{ab} = R_{albl} \cdot R_{akbk} = R_{albl}R_{bkak}$. We get the desired form if $(albk)$ is a permutation. This time we have 16 summands available and we write $16R_{albl}R_{bkak}$ into 16 summands with coefficients ± 1 by transposing the first and second and/or third and fourth index of either factor. Notice if

$$R_{\sigma_1\sigma_2\tau_1\tau_2}R_{\sigma_3\sigma_4\tau_3\tau_4} = R_{albl}R_{bkak},$$

then sign $\sigma \cdot$ sign $\tau = -1$ (which corresponds to the $-16\text{Ric}_{ab}\text{Ric}_{ab}$).

(iii) $R^2 = R_{abab} \cdot R_{cdcd}$. We obtain the desired form if $(abcd)$ is a permutation. We again have a coefficient 4 and write

$$4R_{abab}R_{cdcd} = R_{abab}R_{cdcd} - R_{abab}R_{dccd} + R_{baab}R_{dccd} - R_{baab}R_{cdcd}.$$

Notice that in combining (i), (ii), and (iii) we have picked for any $\sigma \in S_4$ $4 + 16 + 4$ different $\tau \in S_4$, so that in fact we have for any pair $\sigma, \tau \in S_4$ a summand $R_{\sigma_1\sigma_2\tau_1\tau_2}R_{\sigma_3\sigma_4\tau_3\tau_4}$ with the correct sign. To show that the remaining terms add up to 0 we look at the remaining terms of

$$-16\text{Ric}_{ab}\text{Ric}_{ab} = -16R_{albl}R_{bkak},$$

$(albk)$ not a permutation. There are three cases

$$b = a \text{ and } l \neq k \quad \text{with terms } R_{alal}R_{akak}, l \neq k \quad \text{(ci)}$$

$$l = k \text{ and } a \neq b \quad \text{with terms } R_{albl}R_{albl}, a \neq b \quad \text{(cii)}$$

$$l = k, a = b \quad \text{with terms } R_{alal}R_{alal}. \quad \text{(ciii)}$$

The terms in (ci) are cancelled by the 16 terms of $4R^2 = 4R_{abab}R_{cdcd}$ of type

$$(c = a, d \neq b), (c = b, d \neq a), (c \neq a, d = b), (c \neq b, d = a).$$

The terms in (cii) are cancelled by the 16 terms of $4R_{abcd}R_{abcd} = 4R_{abcd}R_{cdab}$ of the same four types for $(abcd)$. So there remains (ciii), from $R_{abcd}R_{cdab}$ the 8 summands

$$4R_{abab}R_{abab} + 4R_{abba}R_{abba} = 8R_{abab}R_{abab},$$

and from $4R^2$ the same 8 summands. □

Apart from the Euler form there is another feature present in the expression of $a_4(x, D^2)$: the Weyl tensor. It shows up when we split the curvature tensor into its irreducible components.

The symmetries of R_{abcd} (for simplicity we use again an orthonormal frame)

$$R_{abcd} = -R_{bacd} \tag{si}$$

$$R_{abcd} = R_{cdab} \tag{sii}$$

$$R_{abcd} + R_{cabd} + R_{bcad} = 0 \quad \text{(algebraic Bianchi)} \tag{siii}$$

show that the curvature tensor is a section of $\mathcal{C}(M) := S^2(\wedge^2 T^*M) \cap \ker b$, where $b : \otimes^4 T^*M \longrightarrow \otimes^4 T^*M$ is the Bianchi map

$$b(T)(X, Y, Z, U) = \frac{1}{3}\left(T(X, Y, Z, U) + T(Z, X, Y, U) + T(Y, Z, X, U)\right).$$

On each fiber $\mathcal{C}_x(M)$ the orthogonal group of the Riemannian metric acts and induces an orthogonal splitting

$$\mathcal{C}(M) = \mathcal{R}(M) \oplus \mathcal{W}(M)$$

where $\mathcal{W}(M) = \ker \rho \cap \mathcal{C}(M)$

$$\rho : S^2(\wedge^2 T^*M) \longrightarrow S^2(T^*M)$$

is the Ricci contraction, i.e., in an orthonormal frame,

$$\rho(T)_{ac} = T_{abcb}.$$

$\mathcal{R}(M)$ can be expressed with the help of the Kulkarni–Nomizu product as the image of

$$-\owedge g : S^2(T^*M) \longrightarrow \mathcal{C}(M).$$

Here, for $T \in S^2(T^*M)$, again with respect to an orthonormal frame,

$$(T \owedge g)_{abcd} = T_{ac}\delta_{bd} + T_{bd}\delta_{ac} - T_{ad}\delta_{bc} - T_{bc}\delta_{ad}.$$

Obviously,

$$\rho(T \owedge g) = (n-2)T + \operatorname{tr}(T) \cdot g,$$

where $n = \dim M$ and $\operatorname{tr}(T) = T_{aa}$ (in an orthonormal frame, or T^μ_μ in general). Thus, if we denote for a tensor T of $\mathcal{C}(M) = \mathcal{R}(M) \oplus \mathcal{W}(M)$ its summands by $\mathcal{R}(T), \mathcal{W}(T)$, then

$$\mathcal{R}(T) = \frac{1}{n-2}\rho(T) \oslash g - \frac{\sigma(T)}{2(n-1)(n-2)} g \oslash g ,$$

where $\sigma(T) = \operatorname{tr}(\rho(T))$ is the "scalar curvature" of T (while $\rho(T)$ is the "Ricci curvature" of T).

Notice: While for each $x \in M$ the $O(n)$ representation $\mathcal{W}_x(M)$ is irreducible, the representation $\mathcal{R}_x(M)$ splits into a 1–dimensional space corresponding to the scalar curvature part and a subspace corresponding to the trace-free Ricci part.

$\mathcal{W}(T)$ is called the Weyl part of T. If $T = K^{LC}$ is the Riemann tensor then $\mathcal{W}(T)$ is called the Weyl tensor of M with components W_{abcd}. We are interested in $W_{abcd} W_{abcd}$ for $n = 4$. Now, if U, T are any tensors in $\mathcal{C}(M)$ a straightforward calculation shows that

$$U_{abcd}\mathcal{R}(T)_{abcd} = -\frac{2\sigma(U)\sigma(T)}{(n-1)(n-2)} + \frac{4}{n-2}\rho(U)_{ab}\rho(T)_{ab} .$$

Since $\rho(T) = \rho(\mathcal{R}(T))$ we obtain, again for any tensor T in $\mathcal{C}(M)$

$$\mathcal{W}(T)_{abcd}\mathcal{W}(T)_{abcd} = T_{abcd}T_{abcd} - \frac{4}{n-2}\rho(T)_{ab}\rho(T)_{ab} + \frac{2}{(n-1)(n-2)}\sigma(T)^2.$$

So, in particular for the Weyl tensor on a 4–manifold we obtain

$$W_{abcd}W_{abcd} = R_{abcd}R_{abcd} - 2\operatorname{Ric}_{ab}\operatorname{Ric}_{ab} + \frac{1}{3}R^2.$$

Setting $R^*R^* = R_{abcd}R_{abcd} - 4\operatorname{Ric}_{ab}\operatorname{Ric}_{ab} + R^2$ so that $\frac{\operatorname{vol}_M}{32\pi^2} R^*R^*$ is the Euler form (by Lemma 4), and plugging this and the formula for the Weyl tensor into our expression for $a_4(x, D^2)$ we finally obtain

$$a_4(x, D^2) = \frac{N}{48\pi^2}\left(-\frac{3}{20}W_{abcd}W_{abcd} + \frac{11}{120}R^*R^* + \frac{1}{10}R_{;aa} + \frac{4}{N}F^j_{ab}F^j_{ab}\right),$$
(1.5a)

or in local coordinates

$$a_4(x, D^2) = \frac{N}{48\pi^2}\left(-\frac{3}{20}W_{\mu\nu\rho\sigma}W^{\mu\nu\rho\sigma} + \frac{11}{120}R^*R^* + \frac{1}{10}R_{;\mu}{}^\mu + \frac{4}{N}F^j_{\mu\nu}F^{\mu\nu j}\right).$$
(1.5μ)

The discrepancy of (1.5μ) with [31, (2.24)], comes from the introduction of a coupling constant g_0 and a factor $-\frac{1}{2}$ in the description of the connection

∇ on $S \otimes M_N(\mathbb{C})$. So writing our connection A on the trivial $M_N(\mathbb{C})$ bundle as $A = -\frac{g_o}{2} \widehat{A}$ (see [31, (2.4)], the last summand in the above formula becomes $\frac{g_0^2}{N} \widehat{F}^j_{\mu\nu} \widehat{F}^{\mu\nu j}$ which is the last summand in (2.24) of [31].

The particular form of equation (1.5) shows that

$$a_4(D^2) = \int_M a_4(x, D^2) \, \text{vol}_M$$

is conformally invariant, where, as before, vol_M is the volume form $\sqrt{g}\, dx$ of M with respect to g. This works only in dimension 4. To see this note that $R^\mu_{;\mu}$ as a divergence can be ignored, and $F^j_{\mu\nu}$ is independent of the metric (but $F^{\mu\nu j}$ depends on g). Furthermore, and this is the important fact, the (3,1)–Weyl tensor $W^\sigma_{\mu\nu\rho}$ is a conformal invariant.

Thus, if $\widetilde{g} = f^2 \cdot g$ for some positive function f, we have $\widetilde{W}_{\mu\nu\rho\sigma} = f^2 \cdot W_{\mu\nu\rho\sigma}$, $\widetilde{W}^{\mu\nu\rho\sigma} = f^{-6} \cdot W_{\mu\nu\rho\sigma}$, $\widetilde{\text{vol}_M} = f^4 \, \text{vol}_M$, $\widetilde{F}^{\mu\nu j} = f^{-4} F^{\mu\nu j}$. Finally

$$\int_M R^* R^* \, \text{vol}_M = 32 \cdot \pi^2 \chi(M) = \int_M \widetilde{R^* R^*} \, \widetilde{\text{vol}_M},$$

where $\chi(M)$ is the Euler characteristic of M. Of course, neither

$$\text{Vol}(M_{,g}) = \int_M \text{vol}_{(M,g)} = \frac{4\pi^2}{N} a_0(D^2)$$

nor the average scalar curvature

$$\int_M R \cdot \text{vol}_{(M,g)} = \frac{48\pi^2}{N} a_2(D^2),$$

which make up the Einstein action, are conformal invariants. Putting it all together, we choose a function χ as in Proposition 1 with $\int_0^\infty u\chi(u)\, du = f_0$, $\int_0^\infty \chi(u) = f_2$, $\chi(0) = f_4$, and $\chi^{(n)}(0) = 0$ for every $n \in \mathbb{N}$. Then the "bare" action $\mathfrak{I}^B_{D_0, \mathcal{A}}$ has an asymptotic expansion in powers of m_0^2 as

$$\mathfrak{I}^B_{D_0, \mathcal{A}} = \text{tr}\, \chi\left(\frac{D^2}{m_0^2}\right) \sim \frac{N}{48\pi^2} \left(a_0 m_0^4 + a_2 m_0^2 + a_4\right) + O(m_0^{-\infty}), \quad \text{as } m_0 \to \infty,$$

where

$$a_0 = 12 f_0 \int_M \text{vol}_{(M,g)}$$

$$a_2 = f_2 \int_M R \, \text{vol}_{(M,g)}$$

$$a_4 = f_4 \int_M \left(-\frac{3}{20} W_{\mu\nu\rho\sigma} W^{\mu\nu\rho\sigma} + \frac{44\pi^2}{15} \chi(M) + \frac{4}{N} F^j_{\mu\nu} F^{\mu\nu j}\right) \text{vol}_{(M,g)}.$$

6.5 Gravity Coupled to the Standard Model

Let us start by recalling some of the basic facts. Besides the carriers of strong and electroweak forces, all elementary particles are in one of two classes: They either feel the strong force — which is the characteristic property of *hadrons* — or they don't — in which case they are called *leptons*. There exist three generations of leptons, which ordered by their respective masses are (e, ν_e), (μ, ν_μ) and (τ, ν_τ), consisting of the electron e, the muon μ and the tauon τ, together with their respective neutrinos (and antiparticles). All leptons are fermions.

The fermions among the hadrons are called *baryons* and the bosons are *mesons*. There is a large number of hadrons which nowadays are successfully classified by their quark content: There are six quarks *up* and *down*, (u, d), *strange* and *charm*, (s, c), as well as *top* and *bottom*, (t, b). The ambitious reader is advised to consult [28] for a complete and up-to-date picture of all known hadrons.

Quarks are fermions and pay due credit to Pauli's exclusion principle by showing an additional degree of freedom, which is called *color*. The three existing colors are called blue, red and green, and to each color exists a complementary anticolor for the respective antiquark. Baryons contain three flavors of quarks, the color degrees of freedom arranged such that they are singlets with respect to $SU(3)_c$. (For the terminology see below.)

Each particle is assigned a number of generalized charges, like *lepton numbers, baryon numbers, (strong) isospin* and several flavor quantum numbers. (The lepton number, e.g. is 1 if the particle is a lepton, -1 if it is an antilepton and 0 in all other cases. Baryon numbers, and flavor are defined in a similar way.) *Charge conjugation* on the level of elementary particles is the transformation that maps a particle to its antiparticle. The antiparticle is obtained by a change of sign of all charge-like quantum numbers, and, in the case of fermions, by an additional change of parity. (If $\psi(x)$ is the state of the fermion then $\psi(-x)$ is the state after a change of parity.)

Fermions can further be distinguished by the orientation their spin has relative to the direction of motion. For *right-handed* particles the spin is parallel to the direction of motion, the spin of *left-handed* particles points in the opposite direction. Mathematically, these states of a particle correspond to the eigenspaces of the chirality operator $\chi = -\gamma^0\gamma^1\gamma^2\gamma^3$ (in 4 dimensions). Note that chirality is a preserved quantity only for massless particles and that χ does not yield a complete symmetry since for neutrinos only the left-handed variant exists.

Mathematically, elementary particles are usually treated in the following way: Quantum numbers are preserved quantities, at least in sufficiently idealized situations, and thus correspond to the infinitesimal generators of a Lie group acting on the set of solutions of the underlying equations. Whatever the picture, elementary particles are certainly distinguished as those states for which none of the known invariants is capable of further distinction. Hence, all elementary particles are usually assigned to irreducible representations of a properly chosen Lie group G. Speaking quantum mechanically, the infinitesimal generators H_α corresponding to the set of all quantum numbers then forms a maximal set of commuting generators of $\mathfrak{g}$, i.e. they generate a Cartan subalgebra $\mathfrak{h}$. The quantum numbers themselves show up in an irreducible representation π as (simultaneous) eigenvalues of the generators H_α. It is customary to organize these tuples into weight vectors pertaining to the basis elements H_α of $\mathfrak{h}$ for $\mathfrak{g}$ in standard form. (That this assignment is independent of the Cartan algebra chosen is a well-known result from Lie group theory). The edges of the emerging weight diagrams then label the different multiplets of elementary particles.

As an example, for $G = SU(3)$ the Lie algebra $\mathfrak{su}(3)$ consists of all antihermitian, traceless matrices, and a choice for $\mathfrak{h}$ is lin $\{Y, I_3\}$, where

$$Y = \begin{pmatrix} \frac{1}{3} & 0 & 0 \\ 0 & \frac{1}{3} & 0 \\ 0 & 0 & -\frac{2}{3} \end{pmatrix} \quad \text{and} \quad I_3 = \begin{pmatrix} \frac{1}{2} & 0 & 0 \\ 0 & -\frac{1}{2} & 0 \\ 0 & 0 & 0 \end{pmatrix}$$

correspond to (strong) hypercharge and the third component of the isospin, respectively. For the particles known at the beginning of the 60's these quantities were connected to the electrical charge Q/e by the Gell-Mann-Nishijima formula $Q/e = I_3 + Y/2$. Also, there seemed to be 6 physically relevant irreducible representations: The trivial one, written **1**, the 'fundamental one' denoted **3** (which has this name since all other irreducible representation show up as summands of tensor powers of **3**), in which $SU(3)$ acts naturally on $\mathbb{C}^3$, the one conjugate to **3** denoted $\overline{\mathbf{3}}$ in which $u \in SU(3)$ acts as u^* on $\mathbb{C}^3$, the adjoint representation **8** where $\mathfrak{su}(3)$ acts on (the vector space) $\mathfrak{su}(3)$ by $A \mapsto [A, X]$ and a 10-dimensional representation, **10**, which is the restriction of $\mathbf{3} \otimes \mathbf{3} \otimes \mathbf{3}$ onto the invariant subspace of symmetric tensors. The credo of this time was that mesons and baryons correspond to the equations

$$\mathbf{3} \otimes \overline{\mathbf{3}} = \mathbf{8} \oplus \mathbf{1} \quad \text{and} \quad \mathbf{3} \otimes \mathbf{3} \otimes \mathbf{3} = \mathbf{10} \oplus \mathbf{8} \oplus \mathbf{8} \oplus \mathbf{1},$$

respectively, and that these assignments could be attributed to three new particles, the quarks u, d and s. The resulting weight diagrams correspond to the quark triplet (u, s, d), octets of pseudoskalar mesons and vectormesons, an octet of spin-1/2-baryons as well as a decuplet of spin-3/2-baryons.

With the advent of further quantum numbers in the following decades, the number of different quarks (not counting antiparticles and disregarding color) has grown to 6, and the multiplets of mesons and baryons had to be generated from the fundamental representation of $SU(6)$. Note, however, that this picture seems to be more and more blurred for higher generations, since the mass differences among the members of the quark duplets increase.

Leptons can be treated in a similar way, with symmetry group $SU(2)$. It turns out, for example, that the six known quarks belong to three duplets that are naturally grouped with the three known generations of leptons. For a more extensive coverage of the basic ideas the reader should consult e.g. [38], [169], [205] or [206].

All groups of the previous section are, in terms of gauge field theory, global gauge groups which yield globally invariant charges. In order to set gauge theory of the standard model to work, it is necessary to specify a gauge group G and to say which particles are to be coupled to the gauge fields by fixing a suitable representation of G. It must also be decided in which way the different particles interact and which will become massive. As an effect, the carrier of forces, which have been excluded from the picture so far, will show up as spin-1 fields associated to the generators of $\mathfrak{g}$.

The standard model uses the (global) gauge group $U(1)_Y \times SU(2)_w \times SU(3)_c$, where the Lie algebra generators correspond to hypercharge, the three carriers of the weak force, the massive bosons $W^\pm$ and Z, as well as the massless eight gluons mediating the strong force. All elementary particles interacting with the gauge fields then appear as basis vectors of a Hilbert space belonging to a suitably chosen irreducible representation of G.

In the following, we will describe a spectral triple that will result in the gauge fields of the standard model with gravity switched on. It will consist of two parts. The first part reflects the internal, noncommutative geometry, the second one is responsible for gravity, and both will be combined within a tensor product. Assigning masses and coupling constants will be the task of the internal Dirac operator and will thus be based on the geometry of the model.

We start with the internal spectral triple Hilbert space, $\mathcal{H}_{\text{int}}$, which turns out to be the space for the underlying gauge theory: Its basis consists of all known elementary fermions. Violation of parity by the weak force makes it necessary to consider left- and right-handed particles separately, so that for the first quark generation of one color, the corresponding basis elements of $\mathcal{H}_{\text{int}}$ are given by

$$u_L \ d_L$$
$$u_R \ d_R$$

and similarly for the remaining two colors. Leptons of the first generation are

$$\nu_L \ e_L$$
$$e_R,$$

where the lack of a right handed neutrino reflects the famous asymmetry of the β-decay. Adding the list of antiparticles brings us to 30 basis elements, and, for the time being, there seems to be no reason to assume existence of more than 3 generations, thus $\dim \mathcal{H}_{\text{int}} = 90$.

The space $\mathcal{H}_{\text{int}}$ splits into lepton, quark, antilepton and antiquark subspaces,

$$\mathcal{H}_{\text{int}} = \mathcal{H}_{\text{int}}^+ \oplus \mathcal{H}_{\text{int}}^- = \mathcal{H}_\ell^+ \oplus \mathcal{H}_q^+ \oplus \mathcal{H}_\ell^- \oplus \mathcal{H}_q^-,$$

and each of these subspaces decomposes according to chirality. Thus, for $\mathcal{H}_\ell^+$, $\mathcal{H}_q^+$, $\mathcal{H}_\ell^-$ and $\mathcal{H}_q^-$ we have $\mathcal{H}_\ell^+ = \mathcal{H}_{\ell R}^+ \oplus \mathcal{H}_{\ell L}^+$ and similarly in the other cases.

(Note, however, that $\dim \mathcal{H}_{\ell L}^+ = \dim \mathcal{H}_{\ell L}^- = 6$ whereas $\dim \mathcal{H}_{\ell R}^+ = \dim \mathcal{H}_{\ell R}^- = 3$.) We will also account for the different particle generations by an internal tensor product with $\mathbb{C}^3$. The internal algebra will reflect the gauge group $U(1) \times SU(2) \times SU(3)$ of the standard model. Let

$$\mathfrak{A}_{\text{int}} = \mathbb{C} \oplus \mathbb{H} \oplus M_3(\mathbb{C}).$$

We note that this algebra is a *real* involutive algebra. As a consequence, those axioms of a real spectral triple that explicitly resort to complex algebras have to be checked for the complexification of $\mathfrak{A}_{\text{int}}$ (as an algebra acting on $\mathcal{H}_{\text{int}}$.) If $q \in \mathbb{H}$ is a quaternion, we will write

$$q = \begin{pmatrix} \alpha & \beta \\ -\bar{\beta} & \bar{\alpha} \end{pmatrix},$$

where α, β are complex numbers.

The representation π_{int} of $\mathfrak{A}_{\text{int}}$ on H_{int} will be defined as a direct sum of representations on the lepton, quark, antilepton and antiquark sectors:

$$\pi_{\text{int}} = \pi_{\text{int}}^+ \oplus \pi_{\text{int}}^- = \pi_\ell^+ \oplus \pi_q^+ \oplus \pi_\ell^- \oplus \pi_q^-.$$

For the quark sector, we have

$$\mathcal{H}_q^+ = \left[\mathbb{C}_{ud}^2 \otimes \mathbb{C}_{\text{gen}}^3 \otimes \mathbb{C}_{\text{col}}^3\right]_R \oplus \left[\mathbb{C}_{ud}^2 \otimes \mathbb{C}_{\text{gen}}^3 \otimes \mathbb{C}_{\text{col}}^3\right]_L$$

and we let, for λ in $\mathbb{C}$, q in $\mathbb{H}$ and m in $M_3(\mathbb{C})$,

$$\pi_q^+(\lambda, q, m) := \begin{pmatrix} \lambda & 0 & 0 & 0 \\ 0 & \bar{\lambda} & 0 & 0 \\ 0 & 0 & \alpha & -\bar{\beta} \\ 0 & 0 & \beta & \bar{\alpha} \end{pmatrix} \otimes 1_3 \otimes 1_3$$

Here, the first matrix is supposed to act on a vector of type $(\mu_R, \nu_R, \mu_L, \nu_L)$, where $\mu \in \{u, c, t\}$ and $\nu \in \{d, s, b\}$ vary according to generation. Note that the color indices have been omitted. For the lepton sector with massless neutrinos, put

$$\pi_\ell^+(\lambda, q) := \begin{pmatrix} \lambda & 0 & 0 \\ 0 & \alpha & \beta \\ 0 & -\bar{\beta} & \bar{\alpha} \end{pmatrix} \otimes 1_3,$$

where the first matrix acts on a vector of type (e_R, e_L, ν_L), and everything takes place on the Hilbert space

$$\mathcal{H}_\ell^+ = \mathcal{H}_{R\ell}^+ \oplus \mathcal{H}_{L\ell}^+ = \left[\mathbb{C}_e \otimes \mathbb{C}_{\text{gen}}^3\right]_R \oplus \left[\mathbb{C}_{e\nu}^2 \otimes \mathbb{C}_{\text{gen}}^3\right]_L.$$

The action of π_{int} for antiparticles is simpler. For the antiquark sector we have

$$\mathcal{H}_q^- = \left[\bar{\mathbb{C}}_{ud}^2 \otimes \bar{\mathbb{C}}_{\text{gen}}^3 \otimes \bar{\mathbb{C}}_{\text{col}}^3\right]_L \oplus \left[\bar{\mathbb{C}}_{ud}^2 \otimes \bar{\mathbb{C}}_{\text{gen}}^3 \otimes \bar{\mathbb{C}}_{\text{col}}^3\right]_R,$$

and for antileptons,
$$\mathcal{H}_\ell^- = \left[\bar{\mathbb{C}}_e \otimes \bar{\mathbb{C}}_{\text{gen}}^3\right]_L \oplus \left[\bar{\mathbb{C}}_{e\nu}^2 \otimes \bar{\mathbb{C}}_{\text{gen}}^3\right]_R,$$

so that we may put
$$\pi_q^-(\lambda, q, m) := 1_4 \otimes 1_3 \otimes m \quad \text{and} \quad \pi_\ell^-(\lambda, q, m) := \lambda \otimes 1_3.$$

The full action of π_{int} can be summarized as follows:
$$\pi_{\text{int}}(\lambda, q, m) \begin{pmatrix} \xi \\ \bar{\eta} \end{pmatrix} = \pi_{\text{int}}(\lambda, q, m) \begin{pmatrix} \xi_\ell \\ \xi_q \\ \bar{\eta}_\ell \\ \bar{\eta}_q \end{pmatrix} = \begin{pmatrix} \pi_\ell^+ \xi_\ell \\ \pi_q^+ \xi_q \\ \lambda \bar{\eta}_\ell \\ m \bar{\eta}_q \end{pmatrix}.$$

For the operator J, let $J = C \otimes J_{\text{int}}$, where C is charge conjugation on $L^2(M, S)$ and J_{int} is charge conjugation on $\mathcal{H}_{\text{int}}$, i.e. for $(\xi, \bar{\eta}) \in \mathcal{H}_{\text{int}}^+ \oplus \mathcal{H}_{\text{int}}^-$ we have
$$J_{\text{int}} \begin{pmatrix} \xi \\ \bar{\eta} \end{pmatrix} = \begin{pmatrix} \eta \\ \bar{\xi} \end{pmatrix}$$

We finally describe the Dirac operator that will be used within the real spectral triple under construction. As in the previous definitions, D will be defined according to the tensor product structure all other components have. So,
$$D_0 = D_S \otimes \text{Id}_{\mathcal{H}_{\text{int}}} + \gamma^5 \otimes D_{\text{int}},$$

where D_{int} is the ingredient that requires some extra effort in the definition. It splits into lepton and quark parts:
$$D_{\text{int}} = \begin{pmatrix} D_\ell & 0 \\ 0 & D_q \otimes 1_3 \end{pmatrix}.$$

With respect to the right-left splitting, each Dirac operator is a Yukawa coupling matrix, has the dimension of a mass, and is of the form:
$$D_{\ell, q} = \begin{pmatrix} 0 & M_{\ell, q} \\ M_{\ell, q}^* & 0 \end{pmatrix}.$$

Specifically,
$$M_q = \begin{pmatrix} m_0^d \otimes H_0 & m_0^u \otimes \tilde{H}_0 \end{pmatrix}$$

and
$$M_\ell = m_0^e \otimes H_0$$

Here, m_0^d, m_0^u and m_0^e are 3×3 positive definite matrices of Yukawa coupling constants, acting on the space of generations. Also,
$$H_0 = \mu \begin{pmatrix} 0 \\ 1 \end{pmatrix} \quad \text{and} \quad \tilde{H}_0 = \mu \begin{pmatrix} 1 \\ 0 \end{pmatrix}.$$

To see the effect of these definitions one has to observe the following facts: The full Dirac operator

$$D = D_0 + A + JAJ^{-1}$$

yields by a calculation which is similar to (though more complicated than) the corresponding one for one gauge field (section 6.2) $U(1)$, $SU(2)$ and $U(3)$ gauge fields as well as a Higgs field. The computation of $A + JAJ^{-1}$ together with $A = A^*$ removes, again as in section 6.2, a $U(1)$ part from the above gauge fields, and the full matrix is traceless. The fermion sector, $\langle \psi, D\psi \rangle$, yields the corresponding action of the standard model, and, a long and cumbersome calculation, again similar to the one we have explained in more detail above, shows that the bosonic part of the Lagrangian has an asymptotic expansion

$$\mathfrak{I}^B_{D_0,\mathcal{A}} = \operatorname{tr} \chi \left(\frac{D^2}{m_0^2} \right) \sim \frac{1}{4\pi^2} \left(a_0 m_0^4 + a_2 m_0^2 + a_4 \right) + O(m_0^{-\infty}), \quad \text{as } m_0 \to \infty,$$

where

$$a_0 = 45 f_0 \int_M \operatorname{vol}_{(M,g)} \cdot$$

$$a_2 = 3 f_2 \int_M \frac{5}{4} R - 2 \operatorname{tr} \left(|m_0^d|^2 + |m_0^u|^2 + \frac{1}{3}|m_0^e|^2 \right) H^* H \operatorname{vol}_{(M,g)},$$

$$a_4 = f_4 \int_M \left(\frac{1}{40} \frac{5}{4} (12 R_{;\mu}{}^\mu + 11 R^* R^* - 18 W_{\mu\nu\rho\sigma} W^{\mu\nu\rho\sigma}) \right.$$

$$+ 3 \operatorname{tr} \left(|m_0^d|^2 + |m_0^u|^2 + \frac{1}{3}|m_0^e|^2 \right) H^* H \left(D_\mu H^* D^\mu H - \frac{1}{6} R H^* H \right)$$

$$+ g_{03}^2 G^i_{\mu\nu} G^{\mu\nu i} + g_{02}^2 F^\alpha_{\mu\nu} F^{\mu\nu\alpha} + \frac{5}{3} g_{01}^2 B_{\mu\nu} B^{\mu\nu}$$

$$+ 3 \operatorname{tr} \left((|m_0^d|^2 + |m_0^u|^2)^2 + \frac{1}{3}|m_0^e|^4 \right) H^* H (H^* H)^2$$

$$\left. - \operatorname{tr} \left(|m_0^d|^2 + |m_0^u|^2 + \frac{1}{3}|m_0^e|^2 \right) H^* H (H^* H)_{;\mu}{}^\mu \right) \operatorname{vol}_{(M,g)}.$$

At this point, second quantization and renormalization would have to enter the scene. Also, all known constants should be inserted and the physical consequences would have to be discussed. We nevertheless stop here and refer to [31], [25], [159] and [196] for a detailed account.

Part II

The Lagrangian of the Standard Model
Derived from Noncommutative Geometry

This part gives a detailed exposition of the Standard Model of elementary particle physics from the point of view of noncommutative geometry. In this sense it is a continuation of the main ideas of Chapter 1, but with emphasis on the full structure (fermionic and bosonic field content) of the Standard Model and other physically realistic theories. The aim is to derive the Lagrange densities for the bosonic and fermionic fields, both in Euclidean and Minkowski space-time, within a uniform notational framework. Although the analytical foundations of noncommutative calculus, such as Dixmier trace and pseudo-differential calculus, are fully established only in the case of compact Riemannian manifolds, the resulting trace formulas are also meaningful in the pseudo-Riemannian case, leading to the detailed expression for the physical Lagrange density.

According to the four known interactions (electro-weak, strong and gravitational interaction) the presentation proceeds in three steps

- electro-weak model,
- Standard Model including the strong interaction,
- Standard Model coupled with gravity.

The first two models are Yang-Mills theories of Connes-Lott type, the third model involves the spectral action of Connes-Chamsedinne. In order to compare the models, the fermions are listed in a way to make the electroweak model a restriction of the complete Standard Model. In this manner, the similarities and characteristic differences of the three approaches become more transparent. The all important Higgs mechanism of symmetry breaking is discussed in detail, both within the noncommutative geometry framework and from the point of view of physics.

7 Dirac Operator and Real Structure on Euclidean and Minkowski Spacetime

Herald Upmeier

University of Marburg

7.1 γ-Matrices on Flat and Curved Spacetime

In this section we collect the basic facts concerning spinors and the Dirac operator on space-time M. We assume that

$$n := \dim M = 2m$$

is even, with $m = 2$ the physically relevant case. Throughout, we fix $\kappa \in \{1, i\}$ in order to treat space-times of Euclidean type ($\kappa = i$) and Minkowski type ($\kappa = 1$) simultaneously. Let

$$\eta = (\eta^{ij}) = \begin{pmatrix} 1 & & & & \\ & -\kappa^2 & & 0 & \\ & & -\kappa^2 & & \\ & 0 & & \ddots & \\ & & & & -\kappa^2 \end{pmatrix} \qquad (7.1)$$

denote the corresponding flat metric. Our first goal is to construct the γ-matrices for η in an explicit way. The γ-matrices $\gamma^j = \gamma_m^j$ ($0 \leq j < 2m$) of size $2^m \times 2^m$ and the associated grading operator

$$\Gamma = \Gamma_m = \kappa i^{m-1} \gamma_m^0 \ldots \gamma_m^{2m-1} \qquad (7.2)$$

are constructed by induction as follows: We put

$$\gamma_1^0 := \begin{pmatrix} 0 & 1 \\ 1 & 0 \end{pmatrix}, \quad \gamma_1^1 := \begin{pmatrix} 0 & \kappa \\ -\kappa & 0 \end{pmatrix}$$

$$\Gamma_1 = \kappa \begin{pmatrix} 0 & 1 \\ 1 & 0 \end{pmatrix} \begin{pmatrix} 0 & \kappa \\ -\kappa & 0 \end{pmatrix} = \begin{pmatrix} -\kappa^2 & 0 \\ 0 & \kappa^2 \end{pmatrix}$$

and for the induction step $1 \leq m \to m+1$ we put

$$\gamma_{m+1}^0 := \begin{pmatrix} 0 & I_m \\ I_m & 0 \end{pmatrix},$$

$$\gamma_{m+1}^1 := \begin{pmatrix} 0 & \kappa \gamma_m^0 \\ -\kappa \gamma_m^0 & 0 \end{pmatrix},$$

$$\gamma_{m+1}^{j+1} := \begin{pmatrix} 0 & -i\gamma_m^j \\ i\gamma_m^j & 0 \end{pmatrix} \quad (1 \leq j < 2m),$$

$$\gamma_{m+1}^{2m+1} := \begin{pmatrix} 0 & -\kappa \Gamma_m \\ \kappa \Gamma_m & 0 \end{pmatrix}.$$

Here $I = I_m$ is the $2^m \times 2^m$ unit matrix. Then

$$\Gamma_{m+1} = \begin{pmatrix} -\kappa^2 I_m & 0 \\ 0 & \kappa^2 I_m \end{pmatrix}. \tag{7.3}$$

From the construction it is clear that we have

$$\Gamma^2 = I \tag{7.4}$$

and

$$\gamma^j \Gamma = -\Gamma \gamma^j \tag{7.5}$$

for all j. This also follows algebraically from

Proposition 1. *The γ-matrices satisfy the anticommutation relations*

$$\gamma^i \gamma^j + \gamma^j \gamma^i = 2\eta^{ij} I. \tag{7.6}$$

Proof. By induction on $m \geq 1$.

In the following we fix m and simply write $\gamma^i := \gamma^i_m$ and

$$\Gamma = \kappa i^{m-1} \gamma^0 \ldots \gamma^{2m-1}. \tag{7.7}$$

Lemma 1. *For $0 \leq i, j, k < n$ we have*

$$\frac{1}{2}(\gamma^k \gamma^i \gamma^j - \gamma^i \gamma^j \gamma^k) = \eta^{ki} \gamma^j - \gamma^i \eta^{jk}. \tag{7.8}$$

Proof.

$$\frac{1}{2}(\gamma^k \gamma^i \gamma^j - \gamma^i \gamma^j \gamma^k) = \frac{\gamma^k \gamma^i + \gamma^i \gamma^k}{2} \gamma^j - \gamma^i \frac{\gamma^j \gamma^k + \gamma^k \gamma^j}{2} = \eta^{ki} \gamma^j - \gamma^i \eta^{jk}.$$

Lemma 2. *For pairwise distinct indices $i_1, \ldots, i_p$ we have*

$$(\gamma^{i_1} \ldots \gamma^{i_p})^2 = (-1)^{p(p-1)/2} (\gamma^{i_1})^2 \ldots (\gamma^{i_p})^2. \tag{7.9}$$

Proof. This follows by induction from Proposition 1.

Lemma 3. *For distinct indices $i_1, \ldots, i_p$ we have*

$$\operatorname{tr} \gamma^{i_1} \ldots \gamma^{i_p} = 0. \tag{7.10}$$

Proof. If p is even, (7.9) implies

$$\operatorname{tr} \gamma^{i_1} \ldots \gamma^{i_p} = (-1)^{p-1} \operatorname{tr} \gamma^{i_2} \ldots \gamma^{i_p} \gamma^{i_1}$$
$$= -\operatorname{tr} \gamma^{i_2} \ldots \gamma^{i_p} \gamma^{i_1} = -\operatorname{tr} \gamma^{i_1} \ldots \gamma^{i_p}.$$

If p is odd, (7.5) implies

$$\Gamma \gamma^{i_1} \ldots \gamma^{i_p} = -\gamma^{i_1} \ldots \gamma^{i_p} \Gamma$$

and hence

$$\operatorname{tr} \gamma^{i_1} \ldots \gamma^{i_p} = \operatorname{tr} \Gamma \gamma^{i_1} \ldots \gamma^{i_p} \Gamma^{-1} = -\operatorname{tr} \gamma^{i_1} \ldots \gamma^{i_p}. \tag{7.11}$$

Let T^* denote the usual hermitian adjoint of a $2^m \times 2^m$ matrix T. Since $\Gamma = \overline{\Gamma} = \Gamma^*$ and $\overline{\kappa} = \kappa^3$ we also get by induction

Proposition 2. *We have*

$$\overset{*}{\gamma}^0 = \gamma^0, \quad \overset{*}{\gamma}^{j+1} = -\kappa^2 \gamma^{j+1} \tag{7.12}$$

and

$$\overline{\gamma}^0 = \gamma^0, \quad \overline{\gamma}^{2j+1} = \kappa^2 \gamma^{2j+1}, \quad \overline{\gamma}^{2j+2} = -\kappa^2 \gamma^{2j+2}. \tag{7.13}$$

As a consequence, it follows that

$$\gamma^j \overset{*}{\gamma}^j = I$$

and

$$\gamma^{2j+\varepsilon} \overline{\gamma}^{2j+\varepsilon} = (-1)^\varepsilon I$$

for all j and $\varepsilon \in \{0, 1\}$.

Remark 1. *The relations (7.12) correspond to the standard involution of real Clifford algebras $C\ell(p, q)$ of signature (p, q), as defined in [39, (47)] or Definition 3 (Chapter I). In fact in the Minkowski case, we have signature $(1, 2m - 1)$ and $\kappa = 1$ so that*

$$\overset{*}{\gamma}^0 = \gamma^0, \quad \overset{*}{\gamma}^{j+1} = -\gamma^{j+1} \quad (j \geq 0). \tag{7.14}$$

In the Euclidean case of signature $(2m, 0)$, we have $\kappa = i$ and hence

$$\overset{*}{\gamma}^0 = \gamma^0, \quad \overset{*}{\gamma}^{j+1} = \gamma^{j+1} \quad (j \geq 0). \tag{7.15}$$

For any column vector $\psi \in \mathbb{C}^{2^m}$ let ψ^* be the adjoint row vector and put

$$\psi^\dagger := \begin{cases} \psi^* & \kappa = i \\ \psi^* \gamma^0 & \kappa = 1. \end{cases} \tag{7.16}$$

Then $\psi^\dagger \psi'$ defines a (non-degenerate) inner product on $\mathbb{C}^{2^m}$ (conjugate-linear in the first variable), and for $T \in \mathbb{C}^{2^m \times 2^m}$ we define $T^\dagger \in \mathbb{C}^{2^m \times 2^m}$ by

$$(T\psi)^\dagger \psi' = \psi^\dagger (T^\dagger \psi') \tag{7.17}$$

for all $\psi, \psi' \in \mathbb{C}^{2^m}$. Thus

$$T^\dagger = \begin{cases} T^* & \kappa = i \\ \gamma^0 T^* \gamma^0 & \kappa = 1. \end{cases} \tag{7.18}$$

Proposition 3. *We have*

$$(\gamma^j)^\dagger = \gamma^j \tag{7.19}$$

for $0 \leq j < n$, and

$$\Gamma^\dagger = -\kappa^2 \Gamma. \tag{7.20}$$

Proof. By Proposition 2, $(\gamma^j)^\dagger = (\gamma^j)^* = \gamma^j$ in case $\kappa = i$, whereas for $j > 0$

$$(\gamma^j)^\dagger = \gamma^0(\gamma^j)^*\gamma^0 = -\gamma^0\gamma^j\gamma^0 = \gamma^j\gamma^0\gamma^0 = \gamma^j$$

in case $\kappa = 1$. Moreover, (7.20) implies

$$\Gamma^\dagger = \Gamma^* = \Gamma$$

for $\kappa = i$ and

$$\Gamma^\dagger = \gamma^0 \Gamma^* \gamma^0 = \gamma^0 \Gamma \gamma^0 = -\gamma^0 \gamma^0 \Gamma = -\Gamma$$

for $\kappa = 1$.

Up to now our considerations were concerned with the case of a flat metric η of Euclidean or Minkowski type. Now we pass to curved space-time M, realized as an n-dimensional manifold. Let TM and $T^\#M$ denote the (real) tangent and cotangent bundle of M, resp., and let

$$\bigwedge(T^\#M) = \bigoplus_{i=0}^{n} \bigwedge^i(T^\#M) \qquad (7.21)$$

denote the bundle of exterior forms over M.

Given a $\mathbb{K}$-vector bundle V over M (with $\mathbb{K} = \mathbb{R}, \mathbb{C}$), the space of smooth sections of V over M will be denoted by $\mathcal{C}^\infty(V)$. It is endowed with the structure of a $\mathcal{C}^\infty(M, \mathbb{K})$-module. In particular, $\mathcal{C}^\infty(T^\#M)$ and $\mathcal{C}^\infty(\bigwedge^2 T^\#M)$ is the space of differential 1-forms and 2-forms, resp. In terms of local coordinates

$$(x^\mu)_{0 \leq \mu < n} \qquad (7.22)$$

of M, every $\omega \in \mathcal{C}^\infty(T^\#M)$ and $\sigma \in \mathcal{C}^\infty(\bigwedge^2 T^\#M)$ can be expressed as

$$\omega = \omega_\mu \, dx^\mu, \qquad (7.23)$$
$$\sigma = \tfrac{1}{2}\sigma_{\mu\nu} \, dx^\mu \wedge dx^\nu, \qquad (7.24)$$

where $\omega_\mu \in \mathcal{C}^\infty(M, \mathbb{R})$, $\sigma_{\mu\nu} = -\sigma_{\nu\mu} \in \mathcal{C}^\infty(M, \mathbb{R})$ and the Einstein summation convention is understood.

Now consider a fibre metric g on TM, of Euclidean or Minkowski type, and let $(\omega \mid \omega')$ denote the induced inner product on $T_x^\#(M)$, for $x \in M$. A similar notation will be used for elements in $\bigwedge T_x^\#(M)$. For sections ω, ω' the inner product is taken fibrewise so that $(\omega \mid \omega')$ becomes a scalar function on M. For $\mathbb{C}$-valued forms, we extend the inner product to be conjugate linear in the first variable. Using local coordinates we define

$$\partial_\mu := \frac{\partial}{\partial x^\mu} \in \mathcal{C}^\infty(TM), \qquad (7.25)$$
$$g_{\mu\nu} := g(\partial_\mu, \partial_\nu), \qquad (7.26)$$
$$g^{\mu\nu} := (dx^\mu \mid dx^\nu). \qquad (7.27)$$

Then
$$(\omega \mid \omega') = g^{\mu\nu}\overline{\omega}_\mu \omega'_\nu \tag{7.28}$$

for all 1-forms $\omega, \omega' \in \mathcal{C}^\infty(T^\# M \otimes \mathbb{C})$. The fibre metric on $T^\# M$ induced by g gives rise to a (smooth) bundle of complex Clifford algebras denoted by $\mathcal{C}\ell(T^\# M)$. For fixed $x \in M$, $\mathcal{C}\ell(T^\#_x M)$ is the 2^n-dimensional algebra generated by $T^\#_x M$ subject to the anti-commutation relations

$$\omega \times \omega' + \omega' \times \omega = 2(\omega \mid \omega') \tag{7.29}$$

for all $\omega, \omega' \in T^\#_x M$. Here $\times$ denotes the Clifford algebra product. In particular, we have

$$dx^\mu \times dx^\nu + dx^\nu \times dx^\mu = 2g^{\mu\nu}. \tag{7.30}$$

In order to construct the spinor bundle and Dirac operator on M one has to assume the existence of a spin structure associated with the metric g. Globally this imposes a topological restriction on M, but locally the complex spinor bundle

$$S \approx M \times \mathbb{C}^{2^m} \tag{7.31}$$

and the associated representation

$$\mathcal{C}^\infty(\mathcal{C}\ell(T^\# M)) \xrightarrow{\gamma} \mathrm{End}(S) \approx \mathcal{C}^\infty(M, \mathbb{C}^{2^m \times 2^m}) \tag{7.32}$$

can be constructed as follows: Using local coordinates x^μ we assume that

$$g^{\mu\nu} = e^\mu_i \eta^{ij} e^\nu_j \tag{7.33}$$

for some smooth $\mathrm{Gl}(n, \mathbb{R})$-valued matrix function $e = (e^\mu_i)$ on M. Let $\overset{-1}{e}$ denote the inverse matrix function satisfying

$$e^i_\mu \overset{-1}{e}{}^\mu_j = \delta^i_j, \quad \overset{-1}{e}{}^\mu_i e^i_\nu = \delta^\mu_\nu.$$

Putting
$$e^i := e^i_\mu dx^\mu \in \mathcal{C}^\infty(T^\# M) \tag{7.34}$$

we have
$$(e^i \mid e^j) = \eta^{ij} \tag{7.35}$$

for all $0 \leq i, j < n$. This implies the relation

$$e^i \times e^j + e^j \times e^i = 2\eta^{ij} \tag{7.36}$$

in $\mathcal{C}^\infty(\mathcal{C}\ell(T^\# M))$. The matrix inverse $\overset{-1}{e}$ defines an η-orthonormal basis

$$\overset{-1}{e}_i := \overset{-1}{e}{}^\mu_i \partial_\mu \in \mathcal{C}^\infty(TM) \tag{7.37}$$

of vector fields. Using the (flat) γ-matrices constructed in Proposition 1, we define

$$\gamma(dx^\mu) := \overset{-1}{e}{}^\mu_i \gamma^i \in \mathcal{C}^\infty(M, \mathbb{C}^{2^m \times 2^m}) \tag{7.38}$$

regarded as endomorphisms of the spinor bundle. Then Proposition 1 implies
$$\gamma(dx^\mu)\gamma(dx^\nu) + \gamma(dx^\nu)\gamma(dx^\mu) = 2g^{\mu\nu}I. \tag{7.39}$$
Now the representation (7.32) is fibrewise defined using the universal property of $C\ell(T^\# M)$. Then we have
$$\gamma(e^i) = e^i_\mu \gamma(dx^\mu) = \gamma^i \tag{7.40}$$
and
$$\gamma(\omega) = \omega_\mu \gamma(dx^\mu) = \omega_\mu \, e^{-1\mu}_{i} \gamma^i \tag{7.41}$$
for $\omega = \omega_\mu \, dx^\mu \in C^\infty(T^\# M \otimes \mathbb{C})$. For functions $f \in C^\infty(M, \mathbb{C}) \subset C^\infty(C\ell(T^\# M))$, $\gamma(f)$ is just multiplication by f and we simply write $\gamma(f) = f$. For an endomorphism
$$T \in \mathrm{End}(S) \approx C^\infty(M, \mathbb{C}^{2^m \times 2^m})$$
of the spinor bundle S, endowed with the fibre metric $\psi^\dagger \psi'$ introduced in (7.16) (using the constant γ^0-matrix in the Minkowski case), we define the adjoint
$$T^\dagger \in \mathrm{End}(S)$$
fibrewise via (7.17). We also let
$$\underline{\mathrm{tr}}_S T := 2^{-m} \mathrm{tr}_S T \tag{7.42}$$
denote the normalized trace, and put
$$(T \mid T')_S := \underline{\mathrm{tr}}_S T^\dagger T'. \tag{7.43}$$
Usually (7.42) and (7.43) are taken fibrewise and therefore depend on $x \in M$. Integration over M is then denoted separately. By Proposition 3, we have
$$\gamma(dx^\mu)^\dagger = \gamma(\overline{dx^\mu}) \tag{7.44}$$
and hence
$$\gamma(\omega)^\dagger = \gamma(\overline{\omega}) \tag{7.45}$$
for all $\omega \in C^\infty(T^\# M \otimes \mathbb{C})$, where $\overline{\omega} := \overline{\omega}_\mu \, dx^\mu$. For $f \in C^\infty(M, \mathbb{C})$ we have
$$f^\dagger = \overline{f} \tag{7.46}$$
and
$$(\Gamma f)^\dagger = -\kappa^2 \Gamma \overline{f} \tag{7.47}$$
since, by Proposition 3,
$$(\Gamma f)^\dagger = \overline{f} \Gamma^\dagger = -\kappa^2 \overline{f} \, \Gamma = -\kappa^2 \Gamma \overline{f}.$$
Also, (7.4) implies
$$\Gamma \gamma(dx^\mu) + \gamma(dx^\mu) \Gamma = 0 \tag{7.48}$$
for all μ, and hence for all $\omega \in C^\infty(T^\# M \otimes \mathbb{C})$
$$\Gamma \gamma(\omega) + \gamma(\omega) \Gamma = 0. \tag{7.49}$$

Lemma 4. For $\omega, \omega' \in \mathcal{C}^\infty(T^\# M \otimes \mathbb{C})$, we have
$$\bigl(\gamma(\omega) \,\bigm|\, \gamma(\omega')\bigr)_S = (\omega \mid \omega'). \tag{7.50}$$

Proof. The relation (7.29) implies for real 1-forms
$$2\operatorname{\underline{tr}}_S \gamma(\omega)\gamma(\omega') = \operatorname{\underline{tr}}_S\bigl(\gamma(\omega)\gamma(\omega') + \gamma(\omega')\gamma(\omega)\bigr) = 2\operatorname{\underline{tr}}_S(\omega \mid \omega')\mathbf{1} = 2(\omega \mid \omega').$$
In view of (7.45), the assertion follows.

Lemma 5. For $\omega, \omega' \in \mathcal{C}^\infty(T^\# M)$, we have
$$\operatorname{\underline{tr}}_S \gamma(\omega_1)\gamma(\omega_2)\gamma(\omega_3)\gamma(\omega_4) =$$
$$(\omega_1 \mid \omega_2)(\omega_3 \mid \omega_4) - (\omega_1 \mid \omega_3)(\omega_2 \mid \omega_4) + (\omega_1 \mid \omega_4)(\omega_2 \mid \omega_3).$$

Proof. The identity
$$\begin{aligned}
&2(\omega_1 \mid \omega_2)\omega_3 \times \omega_4 - 2(\omega_1 \mid \omega_3)\omega_2 \times \omega_4 + 2\omega_2 \times \omega_3(\omega_1 \mid \omega_4)\\
&= (\omega_1 \times \omega_2 + \omega_2 \times \omega_1) \times \omega_3 \times \omega_4\\
&\quad - \omega_2 \times (\omega_1 \times \omega_3 + \omega_3 \times \omega_1) \times \omega_4\\
&\quad + \omega_2 \times \omega_3 \times (\omega_1 \times \omega_4 + \omega_4 \times \omega_1)\\
&= \omega_1 \times \omega_2 \times \omega_3 \times \omega_4 + \omega_2 \times \omega_3 \times \omega_4 \times \omega_1
\end{aligned}$$

within $\mathcal{C}^\infty(\mathcal{Cl}(T^\# M))$ implies
$$\begin{aligned}
&2\operatorname{\underline{tr}}_S \gamma(\omega_1)\gamma(\omega_2)\gamma(\omega_3)\gamma(\omega_4)\\
&= \operatorname{\underline{tr}}_S\bigl(\gamma(\omega_1)\gamma(\omega_2)\gamma(\omega_3)\gamma(\omega_4) + \gamma(\omega_2)\gamma(\omega_3)\gamma(\omega_4)\gamma(\omega_1)\bigr)\\
&= 2(\omega_1 \mid \omega_2)\operatorname{\underline{tr}}_S \gamma(\omega_3)\gamma(\omega_4) - 2(\omega_1 \mid \omega_3)\operatorname{\underline{tr}}_S \gamma(\omega_2)\gamma(\omega_4)\\
&\quad + 2(\omega_2 \mid \omega_3)\operatorname{\underline{tr}}_S \gamma(\omega_1)\gamma(\omega_4).
\end{aligned}$$

Now the assertion follows with Lemma 4.

Corollary 1. We have
$$\operatorname{\underline{tr}}_S \gamma^i \gamma^j \gamma^k \gamma^l = \eta^{ij}\eta^{kl} - \eta^{ik}\eta^{jl} + \eta^{il}\eta^{jk} \tag{7.51}$$

and
$$\operatorname{\underline{tr}}_S \gamma(dx^\mu)\gamma(dx^\nu)\gamma(dx^\alpha)\gamma(dx^\beta) = g^{\mu\nu}g^{\alpha\beta} - g^{\mu\alpha}g^{\nu\beta} + g^{\mu\beta}g^{\nu\alpha}. \tag{7.52}$$

The exterior algebra bundle $\bigwedge(T^\# M)$ has an η-orthonormal basis given by
$$e^{i_1} \wedge \ldots \wedge e^{i_p} \tag{7.53}$$
for all $0 \leq i_1 < \ldots < i_p < n$. Mapping (7.53) to
$$\gamma^{i_1} \ldots \gamma^{i_p} = \gamma(e^{i_1} \times \ldots \times e^{i_p}) \tag{7.54}$$
we may identify $\mathcal{Cl}(T^\# M) \approx \bigwedge(T^\# M) \otimes \mathbb{C}$ as $\mathbb{C}$-vector spaces. The same holds for the $\mathcal{C}^\infty(M, \mathbb{C})$-modules of sections. Accordingly, we write $\gamma(\sigma) \in \operatorname{End}(S)$ for p-forms σ, with $p = 1$ and $p = 2$ the degrees actually used in the sequel. Generalizing Lemma 4 we obtain

Proposition 4. For $\sigma, \sigma' \in \bigwedge(T^\# M) \otimes \mathbb{C}$ we have

$$(\gamma(\sigma) \mid \gamma(\sigma'))_S = (\sigma \mid \sigma'), \tag{7.55}$$

where, as mentioned above, $(\sigma \mid \sigma')$ is defined to be conjugate linear in the first variable.

Proof. It is enough to show that the $\mathbb{C}$-linear basis (7.54) of $C\ell(T^\# M)$ is η-orthonormal under the inner product (7.43). By Proposition 3 we have for $0 \leq i_1 < \ldots < i_p < n$ and $0 \leq j_1 < \ldots < j_q < n$

$$\begin{aligned}\left(\gamma^{i_1} \ldots \gamma^{i_p} \mid \gamma^{j_1} \ldots \gamma^{j_q}\right)_S &= \underline{\text{tr}}_S \left(\gamma^{i_1} \ldots \gamma^{i_p}\right)^\dagger \left(\gamma^{j_1} \ldots \gamma^{j_q}\right) \\ &= \underline{\text{tr}}_S \gamma^{i_p} \ldots \gamma^{i_1} \gamma^{j_1} \ldots \gamma^{j_q}. \end{aligned} \tag{7.56}$$

In case $\{i_1, \ldots, i_p\} \neq \{j_1, \ldots, j_q\}$ we may reduce (7.56) to $\pm \underline{\text{tr}}_S \gamma^{k_1} \ldots \gamma^{k_l}$ where $0 \leq k_1 < \ldots < k_l < n$ and $l \geq 1$. By Lemma 3 this trace vanishes. Now suppose $p = q$ and $i_1 = j_1, \ldots, i_p = j_p$. Then (7.56) equals $\eta^{i_1 i_1} \ldots \eta^{i_p i_p}$ which shows η-orthogonality.

For the construction of the Dirac operator, we consider real matrices $.F^{\cdot} = (_i F^j)$ satisfying the skew-symmetry condition

$$_i F^j \eta_{jj} = -_j F^i \eta_{ii} =: {}_i F_j = -_j F_i. \tag{7.57}$$

Proposition 5. *The inner product of* $.F^{\cdot}$ *and* $.G^{\cdot}$ *is given by*

$$(F \mid G) = -\operatorname{tr}.F^{\cdot}.G^{\cdot}. \tag{7.58}$$

Proof. By definition,

$$\begin{aligned}(F \mid G) &= {}_i F_j \, {}^i G^j = {}_i F^j \, \eta_{jj} \, \eta^{ii} \, {}_i G^j \\ &= -_j F^i \, \eta_{ii} \, \eta^{ii} \, {}_i G^j = -_j F^i \, {}_i G^j = -\operatorname{tr}.F^{\cdot}.G^{\cdot}.\end{aligned}$$

Using the η-orthonormal basis (e^i) of the cotangent bundle $T^\# M$, let

$$.\underset{\sim}{F} = \frac{1}{4} \, {}_i F^j \, \eta_{jj} \, e^i \times e^j \tag{7.59}$$

denote the corresponding element in the Clifford algebra $C\ell(T^\# M)$. Using the spin representation (7.32), we obtain an endomorphism

$$\gamma(.\underset{\sim}{F}) = \frac{1}{4} \, {}_i F^j \, \eta_{jj} \, \gamma^i \gamma^j. \tag{7.60}$$

Proposition 6. *The normalized trace satisfies*

$$\underline{\text{tr}}_S \, \gamma(.\underset{\sim}{F}) \gamma(.\underset{\sim}{G}) = \frac{1}{8} \operatorname{tr}.F^{\cdot}.G^{\cdot} = -\frac{1}{8}(.F^{\cdot} \mid .G^{\cdot}). \tag{7.61}$$

Proof.

$$16\,\underline{\mathrm{tr}}_S\,\gamma(.\underset{\sim}{F}\,\!\!\dot{\,})\gamma(.\underset{\sim}{G}\,\!\!\dot{\,}) = \underline{\mathrm{tr}}_S\,{}_iF^j\eta_{jj}\gamma^i\gamma^j{}_kG^l\eta_{ll}\gamma^k\gamma^l = \underline{\mathrm{tr}}_S\,\gamma^i\gamma^j\gamma^k\gamma^l{}_iF^j\eta_{jj}\,{}_kG^l\eta_{ll}$$
$$= (\eta^{ij}\eta^{kl} - \eta^{ik}\eta^{jl} + \eta^{il}\eta^{jk})_i F^j\eta_{jj}\,{}_kG^l\eta_{ll}$$
$$= -\eta^{ii}\eta^{jj}\underbrace{{}_iF^j\eta_{jj}\,{}_iG^j\eta_{jj}}_{-{}_jF^i\eta_{ii}} + \eta^{ii}\eta^{jj}{}_iF^j\eta_{jj}\,{}_jG^i\eta_{ii} = 2{}_iF^j{}_jG^i.$$

Proposition 7.
$$\gamma(.\underset{\sim}{F}\,\!\!\dot{\,})^\dagger = -\gamma(.\underset{\sim}{F}\,\!\!\dot{\,}). \tag{7.62}$$

Proof.

$$\gamma(.\underset{\sim}{F}\,\!\!\dot{\,})^\dagger = \frac{1}{4}{}_i\overline{F}^j\eta_{jj}\gamma^{j\dagger}\gamma^{i\dagger} = \frac{1}{4}{}_iF^j\eta_{jj}\gamma^j\gamma^i = -\frac{1}{4}{}_jF^i\eta_{ii}\gamma^j\gamma^i = -\gamma(.\underset{\sim}{F}\,\!\!\dot{\,}).$$

Corollary 2.
$$\big(\gamma(.\underset{\sim}{F}\,\!\!\dot{\,})\mid\gamma(.\underset{\sim}{G}\,\!\!\dot{\,})\big)_S = \frac{1}{8}(.F\mid .G). \tag{7.63}$$

Proof. Combining Proposition 7 and Proposition 6, we obtain

$$\big(\gamma(.\underset{\sim}{F}\,\!\!\dot{\,})\mid\gamma(.\underset{\sim}{G}\,\!\!\dot{\,})\big)_S = \underline{\mathrm{tr}}_S\,\gamma(.\underset{\sim}{F}\,\!\!\dot{\,})^\dagger\gamma(.\underset{\sim}{G}\,\!\!\dot{\,})$$
$$= -\underline{\mathrm{tr}}_S\,\gamma(.\underset{\sim}{F}\,\!\!\dot{\,})\gamma(.\underset{\sim}{G}\,\!\!\dot{\,}) = \frac{1}{8}(.F\mid .G).$$

Proposition 8.
$$\underline{\mathrm{tr}}_S\,\gamma(.\underset{\sim}{F}\,\!\!\dot{\,}) = 0. \tag{7.64}$$

Proof. Since $i \ne j$, Lemma 3 implies

$$4\,\underline{\mathrm{tr}}_S\,\gamma(.\underset{\sim}{F}\,\!\!\dot{\,}) = {}_iF^j\eta_{jj}\,\underline{\mathrm{tr}}_S\,\gamma^i\gamma^j = 0.$$

7.2 Levi-Civita Connection and Dirac Operator

We finally come to the construction of the Dirac operator. Consider the Levi-Civita connection ∇ on TM associated with g. Locally, we have

$$\nabla_\mu \partial_\nu = \Gamma^\lambda_{\mu\nu}\partial_\lambda, \tag{7.65}$$

where $\nabla_\mu := \nabla_{\partial_\mu}$ and

$$\Gamma^\lambda_{\mu\nu} = \tfrac{1}{2}g^{\lambda\kappa}(\partial_\mu g_{\nu\kappa} + \partial_\nu g_{\mu\kappa} - \partial_\kappa g_{\mu\nu}) \tag{7.66}$$

are the Christoffel symbols. Now consider the η-orthonormal vector fields $\overset{-1}{e}_i$ defined in (7.37) and define 1-forms

$${}_i\omega^j = {}_i\omega^j_\mu\,dx^\mu \in \mathcal{C}^\infty(T^\#M) \tag{7.67}$$

by putting
$$\nabla_\mu \overset{-1}{e}{}_i = {}_i\omega_\mu^j \overset{-1}{e}{}_j. \tag{7.68}$$

It is easy to show that the matrix $.\omega_\mu := ({}_i\omega_\mu^j)$ satisfies the skew-symmetry condition (7.57). Therefore we may define

$$.\underset{\sim}{\omega}_\mu := \tfrac{1}{4} {}_i\omega_\mu^j \, \eta_{jj} e^i \times e^j \in C\ell(T^\# M) \tag{7.69}$$

and obtain the endomorphism

$$\gamma(.\underset{\sim}{\omega}_\mu) := \tfrac{1}{4} {}_i\omega_\mu^j \, \eta_{jj} \gamma^i \gamma^j \in \mathrm{End}(S). \tag{7.70}$$

Definition 1. *The Dirac operator D associated with g (and the underlying spin structure) is the first order differential operator acting on $C^\infty(S)$ given locally by*

$$D = \gamma(dx^\mu)\big(\partial_\mu - \tfrac{1}{4} {}_i\omega_\mu^j \eta_{jj} \gamma^i \gamma^j\big) = \gamma(dx^\mu)\big(\partial_\mu - \gamma(.\underset{\sim}{\omega}_\mu)\big). \tag{7.71}$$

Proposition 9. *We have*

$$\int_M d^n x \Big((D\psi)^\dagger \psi' + \psi^\dagger (D\psi') \Big) = 0$$

for all compactly supported spinor fields $\psi, \psi' \in C_c^\infty(S)$, i.e.

$$D^\dagger = -D$$

is formally skew-hermitian.

Proof. In terms of the Christoffel symbols we have

$${}_i\omega_\mu^j \, \overset{-1}{e}{}_j^\lambda = \partial_\mu(\overset{-1}{e}{}_i^\lambda) + \overset{-1}{e}{}_i^\lambda \Gamma_{\mu\nu}^\lambda.$$

In fact, the Leibniz rule for ∇ implies

$$\begin{aligned}
{}_i\omega_\mu^j \, \overset{-1}{e}{}_j^\lambda \partial_\lambda &= {}_i\omega_\mu^j \, \overset{-1}{e}{}_j = \nabla_\mu \overset{-1}{e}{}_i = \nabla_\mu(\overset{-1}{e}{}_i^\nu \partial_\nu) \\
&= \partial_\mu(\overset{-1}{e}{}_i^\nu) \partial_\nu + \overset{-1}{e}{}_i^\nu (\nabla_\mu \partial_\nu) = \partial_\mu(\overset{-1}{e}{}_i^\lambda) \partial_\lambda + \overset{-1}{e}{}_i^\nu \Gamma_{\mu\nu}^\lambda \partial_\lambda.
\end{aligned}$$

Using (7.6), (7.38) and Lemma 1, we obtain

$$\begin{aligned}
\tfrac{1}{4} {}_i\omega_\mu^j \eta_{jj} \big(\gamma(dx^\mu)\gamma^i\gamma^j + \gamma^j\gamma^i\gamma(dx^\mu)\big) &= \tfrac{1}{4} {}_i\omega_\mu^j \eta_{jj} \big(\gamma(dx^\mu)\gamma^i\gamma^j - \gamma^i\gamma^j\gamma(dx^\mu)\big) \\
&= \tfrac{1}{4} {}_i\omega_\mu^j \eta_{jj} \overset{-1}{e}{}_k^\mu \big(\gamma^k \gamma^i \gamma^j - \gamma^i\gamma^j\gamma^k\big) = \tfrac{1}{2} {}_i\omega_\mu^j \eta_{jj} \overset{-1}{e}{}_k^\mu \big(\eta^{ki}\gamma^j - \gamma^i\eta^{jk}\big) \\
&= -\tfrac{1}{2} {}_j\omega_\mu^i \eta_{ii} \overset{-1}{e}{}_k^\mu \eta^{ki}\gamma^j - \tfrac{1}{2} {}_i\omega_\mu^j \eta_{jj} \overset{-1}{e}{}_k^\mu \gamma^i \eta^{jk} \\
&= -\tfrac{1}{2} {}_j\omega_\mu^i \overset{-1}{e}{}_i^\mu \gamma^j - \tfrac{1}{2} {}_i\omega_\mu^j \overset{-1}{e}{}_j^\mu \gamma^i = -{}_i\omega_\mu^j \overset{-1}{e}{}_j^\mu \gamma^i \\
&= -\big(\partial_\mu(\overset{-1}{e}{}_i^\mu) + \overset{-1}{e}{}_i^\nu \Gamma_{\mu\nu}^\mu\big)\gamma^i = -\big(\partial_\mu(\gamma(dx^\mu)) + \Gamma_{\mu\nu}^\mu \gamma(dx^\nu)\big).
\end{aligned}$$

Summing over μ, Cramer's rule yields

$$\Gamma^\mu_{\mu\nu} = \tfrac{1}{2}g^{\mu\kappa}(\partial_\mu g_{\nu\kappa} + \partial_\nu g_{\mu\kappa} - \partial_\kappa g_{\mu\nu}) = \frac{1}{2}g^{\mu\kappa}\partial_\nu g_{\mu\kappa} = \frac{1}{2}\operatorname{tr}(\partial_\nu g)g^{-1}$$
$$= \frac{1}{2\operatorname{Det}g}\partial_\nu(\operatorname{Det}g) = \frac{1}{\sqrt{|\operatorname{Det}g|}}\partial_\nu\sqrt{|\operatorname{Det}g|}.$$

Using (7.19) this implies for fixed $\psi, \psi' \in \mathcal{C}^\infty_c(S)$

$$\begin{aligned}(D\psi)^\dagger\psi' &+ \psi^\dagger(D\psi') \\
&= \bigl(\gamma(dx^\mu)(\partial_\mu\psi) - \tfrac{1}{4}i\omega^j_\mu\,\eta_{jj}\gamma(dx^\mu)\gamma^i\gamma^j\psi\bigr)^\dagger\psi' \\
&\quad + \psi^\dagger\bigl(\gamma(dx^\mu)(\partial_\mu\psi') - \tfrac{1}{4}i\omega^j_\mu\,\gamma(dx^\mu)\gamma^i\gamma^j\psi'\bigr) \\
&= (\partial_\mu\psi)^\dagger\gamma(dx^\mu)\psi' + \psi^\dagger\gamma(dx^\mu)(\partial_\mu\psi') \\
&\quad - \tfrac{1}{4}\psi^\dagger\,{}_i\omega^j_\mu\,\eta_{jj}\bigl(\gamma(dx^\mu)\gamma^i\gamma^j + \gamma^j\gamma^i\gamma(dx^\mu)\bigr)\psi' \\
&= (\partial_\mu\psi^\dagger)\gamma(dx^\mu)\psi' + \psi^\dagger\gamma(dx^\mu)(\partial_\mu\psi') \\
&\quad + \psi^\dagger\bigl(\partial_\mu(\gamma(dx^\mu)) + \Gamma^\mu_{\mu\nu}\gamma(dx^\nu)\bigr)\psi' \\
&= \partial_\mu\bigl(\psi^\dagger\gamma(dx^\mu)\psi'\bigr) + \Gamma^\mu_{\mu\nu}\psi^\dagger\gamma(dx^\nu)\psi' \\
&= \frac{1}{\sqrt{\operatorname{Det}g}}\partial_\nu\bigl(\sqrt{\operatorname{Det}g}\,\psi^\dagger\gamma(dx^\nu)\psi'\bigr) \\
&= \frac{1}{\sqrt{\operatorname{Det}g}}\partial_\nu\bigl(\sqrt{\operatorname{Det}g}\,g^{\nu\mu}\alpha_\mu\bigr) = d^*\alpha,\end{aligned}$$

where $\alpha = \alpha_\mu\,dx^\mu \in \mathcal{C}^\infty(T^\# M \otimes \mathbb{C})$ is defined by

$$\alpha_\mu := g_{\mu\nu}\psi^\dagger\gamma(dx^\nu)\psi'$$

for all μ. Since α has compact support, Stokes' Theorem implies

$$\int_M d^*\alpha = 0$$

and the assertion follows.

Proposition 10.
$$\Gamma D = -D\Gamma. \tag{7.72}$$

Proof. Since Γ is independent of x, we have $\Gamma\partial_\mu = \partial_\mu\Gamma$ for all μ. Since $\Gamma\gamma^i\gamma^j = \gamma^i\gamma^j\Gamma$ by (7.5), (7.48) implies

$$\begin{aligned}\Gamma D &= \Gamma\gamma(dx^\mu)\bigl(\partial_\mu - \tfrac{1}{4}i\omega^j_\mu\,\gamma^i\gamma^j\bigr) \\
&= -\gamma(dx^\mu)\Gamma\bigl(\partial_\mu - \tfrac{1}{4}i\omega^j_\mu\,\gamma^i\gamma^j\bigr) \\
&= -\gamma(dx^\mu)\bigl(\partial_\mu - \tfrac{1}{4}i\omega^j_\mu\,\gamma^i\gamma^j\bigr)\Gamma = -D\Gamma.\end{aligned}$$

Proposition 11. *For $f \in C^\infty(M, \mathbb{C})$, we have*

$$[D, f] = \gamma(df). \tag{7.73}$$

Proof. Since $[\partial_\mu, f] = \partial_\mu f$ and f commutes with γ^i and $\gamma(dx^\mu)$, it follows that

$$[D, f] = [\gamma(dx^\mu)\partial_\mu, f] = \gamma(dx^\mu)[\partial_\mu, f] = \gamma(dx^\mu)\,\partial_\mu f = \gamma(\partial_\mu f\, dx^\mu) = \gamma(df).$$

7.3 Real Structure on Spacetime

Apart from the grading operator Γ defined in (7.7), the spinor space $\mathbb{C}^{2^m}$ carries a conjugate linear isometry J which is of importance in Riemannian geometry (real K-theory) and mathematical physics (charge conjugation). Depending on the parity of $m = n/2$, we define $J = J_m$ as follows

$$J_{2k} := \gamma^0 \gamma^2 \ldots \gamma^{4k-2}\,\overline{} \tag{7.74}$$
$$J_{2k+1} := \gamma^1 \gamma^3 \ldots \gamma^{4k+1}\,\overline{}.$$

Here $\overline{}$ denotes the usual conjugation on $\mathbb{C}^{2^m}$. In short, J_m contains the m γ-factors which have the same parity as m.

Example 1. *In 4-dimensional Minkowski space ($m = 2$), note that*

$$J = \gamma^0 \gamma^2\,\overline{} \tag{7.75}$$

is given explicitly by

$$J\psi = \gamma^0\gamma^2\overline{\psi} = i\begin{pmatrix} 0 & 1 & & 0 \\ -1 & 0 & & \\ & & 0 & -1 \\ 0 & & 1 & 0 \end{pmatrix}\begin{pmatrix} \overline{\psi_0} \\ \overline{\psi_1} \\ \overline{\psi_2} \\ \overline{\psi_3} \end{pmatrix} = i\begin{pmatrix} \overline{\psi_1} \\ -\overline{\psi_0} \\ -\overline{\psi_3} \\ \overline{\psi_2} \end{pmatrix}. \tag{7.76}$$

J is closely related to, but not identical with, the well-known charge conjugation operator

$$(PC\psi)(x^0, \boldsymbol{x}) = \gamma^0\gamma^2\overline{\psi}(x^0, -\boldsymbol{x}) \tag{7.77}$$

which involves a reflection in the space coordinates.

Proposition 12. *We have*

$$J_{2k}^2 = (-1)^k I, \quad J_{2k+1}^2 = (-1)^{k+1} I, \tag{7.78}$$
$$J_m \Gamma = (-1)^m \Gamma J_m. \tag{7.79}$$

Proof. The relation (7.79) follows from (7.5) since J_m is given by a product of m γ-factors and $\overline{\Gamma} = \Gamma$. For (7.78) we apply Proposition 2 and Lemma 2 to get

$$
\begin{aligned}
J_{2k}^2 &= \gamma^0 \gamma^2 \ldots \gamma^{4k-2} \overline{\gamma^0 \gamma^2 \ldots \gamma^{4k-2}} \\
&= \gamma^0 \gamma^2 \ldots \gamma^{4k-2} \overline{\gamma}^0 \overline{\gamma}^2 \ldots \overline{\gamma}^{4k-2} \\
&= (-\kappa^2)^{2k-1} \gamma^0 \gamma^2 \ldots \gamma^{4k-2} \gamma^0 \gamma^2 \ldots \gamma^{4k-2} \\
&= -\kappa^2 (-1)^{2k(2k-1)/2} (\gamma^0)^2 (\gamma^2)^2 \ldots (\gamma^{4k-2})^2 \\
&= -\kappa^2 (-1)^{(2k-1)k} (-\kappa^2)^{2k-1} I = (-1)^{(2k-1)k} I = (-1)^k I,
\end{aligned}
$$

$$
\begin{aligned}
J_{2k+1}^2 &= \gamma^1 \gamma^3 \ldots \gamma^{4k+1} \overline{\gamma^1 \gamma^3 \ldots \gamma^{4k+1}} \\
&= \gamma^1 \gamma^3 \ldots \gamma^{4k+1} \overline{\gamma}^1 \overline{\gamma}^3 \ldots \overline{\gamma}^{4k+1} \\
&= (\kappa^2)^{2k+1} \gamma^1 \gamma^3 \ldots \gamma^{4k+1} \gamma^1 \gamma^3 \ldots \gamma^{4k+1} \\
&= \kappa^2 (-1)^{(2k+1)2k/2} (\gamma^1)^2 (\gamma^3)^2 \ldots (\gamma^{4k+1})^2 \\
&= \kappa^2 (-1)^{(2k+1)k} (-\kappa^2)^{2k+1} I = -(-1)^{(2k+1)k} I \\
&= -(-1)^k I = (-1)^{k+1} I.
\end{aligned}
$$

Lemma 6. *For all $0 \leq j < n$, we have*

$$J \gamma^j = -\overset{*j}{\gamma} J. \qquad (7.80)$$

Proof. In view of (7.12), we have to show $J\gamma^0 = -\gamma^0 J$ and $J\gamma^{j+1} = \kappa^2 \gamma^{j+1} J$ for $j \geq 0$. For the various cases, this is shown as follows:

$$
\begin{aligned}
\gamma^0 \gamma^2 \ldots \gamma^{4k-2} \overline{\gamma}^0 &= \gamma^0 \gamma^2 \ldots \gamma^{4k-2} \gamma^0 \\
&= (-1)^{2k-1} \gamma^0 \gamma^0 \gamma^2 \ldots \gamma^{4k-2} = -\gamma^0 \gamma^0 \gamma^2 \ldots \gamma^{4k-2},
\end{aligned}
$$

$$
\begin{aligned}
\gamma^1 \gamma^3 \ldots \gamma^{4k+1} \overline{\gamma}^0 &= \gamma^1 \gamma^3 \ldots \gamma^{4k+1} \gamma^0 \\
&= (-1)^{2k+1} \gamma^0 \gamma^1 \gamma^3 \ldots \gamma^{4k+1} = -\gamma^0 \gamma^1 \gamma^3 \ldots \gamma^{4k+1},
\end{aligned}
$$

$$
\begin{aligned}
\gamma^0 \gamma^2 \ldots \gamma^{4k-2} \overline{\gamma}^{2j+1} &= \kappa^2 \gamma^0 \gamma^2 \ldots \gamma^{4k-2} \gamma^{2j+1} \\
&= \kappa^2 (-1)^{2k} \gamma^{2j+1} \gamma^0 \gamma^2 \ldots \gamma^{4k-2} = \kappa^2 \gamma^{2j+1} \gamma^0 \gamma^2 \ldots \gamma^{4k-2},
\end{aligned}
$$

$$
\begin{aligned}
\gamma^1 \gamma^3 \ldots \gamma^{4k+1} \overline{\gamma}^{2j+1} &= \kappa^2 \gamma^1 \gamma^3 \ldots \gamma^{4k+1} \gamma^{2j+1} \\
&= \kappa^2 (-1)^{2k} \gamma^{2j+1} \gamma^1 \gamma^3 \ldots \gamma^{4k+1} = \kappa^2 \gamma^{2j+1} \gamma^1 \gamma^3 \ldots \gamma^{4k+1},
\end{aligned}
$$

$$
\begin{aligned}
\gamma^0 \gamma^2 \ldots \gamma^{4k-2} \overline{\gamma}^{2j+2} &= -\kappa^2 \gamma^0 \gamma^2 \ldots \gamma^{4k-2} \gamma^{2j+2} \\
&= -\kappa^2 (-1)^{2k-1} \gamma^{2j+2} \gamma^0 \gamma^2 \ldots \gamma^{4k-2} = \kappa^2 \gamma^{2j+2} \gamma^0 \gamma^2 \ldots \gamma^{4k-2},
\end{aligned}
$$

$$
\begin{aligned}
\gamma^1 \gamma^3 \ldots \gamma^{4k+1} \overline{\gamma}^{2j+2} &= -\kappa^2 \gamma^1 \gamma^3 \ldots \gamma^{4k+1} \gamma^{2j+2} \\
&= -\kappa^2 (-1)^{2k+1} \gamma^{2j+2} \gamma^1 \gamma^3 \ldots \gamma^{4k+1} = \kappa^2 \gamma^{2j+2} \gamma^1 \gamma^3 \ldots \gamma^{4k+1}.
\end{aligned}
$$

Corollary 3. *The Dirac operator D satisfies*

$$JD = -\tilde{D}J, \tag{7.81}$$

where

$$\tilde{D} := \gamma(dx^\mu)^*\big(\partial_\mu - \tfrac{1}{4}i\omega_\mu^j\,\eta_{jj}\overset{*}{\gamma}^i\overset{*}{\gamma}^j\big). \tag{7.82}$$

Proof. Since J is independent of $x \in M$ and ${}_i\omega_\mu^j$ is real, Lemma 6 implies

$$\begin{aligned}
-JD &= -J\gamma(dx^\mu)\big(\partial_\mu - \tfrac{1}{4}i\omega_\mu^j\,\eta_{jj}\gamma^i\gamma^j\big)\\
&= \gamma(dx^\mu)^* J\big(\partial_\mu - \tfrac{1}{4}i\omega_\mu^j\,\eta_{jj}\gamma^i\gamma^j\big)\\
&= \gamma(dx^\mu)^*\big(\partial_\mu J + \tfrac{1}{4}i\omega_\mu^j\,\eta_{jj}\overset{*}{\gamma}^i J\gamma^j\big)\\
&= \gamma(dx^\mu)^*\big(\partial_\mu - \tfrac{1}{4}i\omega_\mu^j\,\eta_{jj}\overset{*}{\gamma}^i\overset{*}{\gamma}^j\big)J.
\end{aligned}$$

Remark 2. *On a space-time of signature (p,q), with $p-q$ even, a "real structure" (mod 8) is a conjugate linear isometry J on the spinor bundle satisfying*

$$JD = -DJ \tag{7.83}$$

and

$$J^2 = \iota I, \qquad J\Gamma = \iota'\Gamma J \tag{7.84}$$

where $\iota, \iota' \in \{\pm 1\}$ are given by the following table (cf. Chapter 1.3 and [39, Definition 3])

$p-q \bmod 8$	0	2	4	6
ι	1	-1	-1	1
ι'	1	-1	1	-1

$$\tag{7.85}$$

The preceeding results show that in the Euclidean case ($\kappa = i$) J defines a real structure on space-time of signature $(2m, 0)$ since $\iota' = (-1)^m$ and

$$\iota = \begin{cases} (-1)^{m/2} & m \text{ even} \\ (-1)^{(m+1)/2} & m \text{ odd}. \end{cases}$$

Moreover, (7.83) holds since $\overset{*}{\gamma}^j = \gamma^j$ for all j and hence $\tilde{D} = D$. On the other hand in the physically relevant Minkowski case ($\kappa = 1$) we have signature $(1, 2m-1)$ and J does not fit the table (7.85). Also, $\tilde{D} \neq D$ in this case so that (7.83) is false. In the discussion of the Standard Model (Section 9), a real structure will be defined in both cases using fermion doubling.

Remark 3. *Corollary 3 suggests to define another representation*

$$\tilde{\gamma} : \mathcal{C}^\infty(C\ell(T^\# M)) \to \text{End}(S) \tag{7.86}$$

by using the matrices $\tilde{\gamma}^i$ instead of γ^i. Then $\tilde{D}$ is the Dirac operator associated with $\tilde{\gamma}$. By Proposition 2, we have $\tilde{\gamma} = \gamma$ only in the Euclidean case. Using Lemma 6, we see that for 1-forms $\omega \in \mathcal{C}^\infty(T^\# M)$

$$\tilde{\gamma}(\omega) := \omega_\mu \gamma(dx^\mu)^* \tag{7.87}$$

satisfies

$$\tilde{\gamma}(\omega) = -J\gamma(\omega)J^{-1}. \tag{7.88}$$

On the other hand, for functions $f \in \mathcal{C}^\infty(M, \mathbb{C})$ we have

$$\overline{f} = JfJ^{-1}. \tag{7.89}$$

The basic identities (7.73), (7.49), (7.45) and (7.55) hold analogously for $\tilde{\gamma}$.

7.4 Trace Formulas and Inner Products

As mentioned in the introduction, the local trace formulas for the spinor bundle have a much deeper analytic significance in the case of compact Riemannian manifolds. For any compact Riemannian manifold M of dimension n, a pseudo-differential operator

$$P : \mathcal{C}^\infty(V) \to \mathcal{C}^\infty(V)$$

of order $-n$, acting on a complex vector bundle V over M, has a well-defined "Dixmier trace", denoted by $\mathrm{Tr}_\omega(P)$. The well-known trace formula of A. Connes (Chapter 1.4) expresses $\mathrm{Tr}_\omega(P)$ as an integral involving the principal symbol

$$\sigma_{-n}(P) \in \mathcal{C}^\infty(T^\# M, \mathrm{End}\,(V))$$

of P. This concept applies in particular to the (complex) spinor bundle S over M and the pseudo-differential operator $|D|^{-n}$, where $|D|$ is the absolute value of the Dirac operator acting on S. More generally, any section T of the endomorphism bundle of S gives rise to a pseudo-differential operator of order 0, and by Connes' trace formula, we obtain

$$\mathrm{Tr}_\omega(T|D|^{-n}) = c_n \int_M d^n x \cdot \underline{\mathrm{tr}}_S T_x$$

where $\underline{\mathrm{tr}}_S$ denotes the normalized trace on spinors, $d^n x$ is the Riemannian measure and c_n is a universal constant depending only on the dimension n. Using the spinor representation (7.32) we obtain equivalently

$$\mathrm{Tr}_\omega(\gamma(\vartheta)|D|^{-n}) = c_n \int_M d^n x \cdot \underline{\mathrm{tr}}_S \gamma(\vartheta_x) \tag{7.90}$$

for any section $\vartheta \in \mathcal{C}^\infty(Cl(T^\#M))$. Via the identification (7.54) it follows that

$$\frac{1}{c_n}\mathrm{Tr}_\omega\big(\gamma(\sigma)^\dagger\gamma(\sigma')|D|^{-n}\big) = \int_M d^n x \cdot \big(\gamma(\sigma)|\gamma(\sigma')\big)_S = \int_M d^n x\, (\sigma|\sigma')_x \quad (7.91)$$

for all $\sigma, \sigma' \in \mathcal{C}^\infty(\bigwedge(T^\#M) \otimes \mathbb{C})$. This operator-theoretic interpretation of the metric properties of differential forms is one of the starting points of the noncommutative geometry approach towards the Standard Model.

We end this section by listing the normalizations for traces and inner products used in the sequel. We will consider complex vector bundles V which (locally) are tensor products involving the spinor bundle S over M and also a finite dimensional "generation space" $\mathbb{G}$, endowed with an inner product. Given an endomorphism T of V we define the "partially normalized" trace

$$\underline{\mathrm{tr}}_V T := \frac{1}{c}\,\mathrm{tr}_V T, \qquad (7.92)$$

where the normalization constant c is given as follows:

$$c := \begin{cases} \dim S & \text{if } V \approx S \otimes \mathbb{C}^N \\ \dim \mathbb{G} & \text{if } V \approx \mathbb{C}^N \otimes \mathbb{G} \\ \dim S \cdot \dim G & \text{if } V \approx S \otimes \mathbb{C}^N \otimes \mathbb{G}. \end{cases} \qquad (7.93)$$

In terms of this trace, the inner product of endomorphisms T_1, T_2 of V is defined as

$$(T_1 \mid T_2)_V := \underline{\mathrm{tr}}_V T_1^\dagger T_2 = \frac{1}{c}\,\mathrm{tr}_V T_1^\dagger T_2. \qquad (7.94)$$

We also put

$$\|T\|_V^2 := (T \mid T)_V = \underline{\mathrm{tr}}_V T^\dagger T = \frac{1}{c}\,\mathrm{tr}_V T^\dagger T. \qquad (7.95)$$

8 The Electro-weak Model

Karen Elsner, Holger Neumann, and Harald Upmeier

8.1 Noncommutative Matter Fields

In a first step towards the formulation of the Standard Model of elementary particle physics using noncommutative geometry, we confine ourselves to the description of leptons, which show only electro-weak interaction. Leptons exerimentally appear as pairs in three generations: (e, ν_e), (μ, ν_μ), (τ, ν_τ). We will consider e, μ, τ and ν_e, ν_μ, ν_τ as components of a single object e, ν with three different "generations". Correspondingly, we let $\mathbb{G}_R$ and $\mathbb{G}_L$ be finite-dimensional complex vector spaces called right and left generation spaces. We assume $\dim \mathbb{G}_R = \dim \mathbb{G}_L$ ($= 3$ experimentally). The distinction between right and left (chirality) is motivated as follows:

As is experimentally confirmed the electro-weak interaction can be considered as a gauge interaction with respect to the group $U(1) \times SU(2)$. The electromagnetic interaction only affects the charged leptons e, μ, τ. The weak interaction only affects the left handed components e_L, μ_L, τ_L and the corresponding neutrinos, which are supposed to be massless and to exist only as left-handed components. (An extension of the electro-weak model including massive neutrinos is possible). This "doubling" of the L-components is expressed in our formalism as follows:

We indicate "doubling" of a vector space V or an endomorphism $T \in \mathcal{L}(V)$ by double underlining $\underline{\underline{}}$, i.e.

$$\underline{\underline{V}} = \begin{pmatrix} V \\ V \end{pmatrix} = V \otimes \mathbb{C}^2 \tag{8.1}$$

and

$$\underline{\underline{T}} := \begin{pmatrix} T & 0 \\ 0 & T \end{pmatrix} = T \otimes \mathbf{1}_2. \tag{8.2}$$

Likewise for vector bundles etc. Now consider the complex vector space

$$\mathbb{G}_\ell := \begin{pmatrix} \mathbb{G}_R \\ \underline{\underline{\mathbb{G}_L}} \end{pmatrix} \tag{8.3}$$

($\ell =$ "leptonic") and define

$$S_\ell := S \otimes \mathbb{G}_\ell = \begin{pmatrix} S \otimes \mathbb{G}_R \\ \underline{\underline{S}} \otimes \mathbb{G}_L \end{pmatrix} \tag{8.4}$$

as a complex vector bundle over M. Here S is the spinor bundle of M. Then

$$\mathcal{C}^\infty(S_\ell) = \mathcal{C}^\infty(S \otimes \mathbb{G}_\ell) = \mathcal{C}^\infty(S) \otimes \mathbb{G}_\ell .$$

Definition 1. *The NC matter fields for the electro-weak model are the (compactly supported) smooth sections*

$$\psi_\ell \in \mathcal{C}_c^\infty(S_\ell) \tag{8.5}$$

of the $\mathbb{G}_\ell$-valued spinor bundle S_ℓ. Thus the NC matter fields are (linear combinations of) tensors

$$\psi_\ell = \begin{pmatrix} \psi \otimes e_R \\ \underline{\Psi} \otimes \ell_L \end{pmatrix}, \tag{8.6}$$

where $\psi \in \mathcal{C}_c^\infty(S)$, $\underline{\Psi} \in \mathcal{C}_c^\infty(\underline{S})$ are space-time spinors and $e_R \in \mathbb{G}_R$, $\ell_L \in \mathbb{G}_L$ are arbitrary vectors. The physically relevant fermions satisfy

$$\Gamma\psi = \psi, \quad \underline{\Gamma}\underline{\Psi} = -\underline{\Psi} \tag{8.7}$$

but for now we do not impose this restriction. Thus we overcount the fermionic degrees of freedom two-fold.

Remark 1. *In physics notation, the NC matter fields can be expressed as*

$$\psi_\ell = \begin{pmatrix} \psi_{e_R} \\ \psi_{e_L} \\ \psi_{\nu_L} \end{pmatrix} \tag{8.8}$$

where $\psi_{e_R} \in \mathcal{C}_c^\infty(S \otimes \mathbb{G}_R)$ and $\psi_{e_L}, \psi_{\nu_L} \in \mathcal{C}_c^\infty(S \otimes \mathbb{G}_L)$ correspond to the fermion (spin $\frac{1}{2}$) fields (leptons) associated with the electro-weak model. This follows from the decomposition

$$\mathbb{G}_\ell = \begin{pmatrix} \mathbb{G}_R \\ \mathbb{G}_L \otimes \mathbb{C}^2 \end{pmatrix} = \begin{pmatrix} \mathbb{G}_R \\ \mathbb{G}_L \\ \mathbb{G}_L \end{pmatrix} \begin{array}{l} e_R \\ e_L \\ \nu_L. \end{array} \tag{8.9}$$

The constraint (8.7) takes the form

$$\Gamma \psi_{e_R} = \psi_{e_R}, \quad \Gamma \psi_{e_L/\nu_L} = \psi_{e_L/\nu_L}. \tag{8.10}$$

Let D denote the Dirac operator on $\mathcal{C}^\infty(S)$ induced by the metric g. Put

$$\varepsilon := \begin{pmatrix} 1 \\ 0 \end{pmatrix} \tag{8.11}$$

and define a first-order differential operator D_ℓ acting on (sections of) S_ℓ by

$$D_\ell := \begin{bmatrix} D \otimes 1_R & \kappa^2 \Gamma \varepsilon^* \otimes M_e^* \\ \varepsilon \Gamma \otimes M_e & \underline{D} \otimes 1_L \end{bmatrix} \tag{8.12}$$

where $M_e : \mathbb{G}_R \longrightarrow \mathbb{G}_L$ is an invertible linear map called "mass-matrix". This is the Dirac operator for the electro-weak model. The unit vector ε corresponds to choosing the electron (e_L) part in the double copy $\underline{\mathbb{G}_L}$ of (8.3). Later it will be related to the so-called "Higgs vacuum" (cf. (8.59)).

Proposition 1. *Consider the projection*

$$p = \begin{pmatrix} 0_R & 0 \\ 0 & 1_L \end{pmatrix} \tag{8.13}$$

and put

$$P := \begin{pmatrix} 1 & 0 \\ 0 & p \end{pmatrix} = \begin{pmatrix} 1_R & 0 & & \\ 0 & 1_L & & 0 \\ & & 0_R & 0 \\ & 0 & 0 & 1_L \end{pmatrix} \tag{8.14}$$

as a projection acting on $\underline{\mathbb{G}_R \oplus \mathbb{G}_L}$. *Then we have (as differential operators on* S_ℓ)

$$D_\ell = P \left(\underline{D} \otimes \begin{pmatrix} 1_R & 0 \\ 0 & 1_L \end{pmatrix} + \underline{\Gamma} \otimes \begin{pmatrix} 0 & \kappa^2 M_e^* \\ M_e & 0 \end{pmatrix} \right) P. \tag{8.15}$$

Proof. Since

$$D \otimes \begin{pmatrix} 1_R & 0 \\ 0 & 1_L \end{pmatrix} + \Gamma \otimes \begin{pmatrix} 0 & \kappa^2 M_e^* \\ M_e & 0 \end{pmatrix} = \begin{bmatrix} D \otimes 1_R & \kappa^2 \Gamma \otimes M_e^* \\ \Gamma \otimes M_e & D \otimes 1_L \end{bmatrix}$$

it follows that

$$P \left(\underline{D} \otimes \begin{pmatrix} 1_R & 0 \\ 0 & 1_L \end{pmatrix} + \underline{\Gamma} \otimes \begin{pmatrix} 0 & \kappa^2 M_e^* \\ M_e & 0 \end{pmatrix} \right) P =$$

$$\begin{pmatrix} 1_R & 0 & & \\ 0 & 1_L & & 0 \\ & & 0_R & 0 \\ & 0 & 0 & 1_L \end{pmatrix} \begin{pmatrix} D \otimes 1_R & \kappa^2 \Gamma \otimes M_e^* & & \\ \Gamma \otimes M_e & D \otimes 1_L & & 0 \\ & & D \otimes 1_R & \kappa^2 \Gamma \otimes M_e^* \\ & 0 & \Gamma \otimes M_e & D \otimes 1_L \end{pmatrix} \begin{pmatrix} 1_R & 0 & & \\ 0 & 1_L & & 0 \\ & & 0_R & 0 \\ & 0 & 0 & 1_L \end{pmatrix} =$$

$$\begin{pmatrix} D \otimes 1_R & \kappa^2 \Gamma \otimes M_e^* & 0 & 0 \\ \Gamma \otimes M_e & D \otimes 1_L & 0 & 0 \\ 0 & 0 & 0 & 0 \\ 0 & 0 & 0 & D \otimes 1_L \end{pmatrix} = \begin{bmatrix} D \otimes 1_R & \kappa^2 \Gamma \otimes M_e^* & 0 \\ \Gamma \otimes M_e & D \otimes 1_L & 0 \\ 0 & 0 & D \otimes 1_L \end{bmatrix} =$$

$$\begin{bmatrix} D \otimes 1_R & \kappa^2 (\Gamma, 0) \otimes M_e^* \\ \begin{pmatrix} \Gamma \\ 0 \end{pmatrix} \otimes M_e & \begin{pmatrix} D & 0 \\ 0 & D \end{pmatrix} \otimes 1_L \end{bmatrix} = \begin{bmatrix} D \otimes 1_R & \kappa^2 \Gamma \varepsilon^* \otimes M_e^* \\ \varepsilon \Gamma \otimes M_e & \underline{D} \otimes 1_L \end{bmatrix} = D_\ell.$$

For the next Proposition only, we use a slightly different tensor decomposition (indicated by curly brackets) and write

$$D_\ell = \left\{ \begin{array}{cc} D \otimes 1_R & \kappa^2 \Gamma \otimes M^* \\ \Gamma \otimes M & D \otimes \underline{1_L} \end{array} \right\} \tag{8.16}$$

by putting
$$M = \begin{pmatrix} M_e \\ 0 \end{pmatrix}. \tag{8.17}$$

Proposition 2.
$$D_\ell^\dagger = -D_\ell.$$

Proof. Using the notation (8.16), it follows from Proposition 9 that

$$D_\ell^\dagger = \begin{Bmatrix} D \otimes 1_R & \kappa^2 \Gamma \otimes M^* \\ \Gamma \otimes M & D \otimes \underline{1_L} \end{Bmatrix}^\dagger = \begin{Bmatrix} D^\dagger \otimes 1_R & \Gamma^\star \otimes M^* \\ \kappa^2 \Gamma^\star \otimes M & D^\dagger \otimes \underline{1_L} \end{Bmatrix} =$$

$$\begin{Bmatrix} -D \otimes 1_R & -\kappa^2 \Gamma \otimes M^* \\ -\Gamma \otimes M & -D \otimes \underline{1_L} \end{Bmatrix} = -D_\ell.$$

8.2 Noncommutative Gauge Fields

Having described the Dirac operator associated with the electro-weak model, we now turn to the underlying (noncommutative) algebra. The basic idea of noncommutative geometry [38] is to replace the function algebra $\mathcal{C}^\infty(M, \mathbb{C})$ over a manifold M by a noncommutative algebra $\mathcal{A}$, and to study an appropriate differential calculus (integration of differential forms) over $\mathcal{A}$. In our context, M represents space-time and the noncommutative algebra reflects the interactions between the fermions of the model, i.e., the bosonic (gauge) degrees of freedom. In the electro-weak model, only the leptonic fermions are considered.

Proposition 3. *Let $\mathcal{A}$ denote the (commutative) complex algebra consisting of all endomorphisms of $S \otimes \begin{pmatrix} \mathbb{G}_R \\ \mathbb{G}_L \end{pmatrix}$ of the form*

$$\begin{bmatrix} f_R \otimes 1_R & 0 \\ 0 & f_L \otimes 1_L \end{bmatrix} \tag{8.18}$$

where $f_i \in \mathcal{C}^\infty(M, \mathbb{C})$ for $i \in \{R, L\}$. Then $\mathcal{A}_\ell := P\mathcal{A}^{2\times 2}P$ is the algebra of all endomorphisms f_ℓ of S_ℓ having the form

$$f_\ell = \begin{bmatrix} f \otimes 1_R & 0 \\ 0 & F \otimes 1_L \end{bmatrix} \tag{8.19}$$

where $f \in \mathcal{C}^\infty(M, \mathbb{C})$ and $F \in \mathcal{C}^\infty(M, \mathbb{C}^{2\times 2})$.

Proof. Note that $p \in \mathcal{A}$ and hence $P \in \mathcal{A}^{2\times 2}$. Writing a 2×2-matrix of endomorphisms in $\mathcal{A}$ according to (8.18), we obtain

$$\begin{pmatrix} 1_R & 0 & & \\ 0 & 1_L & & 0 \\ & & 0_R & 0 \\ 0 & & 0 & 1_L \end{pmatrix} \begin{pmatrix} f_R^1 \otimes 1_R & 0 & f_R^2 \otimes 1_R & 0 \\ 0 & f_L^1 \otimes 1_L & 0 & f_L^2 \otimes 1_L \\ f_R^3 \otimes 1_R & 0 & f_R^4 \otimes 1_R & 0 \\ 0 & f_L^3 \otimes 1_L & 0 & f_L^4 \otimes 1_L \end{pmatrix} \begin{pmatrix} 1_R & 0 & & \\ 0 & 1_L & & 0 \\ & & 0_R & 0 \\ 0 & & 0 & 1_L \end{pmatrix}$$

$$= \begin{pmatrix} f_R^1 \otimes 1_R & 0 & 0 & 0 \\ 0 & f_L^1 \otimes 1_L & 0 & f_L^2 \otimes 1_L \\ 0 & 0 & 0 & 0 \\ 0 & f_L^3 \otimes 1_L & 0 & f_L^4 \otimes 1_L \end{pmatrix} = \begin{bmatrix} f_R^1 \otimes 1_R & 0 & 0 \\ 0 & f_L^1 \otimes 1_L & f_L^2 \otimes 1_L \\ 0 & f_L^3 \otimes 1_L & f_L^4 \otimes 1_L \end{bmatrix}$$

$$= \begin{bmatrix} f_R^1 \otimes 1_R & 0 \\ 0 & \begin{pmatrix} f_L^1 & f_L^2 \\ f_L^3 & f_L^4 \end{pmatrix} \otimes 1_L \end{bmatrix}.$$

Putting $f = f_R^1$ and $F = \begin{pmatrix} f_L^1 & f_L^2 \\ f_L^3 & f_L^4 \end{pmatrix}$, the assertion follows.

Let Ω^k denote the space of "universal" k-forms over $\mathcal{A}^{2\times 2}$, with differential $d: \Omega^k \to \Omega^{k+1}$. Using the representation

$$\pi_k : \Omega^k \to \mathcal{L}\left(L^2\left(M, S \otimes \begin{pmatrix} \mathbb{G}_R \\ \mathbb{G}_L \end{pmatrix}\right)\right) \qquad (8.20)$$

we put $\mathcal{A}_\ell^{(k)} := P\pi_k(\Omega^k)P$, considered as a subspace of $\mathcal{L}(L^2(M, S_\ell))$. Note that $\mathcal{A}_\ell^{(0)} = \mathcal{A}_\ell$. We now pass to $k = 1$ and $k = 2$.

Proposition 4. $\mathcal{A}_\ell^{(1)}$ consists of all endomorphisms

$$\omega_\ell = \begin{bmatrix} \gamma(\omega) \otimes 1_R & \kappa^2 \Gamma \psi \otimes M_e^* \\ \varphi \Gamma \otimes M_e & \gamma(\Omega) \otimes 1_L \end{bmatrix} \qquad (8.21)$$

acting on S_ℓ, where $\omega \in \mathcal{C}^\infty(T^\# M \otimes \mathbb{C})$, $\Omega \in \mathcal{C}^\infty(T^\# M \otimes \mathbb{C}^{2\times 2})$, $\varphi \in \mathcal{C}^\infty(M, \mathbb{C}^{2\times 1})$ and $\psi \in \mathcal{C}^\infty(M, \mathbb{C}^{1\times 2})$.

Proof. We have

$$\begin{bmatrix} D \otimes 1_R & \kappa^2 \Gamma \varepsilon^* \otimes M_e^* \\ \varepsilon \Gamma \otimes M_e & \underline{D} \otimes 1_L \end{bmatrix} \begin{bmatrix} f \otimes 1_R & 0 \\ 0 & F \otimes 1_L \end{bmatrix} = \begin{bmatrix} Df \otimes 1_R & \kappa^2 \Gamma \varepsilon^* F \otimes M_e^* \\ \varepsilon f \Gamma \otimes M_e & \underline{D}F \otimes 1_L \end{bmatrix}$$

and
$$\begin{bmatrix} f \otimes 1_R & 0 \\ 0 & F \otimes 1_L \end{bmatrix} \begin{bmatrix} D \otimes 1_R & \kappa^2 \Gamma \varepsilon^* \otimes M_e^* \\ \varepsilon \Gamma \otimes M_e & \underline{D} \otimes 1_L \end{bmatrix} = \begin{bmatrix} fD \otimes 1_R & \kappa^2 \Gamma f \varepsilon^* \otimes M_e^* \\ F\varepsilon\Gamma \otimes M_e & F\underline{D} \otimes 1_L \end{bmatrix}.$$

Since
$$Df - fD = \gamma(df), \quad \underline{D}F - F\underline{D} = \gamma(dF) \quad \text{and} \quad \Gamma f = f\Gamma,$$

for f_ℓ of the form (8.19) it follows that

$$[D_\ell, f_\ell] = \begin{bmatrix} \gamma(df) \otimes 1_R & \kappa^2 \Gamma(\varepsilon^* F - f\varepsilon^*) \otimes M_e^* \\ (\varepsilon f - F\varepsilon)\Gamma \otimes M_e & \gamma(dF) \otimes 1_L \end{bmatrix}.$$

Now consider (finite sums of) tensors $f_\ell^0 \otimes f_\ell^1$ in $\mathcal{A}_\ell$. Then

$$f_\ell^0 [D_\ell, f_\ell^1]$$
$$= \begin{bmatrix} f^0 \otimes 1_R & 0 \\ 0 & F^0 \otimes 1_L \end{bmatrix} \begin{bmatrix} \gamma(df^1) \otimes 1_R & \kappa^2 \Gamma(\varepsilon^* F^1 - f^1 \varepsilon^*) \otimes M_e^* \\ (\varepsilon f^1 - F^1 \varepsilon)\Gamma \otimes M_e & \gamma(dF^1) \otimes 1_L \end{bmatrix}$$
$$= \begin{bmatrix} \gamma(f^0 df^1) \otimes 1_R & \kappa^2 \Gamma f^0(\varepsilon^* F^1 - f^1 \varepsilon^*) \otimes M_e^* \\ F^0(\varepsilon f^1 - F^1 \varepsilon)\Gamma \otimes M_e & \gamma(F^0 dF^1) \otimes 1_L \end{bmatrix}$$
$$= \begin{bmatrix} \gamma(\omega) \otimes 1_R & \kappa^2 \Gamma \psi \otimes M_e^* \\ \varphi \Gamma \otimes M_e & \gamma(\Omega) \otimes 1_L \end{bmatrix}$$

by putting
$$\omega = f^0 df^1, \quad \Omega = F^0 dF^1, \quad \varphi = F^0(\varepsilon f^1 - F^1 \varepsilon), \quad \psi = f^0(\varepsilon^* F^1 - f^1 \varepsilon^*). \tag{8.22}$$

Remark 2. Using (7.73) we also have the operator relations

$$\begin{bmatrix} D \otimes 1_R & \kappa^2 \Gamma \otimes M_e^* \\ \Gamma \otimes M_e & D \otimes 1_L \end{bmatrix} \begin{bmatrix} f_R \otimes 1_R & 0 \\ 0 & f_L \otimes 1_L \end{bmatrix}$$
$$- \begin{bmatrix} f_R \otimes 1_R & 0 \\ 0 & f_L \otimes 1_L \end{bmatrix} \begin{bmatrix} D \otimes 1_R & \kappa^2 \Gamma \otimes M_e^* \\ \Gamma \otimes M_e & D \otimes 1_L \end{bmatrix}$$
$$= \begin{bmatrix} Df_R \otimes 1_R & \kappa^2 \Gamma f_L \otimes M_e^* \\ \Gamma \otimes M_e & D \otimes 1_L \end{bmatrix} - \begin{bmatrix} Df_R \otimes 1_R & \kappa^2 \Gamma f_L \otimes M_e^* \\ \Gamma \otimes M_e & D \otimes 1_L \end{bmatrix}$$
$$= \begin{bmatrix} \gamma(df_R) \otimes 1_R & \kappa^2 \Gamma f_{LR} \otimes M_e^* \\ f_{RL}\Gamma \otimes M_e & \gamma(df_L) \otimes 1_L \end{bmatrix}, \tag{8.23}$$

and

$$\begin{bmatrix} f_R^0 \otimes 1_R & 0 \\ 0 & f_L^0 \otimes 1_L \end{bmatrix} \begin{bmatrix} \gamma(df_R^1) \otimes 1_R & \kappa^2 \Gamma f_{LR}^1 \otimes M_e^* \\ f_{RL}^1 \Gamma \otimes M_e & \gamma(df_L^1) \otimes 1_L \end{bmatrix}$$
$$= \begin{bmatrix} \gamma(f_R^0 df_R^1) \otimes 1_R & \kappa^2 \Gamma f_R^0 f_{LR}^1 \otimes M_e^* \\ f_L^0 f_{RL}^1 \Gamma \otimes M_e & \gamma(f_L^0 df_L^1) \otimes 1_L \end{bmatrix} \quad (8.24)$$

acting on sections of $S \otimes \begin{pmatrix} \mathbb{G}_R \\ \mathbb{G}_L \end{pmatrix}$. Here we put

$$f_{LR} := f_L - f_R = -f_{RL}. \quad (8.25)$$

Proposition 5. $\mathcal{A}_\ell^{(2)}$ consists of all endomorphisms ϑ_ℓ of the form

$$\begin{bmatrix} \gamma(\vartheta) \otimes 1_R + \kappa^2 \lambda \otimes M_e^* M_e & \kappa^2 \Gamma \gamma(\eta) \otimes M_e^* \\ \gamma(\xi)\Gamma \otimes M_e & \gamma(\Theta) \otimes 1_L + \kappa^2 \Lambda \otimes M_e M_e^* \end{bmatrix} \quad (8.26)$$

acting on S_ℓ, where

$$\vartheta \in \mathcal{C}^\infty(\mathcal{C}\ell^{2-\mathrm{ev}}(T^\# M) \otimes \mathbb{C}), \quad \Theta \in \mathcal{C}^\infty(\mathcal{C}\ell^{2-\mathrm{ev}}(T^\# M) \otimes \mathbb{C}^{2\times 2}),$$
$$\xi \in \mathcal{C}^\infty(T^\# M \otimes \mathbb{C}^{2\times 1}), \quad \eta \in \mathcal{C}^\infty(T^\# M \otimes \mathbb{C}^{1\times 2}),$$
$$\lambda \in \mathcal{C}^\infty(M, \mathbb{C}), \quad \Lambda \in \mathcal{C}^\infty(M, \mathbb{C}^{2\times 2}).$$

Here $\mathcal{C}\ell^{2-\mathrm{ev}}$ is the space of all Clifford algebra elements of degree 2 or 0.

Proof. Let $\times$ denote Clifford multiplication. Using the notation of (8.21), we obtain

$$\omega_\ell^0 \omega_\ell^1 = \begin{bmatrix} \gamma(\omega^0) \otimes 1_R & \kappa^2 \Gamma \psi^0 \otimes M_e^* \\ \varphi^0 \Gamma \otimes M_e & \gamma(\Omega^0) \otimes 1_L \end{bmatrix} \begin{bmatrix} \gamma(\omega^1) \otimes 1_R & \kappa^2 \Gamma \psi^1 \otimes M_e^* \\ \varphi^1 \Gamma \otimes M_e & \gamma(\Omega^1) \otimes 1_L \end{bmatrix}$$
$$= \begin{bmatrix} \gamma(\omega^0 \times \omega^1) \otimes 1_R + \kappa^2 \psi^0 \varphi^1 \otimes M_e^* M_e & \kappa^2 \Gamma \gamma(\psi^0 \Omega^1 - \omega^0 \psi^1) \otimes M_e^* \\ \gamma(\Omega^0 \varphi^1 - \varphi^0 \omega^1)\Gamma \otimes M_e & \gamma(\Omega^0 \times \Omega^1) \otimes 1_L + \kappa^2 \varphi^0 \psi^1 \otimes M_e M_e^* \end{bmatrix}$$
$$= \begin{bmatrix} \gamma(\vartheta) \otimes 1_R + \kappa^2 \lambda \otimes M_e^* M_e & \kappa^2 \Gamma \gamma(\eta) \otimes M_e^* \\ \gamma(\xi)\Gamma \otimes M_e & \gamma(\Theta) \otimes 1_L + \kappa^2 \Lambda \otimes M_e M_e^* \end{bmatrix}.$$

This implies the assertion by putting

$$\vartheta = \omega^0 \times \omega^1, \qquad \Theta = \Omega^0 \times \Omega^1,$$
$$\xi = \Omega^0 \varphi^1 - \varphi^0 \omega^1, \qquad \eta = \psi^0 \Omega^1 - \omega^0 \psi^1, \quad (8.27)$$
$$\lambda = \psi^0 \varphi^1, \qquad \Lambda = \varphi^0 \psi^1.$$

The Electro-weak Model   159

Proposition 6. *For $\omega_\ell \in \mathcal{A}_\ell^{(1)}$, given by (8.21), $P(\pi_2 d\pi_1^{-1}\omega_\ell)P$ consists of all endomorphisms*

$$\begin{bmatrix} \gamma(f+d\omega) \otimes 1_R + \kappa^2(\varepsilon^*\varphi + \psi\varepsilon) \otimes M_e^* M_e & -\kappa^2 \Gamma\gamma(d\psi + \omega\varepsilon^* - \varepsilon^*\Omega) \otimes M_e^* \\ \gamma(d\varphi + \Omega\varepsilon - \varepsilon\omega)\Gamma \otimes M_e & \gamma(F+d\Omega) \otimes 1_L + \kappa^2(\varphi\varepsilon^* + \varepsilon\psi) \otimes M_e M_e^* \end{bmatrix}$$

where $f \in \mathcal{C}^\infty(M,\mathbb{C})$ and $F \in \mathcal{C}^\infty(M,\mathbb{C}^{2\times 2})$ are arbitrary.

Proof. According to (8.24), we have

$$\pi_1(f^0 df^1) = \begin{bmatrix} \gamma(\omega_R) \otimes 1_R & \kappa^2 \Gamma \varphi_R \otimes M_e^* \\ \varphi_L \Gamma \otimes M_e & \gamma(\omega_L) \otimes 1_L \end{bmatrix}$$

as endomorphisms acting on $S \otimes \begin{pmatrix} \mathbb{G}_R \\ \mathbb{G}_L \end{pmatrix}$, where $\omega_i = f_i^0 df_i^1$ and $\varphi_i = f_i^0 f_{ji}^1$ for $\{i,j\} = \{R,L\}$. Since $d\omega_i = df_i^0 \wedge df_i^1$, we have $f_i := df_i^0 \times df_i^1 - d\omega_i \in \mathcal{C}^\infty(M,\mathbb{C})$. Moreover $\varphi_i + \varphi_j = f_i^0 f_{ji}^1 + f_j^0 f_{ij}^1 = (f_i^0 - f_j^0)f_{ji}^1 = f_{ij}^0 f_{ji}^1$ and

$$\omega_{ij} + d\varphi_i = f_i^0 df_i^1 - f_j^0 df_j^1 + df_i^0 f_{ji}^1 + f_i^0 df_{ji}^1$$
$$= f_i^0 df_j^1 - f_j^0 df_j^1 + df_i^0 f_{ji}^1 = f_{ij}^0 df_j^1 + df_i^0 f_{ji}^1.$$

Using (8.23) we obtain for $\pi_2\big((df^0)(df^1)\big) = \pi_1(df^0)\pi_1(df^1)$

$$\begin{bmatrix} \gamma(df_R^0) \otimes 1_R & \kappa^2 \Gamma f_{LR}^0 \otimes M_e^* \\ f_{RL}^0 \Gamma \otimes M_e & \gamma(df_L^0) \otimes 1_L \end{bmatrix} \begin{bmatrix} \gamma(df_R^1) \otimes 1_R & \kappa^2 \Gamma f_{LR}^1 \otimes M_e^* \\ f_{RL}^1 \Gamma \otimes M_e & \gamma(df_L^1) \otimes 1_L \end{bmatrix}$$
$$= \begin{bmatrix} \gamma(df_R^0 \times df_R^1) \otimes 1_R + \kappa^2 f_{LR}^0 f_{RL}^1 \otimes M_e^* M_e & \kappa^2 \Gamma\gamma(f_{LR}^0 df_L^1 - df_R^0 f_{LR}^1) \otimes M_e^* \\ \gamma(df_L^0 f_{RL}^1 - f_{RL}^0 df_R^1)\Gamma \otimes M_e & \gamma(df_L^0 \times df_L^1) \otimes 1_L + \kappa^2 f_{RL}^0 f_{LR}^1 \otimes M_e M_e^* \end{bmatrix}$$
$$= \begin{bmatrix} \gamma(f_R + d\omega_R) \otimes 1_R + \kappa^2(\varphi_L + \varphi_R) \otimes M_e^* M_e & -\kappa^2 \Gamma\gamma(\omega_{RL} + d\varphi_R) \otimes M_e^* \\ \gamma(\omega_{LR} + d\varphi_L)\Gamma \otimes M_e & \gamma(f_L + d\omega_L) \otimes 1_L + \kappa^2(\varphi_L + \varphi_R) \otimes M_e M_e^* \end{bmatrix}. \quad (8.28)$$

Now consider a 2×2-matrix of 1-forms

$$A = \begin{pmatrix} \gamma(\omega_R) \otimes 1_R & \kappa^2 \Gamma\psi_1 \otimes M_e^* & 0 & \kappa^2 \Gamma\psi_2 \otimes M_e^* \\ \varphi_1 \Gamma \otimes M_e & \gamma(\omega_L) \otimes 1_L & 0 & \gamma(\omega_L'') \otimes 1_L \\ 0 & 0 & 0 & 0 \\ \varphi_2 \Gamma \otimes M_e & \gamma(\omega_L') \otimes 1_L & 0 & \gamma(\omega_L''') \otimes 1_L \end{pmatrix} = \begin{bmatrix} \gamma(\omega) \otimes 1_R & \kappa^2 \Gamma\psi \otimes M_e^* \\ \varphi \Gamma \otimes M_e & \gamma(\Omega) \otimes 1_L \end{bmatrix}$$

in $P\pi_2(\Omega^2)P$, where $\omega = \omega_R$, $\psi = (\psi_1, \psi_2)$, $\varphi = \begin{pmatrix} \varphi_1 \\ \varphi_2 \end{pmatrix}$, $\Omega = \begin{pmatrix} \omega_L & \omega_L'' \\ \omega_L' & \omega_L''' \end{pmatrix}$.

Applying (8.28) to each entry, we obtain its differential dA as

$$\begin{pmatrix} \gamma(f_R+d\omega_r)\otimes 1_R+ \\ \kappa^2(\varphi_1+\psi_1)\otimes M_e^*M_e \end{pmatrix} \quad -\kappa^2\Gamma\gamma(\omega_{RL}+d\psi_1)\otimes M_e^* \quad \begin{matrix} f_R''\otimes 1_R+ \\ \kappa^2\psi_2\otimes M_e^*M_e \end{matrix} \quad \kappa^2\Gamma\gamma(\omega_L''-d\psi_s)\otimes M_e^* \\ \gamma(\omega_{LR}+d\varphi_1)\Gamma\otimes M_e \quad \begin{matrix} \gamma(f_L+d\omega_L)\otimes 1_L+ \\ \kappa^2(\varphi_1+\psi_1)\otimes M_eM_e^* \end{matrix} \quad \gamma(\omega_L'')\Gamma\otimes M_e \quad \begin{matrix} \gamma(f_L''+d\omega_L'')\otimes 1_L \\ +\kappa^2\psi_2\otimes M_eM_e^* \end{matrix} \\ \begin{matrix} f_R'\otimes 1_R+ \\ \kappa^2\varphi_2\otimes M_e^*M_e \end{matrix} \quad \kappa^2\Gamma\gamma(\omega_L')\otimes M_e^* \quad f_R'''\otimes 1_R \quad \kappa^2\Gamma\gamma(\omega_L''')\otimes M_e^* \\ \gamma(\omega_L'+d\varphi_2)\Gamma\otimes M_e \quad \begin{matrix} \gamma(f_L'+d\omega_L')\otimes 1_L \\ +\kappa^2\varphi_2\otimes M_eM_e^* \end{matrix} \quad \gamma(\omega_L''')\Gamma\otimes M_e \quad \gamma(f_L'''+d\omega_L''')\otimes 1_L \end{pmatrix}$$

where $f_R, f_R', f_R'', f_R''' \in C^\infty(M, \mathbb{C})$ are arbitrary. Deleting the third row and column for $P\,dA\,P$ we get

$$\begin{bmatrix} \gamma(f_R+d\omega_R)\otimes 1_R+ \\ \kappa^2(\varphi_1+\psi_1)\otimes M_e^*M_e \end{bmatrix} \quad -\kappa^2\Gamma\gamma(d\psi_1+\omega_{RL}, d\psi_2-\omega_L'')\otimes M_e^* \\ \gamma\begin{pmatrix} d\varphi_1+\omega_{LR} \\ d\varphi_2+\omega_L' \end{pmatrix}\Gamma\otimes M_e \quad \gamma\begin{pmatrix} f_L+d\omega_L & f_L''+d\omega_L'' \\ f_L'+d\omega_L' & f_L'''+d\omega_L''' \end{pmatrix}\otimes 1_L + \kappa^2\begin{pmatrix} \varphi_1+\psi_1 & \psi_2 \\ \varphi_2 & 0 \end{pmatrix}\otimes M_eM_e^* \end{bmatrix}.$$

Now the assertion follows from the identities

$$d\varphi+\Omega\varepsilon-\varepsilon\omega = \begin{pmatrix} d\varphi_1 \\ d\varphi_2 \end{pmatrix} + \begin{pmatrix} \omega_L & \omega_L'' \\ \omega_L' & \omega_L''' \end{pmatrix}\begin{pmatrix} 1 \\ 0 \end{pmatrix} - \begin{pmatrix} 1 \\ 0 \end{pmatrix}\omega_R = \begin{pmatrix} d\varphi_1+\omega_L-\omega_R \\ d\varphi_2+\omega_L' \end{pmatrix},$$

$$d\psi + \omega\varepsilon^* - \varepsilon^*\Omega = (d\psi_1, d\psi_2) + \omega_R(1, 0) - (1, 0)\begin{pmatrix} \omega_L & \omega_L'' \\ \omega_L' & \omega_L''' \end{pmatrix}$$
$$= (d\psi_1+\omega_R-\omega_L, d\psi_2-\omega_L''),$$

$$\varphi\varepsilon^* + \varepsilon\psi = \begin{pmatrix} \varphi_1 & 0 \\ \varphi_2 & 0 \end{pmatrix} + \begin{pmatrix} \psi_1 & \psi_2 \\ 0 & 0 \end{pmatrix} = \begin{pmatrix} \varphi_1+\psi_1 & \psi_2 \\ \varphi_2 & 0 \end{pmatrix}.$$

Setting $\omega_\ell = 0$ yields the so-called "junk-ideal":

Corollary 1. *We have*

$$P\left[\pi_2 d(\operatorname{Ker}\pi_1)\right]P =$$
$$\left\{\begin{bmatrix} f\otimes 1_R & 0 \\ 0 & F\otimes 1_L \end{bmatrix} : f\in C^\infty(M,\mathbb{C}),\ F\in C^\infty(M,\mathbb{C}^{2\times 2})\right\}.$$

A fibre metric on $S_\ell = S\otimes \mathbb{G}_\ell$ is given by the endomorphism

$$I_\ell = I\otimes \begin{pmatrix} y_R & 0 \\ 0 & y_L \end{pmatrix} = \begin{pmatrix} I\otimes y_R & 0 \\ 0 & I\otimes y_L \end{pmatrix} \tag{8.29}$$

where y_R, y_L are positive definite linear maps on $\mathbb{G}_R$ and $\mathbb{G}_L$, respectively. Clearly, $[f_\ell, I_\ell] = 0$ for all $f_\ell \in \mathcal{A}_\ell$. The condition $[D_\ell, I_\ell] = 0$ is equivalent to

$$M_e\, y_R = y_L\, M_e, \quad y_R\, M_e^* = M_e^*\, y_L. \tag{8.30}$$

Proposition 7. For $\vartheta_\ell, \vartheta'_\ell \in \mathcal{A}_\ell^{(2)}$, given by (8.26), the scalar product has the form

$$(\vartheta_\ell|\vartheta'_\ell I_\ell)_{S_\ell} = s[(\vartheta|\vartheta')_S + (\Theta|\Theta')_{\underline{S}}] \tag{8.31}$$
$$+ \underline{\mathrm{tr}}_{\mathbb{G}_L}(M_e M_e^*)^2 y_L \cdot [(\lambda|\lambda')_S + (\Lambda|\Lambda')_{\underline{S}}]$$
$$+ \kappa^2 r[(\lambda|\vartheta')_S + (\vartheta|\lambda')_S - (\xi|\xi')_S + (\Lambda|\Theta')_{\underline{S}} + (\Theta|\Lambda')_{\underline{S}} - (\eta|\eta')_{\underline{S}}]$$

where

$$\begin{aligned} s &:= \underline{\mathrm{tr}}_{\mathbb{G}_L} y_L & &= \underline{\mathrm{tr}}_{\mathbb{G}_R} y_R, \\ r &:= \underline{\mathrm{tr}}_{\mathbb{G}_L} M_e M_e^* y_L & &= \underline{\mathrm{tr}}_{\mathbb{G}_R} M_e M_e^* y_R, \\ \underline{\mathrm{tr}}_{\mathbb{G}_L}(M_e M_e^*)^2 y_L &= \underline{\mathrm{tr}}_{\mathbb{G}_R}(M_e M_e^*)^2 y_R. \end{aligned} \tag{8.32}$$

Proof. Since M_e is invertible, (8.30) implies

$$\underline{\mathrm{tr}}_{\mathbb{G}_L} y_L = \underline{\mathrm{tr}}_{\mathbb{G}_L} M_e y_R M_e^{-1} = \underline{\mathrm{tr}}_{\mathbb{G}_R} y_R,$$

$$\underline{\mathrm{tr}}_{\mathbb{G}_L} M_e M_e^* y_L = \underline{\mathrm{tr}}_{\mathbb{G}_L} M_e y_R M_e^* = \underline{\mathrm{tr}}_{\mathbb{G}_R} y_R M_e^* M_e = \underline{\mathrm{tr}}_{\mathbb{G}_R} M_e^* M_e y_R,$$

$$\underline{\mathrm{tr}}_{\mathbb{G}_L} M_e M_e^* M_e M_e^* y_L = \underline{\mathrm{tr}}_{\mathbb{G}_L} M_e M_e^* M_e y_R M_e^* = \underline{\mathrm{tr}}_{\mathbb{G}_L} M_e M_e^* y_L M_e M_e^*$$
$$= \underline{\mathrm{tr}}_{\mathbb{G}_L} M_e y_R M_e^* M_e M_e^* = \underline{\mathrm{tr}}_{\mathbb{G}_R} y_R M_e^* M_e M_e^* M_e = \underline{\mathrm{tr}}_{\mathbb{G}_R} M_e^* M_e M_e^* M_e y_R.$$

Using the notation (8.26) $\vartheta_\ell^\dagger \vartheta'_\ell I_\ell$ equals

$$\begin{bmatrix} \gamma(\vartheta) \otimes 1_R + \kappa^2 \lambda \otimes M_e^* M_e & \kappa^2 \Gamma \gamma(\eta) \otimes M_e^* \\ \gamma(\xi)\Gamma \otimes M_e & \gamma(\Theta) \otimes 1_L + \kappa^2 \Lambda \otimes M_e M_e^* \end{bmatrix}^\dagger \begin{bmatrix} \gamma(\vartheta') \otimes 1_R + \kappa^2 \lambda' \otimes M_e^* M_e & \kappa^2 \Gamma \gamma(\eta') \otimes M_e^* \\ \gamma(\xi')\Gamma \otimes M_e & \gamma(\Theta') \otimes 1_L + \kappa^2 \Lambda' \otimes M_e M_e^* \end{bmatrix} \begin{bmatrix} I \otimes y_R & 0 \\ 0 & \underline{I} \otimes y_L \end{bmatrix}$$

$$= \begin{bmatrix} \gamma(\overline{\vartheta}) \otimes 1_R + \kappa^2 \overline{\lambda} \otimes M_e^* M_e & \Gamma^\dagger \gamma(\xi^*) \otimes M_e^* \\ \gamma(\eta^*) \kappa^2 \Gamma^\dagger \otimes M_e & \gamma(\Theta^*) \otimes 1_L + \kappa^2 \Lambda^* \otimes M_e M_e^* \end{bmatrix} \begin{bmatrix} \gamma(\vartheta') \otimes y_R + \kappa^2 \lambda' \otimes M_e^* M_e y_R & \kappa^2 \Gamma \gamma(\eta') \otimes M_e^* y_L \\ \gamma(\xi')\Gamma \otimes M_e y_R & \gamma(\Theta') \otimes y_L + \kappa^2 \Lambda' \otimes M_e M_e^* y_L \end{bmatrix}$$

$$= \begin{bmatrix} \gamma(\overline{\vartheta} \times \vartheta') \otimes y_R + \overline{\lambda}\lambda' \otimes (M_e^* M_e)^2 y_R + \kappa^2 \gamma(\overline{\lambda}\vartheta' + \overline{\vartheta}\lambda' - \xi^* \times \xi') \otimes M_e^* M_e y_R & * \\ * & \gamma(\Theta^* \times \Theta') \otimes y_L + \Lambda^* \Lambda' \otimes (M_e M_e^*)^2 y_L + \kappa^2 \gamma(\Theta^* \Lambda' + \Lambda^* \Theta' - \eta^* \times \eta') \otimes M_e M_e^* y_L \end{bmatrix}.$$

In view of (7.55), the assertion follows by taking the trace.

Remark 3. *For compact Riemannian manifolds M ($\kappa = i$), Proposition 7 gives an explicit expression for the Dixmier trace*

$$\mathrm{Tr}_\omega(\vartheta_\ell^\dagger \vartheta'_\ell I_\ell |D_\ell|^{-n}) = c_n \int_M d^n x \, (\vartheta_\ell|\vartheta'_\ell I_\ell)_{S_\ell} \tag{8.33}$$

by applying (7.91) to $\mathbb{G}_\ell$-valued spinors.

Now consider the inner product (8.31) and let $\perp$ denote the orthogonal complement.

Proposition 8. $P\bigl[\pi_2 d(\mathrm{Ker}\,\pi_1)\bigr]^\perp P$ *consists of endomorphisms* σ_ℓ *of the form*

$$\begin{bmatrix} \gamma(\sigma)\otimes 1_R + \kappa^2\lambda\otimes(M_e^* M_e - \tfrac{r}{s}) & \kappa^2\Gamma\gamma(\eta)\otimes M_e^* \\ \gamma(\xi)\Gamma\otimes M_e & \gamma(\Sigma)\otimes 1_L + \kappa^2\Lambda\otimes(M_e M_e^* - \tfrac{r}{s}) \end{bmatrix} \tag{8.34}$$

with arbitrary

$$\sigma \in \mathcal{C}^\infty(\bigwedge{}^2(T^\# M)\otimes\mathbb{C}), \qquad \Sigma \in \mathcal{C}^\infty(\bigwedge{}^2(T^\# M)\otimes\mathbb{C}^{2\times 2}),$$
$$\xi \in \mathcal{C}^\infty(T^\# M\otimes\mathbb{C}^{2\times 1}), \qquad \eta \in \mathcal{C}^\infty(T^\# M\otimes\mathbb{C}^{1\times 2}).$$

Proof. Writing $\vartheta_\ell \in (d\,\mathrm{Ker}\,\pi_1)^\perp$ and $\vartheta'_\ell \in d\,\mathrm{Ker}\,\pi_1$ according to (8.26), Corollary 1 implies $\vartheta' = f$, $\Theta' = F$, $\lambda' = 0$, $\Lambda' = 0$, $\xi' = 0$ and $\eta' = 0$. Therefore Proposition 7 yields

$$\begin{aligned} 0 &= \bigl(\vartheta_\ell \mid \vartheta'_\ell I_\ell\bigr)_{S_\ell} \\ &= s\bigl[(\vartheta \mid f)_S + (\Theta \mid F)_{\underline{S}}\bigr] + r\bigl[(\lambda \mid f)_S + (\Lambda \mid F)_{\underline{S}}\bigr] \\ &= s\bigl[(\vartheta + \tfrac{r}{s}\lambda \mid f)_S + (\Theta + \tfrac{r}{s}\Lambda \mid F)_{\underline{S}}\bigr]. \end{aligned}$$

Since $f \in \mathcal{C}^\infty(M,\mathbb{C})$ is arbitrary, we have

$$\sigma := \vartheta + \frac{r}{s}\lambda \in \mathcal{C}^\infty(\bigwedge{}^2(T^\# M)\otimes\mathbb{C}),$$

i.e. $\vartheta = \sigma - \tfrac{r}{s}\lambda$. Since $F \in \mathcal{C}^\infty(M,\mathbb{C}^{2\times 2})$ is arbitrary, we have

$$\Sigma := \Theta + \frac{r}{s}\Lambda \in \mathcal{C}^\infty(\bigwedge{}^2(T^\# M)\otimes\mathbb{C}^{2\times 2}),$$

i.e. $\Theta = \Sigma - \tfrac{r}{s}\Lambda$. It follows that

$$\vartheta \otimes 1_R + \lambda \otimes M_e^* M_e = \sigma \otimes 1_R + \lambda \otimes (M_e^* M_e - \frac{r}{s}),$$

$$\Theta \otimes 1_L + \Lambda \otimes M_e M_e^* = \Sigma \otimes 1_L + \Lambda \otimes (M_e M_e^* - \frac{r}{s}).$$

Proposition 9. *The orthogonal projection from* $\mathcal{A}_\ell^{(2)}$ *onto* $P(\pi_2 d\,\mathrm{Ker}\,\pi_1)^\perp P$ *is given by*

$$\begin{bmatrix} \gamma(\vartheta)\otimes 1_R + \kappa^2\lambda\otimes M_e^* M_e & \kappa^2\Gamma\gamma(\eta)\otimes M_e^* \\ \gamma(\xi)\Gamma\otimes M_e & \gamma(\Theta)\otimes 1_L + \kappa^2\Lambda\otimes M_e M_e^* \end{bmatrix}^\perp =$$
$$\begin{bmatrix} \gamma(\vartheta^\perp)\otimes 1_R + \kappa^2\lambda\otimes(M_e^* M_e - \tfrac{r}{s}) & \kappa^2\Gamma\gamma(\eta)\otimes M_e^* \\ \gamma(\xi)\Gamma\otimes M_e & \gamma(\Theta^\perp)\otimes 1_L + \kappa^2\Lambda\otimes(M_e M_e^* - \tfrac{r}{s}) \end{bmatrix}.$$

Proof. For ϑ_ℓ given by (8.26) we have $\vartheta_\ell^\perp \in P[\pi_2 d \operatorname{Ker} \pi_1]^\perp P$ by Proposition 8 and $\vartheta_\ell - \vartheta_\ell^\perp$ equal to

$$\begin{bmatrix} \gamma(\vartheta - \vartheta^\perp + \kappa^2 \frac{r}{s}\lambda) \otimes 1_R & 0 \\ 0 & \gamma(\Theta - \Theta^\perp + \kappa^2 \frac{r}{s}\Lambda) \otimes 1_L \end{bmatrix}$$

and therefore element of $P[\pi_2 d \operatorname{Ker} \pi_1] P$ by Corollary 1. This implies the assertion.

Now consider the differential

$$\boldsymbol{d}: \mathcal{A}_\ell^{(1)} \to \pi_2[P(d\operatorname{Ker}\pi_1)P]^\perp, \quad \pi_1(PAP) \mapsto \pi_2(P(dA)P)^\perp \qquad (8.35)$$

and product

$$\wedge: \mathcal{A}_\ell^{(1)} \otimes \mathcal{A}_\ell^{(1)} \to \pi_2[P(d\operatorname{Ker}\pi_1)P]^\perp, \quad \pi_1(PA_1P) \otimes \pi_1(PA_2P) \mapsto \pi_2(P(A_1A_2)P)^\perp \qquad (8.36)$$

computed in $\pi_2(P(d\operatorname{Ker}\pi_1)P)^\perp$.

Corollary 2. *We have*

$$\boldsymbol{d}\begin{bmatrix} \gamma(\omega) \otimes 1_R & \kappa^2 \Gamma \psi \otimes M_e^* \\ \varphi \Gamma \otimes M_e & \gamma(\Omega) \otimes 1_L \end{bmatrix} =$$

$$\begin{bmatrix} \gamma(d\omega) \otimes 1_R + \kappa^2(\varepsilon^*\varphi + \psi\varepsilon) \otimes (M_e^* M_e - \frac{r}{s}) & -\kappa^2 \Gamma \gamma(d\psi + \omega\varepsilon^* - \varepsilon^* \Omega) \otimes M_e^* \\ \gamma(d\varphi + \Omega\varepsilon - \varepsilon\omega)\Gamma \otimes M_e & \gamma(d\Omega) \otimes 1_R + \kappa^2(\varphi\varepsilon^* + \varepsilon\psi) \otimes (M_e M_e^* - \frac{r}{s}) \end{bmatrix}$$

and

$$\begin{bmatrix} \gamma(\omega^0) \otimes 1_R & \kappa^2 \Gamma \psi^0 \otimes M_e^* \\ \varphi^0 \Gamma \otimes M_e & \gamma(\Omega^0) \otimes 1_L \end{bmatrix} \wedge \begin{bmatrix} \gamma(\omega^1) \otimes 1_R & \kappa^2 \Gamma \psi^1 \otimes M_e^* \\ \varphi^1 \Gamma \otimes M_e & \gamma(\Omega^1) \otimes 1_L \end{bmatrix} =$$

$$\begin{bmatrix} \gamma(\omega^0 \wedge \omega^1) \otimes 1_R + \kappa^2 \psi^0 \varphi^1 \otimes (M_e^* M_e - \frac{r}{s}) & \kappa^2 \Gamma \gamma(\psi^0 \Omega^1 - \omega^0 \psi^1) \otimes M_e^* \\ \gamma(\Omega^0 \varphi^1 - \varphi^0 \omega^1)\Gamma \otimes M_e & \gamma(\Omega^0 \wedge \Omega^1) \otimes 1_L + \kappa^2 \varphi^0 \psi^1 \otimes (M_e M_e^* - \frac{r}{s}) \end{bmatrix}.$$

We are now ready to introduce the "dynamical variable" for the electroweak model, phrased in terms of the noncommutative geometry $\mathcal{A}_\ell$.

Consider, for real 1-forms $\rho = \overline{\rho} \in \mathcal{C}^\infty(T^\# M)$, the elements

$$\hat{\rho} := \begin{bmatrix} \gamma(\rho) \otimes 1_R & 0 \\ 0 & \underline{\gamma(\rho) \otimes 1_L} \end{bmatrix} \in \mathcal{A}_\ell^{(1)} \tag{8.37}$$

which are easily characterized algebraically.

Given any endomorphism

$$\begin{pmatrix} a & b \\ c & d \end{pmatrix} \in \mathcal{A}_\ell^{(1)}$$

we define the *supertrace*

$$\operatorname{str}\begin{pmatrix} a & b \\ c & d \end{pmatrix} := \underline{\operatorname{tr}}_{S \otimes \mathbb{G}_R} a - \underline{\operatorname{tr}}_{S \otimes \mathbb{G}_L} d . \tag{8.38}$$

Definition 2. *The skew-adjoint elements*

$$\omega_\ell = -\omega_\ell^\dagger \in \mathcal{A}_\ell^{(1)} , \tag{8.39}$$

satisfying

$$\operatorname{str} \hat{\rho}\, \omega_\ell\, I_\ell = 0 \tag{8.40}$$

for all $\rho \in \mathcal{C}^\infty(T^\# M)$ will be called NC gauge fields *for the electro-weak model.*

Proposition 10. *The NC gauge fields ω_ℓ for the electro-weak model are given by*

$$\begin{pmatrix} \gamma(\omega) \otimes 1_R & \kappa^2\, \Gamma\, \varphi^* \otimes M_e^* \\ \varphi\, \Gamma \otimes M_e & \gamma(\Omega_0 + \tfrac{1}{2}\underline{\omega}) \otimes 1_L \end{pmatrix} , \tag{8.41}$$

with arbitrary

$$\omega \in \mathcal{C}^\infty(T^\# M \otimes \mathfrak{u}(1)),$$
$$\Omega_0 \in \mathcal{C}^\infty(T^\# M \otimes \mathfrak{su}(2)),$$
$$\varphi \in \mathcal{C}^\infty(M, \mathbb{C}^{2 \times 1}) .$$

Proof. Since

$$\begin{bmatrix} \gamma(\omega) \otimes 1_R & \kappa^2 \Gamma \psi \otimes M_e^* \\ \varphi \Gamma \otimes M_e & \gamma(\Omega) \otimes 1_L \end{bmatrix}^\dagger = \begin{bmatrix} \gamma(\omega)^\dagger \otimes 1_R & \Gamma^\dagger \varphi^* \otimes M_e^* \\ \psi^* \kappa^2 \Gamma^\dagger \otimes M_e & \gamma(\Omega)^\dagger \otimes 1_L \end{bmatrix} = \begin{bmatrix} \gamma(\overline{\omega}) \otimes 1_R & -\kappa^2 \Gamma \varphi^* \otimes M_e^* \\ -\psi^* \Gamma \otimes M_e & \gamma(\Omega^*) \otimes 1_L \end{bmatrix}$$

it follows that $\omega_\ell^\dagger = -\omega_\ell$ if and only if $\overline{\omega} = -\omega$, $\psi = \varphi^*$, and $\Omega^* = -\Omega$. Thus the skew-adjoint elements $\omega_\ell = -\omega_\ell^\dagger \in \mathcal{A}_\ell^{(1)}$ are given by

$$\begin{bmatrix} \gamma(\omega) \otimes 1_R & \kappa^2 \Gamma \varphi^* \otimes M_e^* \\ \varphi \Gamma \otimes M_e & \gamma(\Omega) \otimes 1_L \end{bmatrix} \tag{8.42}$$

where $\omega \in \mathcal{C}^\infty(T^\# M \otimes \mathfrak{u}(1))$, $\Omega \in \mathcal{C}^\infty(T^\# M \otimes \mathfrak{u}(2))$ and $\varphi \in \mathcal{C}^\infty(M, \mathbb{C}^{2\times 1})$. Now let $\rho \in \mathcal{C}^\infty(T^\# M)$ be arbitrary. Then the endomorphism

$$\hat\rho \omega_\ell I_\ell = \begin{pmatrix} \gamma(\rho)\otimes 1_R & 0 \\ 0 & \underline{\underline{\gamma(\rho)\otimes 1_L}} \end{pmatrix} \begin{pmatrix} \gamma(\omega)\otimes 1_R & \kappa^2\Gamma\varphi^*\otimes M_e^* \\ \varphi\Gamma\otimes M_e & \gamma(\underline{\Omega})\otimes 1_L \end{pmatrix} \begin{pmatrix} I\otimes y_R & 0 \\ 0 & \underline{\underline{I\otimes y_L}} \end{pmatrix}$$

$$= \begin{pmatrix} \gamma(\rho\times\omega)\otimes y_R & \kappa^2\gamma(\rho)\Gamma\varphi^*\otimes M_e^* y_R \\ \underline{\underline{\gamma(\rho)\varphi\Gamma \otimes M_e y_R}} & \underline{\underline{\gamma(\underline{\rho}\times\underline{\Omega})\otimes y_L}} \end{pmatrix}$$

has supertrace

$$\operatorname{str}\hat\rho\omega_\ell I_\ell = \underline{\operatorname{tr}}_{S\otimes\mathbb{G}_R}\gamma(\rho\times\omega)\otimes y_R - \underline{\underline{\operatorname{tr}}}_{S\otimes\mathbb{G}_L}\gamma(\underline{\rho}\times\underline{\Omega})\otimes y_L$$
$$= \underline{\operatorname{tr}}_S\gamma(\rho\times\omega)\,\underline{\operatorname{tr}}_{\mathbb{G}_R}y_R - \underline{\underline{\operatorname{tr}}}_{\underline{S}}\gamma(\underline{\rho}\times\underline{\Omega})\,\underline{\underline{\operatorname{tr}}}_{\mathbb{G}_L}y_L$$
$$= [(\rho|\omega)_S - (\underline{\rho}|\underline{\Omega})_{\underline{S}}]s = (\rho|\omega - \operatorname{tr}\Omega)_S \cdot s.$$

Thus the condition (8.40) implies $(\rho|\omega - \operatorname{tr}\Omega)_S = 0$ for all ρ. It follows that $\omega = \operatorname{tr}\Omega$ as 1-forms. Hence

$$\Omega = \Omega_0 + \frac{1}{2}\underline{\underline{\omega}}, \tag{8.43}$$

where $\Omega_0 \in \mathcal{C}^\infty(T^\# M \otimes \mathfrak{u}(2))$ satisfies $\operatorname{tr}\Omega_0 = 0$.

Remark 4. *The NC gauge fields as defined above parametrize exactly all the bosonic fields of the electro-weak model, which therefore occur naturally as components of a single object. This is one of the major conceptual advantages of the noncommutative geometry approach to the electro-weak model. The bosonic (spin 1) gauge fields mediating the electro-weak interaction are given mathematically by Lie algebra valued 1-forms*

$$i\sum_{a=0}^{3} W^{(a)}\tau^{(a)} \in \mathcal{C}^\infty(T^\# M \otimes \mathfrak{u}(2)) \tag{8.44}$$

where $W^{(0)},\ldots,W^{(3)} \in \mathcal{C}^\infty(T^\# M)$ are real spin 1 fields (generating the electro-magnetic field (photon) and the massive vector boson fields $W_\pm, Z$ by taking certain linear combinations), and $\tau^{(0)},\ldots,\tau^{(3)}$ is a suitable basis of $i\cdot\mathfrak{u}(2)$ satisfying $\operatorname{tr}\tau^{(a)}\tau^{(b)} = \frac{1}{2}\delta^{(a)(b)}$.

8.3 Noncommutative Gauge Action Functional

In order to find a gauge-invariant dynamics allowing massive leptons and gauge fields, one also introduces complex scalar (spin 0) fields $\Phi^0, \Phi^+ \in \mathcal{C}^\infty(M, \mathbb{C})$ forming a doublet

$$\begin{pmatrix} \Phi^0 \\ \Phi^+ \end{pmatrix} \in \mathcal{C}^\infty(M, \mathbb{C}^{2\times 1}) \tag{8.45}$$

called the Higgs field, *which experiences a quartic self-interaction in a ϕ^4-potential with a ground state breaking $SU(2)$-symmetry. Putting*

$$\omega = -ig_1 W^{(0)} \tag{8.46}$$

$$\Omega_0 = \frac{ig_2}{2}\begin{pmatrix} -W^{(3)} & W^{(1)} + iW^{(2)} \\ W^{(1)} - iW^{(2)} & W^{(3)} \end{pmatrix} = ig_2 \sum_{a=1}^{3} W^{(a)} \tau^{(a)}$$

and

$$\varphi + \varepsilon = \frac{1}{\sqrt{2r}}\begin{pmatrix} \Phi^0 \\ \Phi^+ \end{pmatrix},$$

it follows that (8.41) encodes all the physical fields $W^{(0)}, \ldots, W^{(3)}, \Phi^0, \Phi^+$ of the electro-weak model. The so-called "coupling constants" $g_1, g_2 > 0$ introduced here will later be expressed in terms of the basic parameter s of the model (cf. (8.55)).

In order to compute the Yang-Mills functional recall that, for a projection $P \in \mathcal{A}^{2\times 2}$, a 1-form $A \in P(\Omega_\mathcal{A}^1)^{2\times 2}P$ defines a covariant derivative

$$\nabla = Pd + A \tag{8.47}$$

on the associated projective module $P\mathcal{A}^2$, whose curvature is given by

$$\nabla(\nabla P) = P(dA + A^2)P + P(dP)^2 \tag{8.48}$$

since

$$\begin{aligned}\nabla(\nabla P) &= (Pd + A)((Pd + A)P) = (Pd + A)(P(dP) + AP) \\ &= P(d(P(dP))) + P(d(AP)) + AP(dP) + AAP \\ &= P(dP)(dP) + P((dA)P - A(dP)) + AP(dP) + AAP \\ &= P(dP)^2 + P(dA + A^2)P.\end{aligned}$$

In our case, with p given by (16.48) and $P := \begin{pmatrix} 1 & 0 \\ 0 & p \end{pmatrix}$, we obtain the connection $Pd + \omega_\ell$ with curvature

$$\left(Pd + \omega_\ell\right)^2 P = P(d\omega_\ell + \omega_\ell^2)P + P(dP)^2.$$

However, we have to compute the curvature in $(d \operatorname{Ker} \pi_1)^\perp$, i.e. modulo the "junk ideal" $d \operatorname{Ker} \pi_1$ characterized in Corollary 1. Accordingly, for any $\omega_\ell \in \mathcal{A}_\ell^{(1)}$ given by (8.41), we define its *curvature* as the element

$$\boldsymbol{F}(\omega_\ell) = \left(P(d\omega_\ell + \omega_\ell^2)P + P(dP)^2\right)^\perp \tag{8.49}$$

of $(d \operatorname{Ker} \pi_1)^\perp$, where $\vartheta_\ell \mapsto \vartheta_\ell^\perp$ denotes the orthogonal projection onto $(d \operatorname{Ker} \pi_1)^\perp$.

Theorem 1. *The curvature $F(\omega_\ell)$ of ω_ℓ is given by*

$$\begin{bmatrix} \gamma(F(\omega))\otimes 1_R + \kappa^2((\varphi+\varepsilon)^*(\varphi+\varepsilon)-I)\otimes(M_e^*M_e-\frac{r}{s}) & -\kappa^2\Gamma\gamma(\nabla^H(\varphi+\varepsilon))^*\otimes M_e^* \\ \gamma(\nabla^H(\varphi+\varepsilon))\Gamma\otimes M_e & \gamma(F(\Omega_0)+\frac{1}{2}\underline{F(\omega)})\otimes 1_L + \kappa^2((\varphi+\varepsilon)(\varphi+\varepsilon)^*-\underline{I})\otimes(M_eM_e^*-\frac{r}{s}) \end{bmatrix}$$

where
$$F(\omega) = d\omega, \qquad F(\Omega_0) = d\Omega_0 + \Omega_0 \wedge \Omega_0 \qquad (8.50)$$
is the curvature (field strength) of ω and Ω_0, resp., and

$$\nabla^H \phi = d\phi + \Omega_0 \phi - \frac{1}{2}\phi\omega \qquad (8.51)$$

denotes the covariant derivative acting on $\mathcal{C}^\infty(M, \mathbb{C}^{2\times 1})$.

Proof. Specializing Corollary 2, we obtain, for $\Omega = \Omega_0 + \frac{1}{2}\underline{\omega}$,

$$d\begin{pmatrix} \gamma(\omega) \otimes 1_R & \kappa^2\Gamma\varphi^* \otimes M_e^* \\ \varphi\Gamma \otimes M_e & \gamma(\Omega) \otimes 1_L \end{pmatrix} =$$

$$\begin{pmatrix} \gamma(d\omega) \otimes 1_R + \kappa^2(\varepsilon^*\varphi + \varphi^*\varepsilon) \otimes (M_e^*M_e - \frac{r}{s}) & -\kappa^2\Gamma\gamma(d\varphi^* + \omega\varepsilon^* - \varepsilon^*\Omega) \otimes M_e^* \\ \gamma(d\varphi + \Omega\varepsilon - \varepsilon\omega)\Gamma \otimes M_e & \gamma(d\Omega) \otimes 1_L + \kappa^2(\varphi\varepsilon^* + \varepsilon\varphi^*) \otimes (M_eM_e^* - \frac{r}{s}) \end{pmatrix}$$

and

$$\begin{pmatrix} \gamma(\omega) \otimes 1_R & \kappa^2\Gamma\varphi^* \otimes M_e^* \\ \varphi\Gamma \otimes M_e & \gamma(\Omega) \otimes 1_L \end{pmatrix} \wedge \begin{pmatrix} \gamma(\omega) \otimes 1_R & \kappa^2\Gamma\varphi^* \otimes M_e^* \\ \varphi\Gamma \otimes M_e & \gamma(\Omega) \otimes 1_L \end{pmatrix} = $$
$$\begin{pmatrix} \kappa^2\varphi^*\varphi \otimes (M_e^*M_e - \frac{r}{s}) & \kappa^2\Gamma\gamma(\varphi^*\Omega - \omega\varphi^*) \otimes M_e^* \\ \gamma(\Omega\varphi - \varphi\omega)\Gamma \otimes M_e & \gamma(\Omega \wedge \Omega) \otimes 1_L + \kappa^2\varphi\varphi^* \otimes (M_eM_e^* - \frac{r}{s}) \end{pmatrix}.$$

By Remark 2, we have

$$dp = \begin{bmatrix} D\otimes 1_R & \kappa^2\Gamma\otimes M_e^* \\ \Gamma\otimes M_e & D\otimes 1_L \end{bmatrix}\begin{bmatrix} 0 & 0 \\ 0 & I\otimes 1_L \end{bmatrix} - \begin{bmatrix} 0 & 0 \\ 0 & I\otimes 1_L \end{bmatrix}\begin{bmatrix} D\otimes 1_R & \kappa^2\Gamma\otimes M_e^* \\ \Gamma\otimes M_e & D\otimes 1_L \end{bmatrix}$$
$$= \begin{bmatrix} 0 & \kappa^2\Gamma \otimes M_e^* \\ -\Gamma\otimes M_e & 0 \end{bmatrix}$$

which implies

$$(dp)^2 = \begin{bmatrix} 0 & \kappa^2\Gamma\otimes M_e^* \\ -\Gamma\otimes M_e & 0 \end{bmatrix}\begin{bmatrix} 0 & \kappa^2\Gamma \otimes M_e^* \\ -\Gamma \otimes M_e & 0 \end{bmatrix} = -\kappa^2\begin{bmatrix} I\otimes M_e^*M_e & 0 \\ 0 & I\otimes M_eM_e^* \end{bmatrix},$$

$$p(dp)^2 = -\kappa^2 \begin{bmatrix} 0 & 0 \\ 0 & I \otimes 1_L \end{bmatrix} \begin{bmatrix} I \otimes M_e^* M_e & 0 \\ 0 & I \otimes M_e M_e^* \end{bmatrix} = \begin{bmatrix} 0 & 0 \\ 0 & -\kappa^2 I \otimes M_e M_e^* \end{bmatrix}.$$

Therefore

$$P(dP)^2 = \begin{pmatrix} 1 & 0 \\ 0 & p \end{pmatrix} (d \begin{pmatrix} 1 & 0 \\ 0 & p \end{pmatrix})^2 = \begin{pmatrix} 1 & 0 \\ 0 & p \end{pmatrix} \begin{pmatrix} 0 & 0 \\ 0 & dp \end{pmatrix}^2 = \begin{pmatrix} 1 & 0 \\ 0 & p \end{pmatrix} \begin{pmatrix} 0 & 0 \\ 0 & (dp)^2 \end{pmatrix} = \begin{pmatrix} 0 & 0 \\ 0 & p(dp)^2 \end{pmatrix}$$

$$= \begin{pmatrix} 0 & 0 & 0 & 0 \\ 0 & 0 & 0 & 0 \\ 0 & 0 & 0 & 0 \\ 0 & 0 & 0 & -\kappa^2 I \otimes M_e M_e^* \end{pmatrix} = \begin{bmatrix} 0 & 0 \\ 0 & -\kappa^2 \begin{pmatrix} 0 & 0 \\ 0 & 1 \end{pmatrix} \otimes M_e M_e^* \end{bmatrix} = \begin{bmatrix} 0 & 0 \\ 0 & -\kappa^2 \underset{\sim}{\varepsilon}\underset{\sim}{\varepsilon}^* \otimes M_e M_e^* \end{bmatrix}$$

and hence

$$(P(dP)^2)^\perp = \begin{pmatrix} 0 & 0 \\ 0 & -\kappa^2 \underset{\sim}{\varepsilon}\underset{\sim}{\varepsilon}^* \otimes (M_e M_e^* - \tfrac{r}{s}) \end{pmatrix}.$$

It follows that $\boldsymbol{F}(\omega_\ell)$ is equal to

$$\begin{pmatrix} \gamma(d\omega) \otimes 1_R + & -\kappa^2 \Gamma \gamma(d\varphi^* - \varphi^*\Omega + \omega\varphi^* + \omega\varepsilon^* - \varepsilon^*\Omega) \otimes M_e^* \\ \kappa^2(\varphi^*\varphi + \varepsilon^*\varphi + \varphi^*\varepsilon) \otimes (M_e^* M_e - \tfrac{r}{s}) & \\ \gamma(d\varphi + \Omega\varepsilon - \varepsilon\omega + \Omega\varphi - \varphi\omega)\Gamma \otimes M_e & \gamma(d\Omega + \Omega \wedge \Omega) \otimes 1_L + \\ & \kappa^2(\varphi\varphi^* + \varphi\varepsilon^* + \varepsilon\varphi^* + \underset{\sim}{\varepsilon}\underset{\sim}{\varepsilon}^*) \otimes (M_e M_e^* - \tfrac{r}{s}) \end{pmatrix}.$$

Now the assertion follows via the identities

$$\nabla^H(\varphi + \varepsilon) = d(\varphi + \varepsilon) + \Omega(\varphi + \varepsilon) - (\varphi + \varepsilon)\omega = d\varphi + \Omega\varphi + \Omega\varepsilon - \varphi\omega - \varepsilon\omega,$$

$$(\nabla^H(\varphi+\varepsilon))^* = (d\varphi)^* + \varphi^*\Omega^* + \varepsilon^*\Omega^* - \overline{\omega}\varphi^* - \omega\varepsilon^* = d\varphi^* - \varphi^*\Omega - \varepsilon^*\Omega + \omega\varphi^* + \omega\varepsilon^*,$$

$$\varphi\varphi^* + \varphi\varepsilon^* + \varepsilon\varphi^* - \underset{\sim}{\varepsilon}\underset{\sim}{\varepsilon}^* = (\varphi+\varepsilon)(\varphi+\varepsilon)^* - \varepsilon\varepsilon^* - \underset{\sim}{\varepsilon}\underset{\sim}{\varepsilon}^* = (\varphi+\varepsilon)(\varphi+\varepsilon)^* - \underline{1},$$

$$\varphi^*\varphi + \varepsilon^*\varphi + \varphi^*\varepsilon = (\varphi+\varepsilon)^*(\varphi+\varepsilon) - \varepsilon^*\varepsilon = \|\varphi+\varepsilon\|^2 - 1.$$

Definition 3. *The NC gauge Lagrangian for the electro-weak model is the functional*

$$\omega_\ell \mapsto \big(\boldsymbol{F}(\omega_\ell) \mid \boldsymbol{F}(\omega_\ell)I_\ell\big)_{S_\ell} \tag{8.52}$$

defined on the NC gauge fields $\omega_\ell \in \mathcal{A}_\ell^{(1)}$. *Note that the inner product is taken fibrewise over* M, *so that (8.52) really defines a Lagrangian density. For the integrated version*

$$\int_M d^n x \cdot \big(\boldsymbol{F}(\omega_\ell) \mid \boldsymbol{F}(\omega_\ell)I_\ell\big)_{S_\ell} \tag{8.53}$$

one has to assume that ω_ℓ *has compact support. For compact Riemannian manifolds* M ($\kappa = i$), *it follows from (7.91) that (8.53) is related to the*

Dixmier trace

$$\mathrm{Tr}_\omega(\boldsymbol{F}(\omega_\ell)^\dagger \boldsymbol{F}(\omega_\ell) I_\ell |D_\ell|^{-n}) = c_n \int_M d^n x \cdot (\boldsymbol{F}(\omega_\ell)|\boldsymbol{F}(\omega_\ell)I_\ell)_{S_\ell}.$$

Theorem 2. *The NC gauge Lagrangian of ω_ℓ is given by*

$$\bigl(\boldsymbol{F}(\omega_\ell) \mid \boldsymbol{F}(\omega_\ell)I_\ell\bigr)_{S_\ell} = \frac{3s}{2}\|F(\omega)\|^2 + s\|F(\Omega_0)\|^2 - 2\kappa^2 r\|\nabla^H(\varphi+\varepsilon)\|^2$$
$$+ 2\underline{\mathrm{tr}}_{\mathbb{G}_L}(M_e M_e^* - \frac{r}{s})^2 y_L \bigl(\|\varphi+\varepsilon\|^2 - 1\bigr)^2 + const,$$

where the constants s, r are defined in (8.32).

Proof. Using Proposition 7 and Theorem 1, it follows that

$$(\boldsymbol{F}(\omega_\ell) \mid \boldsymbol{F}(\omega_\ell)I_\ell)_{S_\ell} =$$

$$s\bigl[\|F(\omega) - \kappa^2 \tfrac{r}{s}((\varphi+\varepsilon)^*(\varphi+\varepsilon) - I)\|_{\underline{S}}^2 + \|F(\Omega_0) + \tfrac{1}{2}\underline{F(\omega)} - \kappa^2 \tfrac{r}{s}((\varphi+\varepsilon)(\varphi+\varepsilon)^* - \underline{I})\|_{\underline{\underline{S}}}^2\bigr]$$

$$+\kappa^2 r\bigl[2\,\mathrm{Re}\bigl(F(\omega) - \kappa^2 \tfrac{r}{s}((\varphi+\varepsilon)^*(\varphi+\varepsilon) - I) \mid (\varphi+\varepsilon)^*(\varphi+\varepsilon) - I\bigr)_S$$

$$+2\,\mathrm{Re}\bigl(F(\Omega_0) + \tfrac{1}{2}\underline{F(\omega)} - \kappa^2 \tfrac{r}{s}((\varphi+\varepsilon)(\varphi+\varepsilon)^* - \underline{I}) \mid (\varphi+\varepsilon)(\varphi+\varepsilon)^* - \underline{I}\bigr)_{\underline{S}}$$

$$-\|\nabla^H(\varphi+\varepsilon)\|_S^2 - \|\nabla^H(\varphi+\varepsilon)^*\|_{\underline{S}}^2\bigr]$$

$$+\underline{\mathrm{tr}}_{\mathbb{G}_L}(M_e M_e^* - \tfrac{r}{s})^2 y_L\bigl[\|(\varphi+\varepsilon)^*(\varphi+\varepsilon) - I\|_S^2 + \|(\varphi+\varepsilon)(\varphi+\varepsilon)^* - \underline{I}\|_{\underline{S}}^2\bigr]$$

$$= s\|F(\omega)\|_S^2 (1 + 2\cdot\tfrac{1}{4}) + s\|F(\Omega_0)\|_{\underline{S}}^2 - 2\kappa^2 r\|\nabla^H(\varphi+\varepsilon)\|_S^2$$

$$+\bigl(\|(\varphi+\varepsilon)^*(\varphi+\varepsilon) - I\|_S^2 + \|(\varphi+\varepsilon)(\varphi+\varepsilon)^* - \underline{I}\|_{\underline{S}}^2\bigr)\cdot\bigl(s\tfrac{r^2}{s^2} - 2\tfrac{r^2}{s} + \underline{\mathrm{tr}}_{\mathbb{G}_L}(M_e M_e^*)^2 y_L\bigr)$$

Since

$$\|\phi\phi^* - \underline{I}\|_{\underline{S}}^2 = \underline{\mathrm{tr}}_{\underline{S}}(\phi\phi^*\phi\phi^* - 2\phi\phi^* + \underline{I})$$
$$= (\phi^*\phi)^2 - 2\phi^*\phi + 2 = (\phi^*\phi - 1)^2 + 1$$

and

$$\underline{\mathrm{tr}}_{\mathbb{G}_L}(M_e M_e^* - \tfrac{r}{s})^2 y_L - \underline{\mathrm{tr}}_{\mathbb{G}_L}(M_e M_e^*)^2 y_L = -\frac{2r}{s}\underline{\mathrm{tr}}_{\mathbb{G}_L} M_e M_e^* y_L + \frac{r^2}{s^2}\underline{\mathrm{tr}}_{\mathbb{G}_L} y_L$$
$$= -\frac{2r^2}{s} + \frac{r^2 s}{s^2}$$

the assertion follows.

Remark 5. *Converting to physics notation (cf. (8.46)), the NC gauge Lagrangian gives the electro-weak Yang-Mills-Higgs functional in the usual form*

$$\frac{1}{g_1^2}\|F(ig_1 W^{(0)}\tau^{(0)})\|^2 + \frac{1}{g_2^2}\|F(ig_2 \sum_{a=1}^{3} W^{(a)}\tau^{(a)})\|^2$$
$$- \kappa^2 \|\nabla^H \begin{pmatrix}\Phi^0\\\Phi^+\end{pmatrix}\|^2 + \frac{m_H^2}{8r}\bigl(\|\begin{pmatrix}\Phi^0\\\Phi^+\end{pmatrix}\|^2 - 2r\bigr)^2 + const \quad (8.54)$$

if the coupling constants g_1, g_2 satisfy

$$\frac{1}{g_1^2} = 3s, \qquad \frac{1}{g_2^2} = s \qquad (8.55)$$

and for the Higgs mass m_H we get

$$m_H^2 = \frac{r}{4}\underline{\text{tr}_{G_L}}(M_e M_e^* - \frac{r}{s})^2 y_L. \qquad (8.56)$$

This follows by combining Theorem 2 and equation (8.46), using tr $\tau^{(0)^2} = \frac{1}{2}$.

8.4 Noncommutative Matter Action Functional

For the remainder of this section we are concerned with the *fermionic action* for the electro-weak model, involving the "perturbed" Dirac operator

$$D_\ell + \omega_\ell = \begin{pmatrix} (D + \gamma(\omega)) \otimes 1_R & \kappa^2 \Gamma(\varphi + \varepsilon)^* \otimes M_e^* \\ (\varphi + \varepsilon)\, \Gamma \otimes M_e & (D + \frac{1}{2}\gamma(\omega) + \gamma(\Omega_0)) \otimes 1_L \end{pmatrix} \qquad (8.57)$$

associated with the NC gauge field ω_ℓ. Note that

$$(D_\ell + \omega_\ell)^\dagger = -(D_\ell + \omega_\ell)$$

since $D, \gamma(\omega), \gamma(\Omega_0)$ are skew-hermitian and $\Gamma^\dagger = -\kappa^2\, \Gamma$.

Definition 4. *The NC matter Lagrangian is the (imaginary) functional*

$$\psi_\ell \mapsto \int_M d^n x \cdot \psi_\ell^\dagger (D_\ell + \omega_\ell)\, \psi_\ell \qquad (8.58)$$

defined on the NC matter fields $\psi_\ell \in \mathcal{C}_c^\infty(S_\ell)$.

Proposition 11. *The NC matter Lagrangian density is given by*

$$\psi_\ell^\dagger (D_\ell + \omega_\ell)\, \psi_\ell = \psi^\dagger (D + \gamma(\omega))\psi\, e_R^*\, e_R$$
$$+\Psi^\dagger (D + \tfrac{1}{2}\gamma(\omega) + \gamma(\Omega_0))\, \Psi\, \overset{*}{\ell_L}\, \ell_L + 2i\, \text{Im}\, \Psi^\dagger (\varphi + \varepsilon)\, \Gamma \psi\, \overset{*}{\ell_L}\, M_e e_R$$

for ψ_ℓ written in the form (8.6).

Proof. Using (8.6) and (8.57) one obtains

$$\psi_\ell^\dagger (D_\ell + \omega_\ell)\, \psi_\ell = \psi^\dagger (D + \gamma(\omega))\, \psi \cdot e_R^* e_R + \Psi^\dagger (D + \tfrac{1}{2}\gamma(\omega) + \gamma(\Omega_0))\, \Psi \cdot \ell_L^* \ell_L$$
$$+\kappa^2\, \psi^\dagger\, \Gamma(\varphi + \varepsilon)^*\, \Psi \cdot e_R^* M_e^* \ell_L + \Psi^\dagger (\varphi + \varepsilon)\, \Gamma \psi \cdot \ell_L^* M_e e_R\ .$$

The last two summands combine to yield the assertion, since

$$\overline{\Psi^\dagger (\varphi+\varepsilon)\Gamma\psi \cdot \ell_L^* M_e e_R} = \psi^\dagger \Gamma^\dagger (\varphi+\varepsilon)^* \Psi \cdot e_R^* M_e^* \ell_L = -\kappa^2 \psi^\dagger \Gamma(\varphi+\varepsilon)^* \Psi \cdot e_R^* M_e^* \ell_L \times$$

Remark 6. *Using physics notation for the NC gauge field ω_ℓ (cf. (8.46)) and the NC matter field ψ_ℓ (cf. (8.8)) we obtain the NC matter Lagrangian in the usual form*

$$\psi_{e_R}^\dagger (D + \gamma(\omega))\psi_{e_R}$$
$$+ (\psi_{e_L}^\dagger, \psi_{\nu_L}^\dagger)(D + \tfrac{1}{2}\gamma(\omega) + \gamma(\Omega_0))\begin{pmatrix}\psi_{e_L}\\ \psi_{\nu_L}\end{pmatrix}$$
$$+ 2i\, Im\, (\psi_{e_L}^\dagger, \psi_{\nu_L}^\dagger)(\varphi + \varepsilon)\,\Gamma M_e \psi_{e_R} =$$

$$\psi_{e_R}^\dagger D\psi_{e_R} + \psi_{e_L}^\dagger D\psi_{e_L} + \psi_{\nu_L}^\dagger D\psi_{\nu_L}$$
$$- ig_1(\psi_{e_R}^\dagger \gamma(W^{(0)})\psi_{e_R} + \tfrac{1}{2}\psi_{e_L}^\dagger \gamma(W^{(0)})\psi_{e_L} + \tfrac{1}{2}\psi_{\nu_L}^\dagger \gamma(W^{(0)})\psi_{\nu_L})$$
$$+ ig_2 \sum_{a=1}^{3}(\psi_{e_L}^\dagger, \psi_{\nu_L}^\dagger)\gamma(W^{(a)}) \otimes \tau^{(a)} \begin{pmatrix}\psi_{e_L}\\ \psi_{\nu_L}\end{pmatrix}$$
$$+ \tfrac{2i}{\sqrt{2r}}\, Im\, (\psi_{e_L}^\dagger \Phi^0 \Gamma M_e \psi_{e_R} + \psi_{\nu_L}^\dagger \Phi^+ \Gamma M_e \psi_{e_R})\,.$$

So far the constraint (8.7) has not been imposed. Restricting the Dirac functional to fields satisfying (8.7) yields a vanishing action in the Euclidean case, but for Minkowski signature we obtain the correct action, in particular the "mass term" in (8.59) reduces to

$$\psi_{e_L}^\dagger \Phi^0 \Gamma M_e \psi_{e_R} + \psi_{\nu_L}^\dagger \Phi^+ \Gamma M_e \psi_{e_R} = \psi_{e_L}^\dagger \Phi^0 M_e \psi_{e_R} + \psi_{\nu_L}^\dagger \Phi^+ \Gamma M_e \psi_{e_R}$$

since $\Gamma\psi_{e_R} = \psi_{e_R}$. In particular, for the "Higgs vacuum"

$$\begin{pmatrix}\Phi^0\\ \Phi^+\end{pmatrix} = \varepsilon = \begin{pmatrix}1\\ 0\end{pmatrix} \tag{8.59}$$

we obtain the term $\psi_{e_L}^\dagger M_e \psi_{e_R}$.

9 The Full Standard Model

Karen Elsner, Holger Neumann, and Harald Upmeier

We will now give a description of the classical field theory underlying the complete Standard model. We allow for curved space-time in the presence of gravity, considered as a non-dynamical background field (in Section 4 the gravitational field will be considered as a dynamical variable as well).

9.1 Noncommutative Matter Fields

The basic fermions are now the leptons introduced in Section 2 and the quarks which experience both electro-weak and strong interaction. Quarks have both left- and right-handed components and again form pairs in three generations, here considered in the order

(d, u) (down/up), (s, c) (strange/charmed) and (b, t) (bottom/top).

This ordering will be convenient for embedding the electro-weak model into the Standard Model. As in Section 3, we denote the different generations by the same letter, i.e., (s, c) and (b, t) are also denoted by (d, u) of different "generation". This is expressed by a complex generation space $\mathbb{G}$ of dimension 3. Since the strong interaction affects both the left- and right-handed components of the quarks on an equal footing, we drop the distinction between $\mathbb{G}_R$ and $\mathbb{G}_L$.

The basic new feature is the strong interaction proved experimentally to be a gauge interaction for the group $SU(3)$. Accordingly all six quarks occur in three colors (red, yellow, blue). As in Section 2, where double underlining $\doubleunderline{}$ indicated the electro-weak interaction with gauge group $SU(2)$, we now use boldface underlining $\underline{\mathbf{}}$ for triple copies, i.e. we put

$$\underline{V} := \begin{pmatrix} V \\ V \\ V \end{pmatrix} \tag{9.1}$$

for any vector space V, and

$$\underline{T} := \begin{pmatrix} T & 0 & 0 \\ 0 & T & 0 \\ 0 & 0 & T \end{pmatrix} \tag{9.2}$$

for any $T \in \mathcal{L}(V)$. This reflects the 3 color degrees of freedom associated with $SU(3)$ (not to be confused with the 3 generations).

Starting with a complex generation space $\mathbb{G}$ (of dimension 3 experimentally), put

$$\underline{\mathbb{G}} := \begin{pmatrix} \mathbb{G} \\ \mathbb{G} \\ \mathbb{G} \end{pmatrix} = \mathbb{G} \otimes \mathbb{C}^3 \tag{9.3}$$

according to the 3 colors. Consider the spinor bundle S over spacetime M and define a complex "quark-lepton" vector bundle $S_{q\ell} = S \otimes \mathbb{G}_{q\ell}$ via tensoring with the complex vector space

$$\mathbb{G}_{q\ell} = \begin{pmatrix} \left(\dfrac{\overline{\mathbb{G}}}{\mathbb{G}}\right) \\ \underline{\underline{\mathbb{G}}} \\ \underline{\underline{\mathbb{C}}} \otimes \left(\dfrac{\overline{\mathbb{G}}}{\mathbb{G}}\right) \end{pmatrix} \approx \mathbb{C}^{45}. \tag{9.4}$$

Thus

$$S_{q\ell} = \begin{pmatrix} S \otimes \left(\dfrac{\overline{\mathbb{G}}}{\mathbb{G}}\right) \\ S \otimes \underline{\underline{\mathbb{G}}} \\ \underline{S} \otimes \left(\dfrac{\overline{\mathbb{G}}}{\mathbb{G}}\right) \end{pmatrix} \tag{9.5}$$

where, as already mentioned, $\underline{}$ means doubling and $\underline{\underline{}}$ means triple copies. Note that the electroweak version (8.4) is a subbundle of (9.5) by deleting all copies of $\underline{\underline{\mathbb{G}}}$ (and putting $\mathbb{G}_R = \mathbb{G} = \mathbb{G}_L$). We have

$$\mathcal{C}^\infty(S_{q\ell}) = \mathcal{C}^\infty(S \otimes \mathbb{G}_{q\ell}) = \mathcal{C}^\infty(S) \otimes \mathbb{G}_{q\ell}.$$

Definition 1. *The NC matter fields for the Standard Model are the (compactly supported) smooth sections*

$$\begin{pmatrix} \psi_{q\ell} \\ \tilde{\psi}_{q\ell} \end{pmatrix} \in \mathcal{C}_c^\infty \begin{pmatrix} S_{q\ell} \\ S_{q\ell} \end{pmatrix} \tag{9.6}$$

of the doubled $\mathbb{G}_{q\ell}$-valued spinor bundle $\begin{pmatrix} S_{q\ell} \\ S_{q\ell} \end{pmatrix}$, corresponding to a fermion doubling with respect to a particle/antiparticle character. (The $\tilde{\psi}_{q\ell}$ correspond to "anti-matter" fields). Thus the NC matter fields are (linear combinations of) tensors of the form

$$\psi_{q\ell} = \begin{pmatrix} \psi_d \otimes \begin{pmatrix} d_R \\ e_R \end{pmatrix} \\ \psi_u \otimes u_R \\ \Psi \otimes \begin{pmatrix} q_L \\ \ell_L \end{pmatrix} \end{pmatrix} \tag{9.7}$$

where $\psi_d, \psi_u \in \mathcal{C}_c^\infty(S)$ and $\Psi \in \mathcal{C}_c^\infty(\underline{S})$ are space-time spinors, and $d_R, u_R, q_L \in \underline{\underline{\mathbb{G}}}$ and $\ell_R, \ell_L \in \mathbb{G}$ are arbitrary vectors. More precisely, there are two copies $\psi, \tilde{\psi}$ of such fermionic fields corresponding to the particle and anti-particle sector. The physically relevant fermionic fields satisfy

$$\Gamma \psi_{d/u} = \psi_{d/u} \qquad \underline{\Gamma} \Psi = -\Psi \tag{9.8}$$
$$J \psi_{d/u} = \tilde{\psi}_{\overline{d}/\overline{u}} \qquad \underline{J} \Psi = -\tilde{\Psi} \tag{9.9}$$

but for now we do not impose this restriction. Thus the fermionic degrees of freedom are overcounted four-fold.

Remark 1. *In physics notation, the NC matter fields can be expressed as*

$$\psi_{q\ell} = \begin{pmatrix} \psi_{d_R} \\ \psi_{e_R} \\ \psi_{u_R} \\ \psi_{d_L} \\ \psi_{e_L} \\ \psi_{u_L} \\ \psi_{\nu_L} \end{pmatrix} \qquad (9.10)$$

where $\psi_{e_R}, \psi_{e_L}, \psi_{\nu_L} \in \mathcal{C}_c^\infty(S \otimes \mathbb{G})$ *are the lepton fields introduced in (8.8), and* $\psi_{u_R}, \psi_{d_L}, \psi_{u_L} \in \mathcal{C}_c^\infty(S \otimes \underline{\mathbb{G}})$ *correspond to the quarks. This follows from the decomposition*

$$\mathbb{G}_{q\ell} = \begin{pmatrix} \begin{pmatrix} \underline{\mathbb{G}} \\ \mathbb{G} \end{pmatrix} \\ \mathbb{G} \\ \underline{\underline{\mathbb{C}}} \otimes \begin{pmatrix} \underline{\mathbb{G}} \\ \mathbb{G} \end{pmatrix} \end{pmatrix} = \begin{pmatrix} \underline{\mathbb{G}} \\ \mathbb{G} \\ \underline{\mathbb{G}} \\ \underline{\mathbb{G}} \\ \mathbb{G} \\ \underline{\mathbb{G}} \\ \mathbb{G} \end{pmatrix} \begin{matrix} d_R \\ e_R \\ u_R \\ d_L \\ e_L \\ u_L \\ \nu_L \end{matrix} . \qquad (9.11)$$

There is also a similar column $\tilde{\psi}_{q\ell}$ *for the antiparticles. The constraints (9.8) take the form*

$$\begin{aligned} \Gamma \psi_{d_R/u_R} &= \psi_{d_R/u_R}, & J \psi_{d_R/u_R} &= \tilde{\psi}_{\overline{d}_R/\overline{u}_R} \\ \Gamma \psi_{d_L/u_L} &= -\psi_{d_L/u_L}, & J \psi_{d_L/u_L} &= -\tilde{\psi}_{\overline{d}_L/\overline{u}_L} \\ \Gamma \psi_{e_L/\nu_L} &= -\psi_{e_L/\nu_L}, & J \psi_{e_L/\nu_L} &= -\tilde{\psi}_{\overline{e}_L/\overline{\nu}_L} . \end{aligned} \qquad (9.12)$$

The charge operator $\boldsymbol{Q}$ on $S_{q\ell}$ is defined by

$$\boldsymbol{Q} = \frac{1}{3} \begin{pmatrix} -\underline{1} & & & & & & \\ & -3 & & & & & \\ & & \underline{2} & & & & \\ & & & -\underline{1} & & & \\ & & & & -3 & & \\ & & & & & \underline{2} & \\ & & & & & & 0 \end{pmatrix} \begin{matrix} d_R \\ e_R \\ u_R \\ d_L \\ e_L \\ u_L \\ \nu_L \end{matrix} . \qquad (9.13)$$

Its eigenvalues are the electric charges of the indicated particle types (down/up-quarks and electron/neutrino-leptons). The doubled vector bundle $\begin{pmatrix} S_{q\ell} \\ S_{q\ell} \end{pmatrix}$ *has charge operator*

$$\begin{pmatrix} \boldsymbol{Q} & 0 \\ 0 & -\boldsymbol{Q} \end{pmatrix} . \qquad (9.14)$$

Define a conjugate linear endomorphism of $\begin{pmatrix} S_{q\ell} \\ S_{q\ell} \end{pmatrix}$ by

$$\begin{pmatrix} 0 & J_{q\ell} \\ -\kappa^2 J_{q\ell} & 0 \end{pmatrix}. \tag{9.15}$$

Here we use the charge conjugation operator J on S defined in (7.74) and set $J_{q\ell}$ equal to

$$\begin{bmatrix} \overline{J \otimes \begin{pmatrix} 1 & 0 \\ 0 & 1 \end{pmatrix}} & 0 & 0 \\ 0 & J \otimes \underline{1}^- & 0 \\ 0 & 0 & \underline{\underline{J}} \otimes \overline{\begin{pmatrix} 1 & 0 \\ 0 & 1 \end{pmatrix}} \end{bmatrix}. \tag{9.16}$$

The vector bundle $\begin{pmatrix} S_{q\ell} \\ S_{q\ell} \end{pmatrix}$ is graded according to

$$\begin{pmatrix} \Gamma_{q\ell} & 0 \\ 0 & -\kappa^2 \Gamma_{q\ell} \end{pmatrix}, \tag{9.17}$$

where $\Gamma_{q\ell}$ is

$$\begin{bmatrix} \Gamma \otimes \begin{pmatrix} 1 & 0 \\ 0 & 1 \end{pmatrix} & 0 & 0 \\ 0 & \Gamma \otimes \underline{1} & 0 \\ 0 & 0 & -\underline{\underline{\Gamma}} \otimes \begin{pmatrix} 1 & 0 \\ 0 & 1 \end{pmatrix} \end{bmatrix} \tag{9.18}$$

and Γ is the grading operator on S defined in (7.7).

For the proof of the next five Propositions it is convenient to use a slightly different matrix decomposition (indicated by curly brackets) by expressing (9.15) and (9.17) in the form

$$\begin{pmatrix} 0 & J_{q\ell} \\ -\kappa^2 J_{q\ell} & 0 \end{pmatrix} = \begin{Bmatrix} 0 & J \otimes 1_R^- & 0 \\ & & 0 & J \otimes 1_L^- \\ -\kappa^2 J \otimes 1_R^- & 0 & \\ 0 & -\kappa^2 J \otimes 1_L^- & & 0 \end{Bmatrix} \tag{9.19}$$

and

$$\begin{pmatrix} \Gamma_{q\ell} & 0 \\ 0 & -\kappa^2 \Gamma_{q\ell} \end{pmatrix} = \begin{Bmatrix} \Gamma \otimes 1_R & 0 & & \\ 0 & -\Gamma \otimes 1_L & & 0 \\ & & -\kappa^2 \Gamma \otimes 1_R & 0 \\ 0 & & 0 & \kappa^2 \Gamma \otimes 1_L \end{Bmatrix}, \tag{9.20}$$

where

$$1_R = \begin{pmatrix} \begin{pmatrix} \frac{1}{0} & 0 \\ 0 & 1 \end{pmatrix} & 0 \\ 0 & \underline{1} \end{pmatrix} = \begin{pmatrix} \frac{1}{0} & 0 & 0 \\ 0 & 1 & 0 \\ 0 & 0 & \underline{1} \end{pmatrix} \begin{matrix} d_R \\ e_R \\ u_R \end{matrix} \qquad (9.21)$$

$$1_L = I \otimes \begin{pmatrix} \frac{1}{0} & 0 \\ 0 & 1 \end{pmatrix} = \begin{pmatrix} \frac{1}{0} & 0 & & 0 \\ 0 & 1 & & \\ & & \frac{1}{0} & 0 \\ 0 & & 0 & 1 \end{pmatrix} \begin{matrix} d_L \\ e_L \\ u_L \\ \nu_L \end{matrix} \qquad (9.22)$$

with a similiar notation for the antiparticle sector.

Proposition 1. *For space-time dimension* $n = 2m$, *we have*

$$\begin{pmatrix} 0 & J_{q\ell} \\ -\kappa^2 J_{q\ell} & 0 \end{pmatrix}^2 = \iota_m \mathrm{Id}$$

where $\iota_{2k} = (-1)^{k+1} \kappa^2$ *and* $\iota_{2k+1} = (-1)^k \kappa^2$.

Proof. Since

$$\left\{ \begin{matrix} 0 & & J \otimes 1_R^- & 0 \\ & 0 & 0 & J \otimes 1_L^- \\ -\kappa^2 J \otimes 1_R^- & 0 & & \\ 0 & -\kappa^2 J \otimes 1_L^- & & \end{matrix} \right\}^2 = -\kappa^2 \left\{ \begin{matrix} J^2 \otimes 1_R & 0 & & 0 \\ 0 & J^2 \otimes 1_L & & \\ & & J^2 \otimes 1_R & 0 \\ 0 & & 0 & J^2 \otimes 1_L \end{matrix} \right\},$$

the assertion follows from Proposition 12.

Proposition 2.

$$\begin{pmatrix} \Gamma_{q\ell} & 0 \\ 0 & -\kappa^2 \Gamma_{q\ell} \end{pmatrix} \begin{pmatrix} 0 & J_{q\ell} \\ -\kappa^2 J_{q\ell} & 0 \end{pmatrix} = -\kappa^2 (-1)^m \begin{pmatrix} 0 & J_{q\ell} \\ -\kappa^2 J_{q\ell} & 0 \end{pmatrix} \begin{pmatrix} \Gamma_{q\ell} & 0 \\ 0 & -\kappa^2 \Gamma_{q\ell} \end{pmatrix}.$$

Proof. Using (9.19) and (9.20), we compute the left product as

$$\left\{ \begin{matrix} \Gamma \otimes 1_R & 0 & & 0 \\ 0 & -\Gamma \otimes 1_L & & \\ & & -\kappa^2 \Gamma \otimes 1_R & 0 \\ 0 & & 0 & \kappa^2 \Gamma \otimes 1_L \end{matrix} \right\} \left\{ \begin{matrix} 0 & & J \otimes 1_R^- & 0 \\ & 0 & 0 & J \otimes 1_L^- \\ -\kappa^2 J \otimes 1_R^- & 0 & & \\ 0 & -\kappa^2 J \otimes 1_L^- & & \end{matrix} \right\}$$

$$= \left\{ \begin{matrix} 0 & & \Gamma J \otimes 1_R^- & 0 \\ & 0 & 0 & -\Gamma J \otimes 1_L^- \\ \Gamma J \otimes 1_R^- & 0 & & \\ 0 & -\Gamma J \otimes 1_L^- & & \end{matrix} \right\}$$

and the right product as

$$\left\{\begin{pmatrix} 0 & & J\otimes 1_R^- & 0 \\ & 0 & 0 & J\otimes 1_L^- \\ -\kappa^2 J\otimes 1_R^- & 0 & & 0 \\ 0 & -\kappa^2 J\otimes 1_L^- & 0 & \end{pmatrix}\right\}\left\{\begin{pmatrix} \Gamma\otimes 1_R & 0 & & 0 \\ 0 & -\Gamma\otimes 1_L & & \\ & 0 & -\kappa^2\Gamma\otimes 1_R & 0 \\ & & 0 & 1\kappa^2\Gamma\otimes 1_L \end{pmatrix}\right\}$$

$$=\left\{\begin{pmatrix} & 0 & -\kappa^2 J\Gamma\otimes^- 1_R & 0 \\ & & 0 & \kappa^2 J\Gamma\otimes^- 1_L \\ -\kappa^2 J\Gamma\otimes^- 1_R & 0 & & 0 \\ 0 & \kappa^2 J\Gamma\otimes^- 1_L & & \end{pmatrix}\right\}.$$

Since $\Gamma J = (-1)^m J\Gamma$ according to (7.79), the assertion follows.

Definition 2. *The* Dirac operator, *acting on sections of* $\begin{pmatrix} S_{q\ell} \\ S_{q\ell} \end{pmatrix}$, *is defined as a* 2×2 *matrix*

$$\begin{pmatrix} D_{q\ell} & 0 \\ 0 & J_{q\ell}D_{q\ell}J_{q\ell} \end{pmatrix} \tag{9.23}$$

where $D_{q\ell}$ *is given as*

$$\begin{bmatrix} D\otimes\begin{pmatrix}1 & 0\\0 & 1\end{pmatrix} & 0 & \kappa^2\Gamma\varepsilon^*\otimes\begin{pmatrix}M_d^* & 0\\0 & M_e^*\end{pmatrix} \\ 0 & D\otimes\underline{1} & \kappa^2\Gamma\underset{\sim}{\varepsilon}^*\otimes(M_u^*,0) \\ \varepsilon\Gamma\otimes\begin{pmatrix}M_d & 0\\0 & M_e\end{pmatrix} & \underset{\sim}{\varepsilon}\Gamma\otimes\begin{pmatrix}M_u\\0\end{pmatrix} & \underline{\underline{D}}\otimes\begin{pmatrix}1 & 0\\0 & 1\end{pmatrix} \end{bmatrix} \tag{9.24}$$

and $J_{q\ell}D_{q\ell}J_{q\ell}$ *equals*

$$\begin{bmatrix} \tilde{D}\otimes\begin{pmatrix}1 & 0\\0 & 1\end{pmatrix} & 0 & -\kappa^2\Gamma\varepsilon^*\otimes\begin{pmatrix}M_d^t & 0\\0 & M_e^t\end{pmatrix} \\ 0 & \tilde{D}\otimes\underline{1} & -\kappa^2\Gamma\underset{\sim}{\varepsilon}^*\otimes(M_u^t,0) \\ -\varepsilon\Gamma\otimes\begin{pmatrix}\overline{M_d} & 0\\0 & \overline{M_e}\end{pmatrix} & -\underset{\sim}{\varepsilon}\Gamma\otimes\begin{pmatrix}\overline{M_u}\\0\end{pmatrix} & \underline{\underline{\tilde{D}}}\otimes\begin{pmatrix}1 & 0\\0 & 1\end{pmatrix} \end{bmatrix}.$$

Here D is the Dirac operator on S induced by the space-time metric g and $\tilde{D}$ is defined by (7.82). Moreover we put

$$\varepsilon := \begin{pmatrix}1\\0\end{pmatrix}, \qquad \underset{\sim}{\varepsilon} := \begin{pmatrix}0\\1\end{pmatrix} \tag{9.25}$$

and $M_d, M_e, M_u : \mathbb{G} \to \mathbb{G}$ are positive definite linear maps called mass-matrices. Generalizing (9.25) we define an operation $\phi \mapsto \underset{\sim}{\phi}$ on 2×1 column vectors by putting

$$\begin{pmatrix} \phi_1 \\ \phi_2 \end{pmatrix}_{\sim} = \begin{pmatrix} -\overline{\phi_2} \\ \overline{\phi_1} \end{pmatrix}. \tag{9.26}$$

Note that the electroweak Dirac operator (8.12) arises as a truncation of (9.24). Another way to write the Dirac operator (9.23) is given by

$$\begin{pmatrix} D_{q\ell} & 0 \\ 0 & J_{q\ell} D_{q\ell} J_{q\ell} \end{pmatrix} = \begin{Bmatrix} D \otimes 1_R & \kappa^2 \Gamma \otimes M^* & & 0 \\ \Gamma \otimes M & D \otimes 1_L & & \\ & & \tilde{D} \otimes 1_R & -\kappa^2 \Gamma \otimes M^t \\ 0 & & -\Gamma \otimes \overline{M} & \tilde{D} \otimes 1_L \end{Bmatrix}, \tag{9.27}$$

where

$$M = \begin{pmatrix} M_d & 0 & 0 \\ 0 & M_e & 0 \\ 0 & 0 & M_u \\ 0 & 0 & 0 \end{pmatrix}. \tag{9.28}$$

Proposition 3.

$$\begin{pmatrix} D_{q\ell} & 0 \\ 0 & J_{q\ell} D_{q\ell} J_{q\ell} \end{pmatrix}^\dagger = -\begin{pmatrix} D_{q\ell} & 0 \\ 0 & J_{q\ell} D_{q\ell} J_{q\ell} \end{pmatrix}.$$

Proof. We treat the two cases separately:

For $\kappa = i$ we have $D^\dagger = -D$ and $\Gamma^\dagger = \Gamma$. Therefore

$$\begin{pmatrix} D_{q\ell} & 0 \\ 0 & J_{q\ell} D_{q\ell} J_{q\ell} \end{pmatrix} = \begin{Bmatrix} D \otimes 1_R & -\Gamma \otimes M^* & & 0 \\ \Gamma \otimes M & D \otimes 1_L & & \\ & & \tilde{D} \otimes 1_R & \Gamma \otimes M^t \\ 0 & & -\Gamma \otimes \overline{M} & \tilde{D} \otimes 1_L \end{Bmatrix} = -\begin{pmatrix} D_{q\ell} & 0 \\ 0 & J_{q\ell} D_{q\ell} J_{q\ell} \end{pmatrix}^\dagger.$$

For $\kappa = 1$ we have $D^\dagger = -D$, $\tilde{D}^\dagger = -\tilde{D}$ and $\Gamma^\dagger = -\Gamma$. This also yields

$$\begin{pmatrix} D_{q\ell} & 0 \\ 0 & J_{q\ell} D_{q\ell} J_{q\ell} \end{pmatrix} = \begin{Bmatrix} D \otimes 1_R & \Gamma \otimes M^* & & 0 \\ \Gamma \otimes M & D \otimes 1_L & & \\ & & \tilde{D} \otimes 1_R & -\Gamma \otimes M^t \\ 0 & & -\Gamma \otimes \overline{M} & \tilde{D} \otimes 1_L \end{Bmatrix} = -\begin{pmatrix} D_{q\ell} & 0 \\ 0 & J_{q\ell} D_{q\ell} J_{q\ell} \end{pmatrix}^\dagger.$$

Proposition 4.

$$\begin{pmatrix} \Gamma_{q\ell} & 0 \\ 0 & -\kappa^2 \Gamma_{q\ell} \end{pmatrix} \begin{pmatrix} D_{q\ell} & 0 \\ 0 & J_{q\ell} D_{q\ell} J_{q\ell} \end{pmatrix} + \begin{pmatrix} D_{q\ell} & 0 \\ 0 & J_{q\ell} D_{q\ell} J_{q\ell} \end{pmatrix} \begin{pmatrix} \Gamma_{q\ell} & 0 \\ 0 & -\kappa^2 \Gamma_{q\ell} \end{pmatrix} = 0.$$

Proof. Since $\Gamma D + D\Gamma = 0$ by (7.72), it follows that

$$\left\{\begin{pmatrix} \Gamma\otimes 1_R & 0 & & \\ 0 & -\Gamma\otimes 1_L & & \\ & & -\kappa^2\Gamma\otimes 1_R & 0 \\ & & 0 & \kappa^2\Gamma\otimes 1_L \end{pmatrix}\begin{pmatrix} D\otimes 1_R & \kappa^2\Gamma\otimes M^* & & \\ \Gamma\otimes M & D\otimes 1_L & & \\ & & \tilde{D}\otimes 1_R & -\kappa^2\Gamma\otimes M^t \\ & & -\Gamma\otimes\overline{M} & \tilde{D}\otimes 1_L \end{pmatrix}\right\}$$

$$+ \left\{\begin{pmatrix} D\otimes 1_R & \kappa^2\Gamma\otimes M^* & & \\ \Gamma\otimes M & D\otimes 1_L & & \\ & & \tilde{D}\otimes 1_R & -\kappa^2\Gamma\otimes M^t \\ & & -\Gamma\otimes\overline{M} & \tilde{D}\otimes 1_L \end{pmatrix}\begin{pmatrix} \Gamma\otimes 1_R & 0 & & \\ 0 & -\Gamma\otimes 1_L & & \\ & & -\kappa^2\Gamma\otimes 1_R & 0 \\ & & 0 & \kappa^2\Gamma\otimes 1_L \end{pmatrix}\right\}$$

$$= \left\{\begin{pmatrix} (\Gamma D + D\Gamma)\otimes 1_R & (\kappa^2-\kappa^2)I\otimes M^* & & \\ (-I+I)\otimes M & -(\Gamma D+D\Gamma)\otimes 1_L & & \\ & & -\kappa^2(\Gamma\tilde{D}+\tilde{D}\Gamma)\otimes 1_R & (I-I)\otimes M^t \\ & & (-\kappa^2+\kappa^2)I\otimes\overline{M} & \kappa^2(\Gamma\tilde{D}+\tilde{D}\Gamma)\otimes 1_L \end{pmatrix}\right\}=0.$$

Proposition 5.

$$\begin{pmatrix} 0 & J_{q\ell} \\ -\kappa^2 J_{q\ell} & 0 \end{pmatrix}\begin{pmatrix} D_{q\ell} & 0 \\ 0 & J_{q\ell}D_{q\ell}J_{q\ell} \end{pmatrix} + \begin{pmatrix} D_{q\ell} & 0 \\ 0 & J_{q\ell}D_{q\ell}J_{q\ell} \end{pmatrix}\begin{pmatrix} 0 & J_{q\ell} \\ -\kappa^2 J_{q\ell} & 0 \end{pmatrix} = 0.$$

Proof. Since $J\Gamma = (-1)^m \Gamma J, JD = -\tilde{D}J$ by (7.79) and (7.81), it follows that the left hand side equals

$$\left\{\begin{pmatrix} 0 & & J\otimes 1_R^- & 0 \\ 0 & & 0 & J\otimes 1_L^- \\ -\kappa^2 J\otimes 1_R^- & 0 & & \\ 0 & -\kappa^2 J\otimes 1_L^- & & \end{pmatrix}\begin{pmatrix} D\otimes 1_R & \kappa^2\Gamma\otimes M^* & & \\ \Gamma\otimes M & D\otimes 1_L & & \\ & & \tilde{D}\otimes 1_R & -\kappa^2\Gamma\otimes M^t \\ & & -\Gamma\otimes\overline{M} & \tilde{D}\otimes 1_L \end{pmatrix}\right\} +$$

$$\left\{\begin{pmatrix} D\otimes 1_R & \kappa^2\Gamma\otimes M^* & & \\ \Gamma\otimes M & D\otimes 1_L & & \\ & & \tilde{D}\otimes 1_R & -\kappa^2\Gamma\otimes M^t \\ & & -\Gamma\otimes\overline{M} & \tilde{D}\otimes 1_L \end{pmatrix}\begin{pmatrix} 0 & & J\otimes 1_R^- & 0 \\ 0 & & 0 & J\otimes 1_L^- \\ -\kappa^2 J\otimes 1_R^- & 0 & & \\ 0 & -\kappa^2 J\otimes 1_L^- & & \end{pmatrix}\right\} =$$

$$\left\{\begin{pmatrix} 0 & & (J\tilde{D}+DJ)\otimes 1_R^- & \kappa^2(\Gamma J\otimes M^{*-}-J\Gamma\otimes^- M^t) \\ & 0 & \Gamma J\otimes M^- -J\Gamma\otimes^-\overline{M} & (J\tilde{D}+DJ)\otimes 1_L^- \\ -\kappa^2(JD+\tilde{D}J)\otimes 1_R^- & \Gamma J\otimes M^{t-}-J\Gamma\otimes^- M^* & 0 & \\ \kappa^2(\Gamma J\otimes\overline{M}^- -J\Gamma\otimes^- M) & -\kappa^2(JD+\tilde{D}J)\otimes 1_L^- & & 0 \end{pmatrix}\right\}$$

$= 0.$

9.2 Noncommutative Gauge Fields

Having described the Dirac operator associated with the Standard Model, we now turn to the underlying (noncommutative) algebra. In this section we often consider 2×2 block matrices of the form $\begin{pmatrix} a & 0 \\ 0 & \tilde{a} \end{pmatrix}$, where $a, \tilde{a}$ are endomorphisms of $S_{q\ell}$. In general, $\tilde{a}$ is independent of a.

Definition 3. Let $\mathcal{A}_{q\ell}$ denote the real algebra of all endomorphisms

$$\begin{pmatrix} f_{q\ell} & 0 \\ 0 & \tilde{f}_{q\ell} \end{pmatrix} \tag{9.29}$$

of $\begin{pmatrix} S_{q\ell} \\ S_{q\ell} \end{pmatrix}$, where $f_{q\ell}$ and $\tilde{f}_{q\ell}$ are endomorphisms of $S_{q\ell} = S \otimes \mathbb{G}_{q\ell}$, having the form

$$\begin{bmatrix} f \otimes \begin{pmatrix} 1 & 0 \\ 0 & 1 \end{pmatrix} & 0 & 0 \\ 0 & \overline{f} \otimes \underline{1} & 0 \\ 0 & 0 & F \otimes \begin{pmatrix} 1 & 0 \\ 0 & 1 \end{pmatrix} \end{bmatrix}, \tag{9.30}$$

and

$$\begin{bmatrix} \overline{\underline{f}}_a \otimes \begin{pmatrix} \overline{a} & 0 \\ 0 & 0 \end{pmatrix} + \overline{f} \otimes \begin{pmatrix} 0 & 0 \\ 0 & 1 \end{pmatrix} & 0 & 0 \\ 0 & \overline{f}_a \otimes \overline{a} & 0 \\ 0 & 0 & \overline{\underline{f}}_a \otimes \begin{pmatrix} \overline{a} & 0 \\ 0 & 0 \end{pmatrix} + \underline{\overline{f}} \otimes \begin{pmatrix} 0 & 0 \\ 0 & 1 \end{pmatrix} \end{bmatrix} \tag{9.31}$$

respectively, with arbitrary

$$f \in \mathcal{C}^\infty(M, \mathbb{C}),$$
$$F \in \mathcal{C}^\infty(M, \mathbb{H}),$$
$$f_a \otimes a \in \mathcal{C}^\infty(M, \mathbb{C}) \otimes \mathbb{C}^{3\times 3} = \mathcal{C}^\infty(M, \mathbb{C}^{3\times 3}).$$

Here $\mathbb{H}$ is the real division algebra of quaternions.

Proposition 6. We have

$$\begin{pmatrix} 0 & J_{q\ell} \\ -\kappa^2 J_{q\ell} & 0 \end{pmatrix} \begin{pmatrix} f_{q\ell} & 0 \\ 0 & \tilde{f}_{q\ell} \end{pmatrix} \begin{pmatrix} 0 & J_{q\ell} \\ -\kappa^2 J_{q\ell} & 0 \end{pmatrix}^{-1} =$$

$$\begin{pmatrix} 0 & J_{q\ell} \\ -\kappa^2 J_{q\ell} & 0 \end{pmatrix} \begin{pmatrix} f_{q\ell} & 0 \\ 0 & \tilde{f}_{q\ell} \end{pmatrix} \begin{pmatrix} 0 & J_{q\ell} \\ -\kappa^2 J_{q\ell} & 0 \end{pmatrix} = \begin{pmatrix} -J_{q\ell}\tilde{f}_{q\ell}J_{q\ell} & 0 \\ 0 & -J_{q\ell}f_{q\ell}J_{q\ell} \end{pmatrix},$$

where $-J_{q\ell}\tilde{f}_{q\ell}J_{q\ell}$ is given by

$$\begin{bmatrix} f_a \otimes \begin{pmatrix} a & 0 \\ 0 & 0 \end{pmatrix} + f \otimes \begin{pmatrix} 0 & 0 \\ 0 & 1 \end{pmatrix} & 0 & 0 \\ 0 & f_a \otimes a & 0 \\ 0 & 0 & \underline{f}_a \otimes \begin{pmatrix} a & 0 \\ 0 & 0 \end{pmatrix} + \underline{f} \otimes \begin{pmatrix} 0 & 0 \\ 0 & 1 \end{pmatrix} \end{bmatrix}$$

and $-J_{q\ell}f_{q\ell}J_{q\ell}$ by

$$\begin{bmatrix} \overline{f} \otimes \begin{pmatrix} 1 & 0 \\ 0 & 1 \end{pmatrix} & 0 & 0 \\ 0 & f \otimes \underline{1} & 0 \\ 0 & 0 & \overline{F} \otimes \begin{pmatrix} 1 & 0 \\ 0 & 1 \end{pmatrix} \end{bmatrix}.$$

Proof. Using (9.16), (9.30) and (9.31) we obtain $J_{q\ell}\tilde{f}_{q\ell}J_{q\ell}$ as

$$\begin{bmatrix} J\overline{f}_a J \otimes {}^-\begin{pmatrix} \overline{a} & 0 \\ 0 & 0 \end{pmatrix}{}^- \\ +J\overline{f}J \otimes {}^-\begin{pmatrix} 0 & 0 \\ \underline{0} & 1 \end{pmatrix}{}^- & 0 & 0 \\ 0 & J\overline{f}_a J \otimes {}^-\overline{a}{}^- & 0 \\ 0 & 0 & \begin{matrix} J\underline{\overline{f}_a}J \otimes {}^-\begin{pmatrix} \overline{a} & 0 \\ 0 & 0 \end{pmatrix}{}^- \\ +J\underline{\overline{f}}J \otimes {}^-\begin{pmatrix} 0 & 0 \\ \underline{0} & 1 \end{pmatrix}{}^- \end{matrix} \end{bmatrix}$$

and $J_{q\ell}f_{q\ell}J_{q\ell}$ as

$$\begin{bmatrix} JfJ \otimes {}^-\begin{pmatrix} 1 & 0 \\ 0 & 1 \end{pmatrix}{}^- & 0 & 0 \\ 0 & J\overline{f}J \otimes {}^-\underline{1}{}^- & 0 \\ 0 & 0 & JFJ \otimes {}^-\begin{pmatrix} 1 & 0 \\ 0 & 1 \end{pmatrix}{}^- \end{bmatrix}.$$

Since $JfJ = -\overline{f}$ and ${}^-a^- = \overline{a}$, the assertion follows.

By Proposition 1, Proposition 2 and Proposition 5, the conjugation (9.15) satisfies the correct sign convention for a "real structure mod 8" with respect to the grading operator (9.17) and the Dirac operator (9.23) (cf. Remark 2). In fact, in the Euclidean setting this is already true of the original space-time involution J. In the Minkowski setting this is achieved after the fermion doubling. Note that also in this case the conjugation (9.15) preserves the right/left parity although it does not commute with the grading operator.

Let Ω^k denote the space of "universal" k-forms over $\mathcal{A}_{q\ell}$, with differential $d : \Omega^k \to \Omega^{k+1}$. Consider the representation

$$\pi_k : \Omega^k \to \mathcal{L}(L^2(M, \begin{pmatrix} S_{q\ell} \\ S_{q\ell} \end{pmatrix})) \tag{9.32}$$

induced by (9.23), and put $\mathcal{A}_{q\ell}^{(k)} := \pi_k(\Omega^k)$. Note that $\mathcal{A}_{q\ell}^{(0)} = \mathcal{A}_{q\ell}$. We now pass to $k = 1$ and $k = 2$.

Proposition 7. $\mathcal{A}_{q\ell}^{(1)}$ consists of all endomorphisms

$$\begin{pmatrix} \omega_{q\ell} & 0 \\ 0 & \tilde{\omega}_{q\ell} \end{pmatrix} \tag{9.33}$$

acting on $\begin{pmatrix} S_{q\ell} \\ S_{q\ell} \end{pmatrix}$, where $\omega_{q\ell}$ and $\tilde{\omega}_{q\ell}$ given by

$$\begin{bmatrix} \gamma(\omega)\otimes\begin{pmatrix}1 & 0\\ 0 & 1\end{pmatrix} & 0 & \kappa^2\Gamma\psi\otimes\begin{pmatrix}\underline{M_d^*} & 0\\ 0 & M_e^*\end{pmatrix} \\ 0 & \gamma(\overline{\omega})\otimes\underline{1} & \kappa^2\Gamma\psi\otimes(\underline{M_u^*},0) \\ \varphi\Gamma\otimes\begin{pmatrix}\underline{M_d} & 0\\ 0 & M_e\end{pmatrix} & \underset{\sim}{\varphi\Gamma}\otimes\begin{pmatrix}M_u\\ 0\end{pmatrix} & \gamma(\Omega)\otimes\begin{pmatrix}1 & 0\\ 0 & 1\end{pmatrix} \end{bmatrix}$$

and

$$\begin{bmatrix} \tilde{\gamma}(\overline{\omega}_a)\otimes\begin{pmatrix}\overline{a} & 0\\ 0 & 0\end{pmatrix}+\tilde{\gamma}(\overline{\omega})\otimes\begin{pmatrix}0 & 0\\ \underline{0} & 1\end{pmatrix} & 0 & 0 \\ 0 & \tilde{\gamma}(\overline{\omega}_a)\otimes\overline{a} & 0 \\ 0 & 0 & \underline{\tilde{\gamma}(\overline{\omega}_a)}\otimes\begin{pmatrix}\overline{a} & 0\\ 0 & 0\end{pmatrix}+\underline{\tilde{\gamma}(\overline{\omega})}\otimes\begin{pmatrix}0 & 0\\ 0 & 1\end{pmatrix} \end{bmatrix}$$

are endomorphisms of $S_{q\ell} = S \otimes \mathbb{G}_{q\ell}$ parametrized by

$$\varphi \in \mathcal{C}^\infty(M, \mathbb{C}^{2\times 1}), \quad \psi \in \mathcal{C}^\infty(M, \mathbb{C}^{1\times 2}),$$
$$\omega \in \mathcal{C}^\infty(T^\#M \otimes \mathbb{C}), \quad \Omega \in \mathcal{C}^\infty(T^\#M \otimes \mathbb{H}),$$
$$\omega_a \otimes a \in \mathcal{C}^\infty(T^\#M \otimes \mathbb{C}^{3\times 3}).$$

Proof. For $\begin{pmatrix} f_{q\ell} & 0 \\ 0 & \tilde{f}_{q\ell} \end{pmatrix} \in \mathcal{A}_{q\ell}$, written as in Definition 3, we calculate $D_{q\ell}f_{q\ell}$ as

$$\begin{bmatrix} Df\otimes\begin{pmatrix}1 & 0\\ 0 & 1\end{pmatrix} & 0 & \kappa^2\Gamma\varepsilon^*F\otimes\begin{pmatrix}\underline{M_d^*} & 0\\ 0 & M_e^*\end{pmatrix} \\ 0 & D\overline{f}\otimes\underline{1} & \kappa^2\Gamma\varepsilon^*F\otimes(\underline{M_u^*},0) \\ \varepsilon f\Gamma\otimes\begin{pmatrix}\underline{M_d} & 0\\ 0 & M_e\end{pmatrix} & \underset{\sim}{\varepsilon\overline{f}\Gamma}\otimes\begin{pmatrix}M_u\\ 0\end{pmatrix} & \underline{DF}\otimes\begin{pmatrix}1 & 0\\ 0 & 1\end{pmatrix} \end{bmatrix}$$

The Full Standard Model

and $f_{q\ell}D_{q\ell}$ as

$$\begin{bmatrix} fD \otimes \begin{pmatrix} 1 & 0 \\ 0 & 1 \end{pmatrix} & 0 & \kappa^2 \Gamma f \varepsilon^* \otimes \begin{pmatrix} M_d^* & 0 \\ 0 & M_e^* \end{pmatrix} \\ 0 & \overline{f}D \otimes \underline{1} & \kappa^2 \Gamma \overline{f} \underset{\sim}{\varepsilon}^* \otimes (\underline{M_u^*}, 0) \\ F\varepsilon\Gamma \otimes \begin{pmatrix} M_d & 0 \\ 0 & M_e \end{pmatrix} & F \underset{\sim}{\varepsilon}\Gamma \otimes \begin{pmatrix} M_u \\ 0 \end{pmatrix} & F\underline{\underline{D}} \otimes \begin{pmatrix} 1 & 0 \\ 0 & 1 \end{pmatrix} \end{bmatrix}.$$

Moreover $(J_{q\ell}D_{q\ell}J_{q\ell})\tilde{f}_{q\ell}$ equals

$$\begin{bmatrix} \overline{Df}_a \otimes \begin{pmatrix} \overline{a} & 0 \\ 0 & 0 \end{pmatrix} + \overline{Df} \otimes \begin{pmatrix} 0 & 0 \\ 0 & 1 \end{pmatrix} & 0 & -\kappa^2\Gamma\varepsilon^*\underline{\overline{f}_a} \otimes \begin{pmatrix} M_d^t\overline{a} & 0 \\ 0 & 0 \end{pmatrix} -\kappa^2\Gamma\varepsilon^*\underline{\overline{f}} \otimes \begin{pmatrix} 0 & 0 \\ 0 & M_e^t \end{pmatrix} \\ 0 & \overline{Df}_a \otimes \overline{a} & -\kappa^2\Gamma\underset{\sim}{\varepsilon}^*\underline{\overline{f}_a} \otimes (\underline{M_u^t\overline{a}}, 0) \\ -\varepsilon\Gamma\overline{f}_a \otimes \begin{pmatrix} \overline{M_d\overline{a}} & 0 \\ 0 & 0 \end{pmatrix} -\varepsilon\Gamma\overline{f} \otimes \begin{pmatrix} 0 & 0 \\ 0 & M_e \end{pmatrix} & -\varepsilon\underset{\sim}{\Gamma}\overline{f}_a \otimes \begin{pmatrix} \overline{M_u\overline{a}} \\ 0 \end{pmatrix} & \underline{\underline{Df_a}} \otimes \begin{pmatrix} \overline{a} & 0 \\ 0 & 0 \end{pmatrix} + \underline{\underline{Df}} \otimes \begin{pmatrix} 0 & 0 \\ 0 & 1 \end{pmatrix} \end{bmatrix}$$

and $\tilde{f}_{q\ell}(J_{q\ell}D_{q\ell}J_{q\ell})$ equals

$$\begin{bmatrix} \overline{f}_a\overline{D} \otimes \begin{pmatrix} \overline{a} & 0 \\ 0 & 0 \end{pmatrix} + \overline{fD} \otimes \begin{pmatrix} 0 & 0 \\ 0 & 1 \end{pmatrix} & 0 & -\kappa^2\overline{f}_a\Gamma\varepsilon^* \otimes \begin{pmatrix} \overline{a}M_d^t & 0 \\ 0 & 0 \end{pmatrix} \\ & & -\kappa^2\overline{f}\Gamma\varepsilon^* \otimes \begin{pmatrix} 0 & 0 \\ 0 & M_e^t \end{pmatrix} \\ 0 & \overline{f}_a\overline{D} \otimes \overline{a} & -\kappa^2\overline{f}_a\underset{\sim}{\Gamma}\varepsilon^* \otimes (\overline{a}M_u^t, 0) \\ -\tilde{\gamma}(\overline{f}_a)\varepsilon\Gamma \otimes \begin{pmatrix} \overline{a}M_d & 0 \\ 0 & 0 \end{pmatrix} & -\tilde{\gamma}(\overline{f}_a)\underset{\sim}{\varepsilon\Gamma} \otimes \begin{pmatrix} \overline{a}M_u \\ 0 \end{pmatrix} & \underline{\underline{f_a\overline{D}}} \otimes \begin{pmatrix} \overline{a} & 0 \\ 0 & 0 \end{pmatrix} + \\ -\tilde{\gamma}(\overline{f})\varepsilon\Gamma \otimes \begin{pmatrix} 0 & 0 \\ 0 & M_e \end{pmatrix} & & \underline{\underline{\overline{fD}}} \otimes \begin{pmatrix} 0 & 0 \\ 0 & 1 \end{pmatrix} \end{bmatrix}.$$

This implies

$$[\begin{pmatrix} D_{q\ell} & 0 \\ 0 & J_{q\ell}D_{q\ell}J_{q\ell} \end{pmatrix}, \begin{pmatrix} f_{q\ell} & 0 \\ 0 & \tilde{f}_{q\ell} \end{pmatrix}] = \begin{pmatrix} [D_{q\ell}, f_{q\ell}] & 0 \\ 0 & [J_{q\ell}D_{q\ell}J_{q\ell}, \tilde{f}_{q\ell}] \end{pmatrix} \quad (9.34)$$

with $[D_{q\ell}, f_{q\ell}]$ equal to

$$\begin{bmatrix} \gamma(df) \otimes \begin{pmatrix} 1 & 0 \\ 0 & 1 \end{pmatrix} & 0 & \kappa^2\Gamma(\varepsilon^*F - f\varepsilon^*) \otimes \begin{pmatrix} M_d^* & 0 \\ 0 & M_e^* \end{pmatrix} \\ 0 & \gamma(d\overline{f}) \otimes \underline{1} & \kappa^2\Gamma(\underset{\sim}{\varepsilon}^*F - \overline{f}\underset{\sim}{\varepsilon}^*) \otimes (\underline{M_u^*}, 0) \\ (\varepsilon f - F\varepsilon)\Gamma \otimes \begin{pmatrix} M_d & 0 \\ 0 & M_e \end{pmatrix} & (\underset{\sim}{\varepsilon}\overline{f} - F\underset{\sim}{\varepsilon})\Gamma \otimes \begin{pmatrix} M_u \\ 0 \end{pmatrix} & \gamma(dF) \otimes \begin{pmatrix} 1 & 0 \\ 0 & 1 \end{pmatrix} \end{bmatrix}$$

and $[J_{q\ell}D_{q\ell}J_{q\ell}, \tilde{f}_{q\ell}]$ equal to

$$\begin{bmatrix} \tilde{\gamma}(d\overline{f}_a)\otimes\begin{pmatrix} \overline{a} & 0 \\ 0 & 0 \end{pmatrix}+\tilde{\gamma}(d\overline{f})\otimes\begin{pmatrix} 0 & 0 \\ 0 & 1 \end{pmatrix} & 0 & 0 \\ 0 & \tilde{\gamma}(d\overline{f}_a)\otimes\overline{a} & 0 \\ 0 & 0 & \underline{\tilde{\gamma}(d\overline{f}_a)\otimes\begin{pmatrix} \overline{a} & 0 \\ 0 & 0 \end{pmatrix}+\tilde{\gamma}(d\overline{f})\otimes\begin{pmatrix} 0 & 0 \\ 0 & 1 \end{pmatrix}} \end{bmatrix}.$$

Taking (finite sums of) tensors $\begin{pmatrix} f_{q\ell}^0 & 0 \\ 0 & \tilde{f}_{q\ell}^0 \end{pmatrix} \otimes \begin{pmatrix} f_{q\ell}^1 & 0 \\ 0 & \tilde{f}_{q\ell}^1 \end{pmatrix}$ it follows that

$$\begin{pmatrix} f_{q\ell}^0 & 0 \\ 0 & \tilde{f}_{q\ell}^0 \end{pmatrix}[\begin{pmatrix} D_{q\ell} & 0 \\ 0 & J_{q\ell}D_{q\ell}J_{q\ell} \end{pmatrix}, \begin{pmatrix} f_{q\ell}^1 & 0 \\ 0 & \tilde{f}_{q\ell}^1 \end{pmatrix}] = \begin{pmatrix} f_{q\ell}^0[D_{q\ell},f_{q\ell}^1] & 0 \\ 0 & \tilde{f}_{q\ell}^0[J_{q\ell}D_{q\ell}J_{q\ell},\tilde{f}_{q\ell}^1] \end{pmatrix}$$

where $f_{q\ell}^0[D_{q\ell}, f_{q\ell}^1]$ is

$$\begin{bmatrix} \gamma(f^0 df^1)\otimes\begin{pmatrix} 1 & 0 \\ 0 & 1 \end{pmatrix} & 0 & \kappa^2 \Gamma f^0(\varepsilon^* F^1 - f^1 \varepsilon^*)\otimes\begin{pmatrix} M_d^* & 0 \\ 0 & M_e^* \end{pmatrix} \\ 0 & \gamma(\overline{f}^0 d\overline{f}^1)\otimes \underline{1} & \kappa^2 \Gamma \overline{f}^0(\varepsilon^* F^1 - \overline{f}^1\underset{\sim}{\varepsilon}^*)\otimes(M_u^*, 0) \\ F^0(\varepsilon f^1 - F^1\varepsilon)\Gamma\otimes\begin{pmatrix} M_d & 0 \\ 0 & M_e \end{pmatrix} & F^0(\varepsilon\overline{f}^1 - F^1\underset{\sim}{\varepsilon})\Gamma\otimes\begin{pmatrix} M_u \\ 0 \end{pmatrix} & \gamma(F^0 dF^1)\otimes\begin{pmatrix} 1 & 0 \\ 0 & 1 \end{pmatrix} \end{bmatrix}$$

and $\tilde{f}_{q\ell}^0[J_{q\ell}D_{q\ell}J_{q\ell}, \tilde{f}_{q\ell}^1]$ given by

$$\begin{bmatrix} \tilde{\gamma}(\overline{f}_b^0 d\overline{f}_c^1)\otimes\begin{pmatrix} \overline{bc} & 0 \\ 0 & 0 \end{pmatrix}+\tilde{\gamma}(\overline{f}^0 d\overline{f}^1)\otimes\begin{pmatrix} 0 & 0 \\ 0 & 1 \end{pmatrix} & 0 & 0 \\ 0 & \tilde{\gamma}(\overline{f}_b^0 d\overline{f}_c^1)\otimes\overline{bc} & 0 \\ 0 & 0 & \underline{\tilde{\gamma}(\overline{f}_b^0 d\overline{f}_c^1)\otimes\begin{pmatrix} \overline{bc} & 0 \\ 0 & 0 \end{pmatrix}+\tilde{\gamma}(\overline{f}^0 d\overline{f}^1)\otimes\begin{pmatrix} 0 & 0 \\ 0 & 1 \end{pmatrix}} \end{bmatrix}.$$

Now the assertion follows by putting

$$\omega = f^0 df^1, \quad \Omega = F^0 dF^1, \quad \omega_a \otimes a = f_b^0 df_c^1 \otimes bc,$$
$$\varphi = F^0(\varepsilon f^1 - F^1\varepsilon), \quad \psi = f^0(\varepsilon^* F^1 - f^1\varepsilon^*),$$

which implies

$$\underset{\sim}{\varphi} = F^0(\varepsilon\overline{f}^1 - F^1\underset{\sim}{\varepsilon}), \quad \underset{\sim}{\psi} = \overline{f}^0(\varepsilon^* F^1 - \overline{f}^1\underset{\sim}{\varepsilon}^*).$$

Proposition 8.

$$-\kappa^2 \begin{pmatrix} 0 & J_{q\ell} \\ -\kappa^2 J_{q\ell} & 0 \end{pmatrix}\begin{pmatrix} \omega_{q\ell} & 0 \\ 0 & \tilde{\omega}_{q\ell} \end{pmatrix}\begin{pmatrix} 0 & J_{q\ell} \\ -\kappa^2 J_{q\ell} & 0 \end{pmatrix}$$

$$= -\begin{pmatrix} 0 & J_{q\ell} \\ -\kappa^2 J_{q\ell} & 0 \end{pmatrix}\begin{pmatrix} \omega_{q\ell} & 0 \\ 0 & \tilde{\omega}_{q\ell} \end{pmatrix}\begin{pmatrix} 0 & J_{q\ell} \\ -\kappa^2 J_{q\ell} & 0 \end{pmatrix}^{-1} = \begin{pmatrix} J_{q\ell}\tilde{\omega}_{q\ell}J_{q\ell} & 0 \\ 0 & J_{q\ell}\omega_{q\ell}J_{q\ell} \end{pmatrix},$$

where $J_{q\ell}\tilde{\omega}_{q\ell}J_{q\ell}$ equals

$$\begin{bmatrix} \gamma(\omega_a)\otimes\begin{pmatrix} a & 0 \\ 0 & 0 \end{pmatrix}+\gamma(\omega)\otimes\begin{pmatrix} 0 & 0 \\ 0 & 1 \end{pmatrix} & 0 & 0 \\ 0 & \gamma(\omega_a)\otimes a & 0 \\ 0 & 0 & \underline{\gamma(\omega_a)}\otimes\begin{pmatrix} a & 0 \\ 0 & 0 \end{pmatrix}+\underline{\gamma(\omega)}\otimes\begin{pmatrix} 0 & 0 \\ 0 & 1 \end{pmatrix} \end{bmatrix}$$

and $J_{q\ell}\omega_{q\ell}J_{q\ell}$ equals

$$\begin{bmatrix} \tilde{\gamma}(\overline{\omega})\otimes\begin{pmatrix} 1 & 0 \\ 0 & 1 \end{pmatrix} & 0 & -\kappa^2\Gamma\overline{\psi}\otimes\begin{pmatrix} M_d^t & 0 \\ 0 & M_e^t \end{pmatrix} \\ 0 & \tilde{\gamma}(\omega)\otimes\underline{1} & -\kappa^2\Gamma\overline{\underset{\sim}{\psi}}\otimes\left(M_u^t,0\right) \\ -\overline{\varphi}\Gamma\otimes\begin{pmatrix} \overline{M_d} & 0 \\ 0 & \overline{M_e} \end{pmatrix} & -\overline{\varphi}\Gamma\otimes\begin{pmatrix} \overline{M_u} \\ 0 \end{pmatrix} & \tilde{\gamma}(\Omega)\otimes\begin{pmatrix} 1 & 0 \\ 0 & 1 \end{pmatrix} \end{bmatrix}.$$

Proof. A straightforward computation shows $J_{q\ell}\tilde{\omega}_{q\ell}J_{q\ell}$ equal to

$$\begin{bmatrix} \begin{array}{l} J\tilde{\gamma}(\overline{\omega}_a)J\otimes\overline{\begin{pmatrix} \overline{a} & 0 \\ 0 & 0 \end{pmatrix}}^- \\ +J\tilde{\gamma}(\overline{\omega})J\otimes\overline{\begin{pmatrix} 0 & 0 \\ 0 & 1 \end{pmatrix}}^- \end{array} & 0 & 0 \\ 0 & J\tilde{\gamma}(\overline{\omega}_a)J\otimes {}^-\overline{a}{}^- & 0 \\ 0 & 0 & \begin{array}{l} J\underline{\tilde{\gamma}(\overline{\omega}_a)}J\otimes\overline{\begin{pmatrix} \overline{a} & 0 \\ 0 & 0 \end{pmatrix}}^- \\ +J\underline{\tilde{\gamma}(\overline{\omega})}J\otimes\overline{\begin{pmatrix} 0 & 0 \\ 0 & 1 \end{pmatrix}}^- \end{array} \end{bmatrix}$$

and $J_{q\ell}\omega_{q\ell}J_{q\ell}$ equal to

$$\begin{bmatrix} J\gamma(\omega)J\otimes\overline{\begin{pmatrix} 1 & 0 \\ 0 & 1 \end{pmatrix}}^- & 0 & \kappa^2 J\Gamma\psi J\otimes\overline{\begin{pmatrix} M_d^* & 0 \\ 0 & M_e^* \end{pmatrix}}^- \\ 0 & J\gamma(\overline{\omega})J\otimes{}^-\underline{1}{}^- & \kappa^2 J\Gamma\underset{\sim}{\psi}J\otimes{}^-\left(M_u^*,0\right)^- \\ J\varphi\Gamma J\otimes\overline{\begin{pmatrix} M_d & 0 \\ 0 & M_e \end{pmatrix}}^- & J\varphi\Gamma J\otimes\overline{\begin{pmatrix} M_u \\ 0 \end{pmatrix}}^- & J\gamma(\Omega)J\otimes\overline{\begin{pmatrix} 1 & 0 \\ 0 & 1 \end{pmatrix}}^- \end{bmatrix}.$$

Since ${}^-\overline{a}{}^- = a$, the assertion follows.

Define an endomorphism $A\mapsto \underset{\sim}{A}$ of $\mathcal{C}^\infty(M,\mathbb{C}^{2\times 2})$ by

$$\begin{pmatrix} a & b \\ c & d \end{pmatrix}_\sim = \begin{pmatrix} \overline{d} & -\overline{c} \\ -\overline{b} & \overline{a} \end{pmatrix} = \begin{pmatrix} 0 & 1 \\ -1 & 0 \end{pmatrix}\begin{pmatrix} \overline{a} & \overline{b} \\ \overline{c} & \overline{d} \end{pmatrix}\begin{pmatrix} 0 & -1 \\ 1 & 0 \end{pmatrix}. \tag{9.35}$$

Then we have
$$A\underset{\sim}{B} = (AB)_{\sim} \tag{9.36}$$
$$\underset{\sim}{A}\underset{\sim}{\phi} = (A\phi)_{\sim} \tag{9.37}$$

for all $A, B \in \mathcal{C}^\infty(M, \mathbb{C}^{2\times 2})$ and $\phi \in \mathcal{C}^\infty(M, \mathbb{C}^{2\times 1})$ (with $\underset{\sim}{\phi}$ defined in (9.26)).

Proposition 9. $\mathcal{A}_{q\ell}^{(2)}$ *consists of all endomorphisms*
$$\begin{pmatrix} \vartheta_{q\ell} & 0 \\ 0 & \tilde{\vartheta}_{q\ell} \end{pmatrix} \tag{9.38}$$

acting on $\begin{pmatrix} S_{q\ell} \\ S_{q\ell} \end{pmatrix}$, *where* $\vartheta_{q\ell}$, *given by*

$$\begin{bmatrix}
\gamma(\vartheta)\otimes\begin{pmatrix}1 & 0 \\ 0 & 1\end{pmatrix} + \kappa^2\varepsilon^*\lambda\varepsilon\otimes\begin{pmatrix}M_d^*M_d & 0 \\ 0 & M_e^*M_e\end{pmatrix} & \kappa^2\varepsilon^*\underset{\sim}{\lambda}\varepsilon\otimes\begin{pmatrix}M_d^*M_u \\ 0\end{pmatrix} & \kappa^2\Gamma\eta\otimes\begin{pmatrix}M_d^* & 0 \\ 0 & M_e^*\end{pmatrix} \\
\kappa^2\varepsilon^*\underset{\sim}{\lambda}\varepsilon\otimes\begin{pmatrix}M_u^*M_d, & 0\end{pmatrix} & \gamma(\overline{\vartheta})\otimes\underline{1}+\kappa^2\varepsilon^*\underset{\sim}{\lambda}\varepsilon\otimes M_u^*M_u & \kappa^2\Gamma\eta\otimes(M_u^*, 0) \\
\xi\Gamma\otimes\begin{pmatrix}M_d & 0 \\ 0 & M_e\end{pmatrix} & \xi\Gamma\otimes\begin{pmatrix}M_u \\ 0\end{pmatrix} & \gamma(\Theta)\otimes\begin{pmatrix}1 & 0 \\ 0 & 1\end{pmatrix}+\kappa^2\Lambda\otimes\begin{pmatrix}M_dM_d^* & 0 \\ 0 & M_eM_e^*\end{pmatrix} \\
& & +\kappa^2\underset{\sim}{\Lambda}\otimes\begin{pmatrix}M_uM_u^* & 0 \\ 0 & 0\end{pmatrix}
\end{bmatrix},$$

and $\tilde{\vartheta}_{q\ell}$, *given by*

$$\begin{bmatrix}
\tilde{\gamma}(\overline{\vartheta}_a)\otimes\begin{pmatrix}\overline{a} & 0 \\ 0 & 0\end{pmatrix}+\tilde{\gamma}(\overline{\vartheta})\otimes\begin{pmatrix}0 & 0 \\ 0 & 1\end{pmatrix} & 0 & 0 \\
0 & \tilde{\gamma}(\overline{\vartheta}_a)\otimes\overline{a} & 0 \\
0 & 0 & \underline{\tilde{\gamma}(\overline{\vartheta}_a)}\otimes\begin{pmatrix}\overline{a} & 0 \\ 0 & 0\end{pmatrix}+\underline{\tilde{\gamma}(\overline{\vartheta})}\otimes\begin{pmatrix}0 & 0 \\ 0 & 1\end{pmatrix}
\end{bmatrix},$$

are endomorphisms of $S_{q\ell}$ *parametrized by*

$$\vartheta \in \mathcal{C}^\infty(Cl^{2-\mathrm{ev}}(T^\# M)\otimes\mathbb{C}), \quad \Theta \in \mathcal{C}^\infty(Cl^{2-\mathrm{ev}}(T^\# M)\otimes\mathbb{H}),$$
$$\lambda \in \mathcal{C}^\infty(M, \mathbb{H}), \quad \vartheta_a\otimes a \in \mathcal{C}^\infty(Cl^{2-\mathrm{ev}}(T^\# M)\otimes\mathbb{C}^{3\times 3}),$$
$$\xi \in \mathcal{C}^\infty(T^\# M\otimes\mathbb{C}^{2\times 1}), \quad \eta \in \mathcal{C}^\infty(T^\# M\otimes\mathbb{C}^{1\times 2}), \quad \Lambda \in \mathcal{C}^\infty(M, \mathbb{C}^{2\times 2}).$$

Proof. Applying Proposition 7 to (finite sums of) products $\begin{pmatrix}\omega_{q\ell}^0 & 0 \\ 0 & \tilde{\omega}_{q\ell}^0\end{pmatrix}\begin{pmatrix}\omega_{q\ell}^1 & 0 \\ 0 & \tilde{\omega}_{q\ell}^1\end{pmatrix}$ in $\mathcal{A}_{q\ell}^{(1)}$ we obtain $\omega_{q\ell}^0\omega_{q\ell}^1$ as

$$\begin{bmatrix} \gamma(\omega^0 \times \omega^1) \otimes \begin{pmatrix} 1 & 0 \\ 0 & 1 \end{pmatrix} + & \kappa^2 \psi^0 \varphi^1 \otimes \begin{pmatrix} M_d^* M_u \\ 0 \end{pmatrix} & \kappa^2 \Gamma \gamma(\psi^0 \Omega^1 - \omega^0 \psi^1) \otimes \begin{pmatrix} M_d^* & 0 \\ 0 & M_e^* \end{pmatrix} \\ \kappa^2 \psi^0 \varphi^1 \otimes \begin{pmatrix} M_d^* M_d & 0 \\ 0 & M_e^* M_e \end{pmatrix} & & \\ \kappa^2 \underset{\sim}{\psi^0} \varphi^1 \otimes (M_u^* M_d, 0) & \begin{matrix} \gamma(\overline{\omega}^0 \times \overline{\omega}^1) \otimes 1 + \\ \kappa^2 \underset{\sim}{\psi^0} \underset{\sim}{\varphi^1} \otimes \underline{M_u^* M_u} \end{matrix} & \kappa^2 \Gamma \gamma(\underset{\sim}{\psi^0} \Omega^1 - \overline{\omega}^0 \underset{\sim}{\psi^1}) \otimes (\underline{M_u^*}, 0) \\ & & \gamma(\Omega^0 \times \Omega^1) \otimes \begin{pmatrix} 1 & 0 \\ 0 & 1 \end{pmatrix} + \\ \gamma(\Omega^0 \varphi^1 - \varphi^0 \omega^1) \Gamma \otimes \begin{pmatrix} M_d & 0 \\ 0 & M_e \end{pmatrix} & \gamma(\Omega^0 \underset{\sim}{\varphi^1} - \underset{\sim}{\varphi^0} \overline{\omega}^1) \Gamma \otimes \begin{pmatrix} M_u \\ 0 \end{pmatrix} & \kappa^2 \varphi^0 \psi^1 \otimes \begin{pmatrix} M_d M_d^* & 0 \\ 0 & M_e M_e^* \end{pmatrix} \\ & & + \kappa^2 \underset{\sim}{\varphi^0} \underset{\sim}{\psi^1} \otimes \begin{pmatrix} M_u M_u^* & 0 \\ 0 & 0 \end{pmatrix} \end{bmatrix},$$

and $\tilde{\omega}_{q\ell}^0 \tilde{\omega}_{q\ell}^1$ as

$$\begin{bmatrix} \tilde{\gamma}(\overline{\omega}_b^0 \times \overline{\omega}_c^1) \otimes \begin{pmatrix} \overline{bc} & 0 \\ 0 & 0 \end{pmatrix} + \tilde{\gamma}(\overline{\omega}^0 \times \overline{\omega}^1) \otimes \begin{pmatrix} 0 & 0 \\ 0 & 1 \end{pmatrix} & 0 & 0 \\ 0 & \tilde{\gamma}(\overline{\omega}_b^0 \times \overline{\omega}_c^1) \otimes x \overline{b} \overline{c} & 0 \\ 0 & 0 & \begin{matrix} \underline{\underline{\tilde{\gamma}(\overline{\omega}_b^0 \times \overline{\omega}_c^1)}} \otimes \begin{pmatrix} \overline{bc} & 0 \\ 0 & 0 \end{pmatrix} \\ + \\ \underline{\underline{\tilde{\gamma}(\overline{\omega}^0 \times \overline{\omega}^1)}} \otimes \begin{pmatrix} 0 & 0 \\ 0 & 1 \end{pmatrix} \end{matrix} \end{bmatrix}.$$

(9.39)

Now the assertion follows by putting

$$\begin{aligned} \vartheta &= \omega^0 \times \omega^1, \quad \Theta = \Omega^0 \times \Omega^1, \quad \vartheta_a \otimes a = \omega_b^0 \times \omega_c^1 \otimes bc, \\ \xi &= \Omega^0 \varphi^1 - \varphi^0 \omega^1, \quad \eta = \psi^0 \Omega^1 - \omega^0 \psi^1, \\ \lambda &= \begin{pmatrix} \psi^0 \\ \underset{\sim}{\psi^0} \end{pmatrix} (\varphi^1, \underset{\sim}{\varphi^1}) = \begin{pmatrix} \psi^0 \varphi^1 & \psi^0 \underset{\sim}{\varphi^1} \\ \underset{\sim}{\psi^0} \varphi^1 & \underset{\sim}{\psi^0} \underset{\sim}{\varphi^1} \end{pmatrix}, \quad \Lambda = \varphi^0 \psi^1 \end{aligned}$$

(9.40)

which implies

$$\underset{\sim}{\xi} = \Omega^0 \underset{\sim}{\varphi^1} - \underset{\sim}{\varphi^0} \overline{\omega}^1, \quad \underset{\sim}{\eta} = \underset{\sim}{\psi^0} \Omega^1 - \overline{\omega}^0 \underset{\sim}{\psi^1}, \quad \underset{\sim}{\lambda} = \lambda, \quad \underset{\sim}{\Lambda} = \underset{\sim}{\varphi^0} \underset{\sim}{\psi^1}.$$

Proposition 10. *For any* $\begin{pmatrix} \omega_{q\ell} & 0 \\ 0 & \tilde{\omega}_{q\ell} \end{pmatrix}$ *in* $\mathcal{A}_{q\ell}^{(1)}$, *written as in Proposition 7,* $\pi_2 d\pi_1^{-1} \begin{pmatrix} \omega_{q\ell} & 0 \\ 0 & \tilde{\omega}_{q\ell} \end{pmatrix}$ *consists of all endomorphisms of the form*

$$\begin{bmatrix} \gamma(f+d\omega)\otimes\begin{pmatrix}1&0\\0&1\end{pmatrix}+ & \kappa^2(\varepsilon^*\varphi+\psi\varepsilon)\otimes\begin{pmatrix}M_d^*M_u&0\\0&\end{pmatrix} & \kappa^2\Gamma\gamma(\varepsilon^*\Omega-\omega\varepsilon^*-d\psi)\otimes\begin{pmatrix}M_d^*&0\\0&M_e^*\end{pmatrix} \\ \kappa^2(\varepsilon^*\varphi+\psi\varepsilon)\otimes\begin{pmatrix}M_d^*M_d&0\\0&M_e^*M_e\end{pmatrix} & & \\ \kappa^2(\varepsilon^*\varphi+\psi\varepsilon)\otimes(M_u^*M_d,0) & \gamma(\overline{f}+d\overline{\omega})\otimes\underline{1}+ & \kappa^2\Gamma\gamma(\varepsilon^*\Omega-\overline{\omega}\varepsilon^*-d\psi)\otimes(M_u^*,0) \\ & \kappa^2(\varepsilon^*\varphi+\psi\varepsilon)\otimes M_u^*M_u & \\ & & \gamma(F+d\Omega)\otimes\begin{pmatrix}1&0\\0&1\end{pmatrix} \\ & & + \\ \gamma(d\varphi+\Omega\varepsilon-\varepsilon\omega)\Gamma\otimes\begin{pmatrix}M_d&0\\0&M_e\end{pmatrix} & \gamma(d\varphi+\Omega\varepsilon-\varepsilon\overline{\omega})\Gamma\otimes\begin{pmatrix}M_u\\0\end{pmatrix} & G\otimes\begin{pmatrix}M_dM_d^*-M_uM_u^*&0\\0&M_eM_e^*\end{pmatrix} \\ & & + \\ & & \frac{\kappa^2}{2}(\varepsilon\psi+\varphi\varepsilon^*+\varepsilon\psi+\varphi\varepsilon^*)\otimes \\ & & \begin{pmatrix}M_dM_d^*-M_uM_u^*&0\\0&M_eM_e^*\end{pmatrix} \end{bmatrix}$$

with arbitrary

$$F \in \mathcal{C}^\infty(M, \mathbb{H}), \quad f \in \mathcal{C}^\infty(M, \mathbb{C}), \quad f_a \otimes a \in \mathcal{C}^\infty(M, \mathbb{C}^{3\times 3}),$$

$$G = \begin{pmatrix} g_1 & g_2 \\ \overline{g}_2 & -\overline{g}_1 \end{pmatrix} \in \mathcal{C}^\infty(M, \mathbb{H}^\perp).$$

Proof. Expressing $\begin{pmatrix}\omega_{q\ell}&0\\0&\tilde{\omega}_{q\ell}\end{pmatrix}$ by (a finite sum of) products $\begin{pmatrix}f^0_{q\ell}&0\\0&\tilde{f}^0_{q\ell}\end{pmatrix}d\begin{pmatrix}f^1_{q\ell}&0\\0&\tilde{f}^1_{q\ell}\end{pmatrix}$, it follows from (9.34) and (??) that $\pi_1(df^0_{q\ell})\pi_1(df^1_{q\ell})$ equals

$$\begin{bmatrix} \gamma(df^0\times df^1)\otimes\begin{pmatrix}1&0\\0&1\end{pmatrix}+ & \kappa^2(\varepsilon^*F^0-f^0\varepsilon^*)(\varepsilon\overline{f}^1-F^1\varepsilon) & \kappa^2\Gamma((\varepsilon^*F^0-f^0\varepsilon^*)dF^1- \\ \kappa^2(\varepsilon^*F^0-f^0\varepsilon^*)(\varepsilon f^1-F^1\varepsilon) & \otimes\begin{pmatrix}M_d^*M_u&0\\0&\end{pmatrix} & df^0(\varepsilon^*F^1-f^1\varepsilon^*)) \\ \otimes\begin{pmatrix}M_d^*M_d&0\\0&M_e^*M_e\end{pmatrix} & & \otimes\begin{pmatrix}M_d^*&0\\0&M_e^*\end{pmatrix} \\ \kappa^2(\varepsilon^*F^0-\overline{f}^0\varepsilon^*)(\varepsilon f^1-F^1\varepsilon) & \gamma(d\overline{f}^0\times d\overline{f}^1)\otimes\underline{1}+ & \kappa^2\Gamma((\varepsilon^*F^0-\overline{f}^0\varepsilon^*)dF^1- \\ \otimes(M_u^*M_d,0) & \kappa^2(\varepsilon^*F^0-\overline{f}^0\varepsilon^*)(\varepsilon\overline{f}^1-F^1\varepsilon) & d\overline{f}^0(\varepsilon^*F^1-\overline{f}^1\varepsilon^*)) \\ & \otimes M_u^*M_u & \otimes(M_u^*,0) \\ & & \gamma(dF^0\times dF^1)\otimes\begin{pmatrix}1&0\\0&1\end{pmatrix}+ \\ (dF^0(\varepsilon f^1-F^1\varepsilon)- & (dF^0(\varepsilon\overline{f}^1-F^1\varepsilon)- & \kappa^2(\varepsilon f^0-F^0\varepsilon)(\varepsilon^*F^1-f^1\varepsilon^*) \\ (\varepsilon f^0-F^0\varepsilon)df^1)\Gamma & (\varepsilon\overline{f}^0-F^0\varepsilon)d\overline{f}^1)\Gamma & \otimes\begin{pmatrix}M_d^*M_d&0\\0&M_eM_e^*\end{pmatrix} \\ \otimes\begin{pmatrix}M_d&0\\0&M_e\end{pmatrix} & \otimes\begin{pmatrix}M_u\\0\end{pmatrix} & +\kappa^2(\varepsilon\overline{f}^0-F^0\varepsilon)(\varepsilon^*F^1-\overline{f}^1\varepsilon^*) \\ & & \otimes\begin{pmatrix}M_uM_u^*&0\\0&0\end{pmatrix} \end{bmatrix},$$

and $\pi_1(d\tilde{f}_{q\ell}^0)\pi_1(d\tilde{f}_{q\ell}^1)$ equals

$$\begin{bmatrix} \tilde{\gamma}(d\overline{f}_b^0 \times d\overline{f}_c^1) \otimes \begin{pmatrix} \overline{b}\overline{c} & 0 \\ 0 & 0 \end{pmatrix} \\ +\tilde{\gamma}(d\overline{f}^0 \times d\overline{f}^1) \otimes \begin{pmatrix} 0 & 0 \\ 0 & 1 \end{pmatrix} & 0 & 0 \\ 0 & \tilde{\gamma}(d\overline{f}_b^0 \times d\overline{f}_c^1) \otimes \overline{b}\overline{c} & 0 \\ 0 & 0 & \tilde{\gamma}(d\overline{f}_b^0 \times d\overline{f}_c^1) \otimes \begin{pmatrix} \overline{b}\overline{c} & 0 \\ 0 & 0 \end{pmatrix} \\ & & +\tilde{\gamma}(d\overline{f}^0 \times d\overline{f}^1) \otimes \begin{pmatrix} 0 & 0 \\ 0 & 1 \end{pmatrix} \end{bmatrix}.$$

Comparing with (9.38), this implies

$$\vartheta = df^0 \times df^1, \quad \omega = f^0 df^1$$
$$\Theta = dF^0 \times dF^1, \quad \Omega = F^0 dF^1$$
$$\vartheta_a \otimes a = (df_b^0 \times df_c^1) \otimes bc, \quad \omega_a \otimes a = (f_b^0 df_c^1) \otimes bc$$

yielding also

$$d\omega = df^0 \wedge df^1, \quad d\Omega = dF^0 \wedge dF^1, \quad d(\omega_a \otimes a) = (df_b^0 \wedge df_c^1) \otimes bc.$$

It follows that

$$f := \vartheta - d\omega \in \mathcal{C}^\infty(M, \mathbb{C}),$$
$$F := \Theta - d\Omega \in \mathcal{C}^\infty(M, \mathbb{H}),$$
$$f_a \otimes a := (\vartheta_a - d\omega_a) \otimes a \in \mathcal{C}^\infty(M, \mathbb{C}^{3\times 3}).$$

Moreover $\varphi = F^0(\varepsilon f^1 - F^1 \varepsilon)$ implies

$$d\varphi + \Omega\varepsilon - \varepsilon\omega = dF^0(\varepsilon f^1 - F^1\varepsilon) + F^0(\varepsilon df^1 - dF^1\varepsilon) + F^0 dF^1\varepsilon - \varepsilon f^0 df^1$$
$$= dF^0(\varepsilon f^1 - F^1\varepsilon) + (F^0\varepsilon - \varepsilon f^0)df^1 = \xi$$

and $\psi = f^0(\varepsilon^* F^1 - f^1\varepsilon^*)$ implies

$$\varepsilon^*\Omega - \omega\varepsilon^* - d\psi = \varepsilon^* F^0 dF^1 - f^0 df^1\varepsilon^* - df^0(\varepsilon^* F^1 - f^1\varepsilon^*) - f^0(\varepsilon^* dF^1 - df^1\varepsilon^*)$$
$$= (\varepsilon^* F^0 - f^0\varepsilon^*)dF^1 - df^0(\varepsilon^* F^1 - f^1\varepsilon^*) = \eta.$$

Finally,

$$\varepsilon^*\lambda\varepsilon - \varepsilon^*\varphi = -f^0 f^1 + f^0\varepsilon^* F^1\varepsilon = \psi\varepsilon,$$

$$\underset{\sim}{\varepsilon}^*\lambda\underset{\sim}{\varepsilon} - \underset{\sim}{\varepsilon}^*\underset{\sim}{\varphi} = \overline{f}^0\underset{\sim}{\varepsilon}^* F^1\underset{\sim}{\varepsilon} = \underset{\sim}{\psi}\underset{\sim}{\varepsilon},$$

$$\varepsilon^*\lambda\underset{\sim}{\varepsilon} - \varepsilon^*\underset{\sim}{\varphi} = f^0\varepsilon^* F^1\underset{\sim}{\varepsilon} = \psi\underset{\sim}{\varepsilon},$$

$$\underset{\sim}{\varepsilon}^*\lambda\underset{\sim}{\varepsilon} - \underset{\sim}{\varepsilon}^*\varphi = -\overline{f}^0\overline{f}^1 + \overline{f}^0\underset{\sim}{\varepsilon}^* F^1\varepsilon = \underset{\sim}{\psi}\varepsilon,$$

$$\Lambda = (\varepsilon f^0 - F^0 \varepsilon)(\varepsilon^* F^1 - f^1\varepsilon^*) = \varepsilon\psi + \varphi\varepsilon^* + F^0(F^1\varepsilon\varepsilon^* - \varepsilon\varepsilon^* F^1),$$

$$\underset{\sim}{\Lambda} = (\underset{\sim}{\varepsilon}\overline{f}^0 - F^0\underset{\sim}{\varepsilon})(\underset{\sim}{\varepsilon}^* F^1 - \overline{f}^1\underset{\sim}{\varepsilon}^*) = \underset{\sim}{\varepsilon}\underset{\sim}{\psi} + \underset{\sim}{\varphi}\underset{\sim}{\varepsilon}^* + F^0(F^1\underset{\sim}{\varepsilon}\underset{\sim}{\varepsilon}^* - \underset{\sim}{\varepsilon}\underset{\sim}{\varepsilon}^* F^1),$$

$$\Lambda + \underset{\sim}{\Lambda} = \varepsilon\psi + \varphi\varepsilon^* + \underset{\sim}{\varepsilon}\underset{\sim}{\psi} + \underset{\sim}{\varphi}\underset{\sim}{\varepsilon}^*.$$

Since $\varepsilon\varepsilon^* + \underset{\sim}{\varepsilon}\underset{\sim}{\varepsilon}^* = 1$, the assertion follows by putting $G = \frac{\kappa^2}{2}(\Lambda - \underset{\sim}{\Lambda})$ and using the identity

$$\Lambda \otimes \begin{pmatrix} M_d M_d^* & 0 \\ 0 & M_e M_e^* \end{pmatrix} + \underset{\sim}{\Lambda} \otimes \begin{pmatrix} M_u M_u^* & 0 \\ 0 & 0 \end{pmatrix} \qquad (9.41)$$

$$= \frac{1}{2}\underbrace{(\Lambda+\underset{\sim}{\Lambda})}_{\in \mathbb{H}} \otimes \begin{pmatrix} M_d^* M_d + M_u M_u^* & 0 \\ 0 & M_e M_e^* \end{pmatrix}$$

$$+ \frac{1}{2}\underbrace{(\Lambda+\underset{\sim}{\Lambda})}_{\in \mathbb{H}^\perp} \otimes \begin{pmatrix} M_d M_d^* - M_u M_u^* & 0 \\ 0 & M_e M_e^* \end{pmatrix}.$$

Setting $\omega_{q\ell} = 0 = \tilde{\omega}_{q\ell}$ we obtain the so-called "junk ideal":

Corollary 1. $\pi_2 d(\mathrm{Ker}\, \pi_1)$ *consists of all endomorphisms*

$$\begin{bmatrix} f \otimes \begin{pmatrix} 1 & 0 \\ 0 & 1 \end{pmatrix} & 0 & 0 \\ 0 & \overline{f} \otimes \underline{1} & 0 \\ 0 & 0 & F \otimes \begin{pmatrix} 1 & 0 \\ 0 & 1 \end{pmatrix} + G \otimes \begin{pmatrix} M_d M_d^* - M_u M_u^* & 0 \\ 0 & M_e M_e^* \end{pmatrix} \end{bmatrix}$$

$$\oplus$$

$$\begin{bmatrix} \overline{f}_a \otimes \begin{pmatrix} \overline{a} & 0 \\ 0 & 0 \end{pmatrix} + \overline{f} \otimes \begin{pmatrix} 0 & 0 \\ 0 & 1 \end{pmatrix} & 0 & 0 \\ 0 & \overline{f}_a \otimes \overline{a} & 0 \\ 0 & 0 & \underline{\overline{f}_a} \otimes \begin{pmatrix} \overline{a} & 0 \\ 0 & 0 \end{pmatrix} + \underline{\overline{f}} \otimes \begin{pmatrix} 0 & 0 \\ 0 & 1 \end{pmatrix} \end{bmatrix}$$

with $f, F, f_a \otimes a$ and G as above.

A fibre metric on $\begin{pmatrix} S_{q\ell} \\ \tilde{S}_{q\ell} \end{pmatrix}$ is defined by

$$\begin{pmatrix} I_{q\ell} & 0 \\ 0 & \tilde{I}_{q\ell} \end{pmatrix}, \tag{9.42}$$

where

$$I_{q\ell} = \begin{bmatrix} I \otimes \begin{pmatrix} x/3 & 0 \\ 0 & y \end{pmatrix} & 0 & 0 \\ 0 & I \otimes x/3 & 0 \\ 0 & 0 & \underline{I} \otimes \begin{pmatrix} x/3 & 0 \\ 0 & y \end{pmatrix} \end{bmatrix}, \tilde{I}_{q\ell} = \begin{bmatrix} I \otimes \begin{pmatrix} \tilde{x} & 0 \\ 0 & \tilde{y} \end{pmatrix} & 0 & 0 \\ 0 & I \otimes \tilde{x} & 0 \\ 0 & 0 & \underline{I} \otimes \begin{pmatrix} \tilde{x} & 0 \\ 0 & \tilde{y} \end{pmatrix} \end{bmatrix}.$$

and I denotes the identity on spinors. Here $x, y, \tilde{x}, \tilde{y}$ are positive definite linear maps on $\mathbb{G}$.

Proposition 11. *For elements in $\mathcal{A}_{q\ell}^{(2)}$ expressed as in Proposition 9 the real scalar product is given by*

$$\operatorname{Re}\left[(\vartheta_{q\ell} \mid \vartheta'_{q\ell} I_{q\ell})_{S_{q\ell}} + (\tilde{\vartheta}_{q\ell} \mid \tilde{\vartheta}'_{q\ell} \tilde{I}_{q\ell})_{S_{q\ell}}\right] = \tag{9.43}$$

$$= s \operatorname{Re}(\vartheta \mid \vartheta')_{\underline{S}} + t \operatorname{Re}(\Theta \mid \Theta')_{\underline{S}} + 4 \underline{\operatorname{tr}}_{\mathbb{G}} \tilde{x} \cdot \operatorname{Re}(\vartheta_b \otimes b \mid \vartheta'_c \otimes c)_{\underline{S}}$$

$$+ r \operatorname{Re}\left[(\vartheta \mid \varepsilon^* \lambda' \varepsilon)_{\underline{S}} + (\varepsilon^* \lambda \varepsilon \mid \vartheta')_{\underline{S}} + (\Theta \mid \Lambda')_{\underline{S}} + (\Lambda \mid \Theta')_{\underline{S}}\right]$$

$$- \kappa^2 \left((\xi \mid \xi')_{\underline{S}} + (\eta \mid \eta')_{\underline{S}}\right)\right] + \underline{\operatorname{tr}}_{\mathbb{G}} \left(((M_d^* M_d)^2 + (M_u^* M_u)^2) x\right)$$

$$+ (M_e^* M_e)^2 y) \operatorname{Re}\left[(\varepsilon^* \lambda \varepsilon \mid \varepsilon^* \lambda' \varepsilon)_{S} + (\Lambda \mid \Lambda')_{\underline{S}}\right]$$

$$+ 2 \underline{\operatorname{tr}}_{\mathbb{G}} (M_u^* M_d M_d^* M_u x) \operatorname{Re}\left[(\varepsilon^* \lambda \varepsilon \mid \varepsilon^* \lambda' \varepsilon)_{S} + (\Lambda \mid \Lambda')_{\underline{S}}\right]$$

where we introduce the (positive) constants

$$\begin{aligned} r &= \underline{\operatorname{tr}}_{\mathbb{G}}((M_d^* M_d \mid M_u^* M_u)x \mid M_e^* M_e y), \\ s &= \underline{\operatorname{tr}}_{\mathbb{G}}(2x + y + 3\tilde{y}), \\ t &= \underline{\operatorname{tr}}_{\mathbb{G}}(x + y). \end{aligned} \tag{9.44}$$

Proof. Using the notation of Proposition 9 it follows that $\vartheta_{q\ell}^\dagger \vartheta'_{q\ell} I_{q\ell}$ has its diagonal blocks composed of the following operators

$$(\gamma(\vartheta) \otimes \begin{pmatrix} 1 & 0 \\ 0 & 1 \end{pmatrix} + \kappa^2 \varepsilon^* \lambda \varepsilon \otimes \begin{pmatrix} M_d^* M_d & 0 \\ 0 & M_e^* M_e \end{pmatrix})^\dagger$$

$$(\gamma(\vartheta') \otimes \begin{pmatrix} 1 & 0 \\ 0 & 1 \end{pmatrix} + \kappa^2 \varepsilon^* \lambda' \varepsilon \otimes \begin{pmatrix} M_d^* M_d & 0 \\ 0 & M_e^* M_e \end{pmatrix}) I \otimes \begin{pmatrix} x/3 & 0 \\ 0 & y \end{pmatrix}$$

$$= (\gamma(\overline{\vartheta}) \otimes \begin{pmatrix} 1 & 0 \\ 0 & 1 \end{pmatrix} + \kappa^2 \varepsilon^* \lambda^* \varepsilon \otimes \begin{pmatrix} M_d^* M_d & 0 \\ 0 & M_e^* M_e \end{pmatrix})$$

$$(\gamma(\vartheta') \otimes \begin{pmatrix} x/3 & 0 \\ 0 & y \end{pmatrix} + \kappa^2 \varepsilon^* \lambda' \varepsilon \otimes \begin{pmatrix} M_d^* M_d \frac{x}{3} & 0 \\ 0 & M_e^* M_e y \end{pmatrix})$$

$$= \gamma(\overline{\vartheta} \times \vartheta') \otimes \begin{pmatrix} x/3 & 0 \\ 0 & y \end{pmatrix} + \kappa^2 \gamma(\varepsilon^* \lambda^* \varepsilon \vartheta' + \overline{\vartheta} \varepsilon^* \lambda' \varepsilon) \otimes \begin{pmatrix} M_d^* M_d \frac{x}{3} & 0 \\ 0 & M_e^* M_e y \end{pmatrix})$$

$$+ \varepsilon^* \lambda^* \varepsilon \varepsilon^* \lambda' \varepsilon \otimes \begin{pmatrix} M_d^* M_d M_d^* M_d \frac{x}{3} & 0 \\ 0 & M_e^* M_e M_e^* M_e y \end{pmatrix},$$

$$(\gamma(\overline{\vartheta}) \otimes \underset{\sim}{1} + \kappa^2 \varepsilon^* \underset{\sim}{\lambda} \varepsilon \otimes \underline{M_u^* M_u})^\dagger (\gamma(\overline{\vartheta'}) \otimes \underset{\sim}{1} + \kappa^2 \varepsilon^* \underset{\sim}{\lambda'} \varepsilon \otimes \underline{M_u^* M_u}) I \otimes x/3$$

$$= (\gamma(\vartheta) \otimes \underset{\sim}{1} + \kappa^2 \varepsilon^* \underset{\sim}{\lambda^*} \varepsilon \otimes \underline{M_u^* M_u})(\gamma(\overline{\vartheta'}) \otimes x/3 + \kappa^2 \varepsilon^* \underset{\sim}{\lambda'} \varepsilon \otimes \underline{M_u^* M_u \frac{x}{3}})$$

$$= \gamma(\vartheta \times \overline{\vartheta'}) \otimes x/3 + \kappa^2 \gamma(\varepsilon^* \underset{\sim}{\lambda^*} \varepsilon \overline{\vartheta'} + \vartheta \varepsilon^* \underset{\sim}{\lambda'} \varepsilon) \otimes \underline{M_u^* M_u \frac{x}{3}}$$

$$+ \varepsilon^* \underset{\sim}{\lambda^*} \varepsilon \varepsilon^* \underset{\sim}{\lambda'} \varepsilon \otimes \underline{M_u^* M_u M_u^* M_u \frac{x}{3}},$$

$$(\gamma(\Theta) \otimes \begin{pmatrix} 1 & 0 \\ 0 & 1 \end{pmatrix} + \kappa^2 \Lambda \otimes \begin{pmatrix} M_d M_d^* & 0 \\ 0 & M_e M_e^* \end{pmatrix} + \kappa^2 \underset{\sim}{\Lambda} \otimes \begin{pmatrix} M_u M_u^* & 0 \\ 0 & 0 \end{pmatrix})^\dagger$$

$$(\gamma(\Theta') \otimes \begin{pmatrix} 1 & 0 \\ 0 & 1 \end{pmatrix} + \kappa^2 \Lambda' \otimes \begin{pmatrix} M_d M_d^* & 0 \\ 0 & M_e M_e^* \end{pmatrix} + \kappa^2 \underset{\sim}{\Lambda'} \otimes \begin{pmatrix} M_u M_u^* & 0 \\ 0 & 0 \end{pmatrix}) \underline{I} \otimes \begin{pmatrix} x/3 & 0 \\ 0 & y \end{pmatrix}$$

$$= (\gamma(\Theta^*) \otimes \begin{pmatrix} 1 & 0 \\ 0 & 1 \end{pmatrix} + \kappa^2 \Lambda^* \otimes \begin{pmatrix} M_d M_d^* & 0 \\ 0 & M_e M_e^* \end{pmatrix} + \kappa^2 \underset{\sim}{\Lambda^*} \otimes \begin{pmatrix} M_u M_u^* & 0 \\ 0 & 0 \end{pmatrix})$$

$$(\gamma(\Theta') \otimes \begin{pmatrix} x/3 & 0 \\ 0 & y \end{pmatrix} + \kappa^2 \Lambda' \otimes \begin{pmatrix} M_d M_d^* \frac{x}{3} & 0 \\ 0 & M_e M_e^* y \end{pmatrix} + \kappa^2 \underset{\sim}{\Lambda'} \otimes \begin{pmatrix} M_u M_u^* \frac{x}{3} & 0 \\ 0 & 0 \end{pmatrix})$$

$$= \gamma(\Theta^* \times \Theta') \otimes \begin{pmatrix} x/3 & 0 \\ 0 & y \end{pmatrix} + \kappa^2 \gamma(\Theta^* \Lambda' + \Lambda^* \Theta') \otimes \begin{pmatrix} M_d M_d^* \frac{x}{3} & 0 \\ 0 & M_e M_e^* y \end{pmatrix}$$

$$+ \kappa^2 \gamma(\Theta^* \underset{\sim}{\Lambda'} + \underset{\sim}{\Lambda^*} \Theta') \otimes \begin{pmatrix} M_u M_u^* \frac{x}{3} & 0 \\ 0 & 0 \end{pmatrix} + \Lambda^* \Lambda' \otimes \begin{pmatrix} M_d M_d^* M_d M_d^* \frac{x}{3} & 0 \\ 0 & M_e M_e^* M_e M_e^* y \end{pmatrix}$$

$$+ \Lambda^* \underset{\sim}{\Lambda'} \otimes \begin{pmatrix} M_d M_d^* M_u M_u^* \frac{x}{3} & 0 \\ 0 & 0 \end{pmatrix} + \underset{\sim}{\Lambda^*} \Lambda' \otimes \begin{pmatrix} M_u M_u^* M_d M_d^* \frac{x}{3} & 0 \\ 0 & 0 \end{pmatrix}$$

$$+ \underset{\sim}{\Lambda^*} \underset{\sim}{\Lambda'} \otimes \begin{pmatrix} M_u M_u^* M_u M_u^* \frac{x}{3} & 0 \\ 0 & 0 \end{pmatrix},$$

$$(\kappa^2 \underset{\sim}{\varepsilon}^* \lambda \varepsilon \otimes (\underline{M_u^* M_d}, 0))^\dagger (\kappa^2 \underset{\sim}{\varepsilon}^* \lambda' \varepsilon \otimes (\underline{M_u^* M_d}, 0)) I \otimes \begin{pmatrix} x/3 & 0 \\ 0 & y \end{pmatrix}$$

$$= (\varepsilon^* \lambda^* \underset{\sim}{\varepsilon} \otimes \begin{pmatrix} \underline{M_d^* M_u} \\ 0 \end{pmatrix})(\varepsilon^* \lambda' \underset{\sim}{\varepsilon} \otimes (\underline{M_u^* M_d \tfrac{x}{3}}, 0))$$

$$= \varepsilon^* \lambda^* \underset{\sim}{\varepsilon} \underset{\sim}{\varepsilon}^* \lambda' \varepsilon \otimes \begin{pmatrix} \underline{M_d^* M_u M_u^* M_d \tfrac{x}{3}} & 0 \\ 0 & 0 \end{pmatrix},$$

$$(\kappa^2 \varepsilon^* \lambda \underset{\sim}{\varepsilon} \otimes \begin{pmatrix} \underline{M_d^* M_u} \\ 0 \end{pmatrix})^\dagger (\kappa^2 \varepsilon^* \lambda' \underset{\sim}{\varepsilon} \otimes \begin{pmatrix} \underline{M_d^* M_u} \\ 0 \end{pmatrix}) I \otimes x/3$$

$$= (\varepsilon^* \lambda^* \varepsilon \otimes (\underline{M_u^* M_d}, 0))(\varepsilon^* \lambda' \underset{\sim}{\varepsilon} \otimes \begin{pmatrix} \underline{M_d^* M_u \tfrac{x}{3}} \\ 0 \end{pmatrix})$$

$$= \varepsilon^* \lambda^* \varepsilon \varepsilon^* \lambda' \underset{\sim}{\varepsilon} \otimes \underline{M_u^* M_d M_d^* M_u} \tfrac{x}{3},$$

$$(\gamma(\xi) \Gamma \otimes \begin{pmatrix} M_d & 0 \\ 0 & M_e \end{pmatrix})^\dagger (\gamma(\xi') \Gamma \otimes \begin{pmatrix} M_d & 0 \\ 0 & M_e \end{pmatrix}) I \otimes \begin{pmatrix} x/3 & 0 \\ 0 & y \end{pmatrix}$$

$$= (\Gamma^\dagger \gamma(\xi^*) \otimes \begin{pmatrix} M_d^* & 0 \\ 0 & M_e^* \end{pmatrix})(\gamma(\xi') \Gamma \otimes \begin{pmatrix} M_d \tfrac{x}{3} & 0 \\ 0 & M_e y \end{pmatrix})$$

$$= -\kappa^2 \gamma(\xi^* \times \xi') \otimes \begin{pmatrix} M_d^* M_d \tfrac{x}{3} & 0 \\ 0 & M_e^* M_e y \end{pmatrix},$$

$$(\gamma(\underset{\sim}{\xi}) \Gamma \otimes \begin{pmatrix} M_u \\ 0 \end{pmatrix})^\dagger (\gamma(\underset{\sim}{\xi}') \Gamma \otimes \begin{pmatrix} M_u \\ 0 \end{pmatrix}) I \otimes x/3$$

$$= (\Gamma^\dagger \gamma(\underset{\sim}{\xi}^*) \otimes (\underline{M_u^*}, 0))(\gamma(\underset{\sim}{\xi}') \Gamma \otimes \begin{pmatrix} \underline{M_u \tfrac{x}{3}} \\ 0 \end{pmatrix}) = -\kappa^2 \gamma(\underset{\sim}{\xi}^* \times \underset{\sim}{\xi}') \otimes \underline{M_u^* M_u} \tfrac{x}{3},$$

$$(\kappa^2 \Gamma \gamma(\eta) \otimes \begin{pmatrix} M_d^* & 0 \\ 0 & M_e^* \end{pmatrix})^\dagger (\kappa^2 \Gamma \gamma(\eta') \otimes \begin{pmatrix} M_d^* & 0 \\ 0 & M_e^* \end{pmatrix}) I \otimes \begin{pmatrix} x/3 & 0 \\ 0 & y \end{pmatrix}$$

$$= (\gamma(\eta^*) \Gamma^\dagger \otimes \begin{pmatrix} M_d & 0 \\ 0 & M_e \end{pmatrix})(\Gamma \gamma(\eta') \otimes \begin{pmatrix} M_d^* \tfrac{x}{3} & 0 \\ 0 & M_e^* y \end{pmatrix})$$

$$= -\kappa^2 \gamma(\eta^* \times \eta') \otimes \begin{pmatrix} M_d M_d^* \tfrac{x}{3} & 0 \\ 0 & M_e M_e^* y \end{pmatrix},$$

$$(\kappa^2 \Gamma \gamma(\underset{\sim}{\eta}) \otimes (\underline{M_u^*}, 0))^\dagger (\kappa^2 \Gamma \gamma(\underset{\sim}{\eta}') \otimes (\underline{M_u^*}, 0)) I \otimes \begin{pmatrix} x/3 & 0 \\ 0 & y \end{pmatrix}$$

$$= (\gamma(\underset{\sim}{\eta}^*) \Gamma^\dagger \otimes \begin{pmatrix} M_u \\ 0 \end{pmatrix})(\Gamma \gamma(\underset{\sim}{\eta}') \otimes (\underline{M_u^* \tfrac{x}{3}}, 0)) = -\kappa^2 \gamma(\underset{\sim}{\eta}^* \times \underset{\sim}{\eta}') \otimes \begin{pmatrix} \underline{M_u M_u^* \tfrac{x}{3}} & 0 \\ 0 & 0 \end{pmatrix}.$$

Similarly, $\tilde{\vartheta}_{q\ell}^\dagger \tilde{\vartheta}_{q\ell}' \tilde{I}_{q\ell}$ consists of the following diagonal blocks

$$(\tilde{\gamma}(\overline{\vartheta}_b) \otimes \begin{pmatrix} \overline{b} & 0 \\ 0 & 0 \end{pmatrix} + \tilde{\gamma}(\overline{\vartheta}) \otimes \begin{pmatrix} 0 & 0 \\ 0 & 1 \end{pmatrix})^\dagger (\tilde{\gamma}(\overline{\vartheta'_c}) \otimes \begin{pmatrix} \overline{c} & 0 \\ 0 & 0 \end{pmatrix} + \tilde{\gamma}(\overline{\vartheta'}) \otimes \begin{pmatrix} 0 & 0 \\ 0 & 1 \end{pmatrix}) I \otimes \begin{pmatrix} \tilde{x} & 0 \\ 0 & \tilde{y} \end{pmatrix}$$

$$= (\tilde{\gamma}(\vartheta_b) \otimes \begin{pmatrix} b^t & 0 \\ 0 & 0 \end{pmatrix} + \tilde{\gamma}(\vartheta) \otimes \begin{pmatrix} 0 & 0 \\ 0 & 1 \end{pmatrix})(\tilde{\gamma}(\overline{\vartheta'_c}) \otimes \begin{pmatrix} \overline{c}\tilde{x} & 0 \\ 0 & 0 \end{pmatrix} + \tilde{\gamma}(\overline{\vartheta'}) \otimes \begin{pmatrix} 0 & 0 \\ 0 & \tilde{y} \end{pmatrix}))$$

$$= \tilde{\gamma}(\vartheta_b \times \overline{\vartheta'_c}) \otimes \begin{pmatrix} b^t \overline{c}\tilde{x} & 0 \\ 0 & 0 \end{pmatrix} + \tilde{\gamma}(\vartheta \times \overline{\vartheta'}) \otimes \begin{pmatrix} 0 & 0 \\ 0 & \tilde{y} \end{pmatrix},$$

$$(\tilde{\gamma}(\overline{\vartheta_b})\otimes\overline{b})^\dagger(\tilde{\gamma}(\overline{\vartheta'_c})\otimes\overline{c})I\otimes\underline{\tilde{x}}=(\tilde{\gamma}(\vartheta_b)\otimes b^t)(\tilde{\gamma}(\overline{\vartheta'_c})\otimes\overline{c}\tilde{x})=\tilde{\gamma}(\vartheta_b\times\overline{\vartheta'_c})\otimes b^t\overline{c}\tilde{x},$$

$$(\tilde{\gamma}(\overline{\underline{\vartheta_b}})\otimes\begin{pmatrix}\overline{b}&0\\0&0\end{pmatrix}+\tilde{\gamma}(\underline{\vartheta})\otimes\begin{pmatrix}0&0\\0&1\end{pmatrix})^\dagger(\tilde{\gamma}(\overline{\underline{\vartheta'_c}})\otimes\begin{pmatrix}\overline{c}&0\\0&0\end{pmatrix}+\tilde{\gamma}(\underline{\vartheta'})\otimes\begin{pmatrix}0&0\\0&1\end{pmatrix})\underline{I}\otimes\begin{pmatrix}\tilde{x}&0\\0&\tilde{y}\end{pmatrix}$$
$$=(\tilde{\gamma}(\underline{\vartheta_b})\otimes\begin{pmatrix}b^t&0\\0&0\end{pmatrix}+\tilde{\gamma}(\underline{\vartheta})\otimes\begin{pmatrix}0&0\\0&1\end{pmatrix})(\tilde{\gamma}(\overline{\underline{\vartheta'_c}})\otimes\begin{pmatrix}\overline{c}\tilde{x}&0\\0&0\end{pmatrix}+\tilde{\gamma}(\overline{\underline{\vartheta'}})\otimes\begin{pmatrix}0&0\\0&\tilde{y}\end{pmatrix})$$
$$=\tilde{\gamma}(\underline{\vartheta_b\times\overline{\vartheta'_c}})\otimes\begin{pmatrix}b^t\overline{c}\tilde{x}&0\\0&0\end{pmatrix}+\tilde{\gamma}(\underline{\vartheta\times\overline{\vartheta'}})\otimes\begin{pmatrix}0&0\\0&\tilde{y}\end{pmatrix}.$$

Here we have used $\Gamma^\dagger=-\kappa^2\Gamma$ and

$$\Gamma^\dagger\gamma(\xi^*)=\gamma(\xi^*)\kappa^2\Gamma,\ \gamma(\xi')\Gamma=-\Gamma\gamma(\xi'),\ \Gamma^\dagger\gamma(\underline{\xi^*})=\kappa^2\gamma(\underline{\xi^*})\Gamma,\ \gamma(\underline{\xi'})\Gamma=-\Gamma\gamma(\underline{\xi'}).$$

Taking the trace we obtain

$$(\tilde{\vartheta}_{q\ell}\mid\tilde{\vartheta}'_{q\ell}I_{q\ell})_{S_{q\ell}}+(\tilde{\vartheta}_{q\ell}\mid\tilde{\vartheta}'_{q\ell}\tilde{I}_{q\ell})_{S_{q\ell}}=$$
$$(\vartheta\mid\vartheta')_S\underline{\mathrm{tr}}_\mathbb{G}(x+y)+\kappa^2((\varepsilon^*\lambda\varepsilon\mid\vartheta')_S+(\vartheta\mid\varepsilon^*\lambda'\varepsilon)_S)\underline{\mathrm{tr}}_\mathbb{G}(M_d^*M_d x+M_e^*M_e y)$$
$$+(\varepsilon^*\lambda\varepsilon\mid\varepsilon^*\lambda'\varepsilon)_S\underline{\mathrm{tr}}_\mathbb{G}((M_d^*M_d)^2 x+(M_e^*M_e)^2 y)+(\overline{\vartheta}\mid\overline{\vartheta'})_S\underline{\mathrm{tr}}_\mathbb{G} x$$
$$+\kappa^2((\underline{\varepsilon^*\lambda\varepsilon}\mid\overline{\vartheta'})_S+(\overline{\vartheta}\mid\underline{\varepsilon^*\lambda'\varepsilon})_S)\underline{\mathrm{tr}}_\mathbb{G} M_u^*M_u x$$
$$+(\underline{\varepsilon^*\lambda\varepsilon}\mid\underline{\varepsilon^*\lambda'\varepsilon})_S\underline{\mathrm{tr}}_\mathbb{G}(M_u^*M_u)^2 x+(\Theta\mid\Theta')_{\underline{S}}\underline{\mathrm{tr}}_\mathbb{G}(x+y)$$
$$+\kappa^2((\Theta\mid\Lambda')_{\underline{S}}+(\Lambda\mid\Theta')_{\underline{S}})\mathrm{tr}_\mathbb{G}(M_d M_d^* x+M_e M_e^* y)$$
$$+\kappa^2((\Theta\mid\underline{\Lambda'})_{\underline{S}}+(\underline{\Lambda}\mid\Theta'))_{\underline{S}}\,\underline{\mathrm{tr}}_\mathbb{G} M_u M_u^* x$$
$$+(\Lambda\mid\Lambda')_{\underline{S}}\underline{\mathrm{tr}}_\mathbb{G}((M_d M_d^*)^2 x+(M_e M_e^*)^2 y)+(\Lambda\mid\underline{\Lambda'})_{\underline{S}}\underline{\mathrm{tr}}_\mathbb{G}(M_d M_d^* M_u M_u^* x)$$
$$+(\underline{\Lambda}\mid\Lambda')_{\underline{S}}\underline{\mathrm{tr}}_\mathbb{G}(M_u M_u^* M_d M_d^* x)+(\underline{\Lambda}\mid\underline{\Lambda'})_{\underline{S}}\underline{\mathrm{tr}}_\mathbb{G}(M_u M_u^*)^2 x$$
$$+(\overline{\vartheta}_b\mid\overline{\vartheta'_c})_S\underline{\mathrm{tr}}_\mathbb{G} b^t\overline{c}\tilde{x}$$
$$+(\overline{\vartheta}\mid\overline{\vartheta'})_S\underline{\mathrm{tr}}_\mathbb{G}\tilde{y}+(\overline{\vartheta}_b\mid\overline{\vartheta'_c})_S\underline{\mathrm{tr}}_\mathbb{G} b^t\overline{c}\tilde{x}+2(\overline{\vartheta}_b\mid\overline{\vartheta'_c})_S\underline{\mathrm{tr}}_\mathbb{G} b^t\overline{c}\tilde{x}+2(\overline{\vartheta}\mid\overline{\vartheta'})_S\underline{\mathrm{tr}}_\mathbb{G}\tilde{y}$$
$$+(\underline{\varepsilon\lambda\varepsilon}\mid\underline{\varepsilon^*\lambda'\varepsilon})_S\underline{\mathrm{tr}}_\mathbb{G}(M_d^*M_u M_u^*M_d x)+(\underline{\varepsilon^*\lambda\varepsilon}\mid\underline{\varepsilon^*\lambda'\varepsilon})_S\underline{\mathrm{tr}}_\mathbb{G}(M_u^*M_d M_d^*M_u x)$$
$$-\kappa^2(\xi\mid\xi')_S\underline{\mathrm{tr}}_\mathbb{G}(M_d^*M_d x+M_e^*M_e y)-\kappa^2(\underline{\xi}\mid\underline{\xi'})_S\underline{\mathrm{tr}}_\mathbb{G}(M_u^*M_u x)$$
$$-\kappa^2(\eta\mid\eta')_{\underline{S}}\underline{\mathrm{tr}}_\mathbb{G}(M_d M_d^* x+M_e M_e^* y)-\kappa^2(\underline{\eta}\mid\underline{\eta'})_{\underline{S}}\underline{\mathrm{tr}}_\mathbb{G}(M_u M_u^* x).$$

Now the assertion follows in view of the identities

$$\mathrm{Re}(\vartheta\mid\vartheta')_S=\mathrm{Re}(\overline{\vartheta}\mid\overline{\vartheta'})_S,$$
$$\mathrm{Re}(\varepsilon^*\lambda\varepsilon\mid\vartheta')_S=\mathrm{Re}(\underline{\varepsilon^*\lambda\varepsilon}\mid\overline{\vartheta'})_S,\quad \mathrm{Re}(\vartheta\mid\varepsilon^*\lambda'\varepsilon)_S=\mathrm{Re}(\overline{\vartheta}\mid\underline{\varepsilon^*\lambda'\varepsilon})_S,$$
$$\mathrm{Re}(\varepsilon^*\lambda\varepsilon\mid\varepsilon^*\lambda'\varepsilon)_S=\mathrm{Re}(\underline{\varepsilon^*\lambda\varepsilon}\mid\underline{\varepsilon^*\lambda'\varepsilon})_S,\mathrm{Re}(\varepsilon^*\lambda\varepsilon\mid\varepsilon^*\lambda'\varepsilon)_S=\mathrm{Re}(\varepsilon^*\lambda\varepsilon\mid\underline{\varepsilon^*\lambda'\varepsilon})_S,$$
$$\mathrm{Re}(\Theta\mid\Lambda')_{\underline{S}}=\mathrm{Re}(\Theta\mid\underline{\Lambda'})_{\underline{S}},\quad \mathrm{Re}(\Lambda\mid\Theta')_{\underline{S}}=\mathrm{Re}(\underline{\Lambda}\mid\Theta')_{\underline{S}},$$

$$\mathrm{Re}(\Lambda\mid\Lambda')_{\underline{S}}=\mathrm{Re}(\underset{\sim}{\Lambda}\mid\underset{\sim}{\Lambda'})_{\underline{S}},\quad \mathrm{Re}(\Lambda\mid\Lambda')_{\underline{\underline{S}}}=\mathrm{Re}(\underset{\sim}{\Lambda}\mid\underset{\sim}{\Lambda'})_{\underline{\underline{S}}},$$
$$\mathrm{Re}(\xi\mid\xi')_{\underline{S}}=\mathrm{Re}(\underset{\sim}{\xi}\mid\underset{\sim}{\xi'})_{S'},\quad \mathrm{Re}(\eta\mid\eta')_{\underline{S}}=\mathrm{Re}(\underset{\sim}{\eta}\mid\underset{\sim}{\eta'})_{\underline{\underline{S}}},$$
$$\mathrm{Re}(\overline{\vartheta}_b\otimes \overline{b}\mid \overline{\vartheta}'_c\otimes \overline{c})_{\underline{S}}=\mathrm{Re}(\vartheta_b\otimes b\mid \vartheta'_c\otimes c)_{\underline{S}}.$$

Remark 2. *For compact Riemannian manifolds M ($\kappa=i$), Proposition 11 gives an explicit expression for the Dixmier trace*

$$\mathrm{Re}\,\mathrm{Tr}_\omega[\vartheta^\dagger_{q\ell}\vartheta'_{q\ell}I_{q\ell}|D_{q\ell}|^{-n} + \tilde{\vartheta}^\dagger_{q\ell}\tilde{\vartheta}'_{q\ell}\tilde{I}_{q\ell}J_{q\ell}|D_{q\ell}|^{-n}J^*_{q\ell}]$$
$$= c_n\int_M d^n x\cdot \mathrm{Re}\big[(\vartheta_{q\ell}|\vartheta'_{q\ell}I_{q\ell})_{S_{q\ell}} + (\tilde{\vartheta}_{q\ell}|\tilde{\vartheta}'_{q\ell}\tilde{I}_{q\ell})_{S_{q\ell}}\big]$$

by applying (7.91) to $\mathbb{G}_{q\ell}$-valued spinors.

Proposition 12. $\pi_2(d\,\mathrm{Ker}\,\pi_1)^\perp$ *consists of all endomorphisms*

$$\begin{pmatrix}\sigma_{q\ell} & 0 \\ 0 & \tilde{\sigma}_{q\ell}\end{pmatrix} \tag{9.45}$$

such that $\sigma_{q\ell}$ equals

$$\begin{bmatrix}
\gamma(\sigma)\otimes\begin{pmatrix}1 & 0\\ 0 & 1\end{pmatrix}+ & \varepsilon^*\underset{\sim}{\lambda}\varepsilon\otimes\begin{pmatrix}M^*_d M_u\\ 0\end{pmatrix} & \kappa^2\Gamma\gamma(\eta)\otimes\begin{pmatrix}M^*_d & 0\\ 0 & M^*_e\end{pmatrix} \\
\varepsilon^*\lambda\varepsilon\otimes\begin{pmatrix}M^*_d M_d-\frac{r}{s} & 0\\ 0 & M^*_e M_e-\frac{r}{s}\end{pmatrix} & & \\
\varepsilon^*\underset{\sim}{\lambda}\varepsilon\otimes(M^*_u M_d,\,0) & \gamma(\overline{\sigma})\otimes 1+ \varepsilon^*\underset{\sim}{\lambda}\underset{\sim}{\varepsilon}\otimes\overline{M^*_u M_u-\frac{r}{s}} & \kappa^2\Gamma\gamma(\eta)\otimes(M^*_u,\,0) \\
\gamma(\xi)\Gamma\otimes\begin{pmatrix}M_d & 0\\ 0 & M_e\end{pmatrix} & \gamma(\underset{\sim}{\xi})\Gamma\otimes\begin{pmatrix}M_u\\ 0\end{pmatrix} & \gamma(\Sigma)\otimes\begin{pmatrix}1 & 0\\ 0 & 1\end{pmatrix}+ \Lambda\otimes\begin{pmatrix}M_dM^*_d+M_uM^*_u-\frac{r}{t} & 0\\ 0 & M_eM^*_e-\frac{r}{t}\end{pmatrix}
\end{bmatrix}$$

and $\tilde{\sigma}_{q\ell}$ equals

$$\begin{bmatrix}
\tilde{\gamma}(\overline{\sigma}_a)\otimes\begin{pmatrix}\overline{a} & 0\\ 0 & 0\end{pmatrix}+\tilde{\gamma}(\overline{\sigma}-\frac{r}{s}\overline{\varepsilon^* a\varepsilon})\otimes\begin{pmatrix}0 & 0\\ 0 & 1\end{pmatrix} & 0 & 0 \\
0 & \tilde{\gamma}(\overline{\sigma}_a)\otimes\overline{a} & 0 \\
0 & 0 & \underline{\tilde{\gamma}(\overline{\sigma}_a)}\otimes\begin{pmatrix}\overline{a} & 0\\ 0 & 0\end{pmatrix}+\underline{\tilde{\gamma}(\overline{\sigma}-\frac{r}{s}\overline{\varepsilon^* a\varepsilon})}\otimes\begin{pmatrix}0 & 0\\ 0 & 1\end{pmatrix}
\end{bmatrix}$$

with arbitrary

$$\sigma\in\mathcal{C}^\infty(\textstyle\bigwedge^2 T^\# M\otimes\mathbb{C}),\quad \sigma_a\otimes a\in\mathcal{C}^\infty(\textstyle\bigwedge^2 T^\# M\otimes\mathbb{C}^{3\times 3}),$$
$$\Sigma\in\mathcal{C}^\infty(\textstyle\bigwedge^2 T^\# M\otimes\mathbb{H}),\quad \lambda\in\mathcal{C}^\infty(M,\mathbb{H}),\quad \xi\in\mathcal{C}^\infty(T^\# M\otimes\mathbb{C}^{2\times 1}),$$
$$\eta\in\mathcal{C}^\infty(T^\# M\otimes\mathbb{C}^{1\times 2}),\quad \Lambda\in\mathcal{C}^\infty(M,\mathbb{C}^{2\times 2}).$$

Proof. Writing $\begin{pmatrix} \vartheta_{q\ell} & 0 \\ 0 & \tilde{\vartheta}_{q\ell} \end{pmatrix} \in \pi_2(d\operatorname{Ker}\pi_1)^\perp$ and $\begin{pmatrix} \vartheta'_{q\ell} & 0 \\ 0 & \tilde{\vartheta}'_{q\ell} \end{pmatrix} \in \pi_2(d\operatorname{Ker}\pi_1)$ according to Proposition 9, it follows from Corollary 1 that

$$\vartheta' = f, \quad \Theta' = F, \quad \vartheta'_c \otimes c = f_c \otimes c,$$
$$\lambda' = 0, \quad \xi' = 0, \quad \eta' = 0, \quad \Lambda' = -\underset{\sim}{\Lambda'} = G \in \mathcal{C}^\infty(M, \mathbb{H}^\perp).$$

Therefore Proposition 11 yields

$$0 = \operatorname{Re}\big[\big(\vartheta_{q\ell} \mid \vartheta'_{q\ell} I_{q\ell}\big)_{S_{q\ell}} + \big(\tilde{\vartheta}_{q\ell} \mid \tilde{\vartheta}'_{q\ell} \tilde{I}_{q\ell}\big)_{S_{q\ell}}\big]$$
$$= s\operatorname{Re}(\vartheta \mid f)_{\underline{S}} + t\operatorname{Re}(\Theta \mid F)_{\underline{S}} + 4\underline{\operatorname{tr}}_{\mathbb{G}}\tilde{x}\operatorname{Re}(\vartheta_b \otimes b \mid f_c \otimes c)_{\underline{S}}$$
$$+ r\operatorname{Re}\big[(\varepsilon^*\lambda\varepsilon \mid f)_{\underline{S}} + (\Theta \mid G)_{\underline{S}} + (\Lambda \mid F)_{\underline{S}}\big]$$
$$+ \underline{\operatorname{tr}}_{\mathbb{G}}(((M_d^* M_d)^2 + (M_u^* M_u)^2 - 2M_u^* M_d M_d^* M_u)x + (M_e^* M_e)^2 y)\operatorname{Re}(\Lambda \mid G)_{\underline{S}}$$
$$= s\operatorname{Re}\big(\vartheta + \tfrac{r}{s}\varepsilon^*\lambda\varepsilon \mid f\big)_{\underline{S}} + t\operatorname{Re}\big(\Theta + \tfrac{r}{t}\Lambda \mid F\big)_{\underline{S}} + 4\underline{\operatorname{tr}}_{\mathbb{G}}\tilde{x}\operatorname{Re}(\vartheta_b \otimes b \mid f_c \otimes c)_{\underline{S}}$$
$$+ \underline{\operatorname{tr}}_{\mathbb{G}}(((M_d^* M_d)^2 + (M_u^* M_u)^2)x + (M_e^* M_e)^2 y)(\Lambda \mid G)_{\underline{S}}.$$

Since $f \in \mathcal{C}^\infty(M, \mathbb{C})$ is arbitrary, we have

$$\sigma := \vartheta + \tfrac{r}{s}\varepsilon^*\lambda\varepsilon \in \mathcal{C}^\infty(\bigwedge^2 T^\# M \otimes \mathbb{C}),$$

i.e. $\vartheta = \sigma - \tfrac{r}{s}\varepsilon^*\lambda\varepsilon$. Since $F \in \mathcal{C}^\infty(M, \mathbb{H})$ is arbitrary, we have

$$\Sigma := \Theta + \tfrac{r}{t}\Lambda \in \mathcal{C}^\infty(\bigwedge^2 T^\# M \otimes \mathbb{H}),$$

i.e. $\Theta = \Sigma - \tfrac{r}{t}\Lambda$. Since $f_c \otimes c \in \mathcal{C}^\infty(M, \mathbb{C}^{3\times 3})$ is arbitrary, we have

$$\vartheta_b \otimes b = \sigma_b \otimes b \in \mathcal{C}^\infty(\bigwedge^2 T^\# M \otimes \mathbb{C}^{3\times 3}).$$

Since $G \in \mathcal{C}^\infty(M, \mathbb{H}^\perp)$ is arbitrary, we have

$$\Lambda = \underset{\sim}{\Lambda} \in \mathcal{C}^\infty(M, \mathbb{H}).$$

Proposition 13. *The orthogonal projection from $\mathcal{A}^{(2)}_{q\ell}$ onto $\pi_2(d\operatorname{Ker}\pi_1)^\perp$ is given by*

$$\begin{pmatrix} \vartheta_{q\ell} & 0 \\ 0 & \tilde{\vartheta}_{q\ell} \end{pmatrix}^\perp = \begin{pmatrix} \vartheta^\perp_{q\ell} & 0 \\ 0 & \tilde{\vartheta}^\perp_{q\ell} \end{pmatrix} \tag{9.46}$$

where $\vartheta_{q\ell}^\perp$ is given by

$$\begin{bmatrix} \varepsilon^*\underset{\sim}{\lambda}\varepsilon \otimes \begin{pmatrix} \gamma(\vartheta^\perp) \otimes \begin{pmatrix} 1 & 0 \\ 0 & 1 \end{pmatrix} + \\ \begin{pmatrix} M_d^*M_d - \frac{r}{s} & 0 \\ 0 & M_e^*M_e - \frac{r}{s} \end{pmatrix} \end{pmatrix} & \varepsilon^*\underset{\sim}{\lambda}\underset{\sim}{\varepsilon} \otimes \begin{pmatrix} M_d^*M_u \\ 0 \end{pmatrix} & \Gamma\gamma(\eta) \otimes \begin{pmatrix} M_d^* & 0 \\ 0 & M_e^* \end{pmatrix} \\ \varepsilon^*\underset{\sim}{\lambda}\underset{\sim}{\varepsilon} \otimes (M_u^*M_d, 0) & \gamma(\overline{\vartheta}^\perp) \otimes \underline{1} + \varepsilon^*\underset{\sim}{\lambda}\underset{\sim}{\varepsilon} \otimes \underline{M_u^*M_u - \frac{r}{s}} & \Gamma\gamma(\eta) \otimes (M_u^*, 0) \\ \gamma(\xi)\Gamma \otimes \begin{pmatrix} M_d & 0 \\ 0 & M_e \end{pmatrix} & \gamma(\xi)\Gamma \otimes \begin{pmatrix} M_u \\ 0 \end{pmatrix} & \begin{array}{c} \gamma(\Theta^\perp) \otimes \begin{pmatrix} 1 & 0 \\ 0 & 1 \end{pmatrix} + \\ \frac{\Lambda+\underset{\sim}{\Lambda}}{2} \otimes \begin{pmatrix} M_dM_d^* + M_uM_u^* - \frac{r}{t} & 0 \\ 0 & M_eM_e^* - \frac{r}{t} \end{pmatrix} \end{array} \end{bmatrix}$$

and $\tilde{\vartheta}_{q\ell}^\perp$ is given by

$$\begin{bmatrix} \tilde{\gamma}(\overline{\vartheta}_a^\perp) \otimes \begin{pmatrix} \overline{a} & 0 \\ 0 & 0 \end{pmatrix} + \tilde{\gamma}(\overline{\vartheta}^\perp - \frac{r}{s}\overline{\varepsilon^*\lambda\varepsilon}) \otimes \begin{pmatrix} 0 & 0 \\ 0 & 1 \end{pmatrix} & 0 & 0 \\ 0 & \tilde{\gamma}(\overline{\vartheta}_a^\perp) \otimes \overline{a} & 0 \\ 0 & 0 & \underline{\tilde{\gamma}(\overline{\vartheta}_a^\perp) \otimes \begin{pmatrix} \overline{a} & 0 \\ 0 & 0 \end{pmatrix} + \tilde{\gamma}(\overline{\vartheta}^\perp - \frac{r}{s}\overline{\varepsilon^*\lambda\varepsilon}) \otimes \begin{pmatrix} 0 & 0 \\ 0 & 1 \end{pmatrix}} \end{bmatrix}$$

Proof. For $\begin{pmatrix} \vartheta_{q\ell} & 0 \\ 0 & \tilde{\vartheta}_{q\ell} \end{pmatrix}$, given as in Proposition 9, we have $\begin{pmatrix} \vartheta_{q\ell}^\perp & 0 \\ 0 & \tilde{\vartheta}_{q\ell}^\perp \end{pmatrix} \in \pi_2(d \operatorname{Ker} \pi_1)^\perp$ by Proposition 12 and

$$\begin{pmatrix} \vartheta_{q\ell} & 0 \\ 0 & \tilde{\vartheta}_{q\ell} \end{pmatrix} - \begin{pmatrix} \vartheta_{q\ell}^\perp & 0 \\ 0 & \tilde{\vartheta}_{q\ell}^\perp \end{pmatrix} = \begin{pmatrix} \vartheta_{q\ell} - \vartheta_{q\ell}^\perp & 0 \\ 0 & \tilde{\vartheta}_{q\ell} - \tilde{\vartheta}_{q\ell}^\perp \end{pmatrix} \in \pi_2(d \operatorname{Ker} \pi_1)$$

by Corollary 1, since $\vartheta_{q\ell} - \vartheta_{q\ell}^\perp$ equals

$$\begin{bmatrix} (\vartheta - \vartheta^\perp + \frac{r}{s}\varepsilon^*\underset{\sim}{\lambda}\varepsilon) \otimes \begin{pmatrix} 1 & 0 \\ 0 & 1 \end{pmatrix} & 0 & 0 \\ 0 & (\overline{\vartheta} - \overline{\vartheta}^\perp + \frac{r}{s}\underset{\sim}{\varepsilon}^*\underset{\sim}{\lambda}\underset{\sim}{\varepsilon}) \otimes \underline{1} & 0 \\ 0 & 0 & \begin{array}{c}(\Theta - \Theta^\perp + \frac{r}{t}\frac{\Lambda+\underset{\sim}{\Lambda}}{2}) \otimes \begin{pmatrix} 1 & 0 \\ 0 & 0 \end{pmatrix} + \\ \frac{\Lambda+\underset{\sim}{\Lambda}}{2} \otimes \begin{pmatrix} M_dM_d^* - M_uM_u^* & 0 \\ 0 & M_eM_e^* \end{pmatrix}\end{array} \end{bmatrix}$$

and $\tilde{\vartheta}_{q\ell} - \tilde{\vartheta}_{q\ell}^\perp$ equals

$$\begin{bmatrix} (\overline{\vartheta}_a - \overline{\vartheta}_a^\perp) \otimes \begin{pmatrix} \overline{a} & 0 \\ 0 & 0 \end{pmatrix} + & 0 & 0 \\ (\overline{\vartheta} - \overline{\vartheta}^\perp + \frac{r}{s}\overline{\varepsilon^*\lambda\varepsilon}) \otimes \begin{pmatrix} 0 & 0 \\ 0 & 1 \end{pmatrix} & & \\ 0 & (\overline{\vartheta}_a - \overline{\vartheta}_a^\perp) \otimes \overline{a} & 0 \\ 0 & 0 & \overline{\vartheta}_a - \overline{\vartheta}_a^\perp \otimes \begin{pmatrix} \overline{a} & 0 \\ 0 & 0 \end{pmatrix} + \\ & & \overline{\vartheta} - \overline{\vartheta}^\perp + \frac{r}{s}\overline{\varepsilon^*\lambda\varepsilon} \otimes \begin{pmatrix} 0 & 0 \\ 0 & 1 \end{pmatrix} \end{bmatrix}.$$

This implies the assertion since

$$\Lambda + \underset{\sim}{\Lambda} \in \mathcal{C}^\infty(M, \mathbb{H}), \quad \Lambda - \underset{\sim}{\Lambda} \in \mathcal{C}^\infty(M, \mathbb{H}^\perp).$$

Now consider the differential

$$\boldsymbol{d} : \mathcal{A}_{q\ell}^{(1)} \to \pi_2(d\operatorname{Ker}\pi_1)^\perp, \quad \pi_1(A) \mapsto \pi_2(dA)^\perp \tag{9.47}$$

and product

$$\wedge : \mathcal{A}_{q\ell}^{(1)} \otimes \mathcal{A}_{q\ell}^{(1)} \to \pi_2(d\operatorname{Ker}\pi_1)^\perp, \quad \pi_1(A_1) \otimes \pi_1(A_2) \mapsto \pi_2(A_1 A_2)^\perp \tag{9.48}$$

computed in $\pi_2(d\operatorname{Ker}\pi_1)^\perp$. Applying Proposition 10, (9.39) and Proposition 13, we obtain

Corollary 2.

$$\boldsymbol{d}\begin{pmatrix} \omega_{q\ell} & 0 \\ 0 & \tilde{\omega}_{q\ell} \end{pmatrix} = \begin{pmatrix} \boldsymbol{d}\omega_{q\ell} & 0 \\ 0 & \boldsymbol{d}\tilde{\omega}_{q\ell} \end{pmatrix},$$

where $\boldsymbol{d}\omega_{q\ell}$ equals

$$\begin{bmatrix} \gamma(d\omega) \otimes \begin{pmatrix} 1 & 0 \\ 0 & 1 \end{pmatrix} + & \kappa^2(\varepsilon^*\underset{\sim}{\varphi} + \underset{\sim}{\psi}\varepsilon) \otimes \begin{pmatrix} M_d^* M_u \\ 0 \end{pmatrix} & \kappa^2\Gamma\gamma(\varepsilon^*\Omega - \omega\varepsilon^* - d\psi) \otimes \begin{pmatrix} M_d^* & 0 \\ 0 & M_e^* \end{pmatrix} \\ \kappa^2(\varepsilon^*\underset{\sim}{\varphi} + \underset{\sim}{\psi}\varepsilon) \otimes \begin{pmatrix} M_d^* M_d - \frac{r}{s} & 0 \\ 0 & M_e^* M_e - \frac{r}{s} \end{pmatrix} & & \\ \kappa^2(\varepsilon^*\underset{\sim}{\varphi} + \underset{\sim}{\psi}\varepsilon) \otimes \begin{pmatrix} M_u^* M_d, 0 \end{pmatrix} & \begin{matrix} \gamma(d\overline{\omega}) \otimes \underset{\sim}{1} + \\ \kappa^2(\varepsilon^*\underset{\sim}{\varphi} + \underset{\sim}{\psi}\varepsilon) \otimes M_u^* M_u - \frac{r}{s} \end{matrix} & \kappa^2\Gamma\gamma(\varepsilon^*\Omega - \overline{\omega}\varepsilon^* - d\psi) \otimes \begin{pmatrix} M_u^*, 0 \end{pmatrix} \\ & & \gamma(d\Omega) \otimes \begin{pmatrix} 1 & 0 \\ 0 & 1 \end{pmatrix} \\ \gamma(d\varphi + \Omega\varepsilon - \varepsilon\omega)\Gamma \otimes \begin{pmatrix} M_d & 0 \\ 0 & M_e \end{pmatrix} & \gamma(d\varphi + \Omega\underset{\sim}{\varepsilon} - \varepsilon\overline{\omega})\Gamma \otimes \begin{pmatrix} M_u \\ 0 \end{pmatrix} & \begin{matrix} + \\ \frac{\kappa^2}{2}(\varepsilon\underset{\sim}{\psi} + \varphi\varepsilon^* + \underset{\sim}{\varepsilon}\underset{\sim}{\psi} + \underset{\sim}{\varphi}\varepsilon^*) \otimes \\ \begin{pmatrix} M_d M_d^* + M_u M_u^* - \frac{r}{t} & 0 \\ 0 & M_e M_e^* - \frac{r}{t} \end{pmatrix} \end{matrix} \end{bmatrix}$$

and $\boldsymbol{d}\tilde{\omega}_{q\ell}$ equals

$$\begin{bmatrix} \tilde{\gamma}(d\overline{\omega}_a) \otimes \begin{pmatrix} \overline{a} & 0 \\ 0 & 0 \end{pmatrix} + \tilde{\gamma}(d\overline{\omega} - \kappa^2\frac{r}{s}(\varepsilon^*\underset{\sim}{\varphi} + \underset{\sim}{\psi}\varepsilon)) \otimes \begin{pmatrix} 0 & 0 \\ 0 & 1 \end{pmatrix} & 0 & 0 \\ 0 & \tilde{\gamma}(d\overline{\omega}_a) \otimes \overline{a} & 0 \\ 0 & 0 & \tilde{\gamma}(d\overline{\omega}_a) \otimes \begin{pmatrix} \overline{a} & 0 \\ 0 & 0 \end{pmatrix} \\ & & +\tilde{\gamma}(d\overline{\omega} - \kappa^2\frac{r}{s}(\varepsilon^*\underset{\sim}{\varphi} + \underset{\sim}{\psi}\varepsilon)) \otimes \begin{pmatrix} 0 & 0 \\ 0 & 1 \end{pmatrix} \end{bmatrix}$$

Similarly,
$$\begin{pmatrix} \omega_{q\ell}^0 & 0 \\ 0 & \tilde{\omega}_{q\ell}^0 \end{pmatrix} \wedge \begin{pmatrix} \omega_{q\ell}^1 & 0 \\ 0 & \tilde{\omega}_{q\ell}^1 \end{pmatrix} = \begin{pmatrix} \omega_{q\ell}^0 \wedge \omega_{q\ell}^1 & 0 \\ 0 & \tilde{\omega}_{q\ell}^0 \wedge \tilde{\omega}_{q\ell}^1 \end{pmatrix},$$

where $\omega_{q\ell}^0 \wedge \omega_{q\ell}^1$ equals

$$\begin{bmatrix} \gamma(\omega^0 \wedge \omega^1) \otimes \begin{pmatrix} 1 & 0 \\ 0 & 1 \end{pmatrix} + & \kappa^2 \psi^0 \varphi^1 \otimes \begin{pmatrix} M_d^* M_u \\ 0 \end{pmatrix} & \kappa^2 \Gamma \gamma(\psi^0 \Omega^1 - \omega^0 \psi^1) \otimes \begin{pmatrix} M_d^* & 0 \\ 0 & M_e^* \end{pmatrix} \\ \kappa^2 \psi^0 \varphi^1 \otimes \begin{pmatrix} M_d^* M_d - \frac{r}{s} & 0 \\ 0 & M_e^* M_e - \frac{r}{s} \end{pmatrix} & & \\ \kappa^2 \psi^0 \varphi^1 \otimes (M_u^* M_d, 0) & \begin{matrix} \gamma(\overline{\omega}^0 \wedge \overline{\omega}^1) \otimes \underline{1} + \\ \kappa^2 \psi^0 \varphi^1 \otimes \underline{M_u^* M_u - \frac{r}{s}} \end{matrix} & \kappa^2 \Gamma \gamma(\psi^0 \Omega^1 - \overline{\omega}^0 \psi^1) \otimes (M_u^*, 0) \\ & & \gamma(\Omega^0 \wedge \Omega^1) \otimes \begin{pmatrix} 1 & 0 \\ 0 & 1 \end{pmatrix} + \\ \gamma(\Omega^0 \varphi^1 - \varphi^0 \omega^1) \Gamma \otimes \begin{pmatrix} M_d & 0 \\ 0 & M_e \end{pmatrix} & \gamma(\Omega^0 \varphi^1 - \varphi^0 \overline{\omega}^1) \Gamma \otimes \begin{pmatrix} M_u \\ 0 \end{pmatrix} & \frac{\kappa^2}{2}(\varphi^0 \psi^1 + \varphi^0 \psi^1) \\ & & \otimes \begin{pmatrix} M_d M_d^* + M_u M_u^* - \frac{r}{t} & 0 \\ 0 & M_e M_e^* - \frac{r}{t} \end{pmatrix} \end{bmatrix}$$

and $\tilde{\omega}_{q\ell}^0 \wedge \tilde{\omega}_{q\ell}^1$ equals

$$\begin{bmatrix} \tilde{\gamma}(\overline{\omega}_b^0 \wedge \overline{\omega}_c^1) \otimes \begin{pmatrix} \overline{bc} & 0 \\ 0 & 0 \end{pmatrix} + & 0 & 0 \\ \tilde{\gamma}(\overline{\omega}^0 \wedge \overline{\omega}^1 - \kappa^2 \frac{r}{s} \psi^0 \varphi^1) \otimes \begin{pmatrix} 0 & 0 \\ 0 & 1 \end{pmatrix} & & \\ 0 & \tilde{\gamma}(\overline{\omega}_b^0 \wedge \overline{\omega}_c^1) \otimes \overline{bc} & 0 \\ 0 & 0 & \tilde{\gamma}(\overline{\omega}_b^0 \wedge \overline{\omega}_c^1) \otimes \begin{pmatrix} \overline{bc} & 0 \\ 0 & 0 \end{pmatrix} + \\ & & \tilde{\gamma}(\overline{\omega}^0 \wedge \overline{\omega}^1 - \kappa^2 \frac{r}{s} \psi^0 \varphi^1) \otimes \begin{pmatrix} 0 & 0 \\ 0 & 1 \end{pmatrix} \end{bmatrix}.$$

We are now ready to introduce the "dynamical variable" for the Standard Model, phrased in terms of the noncommutative geometry $\mathcal{A}_{q\ell}$.

Proposition 14. *For* $\begin{pmatrix} \omega_{q\ell} & 0 \\ 0 & \tilde{\omega}_{q\ell} \end{pmatrix} \in \mathcal{A}_{q\ell}^{(1)}$ *we have*

$$\begin{pmatrix} \omega_{q\ell} & 0 \\ 0 & \tilde{\omega}_{q\ell} \end{pmatrix} = -\begin{pmatrix} 0 & J_{q\ell} \\ -\kappa^2 J_{q\ell} & 0 \end{pmatrix} \begin{pmatrix} \omega_{q\ell} & 0 \\ 0 & \tilde{\omega}_{q\ell} \end{pmatrix} \begin{pmatrix} 0 & J_{q\ell} \\ -\kappa^2 J_{q\ell} & 0 \end{pmatrix}^{-1} \quad (9.49)$$

if and only if

$$\begin{pmatrix} \omega_{q\ell} & 0 \\ 0 & \tilde{\omega}_{q\ell} \end{pmatrix} = \hat{\rho}, \quad (9.50)$$

where $\rho = \overline{\rho} \in \mathcal{C}^\infty(T^\# M)$ and we define $\hat{\rho}$ as

$$\begin{bmatrix} \gamma(\rho)\otimes \begin{pmatrix} \frac{1}{\bullet} & 0 \\ 0 & 1 \end{pmatrix} & & & & & \\ & \gamma(\rho)\otimes\underline{1} & & & & \\ & & \underline{\underline{\gamma(\rho)}}\otimes \begin{pmatrix} \frac{1}{\bullet} & 0 \\ 0 & 1 \end{pmatrix} & & & \\ & & & \tilde{\gamma}(\rho)\otimes \begin{pmatrix} \frac{1}{\bullet} & 0 \\ 0 & 1 \end{pmatrix} & & \\ & & & & \tilde{\gamma}(\rho)\otimes\underline{1} & \\ & & & & & \underline{\underline{\tilde{\gamma}(\rho)}}\otimes \begin{pmatrix} \frac{1}{\bullet} & 0 \\ 0 & 1 \end{pmatrix} \end{bmatrix}.$$

Proof. Using Proposition 8, it is clear that a 1-form given as in Proposition 7 satisfies (9.49) if and only if the following conditions hold

$$\omega_a \otimes \begin{pmatrix} a & 0 \\ 0 & 0 \end{pmatrix} + \omega \otimes \begin{pmatrix} 0 & 0 \\ 0 & 1 \end{pmatrix} = \omega \otimes \begin{pmatrix} 1 & 0 \\ 0 & 1 \end{pmatrix}, \quad \omega_a \otimes a = \overline{\omega} \otimes \underline{1},$$

$$\underline{\underline{\omega_a}} \otimes \begin{pmatrix} a & 0 \\ 0 & 0 \end{pmatrix} + \underline{\underline{\omega}} \otimes \begin{pmatrix} 0 & 0 \\ 0 & 1 \end{pmatrix} = \Omega \otimes \begin{pmatrix} 1 & 0 \\ 0 & 1 \end{pmatrix}, \quad \varphi = 0, \quad \psi = 0.$$

As a consequence, we obtain

$$\omega_a \otimes a = \omega \otimes \underline{1} = \overline{\omega} \otimes \underline{1}, \quad \Omega = \underline{\underline{\omega}}.$$

Therefore $\omega = \overline{\omega} =: \rho$ and we get

$$\omega \otimes \begin{pmatrix} 1 & 0 \\ 0 & 1 \end{pmatrix} = \rho \otimes \begin{pmatrix} 1 & 0 \\ 0 & 1 \end{pmatrix}, \quad \overline{\omega}\otimes\underline{1} = \rho\otimes\underline{1}, \quad \Omega \otimes \begin{pmatrix} 1 & 0 \\ 0 & 1 \end{pmatrix} = \underline{\underline{\rho}} \otimes \begin{pmatrix} 1 & 0 \\ 0 & 1 \end{pmatrix},$$

$$\overline{\omega}_a\otimes \begin{pmatrix} \overline{a} & 0 \\ 0 & 0 \end{pmatrix} + \overline{\omega}\otimes \begin{pmatrix} 0 & 0 \\ 0 & 1 \end{pmatrix} = \overline{\omega}\otimes \begin{pmatrix} 1 & 0 \\ 0 & 1 \end{pmatrix} = \rho\otimes \begin{pmatrix} 1 & 0 \\ 0 & 1 \end{pmatrix}, \quad \overline{\omega}_a\otimes\overline{a}=\omega\otimes\underline{1}=\rho\otimes\underline{1},$$

$$\underline{\underline{\overline{\omega}_a}} \otimes \begin{pmatrix} \overline{a} & 0 \\ 0 & 0 \end{pmatrix} + \underline{\underline{\overline{\omega}}} \otimes \begin{pmatrix} 0 & 0 \\ 0 & 1 \end{pmatrix} = \overline{\Omega} \otimes \begin{pmatrix} 0 & 0 \\ 0 & 1 \end{pmatrix} = \underline{\underline{\overline{\rho}}} \otimes \begin{pmatrix} 1 & 0 \\ 0 & 1 \end{pmatrix} = \underline{\underline{\rho}} \otimes \begin{pmatrix} 1 & 0 \\ 0 & 1 \end{pmatrix}.$$

Definition 4. *The skew-adjoint elements* $\begin{pmatrix} \omega_{q\ell} & 0 \\ 0 & \tilde{\omega}_{q\ell} \end{pmatrix} \in \mathcal{A}^{(1)}_{q\ell}$ *which satisfy the identity*

$$\left(\hat{\rho} \mid \omega_{q\ell} I_{q\ell}\right)_{S_{q\ell}} = \left(\hat{\rho} \mid \tilde{\omega}_{q\ell} \tilde{I}_{q\ell}\right)_{S_{q\ell}} \tag{9.51}$$

for all $\hat{\rho} \in \mathcal{A}^{(1)}_{q\ell}$ *are called* NC gauge fields *for the Standard Model.*

In the following we make the crucial assumption

$$4\,\underline{\text{tr}}_\mathbb{G}\,\tilde{x} = \underline{\text{tr}}_\mathbb{G}(y + 3\tilde{y}). \tag{9.52}$$

Proposition 15. *The NC gauge fields for the Standard Model are given by* $\omega_{q\ell}$ *equal to*

$$\begin{bmatrix} \gamma(\omega) \otimes \begin{pmatrix} 1 & 0 \\ 0 & 1 \end{pmatrix} & 0 & \kappa^2 \Gamma \varphi^* \otimes \begin{pmatrix} M_d^* & 0 \\ 0 & M_e^* \end{pmatrix} \\ 0 & -\gamma(\omega) \otimes \underline{1} & \kappa^2 \Gamma \varphi^* \otimes (\underset{\sim}{M_u^*}, 0) \\ \varphi\Gamma \otimes \begin{pmatrix} M_d & 0 \\ 0 & M_e \end{pmatrix} & \varphi\Gamma \otimes \begin{pmatrix} M_u \\ 0 \end{pmatrix} & \gamma(\Omega) \otimes \begin{pmatrix} 1 & 0 \\ 0 & 1 \end{pmatrix} \end{bmatrix}$$

and $\tilde{\omega}_{q\ell}$ *equal to*

$$\begin{bmatrix} \tilde{\gamma}(\overline{\omega}_a) \otimes \begin{pmatrix} \overline{a} & 0 \\ 0 & 0 \end{pmatrix} + \\ \tilde{\gamma}(\omega) \otimes \begin{pmatrix} \frac{1}{3} & 0 \\ 0 & -1 \end{pmatrix} & 0 & 0 \\ 0 & \tilde{\gamma}(\overline{\omega}_a) \otimes \overline{a} + \tilde{\gamma}(\omega) \otimes \frac{1}{3} & 0 \\ 0 & 0 & \underline{\tilde{\gamma}(\overline{\omega}_a)} \otimes \begin{pmatrix} \overline{a} & 0 \\ 0 & 0 \end{pmatrix} + \\ & & \underline{\tilde{\gamma}(\omega)} \otimes \begin{pmatrix} \frac{1}{3} & 0 \\ 0 & -1 \end{pmatrix} \end{bmatrix}$$

with arbitrary

$$\begin{aligned} \omega \in \mathcal{C}^\infty(T^\# M \otimes \mathfrak{u}(1)), \quad \Omega \in \mathcal{C}^\infty(T^\# M \otimes \mathfrak{su}(2)), \\ \omega_a \otimes a \in \mathcal{C}^\infty(T^\# M \otimes \mathfrak{su}(3)), \quad \varphi \in \mathcal{C}^\infty(M, \mathbb{C}^{2\times 1}). \end{aligned} \qquad (9.53)$$

Proof. Since

$$\begin{bmatrix} \gamma(\omega) \otimes \begin{pmatrix} 1 & 0 \\ 0 & 1 \end{pmatrix} & 0 & \kappa^2 \Gamma \psi \otimes \begin{pmatrix} M_d^* & 0 \\ 0 & M_e^* \end{pmatrix} \\ 0 & \gamma(\overline{\omega}) \otimes \underline{1} & \kappa^2 \Gamma \psi \otimes (\underset{\sim}{M_u^*}, 0) \\ \varphi\Gamma \otimes \begin{pmatrix} M_d & 0 \\ 0 & M_e \end{pmatrix} & \varphi\Gamma \otimes \begin{pmatrix} M_u \\ 0 \end{pmatrix} & \gamma(\Omega) \otimes \begin{pmatrix} 1 & 0 \\ 0 & 1 \end{pmatrix} \end{bmatrix}^\dagger$$

$$= \begin{bmatrix} \gamma(\omega)^\dagger \otimes \begin{pmatrix} 1 & 0 \\ 0 & 1 \end{pmatrix} & 0 & \Gamma^\dagger \varphi^\dagger \otimes \begin{pmatrix} M_d^* & 0 \\ 0 & M_e^* \end{pmatrix} \\ 0 & \gamma(\overline{\omega})^\dagger \otimes \underline{1} & \Gamma^\dagger \varphi^\dagger \otimes (\underset{\sim}{M_u^*}, 0) \\ \kappa^2 \psi^\dagger \Gamma^\dagger \otimes \begin{pmatrix} M_d & 0 \\ 0 & M_e \end{pmatrix} & \kappa^2 \psi^\dagger \Gamma^\dagger \otimes \begin{pmatrix} M_u \\ 0 \end{pmatrix} & \gamma(\Omega)^\dagger \otimes \begin{pmatrix} 1 & 0 \\ 0 & 1 \end{pmatrix} \end{bmatrix}$$

$$= \begin{bmatrix} \gamma(\overline{\omega}) \otimes \begin{pmatrix} 1 & 0 \\ 0 & 1 \end{pmatrix} & 0 & -\kappa^2 \Gamma \varphi^* \otimes \begin{pmatrix} M_d^* & 0 \\ 0 & M_e^* \end{pmatrix} \\ 0 & \gamma(\omega) \otimes \underline{1} & -\kappa^2 \Gamma \varphi^* \otimes (\underline{M_u^*}, 0) \\ -\psi^* \Gamma \otimes \begin{pmatrix} M_d & 0 \\ 0 & M_e \end{pmatrix} & -\psi^* \Gamma \otimes \begin{pmatrix} \underline{M_u} \\ 0 \end{pmatrix} & \gamma(\Omega^*) \otimes \begin{pmatrix} 1 & 0 \\ 0 & 1 \end{pmatrix} \end{bmatrix}$$

and

$$\begin{bmatrix} \tilde{\gamma}(\overline{\omega}_a) \otimes \begin{pmatrix} \overline{a} & 0 \\ 0 & 0 \end{pmatrix} + \tilde{\gamma}(\overline{\omega}) \otimes \begin{pmatrix} 0 & 0 \\ 0 & 1 \end{pmatrix} & 0 & 0 \\ 0 & \tilde{\gamma}(\overline{\omega}_a) \otimes \overline{a} & 0 \\ 0 & 0 & \underline{\tilde{\gamma}(\overline{\omega}_a)} \otimes \begin{pmatrix} \overline{a} & 0 \\ 0 & 0 \end{pmatrix} + \underline{\tilde{\gamma}(\overline{\omega})} \otimes \begin{pmatrix} 0 & 0 \\ 0 & 1 \end{pmatrix} \end{bmatrix}^\dagger$$

$$= \begin{bmatrix} \tilde{\gamma}(\overline{\omega}_a)^\dagger \otimes \begin{pmatrix} \overline{a}^* & 0 \\ 0 & 0 \end{pmatrix} + \tilde{\gamma}(\overline{\omega})^\dagger \otimes \begin{pmatrix} 0 & 0 \\ 0 & 1 \end{pmatrix} & 0 & 0 \\ 0 & \tilde{\gamma}(\overline{\omega}_a)^\dagger \otimes \overline{a}^* & 0 \\ 0 & 0 & \underline{\tilde{\gamma}(\overline{\omega}_a)^\dagger} \otimes \begin{pmatrix} \overline{a}^* & 0 \\ 0 & 0 \end{pmatrix} + \underline{\tilde{\gamma}(\overline{\omega})^\dagger} \otimes \begin{pmatrix} 0 & 0 \\ 0 & 1 \end{pmatrix} \end{bmatrix}$$

$$= \begin{bmatrix} \tilde{\gamma}(\omega_a) \otimes \begin{pmatrix} a^t & 0 \\ 0 & 0 \end{pmatrix} + \tilde{\gamma}(\omega) \otimes \begin{pmatrix} 0 & 0 \\ 0 & 1 \end{pmatrix} & 0 & 0 \\ 0 & \tilde{\gamma}(\omega_a) \otimes a^t & 0 \\ 0 & 0 & \underline{\tilde{\gamma}(\omega_a)} \otimes \begin{pmatrix} a^t & 0 \\ 0 & 0 \end{pmatrix} + \underline{\tilde{\gamma}(\omega)} \otimes \begin{pmatrix} 0 & 0 \\ 0 & 1 \end{pmatrix} \end{bmatrix}$$

it follows that $\begin{pmatrix} \omega_{q\ell} & 0 \\ 0 & \tilde{\omega}_{q\ell} \end{pmatrix}^\dagger = -\begin{pmatrix} \omega_{q\ell} & 0 \\ 0 & \tilde{\omega}_{q\ell} \end{pmatrix}$ if and only if

$$\psi = \varphi^*, \quad \overline{\omega} = -\omega, \quad \Omega^* = -\Omega, \quad \overline{\omega}_a \otimes \overline{a} = -\omega_a \otimes a^t.$$

Thus the skew-adjoint elements in $\mathcal{A}_{q\ell}^{(1)}$ have the form

$$\begin{pmatrix} \omega_{q\ell} & 0 \\ 0 & \tilde{\omega}_{q\ell} \end{pmatrix},$$

The Full Standard Model

where $\omega_{q\ell}$ and $\tilde{\omega}_{q\ell}$ are given by

$$\begin{bmatrix} \gamma(\omega) \otimes \begin{pmatrix} 1 & 0 \\ 0 & 1 \end{pmatrix} & 0 & \kappa^2 \Gamma \varphi^* \otimes \begin{pmatrix} M_d^* & 0 \\ 0 & M_e^* \end{pmatrix} \\ 0 & -\gamma(\omega) \otimes \underline{1} & \kappa^2 \Gamma \varphi^* \otimes (\underline{M_u^*}, 0) \\ \varphi \Gamma \otimes \begin{pmatrix} M_d & 0 \\ 0 & M_e \end{pmatrix} & \underset{\sim}{\varphi \Gamma} \otimes \begin{pmatrix} M_u \\ 0 \end{pmatrix} & \gamma(\Omega) \otimes \begin{pmatrix} 1 & 0 \\ 0 & 1 \end{pmatrix} \end{bmatrix}$$

and

$$\begin{bmatrix} \tilde{\gamma}(\overline{\omega}_a) \otimes \begin{pmatrix} \overline{a} & 0 \\ 0 & 0 \end{pmatrix} - \tilde{\gamma}(\omega) \otimes \begin{pmatrix} 0 & 0 \\ 0 & 1 \end{pmatrix} & 0 & 0 \\ 0 & \tilde{\gamma}(\overline{\omega}_a) \otimes \overline{a} & 0 \\ 0 & 0 & \underline{\tilde{\gamma}(\overline{\omega}_a)} \otimes \begin{pmatrix} \overline{a} & 0 \\ 0 & 0 \end{pmatrix} - \underline{\tilde{\gamma}(\omega)} \otimes \begin{pmatrix} 0 & 0 \\ 0 & 1 \end{pmatrix} \end{bmatrix}$$

respectively, with arbitrary

$$\varphi \in \mathcal{C}^\infty(M, \mathbb{C}^{2\times 1}), \quad \omega \in \mathcal{C}^\infty(T^\# M \otimes \mathfrak{u}(1)),$$
$$\Omega \in \mathcal{C}^\infty(T^\# M \otimes \mathfrak{su}(2)), \quad \omega_a \otimes a \in \mathcal{C}^\infty(T^\# M \otimes \mathfrak{u}(3)).$$

Now let $\rho \in \mathcal{C}^\infty(T^\# M)$ be arbitrary. Then $\hat{\rho}\,\omega_{q\ell}\,I_{q\ell}$ equals

$$\begin{bmatrix} \gamma(\rho) \otimes \begin{pmatrix} 1 & 0 \\ 0 & 1 \end{pmatrix} & 0 & 0 \\ 0 & \gamma(\rho) \otimes \underline{1} & 0 \\ 0 & 0 & \gamma(\rho) \otimes \begin{pmatrix} 1 & 0 \\ 0 & 1 \end{pmatrix} \end{bmatrix} \begin{bmatrix} \gamma(\omega) \otimes \begin{pmatrix} 1 & 0 \\ 0 & 1 \end{pmatrix} & 0 & \kappa^2 \Gamma \varphi^* \otimes \begin{pmatrix} M_d^* & 0 \\ 0 & M_e^* \end{pmatrix} \\ 0 & -\gamma(\omega) \otimes \underline{1} & \kappa^2 \Gamma \varphi^* \otimes (\underline{M_u^*}, 0) \\ \varphi \Gamma \otimes \begin{pmatrix} M_d & 0 \\ 0 & M_e \end{pmatrix} & \underset{\sim}{\varphi \Gamma} \otimes \begin{pmatrix} M_u \\ 0 \end{pmatrix} & \gamma(\Omega) \otimes \begin{pmatrix} 1 & 0 \\ 0 & 1 \end{pmatrix} \end{bmatrix}$$

$$\times \begin{bmatrix} I \otimes \begin{pmatrix} \frac{x}{3} & 0 \\ 0 & y \end{pmatrix} & 0 & 0 \\ 0 & I \otimes \underline{\frac{x}{3}} & 0 \\ 0 & 0 & \underline{I} \otimes \begin{pmatrix} \frac{x}{3} & 0 \\ 0 & y \end{pmatrix} \end{bmatrix}$$

$$= \begin{bmatrix} \gamma(\rho\times\omega) \otimes \begin{pmatrix} \frac{x}{3} & 0 \\ 0 & y \end{pmatrix} & 0 & \kappa^2 \Gamma\rho\varphi^* \otimes \begin{pmatrix} M_d^* \frac{x}{3} & 0 \\ 0 & M_e^* y \end{pmatrix} \\ 0 & -\gamma(\rho\times\omega) \otimes \frac{x}{3} & \kappa^2 \Gamma\rho\varphi^* \otimes \left(M_u^* \frac{x}{3}, 0\right) \\ \underline{\rho\varphi}\Gamma \otimes \begin{pmatrix} M_d \frac{x}{3} & 0 \\ 0 & M_e y \end{pmatrix} & \underline{\rho\varphi}\Gamma \otimes \begin{pmatrix} M_u \frac{x}{3} \\ 0 \end{pmatrix} & \gamma(\underline{\rho}\times\Omega) \otimes \begin{pmatrix} \frac{x}{3} & 0 \\ 0 & y \end{pmatrix} \end{bmatrix}$$

and $\hat{\rho}\tilde{\omega}_{q\ell}\tilde{I}_{q\ell}$ equals

$$\begin{bmatrix} \tilde{\gamma}(\rho) \otimes \begin{pmatrix} 1 & 0 \\ 0 & 1 \end{pmatrix} & 0 & 0 \\ 0 & \tilde{\gamma}(\rho) \otimes \underline{1} & 0 \\ 0 & 0 & \tilde{\gamma}(\underline{\rho}) \otimes \begin{pmatrix} 1 & 0 \\ 0 & 1 \end{pmatrix} \end{bmatrix} \begin{bmatrix} \tilde{\gamma}(\overline{\omega}_a) \otimes \begin{pmatrix} \overline{a} & 0 \\ 0 & 0 \end{pmatrix} & 0 & 0 \\ -\tilde{\gamma}(\omega) \otimes \begin{pmatrix} 0 & 0 \\ 0 & 1 \end{pmatrix} & & \\ 0 & \tilde{\gamma}(\overline{\omega}_a) \otimes \overline{a} & 0 \\ 0 & 0 & \tilde{\gamma}(\underline{\overline{\omega}_a}) \otimes \begin{pmatrix} \overline{a} & 0 \\ 0 & 0 \end{pmatrix} \\ & & -\tilde{\gamma}(\underline{\omega}) \otimes \begin{pmatrix} 0 & 0 \\ 0 & 1 \end{pmatrix} \end{bmatrix}$$

$$\times \begin{bmatrix} I \otimes \begin{pmatrix} \tilde{x} & 0 \\ 0 & \tilde{y} \end{pmatrix} & 0 & 0 \\ 0 & I \otimes \tilde{\underline{x}} & 0 \\ 0 & 0 & \underline{I} \otimes \begin{pmatrix} \tilde{x} & 0 \\ 0 & \tilde{y} \end{pmatrix} \end{bmatrix}$$

$$= \begin{bmatrix} \tilde{\gamma}(\rho\times\overline{\omega}_a) \otimes \begin{pmatrix} \overline{a}\tilde{x} & 0 \\ 0 & 0 \end{pmatrix} & 0 & 0 \\ -\tilde{\gamma}(\rho\times\omega) \otimes \begin{pmatrix} 0 & 0 \\ 0 & \tilde{y} \end{pmatrix} & & \\ 0 & \tilde{\gamma}(\rho\times\overline{\omega}_a) \otimes \overline{a}\tilde{x} & 0 \\ 0 & 0 & \tilde{\gamma}(\underline{\rho\times\overline{\omega}_a}) \otimes \begin{pmatrix} \overline{a}\tilde{x} & 0 \\ 0 & 0 \end{pmatrix} \\ & & -\tilde{\gamma}(\underline{\rho\times\omega}) \otimes \begin{pmatrix} 0 & 0 \\ 0 & \tilde{y} \end{pmatrix} \end{bmatrix}.$$

Taking the trace, we obtain

$$\begin{aligned} \left(\hat{\rho} \mid \omega_{q\ell} I_{q\ell}\right)_{S_{q\ell}} &= \underline{\mathrm{tr}}_{S_{q\ell}} \hat{\rho} \omega_{q\ell} I_{q\ell} \\ &= \underline{\mathrm{tr}}_S \gamma(\rho\times\omega) \, \underline{\mathrm{tr}}_{\mathbb{G}} y + \underline{\mathrm{tr}}_{\underline{S}} \gamma(\underline{\rho}\times\Omega) \, \underline{\mathrm{tr}}_{\mathbb{G}} (x+y) \\ &= \left(\rho \mid \omega\right)_S \underline{\mathrm{tr}}_{\mathbb{G}} y \end{aligned}$$

since $(\rho \mid \Omega)_{\underline{S}} = 0$, and

$$\begin{aligned}
(\hat{\rho} \mid \tilde{\omega}_{q\ell}\tilde{I}_{q\ell})_{S_{q\ell}} &= \underline{\text{tr}}_{S_{q\ell}}\hat{\rho}\tilde{\omega}_{q\ell}\tilde{I}_{q\ell} \\
&= 4\underline{\text{tr}}_S\tilde{\gamma}(\rho \times \overline{\omega}_a) \underset{\underline{\mathbb{C}}}{\text{tr}}\,\overline{a}\,\underline{\text{tr}}_\mathbb{G}\tilde{x} - 3\underline{\text{tr}}_S\tilde{\gamma}(\rho \times \omega)\,\underline{\text{tr}}_\mathbb{G}\tilde{y} \\
&= 4(\rho \mid \overline{\omega}_a)_S \underset{\underline{\mathbb{C}}}{\text{tr}}\,\overline{a}\,\underline{\text{tr}}_\mathbb{G}\tilde{x} - 3(\rho \mid \omega)_S\,\underline{\text{tr}}_\mathbb{G}\tilde{y}.
\end{aligned}$$

Assuming the equality (9.51), this implies

$$4(\rho \mid \overline{\omega}_a)_S\,\underset{\underline{\mathbb{C}}}{\text{tr}}\,\overline{a}\,\underline{\text{tr}}_\mathbb{G}\tilde{x} = (\rho \mid \omega)_S\,\underline{\text{tr}}_\mathbb{G}(y+3\tilde{y}).$$

Now the basic assumption (9.52) implies

$$(\rho \mid \overline{\omega}_a)_S\,\underset{\underline{\mathbb{C}}}{\text{tr}}\,\overline{a} = (\rho \mid \omega)_S$$

for all $\rho = \overline{\rho} \in \mathcal{C}^\infty(T^\# M)$. This yields $\overline{\omega}_a\,\text{tr}_{\underline{\mathbb{C}}}\,\overline{a} = \omega$, $\omega_a\,\text{tr}_{\underline{\mathbb{C}}}\,a = \overline{\omega} = -\omega$ and therefore

$$\omega' := \omega_a \otimes a + \omega \otimes \tfrac{1}{3} \in \mathcal{C}^\infty(T^\# M \otimes \mathfrak{su}(3)).$$

Thus we may write

$$\omega' = \omega_{a'} \otimes a'$$

with $a' \in \mathfrak{su}(3)$.

Remark 3. *The NC gauge fields as defined above parametrize exactly all the bosonic fields of the Standard Model, which therefore occur naturally as components of a single object. This is one of the major conceptual advantages of the noncommutative geometry approach to the Standard Model. The bosonic (spin 1) gauge fields mediating the strong interaction are given mathematically by Lie algebra valued 1-forms*

$$i\sum_{a=1}^{8} G^{(a)}\lambda^{(a)} \in \mathcal{C}^\infty(T^\# M \otimes \mathfrak{su}(3)), \tag{9.54}$$

where $G^{(1)}, \ldots, G^{(8)} \in \mathcal{C}^\infty(T^\# M)$ are real spin 1 fields (gluons) and $\lambda^{(1)}, \ldots, \lambda^{(8)}$ are the so-called Gell-Mann matrices forming a basis of $i \cdot \mathfrak{su}(3)$ satisfying $\text{tr}\,\lambda^{(a)}\lambda^{(b)} = \tfrac{1}{2}\delta^{(a)(b)}$. In addition, one has the bosonic fields (8.44) and (8.45) which experience only electro-weak interactions. Putting

$$\omega = \frac{ig_1}{2}W^{(0)}$$

$$\Omega = \frac{ig_2}{2}\begin{pmatrix} -W^{(3)} & W^{(1)}+iW^{(2)} \\ W^{(1)}-iW^{(2)} & W^{(3)} \end{pmatrix} = ig_2\sum_{a=1}^{3}W^{(a)}\tau^{(a)}$$

$$\omega_a \otimes a = ig_3\sum_{a=1}^{8}G^{(a)}\lambda^{(a)}, \tag{9.55}$$

and
$$\varphi + \varepsilon = \frac{1}{\sqrt{2r}} \begin{pmatrix} \Phi^0 \\ \Phi^+ \end{pmatrix},$$

it follows that (9.53) encodes all the physical fields $W^{(0)},...,W^{(3)}$, $G^{(1)},...,G^{(8)}$, Φ^0, Φ^+ of the Standard Model. The so-called "coupling constants" $g_1, g_2, g_3 > 0$ introduced here will later be expressed in terms of the basic parameters s, t and $\tilde{x}$ of the model (cf. (9.62)).

9.3 Noncommutative Gauge Action Functional

In order to compute the Yang-Mills functional recall that the *curvature* of A is algebraically given by $dA + A^2$. However, since we are working in $\pi_2(d\operatorname{Ker}\pi_1)^\perp$ (i.e., modulo the "junk ideal" $d\operatorname{Ker}\pi_1$), we have to compute the differential and square in $\pi_2(d\operatorname{Ker}\pi_1)^\perp$. Accordingly, we define the *curvature* of $\begin{pmatrix} \omega_{q\ell} & 0 \\ 0 & \tilde{\omega}_{q\ell} \end{pmatrix}$ as

$$F\begin{pmatrix} \omega_{q\ell} & 0 \\ 0 & \tilde{\omega}_{q\ell} \end{pmatrix} = d\begin{pmatrix} \omega_{q\ell} & 0 \\ 0 & \tilde{\omega}_{q\ell} \end{pmatrix} + \begin{pmatrix} \omega_{q\ell} & 0 \\ 0 & \tilde{\omega}_{q\ell} \end{pmatrix} \wedge \begin{pmatrix} \omega_{q\ell} & 0 \\ 0 & \tilde{\omega}_{q\ell} \end{pmatrix} \in \pi_2(d\operatorname{Ker}\pi_1)^\perp \tag{9.56}$$

Theorem 1. *Let* $\begin{pmatrix} \omega_{q\ell} & 0 \\ 0 & \tilde{\omega}_{q\ell} \end{pmatrix} \in \mathcal{A}_{q\ell}^{(1)}$ *be a NC gauge field. Then*

$$F\begin{pmatrix} \omega_{q\ell} & 0 \\ 0 & \tilde{\omega}_{q\ell} \end{pmatrix} = \begin{pmatrix} F(\omega_{q\ell}) & 0 \\ 0 & F(\tilde{\omega}_{q\ell}) \end{pmatrix}, \tag{9.57}$$

where $F(\omega_{q\ell})$ *equals*

$$\begin{bmatrix} \gamma(F(\omega)) \otimes \begin{pmatrix} 1 & 0 \\ 0 & 1 \end{pmatrix} \\ + \\ \kappa^2(\|\varphi+\varepsilon\|^2 - 1)\otimes \\ \begin{pmatrix} M_d^2 - \frac{r}{s} & 0 \\ 0 & M_e^2 - \frac{r}{s} \end{pmatrix} & 0 & -\kappa^2 \Gamma\gamma(\nabla^H(\varphi+\varepsilon))^* \otimes \begin{pmatrix} M_d^* & 0 \\ 0 & M_e^* \end{pmatrix} \\ 0 & \begin{matrix} -\gamma(F(\omega)) \otimes \underline{1} \\ + \\ \kappa^2(\|\varphi+\varepsilon\|^2 - 1) \otimes \underline{M_u^2 - \frac{r}{s}} \end{matrix} & -\kappa^2 \Gamma\gamma(\nabla^H(\varphi+\varepsilon))^* \otimes (\underline{M_u^*}, 0) \\ \gamma(\nabla^H(\varphi+\varepsilon))\Gamma \otimes \begin{pmatrix} M_d & 0 \\ 0 & M_e \end{pmatrix} & \gamma((\nabla^H(\varphi+\varepsilon)_\sim)\Gamma \otimes \begin{pmatrix} M_u \\ 0 \end{pmatrix} & \begin{matrix} \gamma(F(\Omega)) \otimes \begin{pmatrix} 1 & 0 \\ 0 & 1 \end{pmatrix} \\ + \\ \frac{\kappa^2}{2}(\|\varphi+\varepsilon\|^2 - 1)\otimes \\ \begin{pmatrix} M_d^2 + M_u^2 - \frac{r}{t} & 0 \\ 0 & M_e^2 - \frac{r}{t} \end{pmatrix} \end{matrix} \end{bmatrix}$$

and $\boldsymbol{F}(\tilde{\omega}_{q\ell})$ equals

$$\begin{bmatrix} \tilde{\gamma}(d\overline{\omega}_a) \otimes \begin{pmatrix} \overline{a} & 0 \\ 0 & 0 \end{pmatrix} + \tilde{\gamma}(d\omega) \otimes \begin{pmatrix} \frac{1}{3} & 0 \\ 0 & -1 \end{pmatrix} \\ + \\ \tilde{\gamma}(\overline{\omega}_a \wedge \overline{\omega}_b) \otimes \begin{pmatrix} \overline{ab} & 0 \\ 0 & 0 \end{pmatrix} & 0 & 0 \\ + \\ -\kappa^2 \frac{r}{s}(\|\varphi + \varepsilon\|^2 - 1) \otimes \begin{pmatrix} 0 & 0 \\ 0 & 1 \end{pmatrix} \\[1em] & \tilde{\gamma}(\overline{\omega}_a) \otimes \overline{a} \\ & + \\ 0 & \tilde{\gamma}(d\omega) \otimes \frac{1}{3} & 0 \\ & + \\ & \tilde{\gamma}(\overline{\omega}_a \wedge \overline{\omega}_b) \otimes \overline{ab} \\[1em] & & \underline{\tilde{\gamma}(d\overline{\omega}_a)} \otimes \begin{pmatrix} \overline{a} & 0 \\ 0 & 0 \end{pmatrix} + \underline{\tilde{\gamma}(d\omega)} \otimes \begin{pmatrix} \frac{1}{3} & 0 \\ 0 & -1 \end{pmatrix} \\ & & + \\ 0 & 0 & \underline{\tilde{\gamma}(\overline{\omega}_a \wedge \overline{\omega}_b)} \otimes \begin{pmatrix} \overline{ab} & 0 \\ 0 & 0 \end{pmatrix} \\ & & + \\ & & -\kappa^2 \frac{r}{s}\underline{(\|\varphi + \varepsilon\|^2 - 1)} \otimes \begin{pmatrix} 0 & 0 \\ 0 & 1 \end{pmatrix} \end{bmatrix}.$$

Here

$$F(\omega) = d\omega, \quad F(\Omega) = d\Omega + \Omega \wedge \Omega,$$
$$F(\omega_a \otimes a) = d\omega_a \otimes a + \omega_a \wedge \omega_b \otimes ab \tag{9.58}$$

are the respective curvature 2-forms (field strength), and

$$\nabla^H \phi = d\phi + \Omega\phi - \phi\omega \tag{9.59}$$

denotes the covariant derivative acting on $\mathcal{C}^\infty(M, \mathbb{C}^{2\times 1})$.

Proof. Specializing Corollary 2, we obtain

$$d\begin{pmatrix} \omega_{q\ell} & 0 \\ 0 & \tilde{\omega}_{q\ell} \end{pmatrix} = \begin{pmatrix} \boldsymbol{d\omega_{q\ell}} & 0 \\ 0 & \boldsymbol{d\tilde{\omega}_{q\ell}} \end{pmatrix}$$

with $\boldsymbol{d\omega_{q\ell}}$ equal to

$$\begin{bmatrix} \gamma(d\omega) \otimes \begin{pmatrix} 1 & 0 \\ 0 & 1 \end{pmatrix} + & 0 & \kappa^2 \Gamma\gamma(\varepsilon^*\Omega - \omega\varepsilon^* - d\varphi^*) \otimes \begin{pmatrix} M_d^* & 0 \\ 0 & M_e^* \end{pmatrix} \\ \kappa^2(\varphi_1 + \overline{\varphi}_1) \otimes \begin{pmatrix} M_d^2 - \frac{r}{s} & 0 \\ 0 & M_e^2 - \frac{r}{s} \end{pmatrix} & & \\[1em] 0 & \begin{array}{c} -\gamma(d\omega) \otimes \underline{1} + \\ \kappa^2(\varphi_1 + \overline{\varphi}_1) \otimes \underline{M_u^2 - \frac{r}{s}} \end{array} & \kappa^2 \Gamma\gamma(\underline{\varepsilon}^*\Omega - \omega\underline{\varepsilon}^* - d\underline{\varphi}^*) \otimes \begin{pmatrix} M_u^* , 0 \end{pmatrix} \\[1em] \gamma(d\varphi + \Omega\varepsilon - \varepsilon\omega)\Gamma \otimes \begin{pmatrix} M_d & 0 \\ 0 & M_e \end{pmatrix} & \gamma(d\underline{\varphi} + \Omega\underline{\varepsilon} - \underline{\varepsilon}\omega)\Gamma \otimes \begin{pmatrix} M_u \\ 0 \end{pmatrix} & \begin{array}{c} \gamma(d\Omega) \otimes \begin{pmatrix} 1 & 0 \\ 0 & 1 \end{pmatrix} + \\ \frac{\kappa^2}{2}(\varphi_1 + \overline{\varphi}_1) \otimes \begin{pmatrix} M_d^2 + M_u^2 - \frac{r}{t} & 0 \\ 0 & M_e^2 - \frac{r}{t} \end{pmatrix} \end{array} \end{bmatrix}$$

and $d\tilde{\omega}_{q\ell}$ equal to

$$\begin{bmatrix} \tilde{\gamma}(d\overline{\omega}_a)\otimes\begin{pmatrix}\overline{a}&0\\0&0\end{pmatrix}+\tilde{\gamma}(d\omega)\otimes\begin{pmatrix}\frac{1}{3}&0\\0&0\end{pmatrix} & 0 & 0 \\ -\tilde{\gamma}(d\omega+\kappa^2\frac{r}{s}(\varphi_1+\overline{\varphi}_1))\otimes\begin{pmatrix}0&0\\0&1\end{pmatrix} & & \\ 0 & \tilde{\gamma}(d\overline{\omega}_a)\otimes\overline{a}+\tilde{\gamma}(d\omega)\otimes\frac{1}{3} & 0 \\ 0 & 0 & \underline{\tilde{\gamma}(d\overline{\omega}_a)}\otimes\begin{pmatrix}\overline{a}&0\\0&0\end{pmatrix}+\underline{\tilde{\gamma}(d\omega)}\otimes\begin{pmatrix}\frac{1}{3}&0\\0&0\end{pmatrix} \\ & & -\underline{\tilde{\gamma}(d\omega+\kappa^2\frac{r}{s}(\varphi_1+\overline{\varphi}_1))}\otimes\begin{pmatrix}0&0\\0&1\end{pmatrix} \end{bmatrix}.$$

Similarly

$$\begin{pmatrix}\omega_{q\ell}&0\\0&\tilde{\omega}_{q\ell}\end{pmatrix}\wedge\begin{pmatrix}\omega_{q\ell}&0\\0&\tilde{\omega}_{q\ell}\end{pmatrix}=\begin{pmatrix}\omega_{q\ell}\wedge\omega_{q\ell}&0\\0&\tilde{\omega}_{q\ell}\wedge\tilde{\omega}_{q\ell}\end{pmatrix}$$

where $\omega_{q\ell}\wedge\omega_{q\ell}$ equals

$$\begin{bmatrix} \kappa^2\|\varphi\|^2\otimes\begin{pmatrix}M_d^2-\frac{r}{s}&0\\0&M_e^2-\frac{r}{s}\end{pmatrix} & 0 & \kappa^2\Gamma\gamma(\varphi^*\Omega-\omega\varphi^*)\otimes\begin{pmatrix}M_d^*&0\\0&M_e^*\end{pmatrix} \\ 0 & \kappa^2\|\varphi\|^2\otimes\underline{M_u^2-\frac{r}{s}} & \kappa^2\Gamma\gamma(\underset{\sim}{\varphi^*\Omega}+\underset{\sim}{\omega\varphi^*})\otimes(\underline{M_u^*},0) \\ \gamma(\Omega\varphi-\varphi\omega)\Gamma\otimes\begin{pmatrix}M_d&0\\0&M_e\end{pmatrix} & \gamma(\underset{\sim}{\Omega\varphi}+\underset{\sim}{\varphi\omega})\Gamma\otimes\begin{pmatrix}M_u\\0\end{pmatrix} & \gamma(\Omega\wedge\Omega)\otimes\begin{pmatrix}1&0\\0&1\end{pmatrix}+ \\ & & \frac{\kappa^2}{2}\|\varphi\|^2\otimes\begin{pmatrix}M_d^2+M_u^2-\frac{r}{t}&0\\0&M_e^2-\frac{r}{t}\end{pmatrix} \end{bmatrix}$$

and $\tilde{\omega}_{q\ell}\wedge\tilde{\omega}_{q\ell}$ equals

$$\begin{bmatrix} \tilde{\gamma}(\overline{\omega}_a\wedge\overline{\omega}_b)\otimes\begin{pmatrix}\overline{ab}&0\\0&0\end{pmatrix} & 0 & 0 \\ -\kappa^2\frac{r}{s}\|\varphi\|^2\otimes\begin{pmatrix}0&0\\0&1\end{pmatrix} & & \\ 0 & \tilde{\gamma}(\overline{\omega}_a\wedge\overline{\omega}_b)\otimes\overline{ab} & 0 \\ 0 & 0 & \underline{\tilde{\gamma}(\overline{\omega}_a\wedge\overline{\omega}_b)}\otimes\begin{pmatrix}\overline{ab}&0\\0&0\end{pmatrix} \\ & & -\kappa^2\frac{r}{s}\underline{\|\varphi\|^2}\otimes\begin{pmatrix}0&0\\0&1\end{pmatrix} \end{bmatrix}.$$

Therefore the assertion follows via the identities

$$\nabla^H(\varphi+\varepsilon)=d\varphi+\Omega\varepsilon+\Omega\varphi-\varepsilon\omega-\varphi\omega,$$
$$\underset{\sim}{\varepsilon^*}\varphi+\underset{\sim}{\varphi^*}\varepsilon=\overline{\varphi}_1+\varphi_1=\varepsilon^*\varphi+\varphi^*\varepsilon,\quad \underset{\sim}{\varepsilon^*}\varphi+\underset{\sim}{\varphi^*}\varepsilon=0=\varepsilon^*\underset{\sim}{\varphi}+\varphi^*\underset{\sim}{\varepsilon},$$

$$\varepsilon\varphi^*+\varphi\varepsilon^*+\underset{\sim\sim}{\varepsilon\varphi^*}+\underset{\sim\sim}{\varphi\varepsilon^*}=\underline{\underline{\varphi_1+\overline{\varphi}_1}},$$

$$\varphi^*\varphi=\|\varphi\|^2=\underset{\sim}{\varphi^*}\underset{\sim}{\varphi},\quad \varphi^*\underset{\sim}{\varphi}=0=\underset{\sim}{\varphi^*}\varphi,\quad \varphi\varphi^*+\underset{\sim}{\varphi}\underset{\sim}{\varphi^*}=\underline{\underline{\|\varphi\|^2}},$$

$$\omega\wedge\omega=0,$$

$$\overline{\omega}_b\wedge\overline{\omega}_c\otimes\overline{b}\overline{c}=(\overline{\omega}_a\otimes\overline{a}+\omega\otimes\tfrac{1}{3})\wedge(\overline{\omega}_b\otimes\overline{b}+\omega\otimes\tfrac{1}{3})$$

$$=(\overline{\omega}_a\wedge\overline{\omega}_b)\otimes\overline{a}\overline{b}+\underbrace{\omega\wedge\omega}_{=0}\otimes\tfrac{1}{9}+\underbrace{\overline{\omega}_a\wedge\omega+\omega\wedge\overline{\omega}_b}_{=0}(\overline{\omega}_a\tfrac{\overline{a}}{3}\wedge\overline{\omega}_b)\otimes\overline{a}\overline{b}.$$

Definition 5. *The NC gauge Lagrangian for the Standard Model is the functional*

$$\begin{pmatrix} \omega_{q\ell} & 0 \\ 0 & \tilde{\omega}_{q\ell} \end{pmatrix} \mapsto \big(\boldsymbol{F}(\omega_{q\ell}) \mid \boldsymbol{F}(\omega_{q\ell})I_{q\ell}\big)_{S_{q\ell}} + \big(\boldsymbol{F}(\tilde{\omega}_{q\ell}) \mid \boldsymbol{F}(\tilde{\omega}_{q\ell})\tilde{I}_{q\ell}\big)_{S_{q\ell}} \quad (9.60)$$

defined on the NC gauge fields $\begin{pmatrix} \omega_{q\ell} & 0 \\ 0 & \tilde{\omega}_{q\ell} \end{pmatrix} \in \mathcal{A}^{(1)}_{q\ell}$. *Note that the inner product is taken fibrewise over* M, *so that (9.60) really defines a Lagrangian density. For the integrated version*

$$\int_M d^n x \cdot [\big(\boldsymbol{F}(\omega_{q\ell}) \mid \boldsymbol{F}(\omega_{q\ell})I_{q\ell}\big)_{S_{q\ell}} + \big(\boldsymbol{F}(\tilde{\omega}_{q\ell}) \mid \boldsymbol{F}(\tilde{\omega}_{q\ell})\tilde{I}_{q\ell}\big)_{S_{q\ell}}] \quad (9.61)$$

one has to assume that $\omega_{q\ell}$ *and* $\tilde{\omega}_{q\ell}$ *have compact support. For compact Riemannian manifolds* M *(*$\kappa = i$*), it follows from (7.91) that (9.61) is related to the Dixmier trace*

$$\mathrm{Tr}_\omega\big[\boldsymbol{F}(\omega_{q\ell})^\dagger \boldsymbol{F}(\omega_{q\ell}) I_{q\ell}|D_{q\ell}|^{-n} + \boldsymbol{F}(\tilde{\omega}_{q\ell})^\dagger \boldsymbol{F}(\tilde{\omega}_{q\ell})\tilde{I}_{q\ell} J_{q\ell}|D_{q\ell}|^{-n} J^*_{q\ell}\big]$$

$$= c_n \int_M d^n x \cdot [(\boldsymbol{F}(\omega_{q\ell})|\boldsymbol{F}(\omega_{q\ell})I_{q\ell})_{S_{q\ell}} + (\boldsymbol{F}(\tilde{\omega}_{q\ell})|\boldsymbol{F}(\tilde{\omega}_{q\ell})\tilde{I}_{q\ell})_{S_{q\ell}}].$$

Theorem 2. *The NC gauge Lagrangian of* $\begin{pmatrix} \omega_{q\ell} & 0 \\ 0 & \tilde{\omega}_{q\ell} \end{pmatrix}$ *is given by*

$$\big(\boldsymbol{F}(\omega_{q\ell}) \mid \boldsymbol{F}(\omega_{q\ell})I_{q\ell}\big)_{S_{q\ell}} + \big(\boldsymbol{F}(\tilde{\omega}_{q\ell}) \mid \boldsymbol{F}(\tilde{\omega}_{q\ell})\tilde{I}_{q\ell}\big)_{S_{q\ell}} =$$
$$(s+\tfrac{4}{3}\underline{\mathrm{tr}}_\mathbb{G}\tilde{x})\|F(\omega)\|^2 + t\|F(\Omega)\|^2 + 4\underline{\mathrm{tr}}_\mathbb{G}\tilde{x}\|F(\omega_a\otimes a)\|^2 - 2\kappa^2 r\|\nabla^H(\varphi+\varepsilon)\|^2$$
$$+ \big(\|\varphi+\varepsilon\|^2 - 1\big)^2 \cdot \{\underline{\mathrm{tr}}_\mathbb{G}\big[(M_d^2 - \tfrac{r}{s})^2 + (M_u^2 - \tfrac{r}{s})^2 + \tfrac{1}{2}(M_d^2 + M_u^2 - \tfrac{r}{t})^2\big]x$$
$$+ \underline{\mathrm{tr}}_\mathbb{G}\big[(M_e^2 - \tfrac{r}{s})^2 + \tfrac{1}{2}(M_e^2 - \tfrac{r}{t})^2\big]y + \tfrac{3r^2}{s^2}\underline{\mathrm{tr}}_\mathbb{G}\tilde{y}\} + const$$

where the constants r, s, t *are defined in (9.44).*

Proof. Using Proposition 11 and Theorem 1 it follows that

$$\big(F(\omega_{q\ell}) \mid F(\omega_{q\ell})I_{q\ell}\big)_{S_{q\ell}} + \big(F(\tilde{\omega}_{q\ell}) \mid F(\tilde{\omega}_{q\ell})\tilde{I}_{q\ell}\big)_{S_{q\ell}} =$$
$$s\|F(\omega) - \kappa^2 \tfrac{r}{s}((\varphi+\varepsilon)^*(\varphi+\varepsilon) - I)\|_S^2$$
$$+ t\|F(\Omega) - \kappa^2 \tfrac{r}{2t}(\varphi+\varepsilon)^*(\varphi+\varepsilon) - I\|_{\underline{S}}^2$$
$$+ 4\underline{\text{tr}}_{\mathbb{G}} \tilde{x} \|F(\omega_a \otimes a) + F(\omega) \otimes \tfrac{1}{3}\|_{\underline{S}}^2$$
$$+ 2\kappa^2 r \big\{ \text{Re}\big(F(\omega) - \kappa^2 \tfrac{r}{s}(\varphi+\varepsilon)^*(\varphi+\varepsilon) - I\big) \mid (\varphi+\varepsilon)^*(\varphi+\varepsilon) - I\big)_S$$
$$+ \text{Re}\big(F(\Omega) - \kappa^2 \tfrac{r}{2t}(\varphi+\varepsilon)^*(\varphi+\varepsilon) - I \mid \tfrac{1}{2}(\varphi+\varepsilon)^*(\varphi+\varepsilon) - I\big)_{\underline{S}}$$
$$- \kappa^2 r \big\{ \|\nabla^H(\varphi+\varepsilon)\|_S^2 + \|\nabla^H(\varphi+\varepsilon)^*\|_{\underline{S}}^2 \big\}$$
$$+ \underline{\text{tr}}_{\mathbb{G}}\big((M_d^4 + M_u^4)x + M_e^4 y\big)$$
$$\cdot \big\{ \|(\varphi+\varepsilon)^*(\varphi+\varepsilon) - I\|\big)_S^2 + \tfrac{1}{4}\|(\varphi+\varepsilon)^*(\varphi+\varepsilon) - I\|_{\underline{S}}^2 \big\}$$
$$+ 2\underline{\text{tr}}_{\mathbb{G}}(M_u^* M_d M_d^* M_u x)\tfrac{1}{4}\|(\varphi+\varepsilon)^*(\varphi+\varepsilon) - I\|_{\underline{S}}^2$$
$$= \|F(\omega)\|_S^2 (s + 4\underline{\text{tr}}_{\mathbb{G}} \tilde{x} \cdot 3 \cdot \tfrac{1}{9}) + t\|F(\Omega)\|_{\underline{S}}^2 + 4\underline{\text{tr}}_{\mathbb{G}} \tilde{x} \|F(\omega_a \otimes a)\|_{\underline{S}}^2$$
$$+ \|(\varphi+\varepsilon)^*(\varphi+\varepsilon) - I\|_S^2$$
$$\cdot \big\{ s\tfrac{r^2}{s^2} + 2t\tfrac{r^2}{4t^2} - 2r\tfrac{r}{s} - 2r2\tfrac{r}{2t} \cdot \tfrac{1}{2} + \underline{\text{tr}}_{\mathbb{G}}((M_d^4 + M_u^4)x + M_e^4 y)(1 + \tfrac{1}{4} \cdot 2)$$
$$+ 2\underline{\text{tr}}_{\mathbb{G}}(M_u^* M_d M_d^* M_u x)\tfrac{1}{4} \cdot 2 \big\} - 2\kappa^2 r \|\nabla^H(\varphi+\varepsilon)\|_S^2.$$

Now the assertion follows from the calculation

$$\underline{\text{tr}}_{\mathbb{G}}\big[(M_d^2 - \tfrac{r}{s})^2 + (M_u^2 - \tfrac{r}{s})^2 + \tfrac{1}{2}(M_d^2 + M_u^2 - \tfrac{r}{t})^2\big]x$$
$$+ \underline{\text{tr}}_{\mathbb{G}}\big[(M_e^2 - \tfrac{r}{s})^2 + \tfrac{1}{2}(M_e^2 - \tfrac{r}{t})^2\big]y + \tfrac{3r^2}{s^2}\underline{\text{tr}}_{\mathbb{G}}\tilde{y}$$
$$- \tfrac{3}{2}\underline{\text{tr}}_{\mathbb{G}}\big[(M_d^4 + M_u^4)x + M_e^4 y\big] - \underline{\text{tr}}_{\mathbb{G}} M_u^* M_d M_d^* M_u x$$

$$= -(\tfrac{2r}{s} + \tfrac{r}{t})\underline{\text{tr}}_{\mathbb{G}}\big[(M_d^2 + M_u^2)x + M_e^2 y\big]$$
$$+ \tfrac{3r^2}{s^2}\underline{\text{tr}}_{\mathbb{G}}\tilde{y} + (\tfrac{2r^2}{s^2} + \tfrac{r^2}{2t^2})\underline{\text{tr}}_{\mathbb{G}} x + (\tfrac{r^2}{s^2} + \tfrac{r^2}{2t^2})\underline{\text{tr}}_{\mathbb{G}} y$$
$$= -(\tfrac{2r}{s} + \tfrac{r}{t})r + \tfrac{r^2}{s^2}\underline{\text{tr}}_{\mathbb{G}}(3\tilde{y} + 2x + y) + \tfrac{r^2}{2t^2}\underline{\text{tr}}_{\mathbb{G}}(x + y)$$
$$= -\tfrac{2r^2}{s} - \tfrac{r^2}{t} + \tfrac{r^2 s}{s^2} + \tfrac{r^2 t}{2t^2} = -\tfrac{r^2}{s} - \tfrac{r^2}{2t}.$$

Remark 4. *Converting to physics notation (9.55), the NC gauge Lagrangian gives the Standard Model Yang-Mills-Higgs functional in the usual form*

$$\tfrac{1}{g_1^2}\|F(ig_1 W^{(0)} \tau^{(0)})\|^2 + \tfrac{1}{g_2^2}\|F(ig_2 \sum_{a=1}^{3} W^{(a)}\tau^{(a)})\|^2 + \tfrac{1}{g_3^2}\|F(ig_3 \sum_{a=1}^{8} G^{(a)}\lambda^{(a)})\|^2$$
$$- \kappa^2 \|\nabla^H(\begin{pmatrix}\Phi^0\\\Phi^+\end{pmatrix})\|^2 + \tfrac{m_H^2}{8r}(\|\begin{pmatrix}\Phi^0\\\Phi^+\end{pmatrix}\|^2 - 2r)^2 + const$$

if the coupling constants g_1, g_2, g_3 satisfy

$$\frac{1}{g_1^2} = s + \frac{4}{3}\underline{\mathrm{tr}}_G \tilde{x}, \qquad \frac{1}{g_2^2} = t, \qquad \frac{1}{g_3^2} = 4\underline{\mathrm{tr}}_G \tilde{x} \tag{9.62}$$

and for the Higgs mass m_H we get

$$\frac{r}{2} m_H^2 = \underline{\mathrm{tr}}_G\left[(M_d^2 - \tfrac{r}{s})^2 + (M_u^2 - \tfrac{r}{s})^2 + \tfrac{1}{2}(M_d^2 + M_u^2 - \tfrac{r}{t})^2\right]x \tag{9.63}$$
$$+ \underline{\mathrm{tr}}_G\left[(M_e^2 - \tfrac{r}{s})^2 + \tfrac{1}{2}(M_e^2 - \tfrac{r}{t})^2\right]y + \tfrac{3r^2}{s^2}\underline{\mathrm{tr}}_G \tilde{y}.$$

9.4 Noncommutative Matter Action Functional

For the remainder of this section we are concerned with the *fermionic action* for the Standard Model, involving the "perturbed" Dirac operator

$$\begin{pmatrix} D_{gq\ell} & 0 \\ 0 & J_{q\ell} D_{gq\ell} J_{q\ell} \end{pmatrix} := \begin{pmatrix} D_{q\ell} & 0 \\ 0 & J_{q\ell} D_{q\ell} J_{q\ell} \end{pmatrix} \tag{9.64}$$
$$+ \begin{pmatrix} \omega_{q\ell} & 0 \\ 0 & \tilde{\omega}_{q\ell} \end{pmatrix} - \begin{pmatrix} 0 & J_{q\ell} \\ -\kappa^2 J_{q\ell} & 0 \end{pmatrix} \begin{pmatrix} \omega_{q\ell} & 0 \\ 0 & \tilde{\omega}_{q\ell} \end{pmatrix} \begin{pmatrix} 0 & J_{q\ell} \\ -\kappa^2 J_{q\ell} & 0 \end{pmatrix}^{-1}$$

associated with the NC gauge field $\begin{pmatrix} \omega_{q\ell} & 0 \\ 0 & \tilde{\omega}_{q\ell} \end{pmatrix}$. Note that

$$\begin{pmatrix} D_{gq\ell} & 0 \\ 0 & J_{q\ell} D_{gq\ell} J_{q\ell} \end{pmatrix}^\dagger = -\begin{pmatrix} D_{gq\ell} & 0 \\ 0 & J_{q\ell} D_{gq\ell} J_{q\ell} \end{pmatrix}$$

since $D, \gamma(\omega), \gamma(\Omega), \gamma(\omega_a \otimes a)$ and the analogous operators in the anti-particle sector are skew-hermitian.

Proposition 16. *We have $D_{gq\ell} = D_{q\ell} + \omega_{q\ell} + J_{q\ell}\tilde{\omega}_{q\ell} J_{q\ell}$ equal to*

$$\begin{bmatrix} D \otimes \begin{pmatrix} 1 & 0 \\ 0 & 1 \end{pmatrix} + & 0 & \kappa^2 \Gamma(\varphi+\varepsilon)^* \otimes \begin{pmatrix} M_d^* & 0 \\ 0 & M_e^* \end{pmatrix} \\ \gamma(\omega_a) \otimes \begin{pmatrix} a & 0 \\ 0 & 0 \end{pmatrix} + \gamma(\omega) \otimes \begin{pmatrix} \frac{2}{3} & 0 \\ 0 & 2 \end{pmatrix} & & \\ 0 & \begin{matrix} D \otimes \underline{1} + \\ \gamma(\omega_a)\otimes a - \gamma(\omega)\otimes \tfrac{4}{3}\end{matrix} & \kappa^2 \Gamma(\underset{\sim}{\varphi+\varepsilon})^* \otimes (\underline{M_u^*}, 0) \\ (\varphi+\varepsilon)\Gamma \otimes \begin{pmatrix} M_d & 0 \\ 0 & M_e \end{pmatrix} & (\underset{\sim}{\varphi+\varepsilon})\Gamma \otimes \begin{pmatrix} M_u \\ 0 \end{pmatrix} & \begin{matrix} (\underline{D}+\gamma(\underline{\Omega}))\otimes\begin{pmatrix}1 & 0 \\ 0 & 1\end{pmatrix}+ \\ \underline{\gamma(\omega_a)}\otimes\begin{pmatrix}a & 0 \\ 0 & 0\end{pmatrix}-\underline{\gamma(\omega)}\otimes\begin{pmatrix}\tfrac{1}{3} & 0 \\ 0 & -1\end{pmatrix}\end{matrix} \end{bmatrix}$$

and $J_{q\ell} D_{gq\ell} J_{q\ell}$ equal to

$$\begin{bmatrix} \overline{D} \otimes \begin{pmatrix} 1 & 0 \\ 0 & 1 \end{pmatrix} + \\ \tilde{\gamma}(\overline{\omega}_a) \otimes \begin{pmatrix} \overline{a} & 0 \\ 0 & 0 \end{pmatrix} - \tilde{\gamma}(\omega) \otimes \begin{pmatrix} \frac{2}{3} & 0 \\ 0 & 2 \end{pmatrix} & 0 & -\kappa^2 \Gamma (\varphi + \varepsilon)^t \otimes \begin{pmatrix} M_d^t & 0 \\ 0 & M_e^t \end{pmatrix} \\ 0 & \overline{D} \otimes \underset{\sim}{1} + \tilde{\gamma}(\overline{\omega}_a) \otimes \overline{a} + \tilde{\gamma}(\omega) \otimes \frac{4}{3} & -\kappa^2 \Gamma (\varphi + \varepsilon)^t \otimes (M_u^t, 0) \\ -(\overline{\varphi} + \varepsilon) \Gamma \otimes \begin{pmatrix} \overline{M_d} & 0 \\ 0 & \overline{M_e} \end{pmatrix} & -(\overline{\varphi} + \varepsilon) \Gamma \otimes \begin{pmatrix} \overline{M_u} \\ 0 \end{pmatrix} & (\underline{\underline{\overline{D}}} + \tilde{\gamma}(\overline{\Omega})) \otimes \begin{pmatrix} 1 & 0 \\ 0 & 1 \end{pmatrix} + \\ & & \tilde{\gamma}(\overline{\omega}_a) \otimes \begin{pmatrix} \overline{a} & 0 \\ 0 & 0 \end{pmatrix} + \tilde{\gamma}(\omega) \otimes \begin{pmatrix} \frac{1}{3} & 0 \\ 0 & -1 \end{pmatrix} \end{bmatrix}.$$

Proof. Writing $\begin{pmatrix} \omega_{q\ell} & 0 \\ 0 & \tilde{\omega}_{q\ell} \end{pmatrix}$ according to Proposition 15 we have

$$\begin{pmatrix} \omega_{q\ell} & 0 \\ 0 & \tilde{\omega}_{q\ell} \end{pmatrix} - \begin{pmatrix} 0 & J_{q\ell} \\ -\kappa^2 J_{q\ell} & 0 \end{pmatrix} \begin{pmatrix} \omega_{q\ell} & 0 \\ 0 & \tilde{\omega}_{q\ell} \end{pmatrix} \begin{pmatrix} 0 & \kappa^2 J_{q\ell} \\ -J_{q\ell} & 0 \end{pmatrix}$$
$$= \begin{pmatrix} \omega_{q\ell} + J_{q\ell} \tilde{\omega}_{q\ell} J_{q\ell} & 0 \\ 0 & \tilde{\omega}_{q\ell} + J_{q\ell} \omega_{q\ell} J_{q\ell} \end{pmatrix}.$$

Since $J_{q\ell} \tilde{\omega}_{q\ell} J_{q\ell}$ calculates as

$$\begin{bmatrix} \gamma(\omega_a) \otimes \begin{pmatrix} a & 0 \\ 0 & 0 \end{pmatrix} - \gamma(\omega) \otimes \begin{pmatrix} \frac{1}{3} & 0 \\ 0 & -1 \end{pmatrix} & 0 & 0 \\ 0 & \gamma(\omega_a) \otimes a - \gamma(\omega) \otimes \frac{1}{3} & 0 \\ 0 & 0 & \underline{\underline{\gamma(\omega_a)}} \otimes \begin{pmatrix} a & 0 \\ 0 & 0 \end{pmatrix} - \underline{\underline{\gamma(\omega)}} \otimes \begin{pmatrix} \frac{1}{3} & 0 \\ 0 & -1 \end{pmatrix} \end{bmatrix}$$

it follows that $\omega_{q\ell} + J_{q\ell} \tilde{\omega}_{q\ell} J_{q\ell}$ equals

$$\begin{bmatrix} \gamma(\omega_a) \otimes \begin{pmatrix} a & 0 \\ 0 & 0 \end{pmatrix} + \gamma(\omega) \otimes \begin{pmatrix} \frac{2}{3} & 0 \\ 0 & 2 \end{pmatrix} & 0 & \kappa^2 \Gamma \varphi^* \otimes \begin{pmatrix} M_d^* & 0 \\ 0 & M_e^* \end{pmatrix} \\ 0 & \gamma(\omega_a) \otimes a - \gamma(\omega) \otimes \frac{4}{3} & \kappa^2 \Gamma \varphi^* \otimes (M_u^*, 0) \\ \varphi \Gamma \otimes \begin{pmatrix} M_d & 0 \\ 0 & M_e \end{pmatrix} & \varphi \Gamma \otimes \begin{pmatrix} M_u \\ 0 \end{pmatrix} & \gamma(\Omega) \otimes \begin{pmatrix} 1 & 0 \\ 0 & 1 \end{pmatrix} + \\ & & \underline{\underline{\gamma(\omega_a)}} \otimes \begin{pmatrix} a & 0 \\ 0 & 0 \end{pmatrix} - \underline{\underline{\gamma(\omega)}} \otimes \begin{pmatrix} \frac{1}{3} & 0 \\ 0 & -1 \end{pmatrix} \end{bmatrix}$$

and $\tilde{\omega}_{q\ell} + J_{q\ell}\omega_{q\ell}J_{q\ell}J_{q\ell}(\omega_{q\ell} + J_{q\ell}\tilde{\omega}_{q\ell}J_{q\ell})J_{q\ell}$ equals

$$\left[\begin{array}{ccc} \tilde{\gamma}(\overline{\omega}_a) \otimes \begin{pmatrix} \overline{a} & 0 \\ 0 & 0 \end{pmatrix} - \tilde{\gamma}(\omega) \otimes \begin{pmatrix} \frac{2}{3} & 0 \\ 0 & 2 \end{pmatrix} & 0 & -\kappa^2 \Gamma \varphi^t \otimes \begin{pmatrix} M_d^t & 0 \\ 0 & M_e^t \end{pmatrix} \\ 0 & \tilde{\gamma}(\overline{\omega}_a) \otimes \overline{a} + \tilde{\gamma}(\omega) \otimes \underline{\underline{\frac{4}{3}}} & -\kappa^2 \Gamma \varphi^t \otimes (\underline{M_u^t}, 0) \\ -\overline{\varphi} \Gamma \otimes \begin{pmatrix} \overline{M_d} & 0 \\ 0 & \overline{M_e} \end{pmatrix} & -\overline{\varphi} \Gamma \otimes \begin{pmatrix} \overline{M_u} \\ 0 \end{pmatrix} & \tilde{\gamma}(\overline{\Omega}) \otimes \begin{pmatrix} 1 & 0 \\ 0 & 1 \end{pmatrix} + \\ & & \tilde{\gamma}(\overline{\omega}_a) \otimes \begin{pmatrix} \overline{a} & 0 \\ 0 & 0 \end{pmatrix} + \underline{\underline{\tilde{\gamma}(\omega)}} \otimes \begin{pmatrix} \frac{1}{3} & 0 \\ 0 & -1 \end{pmatrix} \end{array}\right].$$

Definition 6. *The* NC *matter Lagrangian for the Standard Model is the (imaginary) functional*

$$\begin{pmatrix} \psi_{q\ell} \\ \tilde{\psi}_{q\ell} \end{pmatrix} \mapsto \int_M (\psi_{q\ell}^\dagger, \tilde{\psi}_{q\ell}^\dagger) \begin{pmatrix} D_{gq\ell} & 0 \\ 0 & J_{q\ell} D_{gq\ell} J_{q\ell} \end{pmatrix} \begin{pmatrix} \psi_{q\ell} \\ \tilde{\psi}_{q\ell} \end{pmatrix} \quad (9.65)$$

$$= \int_M \left(\psi_{q\ell}^\dagger D_{gq\ell} \psi_{q\ell} + \tilde{\psi}_{q\ell}^\dagger J_{q\ell} D_{gq\ell} J_{q\ell} \tilde{\psi}_{q\ell} \right)$$

defined on the NC *matter fields* $\begin{pmatrix} \psi_{q\ell} \\ \tilde{\psi}_{q\ell} \end{pmatrix}$.

Proposition 17. *The* NC *matter Lagrangian (density) is given by*

$$\psi_{q\ell}^\dagger D_{gq\ell} \psi_{q\ell} = \psi_d^\dagger D \psi_d \cdot (d_R^* d_R + e_R^* e_R) +$$
$$\psi_d^\dagger \gamma(\omega_a) \psi_d \cdot d_R^* a d_R + \psi_d^\dagger \gamma(\omega) \psi_d \left(\tfrac{2}{3} d_R^* d_R + 2 e_R^* e_R \right)$$
$$+ \psi_u^\dagger D \psi_u \cdot u_R^* u_R + \psi_u^\dagger \gamma(\omega_a) \psi_u \cdot u_R^* a u_R - \psi_u^\dagger \gamma(\omega) \psi_u \cdot \tfrac{4}{3} u_R^* u_R$$
$$+ \Psi^\dagger (\underline{D} + \gamma(\Omega)) \Psi \cdot (q_L^* q_L + \ell_L^* \ell_L) + \Psi^\dagger \underline{\underline{\gamma(\omega_a)}} \Psi \cdot q_L^* a q_L$$
$$- \Psi^\dagger \underline{\underline{\gamma(\omega)}} \Psi \cdot \left(\tfrac{1}{3} q_L^* q_L - \ell_L^* \ell_L \right)$$
$$+ 2i \, \mathrm{Im}\, \Psi^\dagger (\varphi + \varepsilon) \Gamma \psi_d \cdot (q_L^* M_d d_R + \ell_L^* M_e e_R)$$
$$+ 2i \, \mathrm{Im}\, \Psi^\dagger (\underset{\sim}{\varphi} + \underset{\sim}{\varepsilon}) \Gamma \psi_u \cdot q_L^* \underline{M_u} u_R$$

for $\psi_{q\ell}$ written in the form (9.7), plus a similar expression for the anti-particle sector.

Proof. Using (9.7) and (9.64) one obtains

$$\psi_{q\ell}^\dagger D_{gq\ell}\psi_{q\ell} = \psi_d^\dagger D\psi_d \cdot (d_R^* d_R + e_R^* e_R) +$$
$$\psi_d^\dagger \gamma(\omega_a) \psi_d \cdot d_R^* a d_R + \psi_d^\dagger \gamma(\omega) \psi_d \left(\tfrac{2}{3} d_R^* d_R + 2 e_R^* e_R\right)$$
$$+ \psi_u^\dagger D\psi_u \cdot u_R^* u_R + \psi_u^\dagger \gamma(\omega_a) \psi_u \cdot u_R^* a u_R - \psi_u^\dagger \gamma(\omega) \psi_u \cdot \tfrac{4}{3} u_R^* u_R$$
$$+ \Psi^\dagger (\underline{D} + \gamma(\Omega)) \Psi \cdot (q_L^* q_L + \ell_L^* \ell_L) + \Psi^\dagger \underline{\gamma(\omega_a)} \Psi \cdot q_L^* a q_L$$
$$- \Psi^\dagger \underline{\gamma(\omega)} \Psi \cdot (\tfrac{1}{3} q_L^* q_L - \ell_L^* \ell_L)$$
$$+ \Psi^\dagger (\varphi + \varepsilon) \Gamma \psi_d \cdot (q_L^* \underline{M_d} d_R + \ell_L^* M_e e_R) + \Psi^\dagger (\varphi + \underset{\sim}{\varepsilon}) \Gamma \psi_u \cdot q_L^* \underline{M_u} u_R$$
$$+ \kappa^2 \psi_d^\dagger \Gamma (\varphi + \varepsilon)^* \Psi \cdot (d_R^* \underline{M_d}^* q_L + e_R^* M_e^* \ell_L) + \kappa^2 \psi_u^\dagger \Gamma (\varphi + \underset{\sim}{\varepsilon})^* \Psi \cdot u_R^* \underline{M_u}^* q_L .$$

The last four summands combine to yield the assertion, since

$$\overline{\Psi^\dagger (\varphi + \varepsilon) \Gamma \Psi_d \cdot (q_L^* \underline{M_d} d_R + \ell_L^* M_e e_R)}$$
$$= \psi_d^\dagger \Gamma^\dagger (\varphi + \varepsilon)^* \Psi \cdot (d_R^* \underline{M_d}^* q_L + e_R^* M_e^* \ell_L)$$
$$= -\kappa^2 \psi_d^\dagger \Gamma (\varphi + \varepsilon)^* \Psi \cdot (d_R^* \underline{M_d}^* q_L + e_R^* M_e^* \ell_L)$$

and

$$\overline{\Psi^\dagger (\varphi + \underset{\sim}{\varepsilon}) \Gamma \Psi_u \cdot q_L^* \underline{M_u} u_R} = \psi_u^\dagger \Gamma^\dagger (\varphi + \underset{\sim}{\varepsilon})^* \Psi \cdot u_R^* \underline{M_u}^* q_L$$
$$= -\kappa^2 \psi_u^\dagger \Gamma (\varphi + \underset{\sim}{\varepsilon})^* \Psi \cdot u_R^* \underline{M_u}^* q_L .$$

Remark 5. *Using physics notation for the NC gauge fields* $\begin{pmatrix} \omega_{q\ell} & 0 \\ 0 & \tilde{\omega}_{q\ell} \end{pmatrix}$ *(cf. (9.55)) and the NC matter fields* $\begin{pmatrix} \psi_{q\ell} \\ \tilde{\psi}_{q\ell} \end{pmatrix}$ *(cf. (9.10)) we obtain the NC matter Lagrangian in the usual form*

$$\psi_{d_R}^\dagger D\psi_{d_R} + \psi_{e_R}^\dagger D\psi_{e_R} + \psi_{u_R}^\dagger D\psi_{u_R}$$

$$+ (\psi_{d_L}^\dagger, \psi_{u_L}^\dagger)(\underline{D} + \gamma(\Omega)) \begin{pmatrix} \psi_{d_L} \\ \psi_{u_L} \end{pmatrix} + (\psi_{e_L}^\dagger, \psi_{\nu_L}^\dagger)(\underline{D} + \gamma(\Omega)) \begin{pmatrix} \psi_{e_L} \\ \psi_{\nu_L} \end{pmatrix}$$

$$+ \tfrac{2}{3} \psi_{d_R}^\dagger \gamma(\omega) \psi_{d_R} + 2 \psi_{e_R}^\dagger \gamma(\omega) \psi_{e_R} - \tfrac{4}{3} \psi_{u_R}^\dagger \gamma(\omega) \psi_{u_R}$$

$$- \tfrac{1}{3} (\psi_{d_L}^\dagger, \psi_{u_L}^\dagger) \underline{\gamma(\omega)} \begin{pmatrix} \psi_{d_L} \\ \psi_{u_L} \end{pmatrix} + (\psi_{e_L}^\dagger, \psi_{\nu_L}^\dagger) \underline{\gamma(\omega)} \begin{pmatrix} \psi_{e_L} \\ \psi_{\nu_L} \end{pmatrix}$$

$$+ \psi_{d_R}^\dagger \gamma(\omega_a) \otimes a \, \psi_{d_R} + \psi_{u_R}^\dagger \gamma(\omega_a) \otimes a \, \psi_{u_R}$$

$$+ (\psi_{d_L}^\dagger, \psi_{u_L}^\dagger) \underline{\gamma(\omega_a)} \otimes a \begin{pmatrix} \psi_{d_L} \\ \psi_{u_L} \end{pmatrix} + 2i \, Im \left[\left(\psi_{d_L}^\dagger, \psi_{u_L}^\dagger \right) (\varphi + \varepsilon) \Gamma \underline{M_d} \psi_{d_R} \right.$$

$$+(\psi_{e_L}^\dagger,\psi_{\nu_L}^\dagger)(\varphi+\varepsilon)\,\Gamma M_e\psi_{e_R}+(\psi_{d_L}^\dagger,\psi_{u_L}^\dagger)(\varphi+\underset{\sim}{\varepsilon})\,\Gamma\underline{M_u}\,\psi_{u_R}\Big]\;=$$

$$\psi_{d_R}^\dagger D\psi_{d_R}+\psi_{e_R}^\dagger D\psi_{e_R}+\psi_{u_R}^\dagger D\psi_{u_R}+\psi_{d_L}^\dagger D\psi_{d_L}+\psi_{u_L}^\dagger D\psi_{u_L}+\psi_{e_L}^\dagger D\psi_{e_L}$$

$$+\psi_{\nu_L}^\dagger D\psi_{\nu_L}+\frac{ig_1}{2}\Big(\frac{2}{3}\;\psi_{d_R}^\dagger\gamma(W^{(0)})\,\psi_{d_R}+2\psi_{e_R}^\dagger\gamma(W^{(0)})\,\psi_{e_R}$$

$$-\frac{4}{3}\;\psi_{u_R}^\dagger\gamma(W^{(0)})\,\psi_{u_R}-\frac{1}{3}\;\psi_{d_L}^\dagger\gamma(W^{(0)})\,\psi_{d_L}-\frac{1}{3}\;\psi_{u_L}^\dagger\gamma(W^{(0)})\,\psi_{u_L}$$

$$+\psi_{e_L}^\dagger\gamma(W^{(0)})\psi_{e_L}+\psi_{\nu_L}^\dagger\gamma(W^{(0)})\,\psi_{\nu_L})$$

$$+ig_2\sum_{a=1}^{3}\Big[(\psi_{d_L}^\dagger,\psi_{u_L}^\dagger)\gamma(W^{(a)})\otimes\tau^{(a)}\begin{pmatrix}\psi_{d_L}\\ \psi_{u_L}\end{pmatrix}+(\psi_{e_L}^\dagger,\psi_{\nu_L}^\dagger)\gamma(W^{(a)})\otimes\tau^{(a)}\begin{pmatrix}\psi_{e_L}\\ \psi_{\nu_L}\end{pmatrix}\Big]$$

$$+ig_3\sum_{a=1}^{8}\Big[\psi_{d_R}^\dagger\,\gamma(G^{(a)})\otimes\lambda^{(a)}\,\psi_{d_R}+\psi_{u_R}^\dagger\,\gamma(G^{(a)})\otimes\lambda^{(a)}\,\psi_{u_R}+$$

$$\psi_{d_L}^\dagger\,\gamma(G^{(a)})\otimes\lambda^{(a)}\,\psi_{d_L}+\psi_{u_L}^\dagger\,\gamma(G^{(a)})\otimes\lambda^{(a)}\,\psi_{u_L}\Big]$$

$$+\frac{2i\,\mathrm{Im}}{\sqrt{2r}}\;(\psi_{d_L}^\dagger\,\Phi^0\,\Gamma\,\underline{M_d}\,\psi_{d_R}+\psi_{u_L}^\dagger\,\Phi^+\,\Gamma\,\underline{M_d}\,\psi_{d_R}+\psi_{e_L}^\dagger\,\Phi^0\,\Gamma\,M_e\,\psi_{e_R}$$

$$+\psi_{\nu_L}^\dagger\,\Phi^+\,\Gamma\,M_e\,\psi_{e_R}-\psi_{d_L}^\dagger\,\overline{\Phi}^+\,\Gamma\,\underline{M_u}\,\psi_{u_R}+\psi_{u_L}^\dagger\,\overline{\Phi}^0\,\Gamma\,\underline{M_u}\,\psi_{u_R})$$

plus an analogous expression for the anti-particle sector. Up to now the constraints (9.12) have not been imposed. Restricting the Lagrangian to fields satisfying (9.12) yields a vanishing action in the Euclidean case, but for Minkowski signature one obtains the correct action (in particular for the mass terms) as discussed at the end of Section 8.

10 Standard Model Coupled with Gravity

Holger Neumann and Harald Upmeier

In this section we combine the Yang-Mills type interactions (electromagnetic, weak and strong force) discussed in Section 9 with gravity, by treating the space-time metric g together with the 1-forms described in Proposition 15 as an independent dynamical variable. In this section we keep the notation introduced in Section 9 but for simplicity we put $I_{q\ell} = 1 = \tilde{I}_{q\ell}$. According to the "Spectral Action Principle" (Section 6, [31]), the Lagrangian of the Standard Model coupled with gravity is expressed in terms of the spectral properties of the "perturbed" Dirac operator

$$\begin{pmatrix} D_{gq\ell} & 0 \\ 0 & J_{q\ell} D_{gq\ell} J_{q\ell} \end{pmatrix} := \begin{pmatrix} D_{q\ell} & 0 \\ 0 & J_{q\ell} D_{q\ell} J_{q\ell} \end{pmatrix} \tag{10.1}$$

$$+ \begin{pmatrix} \omega_{q\ell} & 0 \\ 0 & \tilde{\omega}_{q\ell} \end{pmatrix} - \begin{pmatrix} 0 & J_{q\ell} \\ -\kappa^2 J_{q\ell} & 0 \end{pmatrix} \begin{pmatrix} \omega_{q\ell} & 0 \\ 0 & \tilde{\omega}_{q\ell} \end{pmatrix} \begin{pmatrix} 0 & J_{q\ell} \\ -\kappa^2 J_{q\ell} & 0 \end{pmatrix}^{-1}$$

already discussed in Section 9. In this section we regard $\begin{pmatrix} \omega_{q\ell} & 0 \\ 0 & \tilde{\omega}_{q\ell} \end{pmatrix}$, more precisely $\begin{pmatrix} \omega_{q\ell} & 0 \\ 0 & \tilde{\omega}_{q\ell} \end{pmatrix} - \begin{pmatrix} 0 & J_{q\ell} \\ -\kappa^2 J_{q\ell} & 0 \end{pmatrix} \begin{pmatrix} \omega_{q\ell} & 0 \\ 0 & \tilde{\omega}_{q\ell} \end{pmatrix} \begin{pmatrix} 0 & J_{q\ell} \\ -\kappa^2 J_{q\ell} & 0 \end{pmatrix}^{-1}$, as an *internal fluctuation* of the metric considered as a dynamical variable as well. The spectral action is concerned with the spectral properties of the *square*

$$\begin{pmatrix} D_{gq\ell} & 0 \\ 0 & J_{q\ell} D_{gq\ell} J_{q\ell} \end{pmatrix}^2 = \begin{pmatrix} D_{gq\ell}^2 & 0 \\ 0 & -J_{q\ell} D_{gq\ell}^2 J_{q\ell} \end{pmatrix} = \begin{pmatrix} D_{gq\ell}^2 & 0 \\ 0 & J_{q\ell} D_{gq\ell}^2 J_{q\ell}^{-1} \end{pmatrix}.$$

It is clearly enough to study $D_{gq\ell}$ viewed as a differential operator on $S_{q\ell}$.

10.1 Generalized Dirac Operators

In order to realize $D_{gq\ell}$ as a generalized Dirac operator in the sense of differential geometry over space-time M [91], we consider the (trivial) complex vector bundle

$$E = M \times \mathbb{G}_{q\ell}$$

and realize $S_{q\ell}$ as the vector bundle tensor product

$$S_{q\ell} = S \otimes \mathbb{G}_{q\ell} = S \otimes_M E \tag{10.2}$$

so that

$$\mathcal{C}^\infty(S_{q\ell}) = \mathcal{C}^\infty(S) \otimes \mathbb{G}_{q\ell} = \mathcal{C}^\infty(S) \otimes_A \mathcal{C}^\infty(E), \tag{10.3}$$

where $A := \mathcal{C}^\infty(M, \mathbb{C})$. In local coordinates (7.22), the μ-th covariant derivative of the spin connection ∇^S on the spinor bundle S is given by

$$\nabla^S_\mu = \partial_\mu - \gamma(\underset{\sim}{\omega}_\mu) = \partial_\mu - \tfrac{1}{4} i\omega^j_\mu \eta_{jj} \gamma^i \gamma^j \qquad (10.4)$$

where ${}_i\omega^j_\mu$ is defined in (7.68). Similarly, a Yang-Mills connection ∇^E on E has covariant derivatives

$$\nabla^E_\mu = \partial_\mu + \omega^E_\mu, \qquad (10.5)$$

where

$$\omega^E = dx^\mu \otimes \omega^E_\mu \in \mathcal{C}^\infty(T^\# M \otimes \mathfrak{u}(\mathbb{G}_{q\ell})) \qquad (10.6)$$

is the matrix-valued connection 1-form. Note that (10.4) and (10.5) define first order differential operators on $\mathcal{C}^\infty(S)$ and $\mathcal{C}^\infty(E)$, resp. Now consider the tensor product connection ∇ on $S \otimes_M E$. Using the decomposition (10.3) we obtain covariant derivatives

$$\nabla_\mu = \nabla^S_\mu \otimes_A 1 + 1 \otimes_A \nabla^E_\mu \qquad (10.7)$$

or, more explicitly,

$$\begin{aligned}\nabla_\mu(\psi \otimes_A \xi) &= (\nabla^S_\mu \psi) \otimes_A \xi + \psi \otimes_A (\nabla^E_\mu \xi) \\ &= (\partial_\mu \psi - \gamma(\underset{\sim}{\omega}_\mu)\psi) \otimes_A \xi + \psi \otimes_A (\partial_\mu \xi + \omega^E_\mu \xi) \\ &= \partial_\mu(\psi \otimes_A \xi) + \psi \otimes_A (\omega^E_\mu \xi) - (\gamma(\underset{\sim}{\omega}_\mu)\psi) \otimes_A \xi\end{aligned}$$

for all $\psi \in \mathcal{C}^\infty(S)$ and $\xi \in \mathcal{C}^\infty(E)$. The "$E$-valued" Dirac operator associated with the Clifford connection ∇ is the first order differential operator

$$\begin{aligned}D_\nabla &= (\gamma(dx^\mu) \otimes_A 1)\nabla_\mu \\ &= (\gamma(dx^\mu) \otimes_A 1)(\nabla^S_\mu \otimes_A 1 + 1 \otimes_A \nabla^E_\mu) \qquad (10.8) \\ &= (\gamma(dx^\mu)\nabla^S_\mu) \otimes_A 1 + \gamma(dx^\mu) \otimes_A \nabla^E_\mu \\ &= D \otimes_A 1 + \gamma(dx^\mu) \otimes_A \nabla^E_\mu\end{aligned}$$

acting on the Clifford module $\mathcal{C}^\infty(S \otimes_M E)$. Here D is the Dirac operator induced by g. More explicitly,

$$\begin{aligned}D_\nabla(\psi \otimes_A \xi) &= (D\psi) \otimes_A \xi + \gamma(dx^\mu)\psi \otimes_A (\nabla^E_\mu \xi) \qquad (10.9) \\ &= (D\psi) \otimes_A \xi + \gamma(dx^\mu)\psi \otimes_A (\partial_\mu \xi + \omega^E_\mu \xi)\end{aligned}$$

for all $\psi \in \mathcal{C}^\infty(S)$, $\xi \in \mathcal{C}^\infty(E)$.

Theorem 1. $D_{gq\ell}$ *is a generalized Dirac operator on* $S_{q\ell}$, *i.e. there exist a tensor connection*

$$\nabla = \nabla^S \otimes_A 1 + 1 \otimes_A \nabla^E \qquad (10.10)$$

with Dirac operator D_∇ and a vector bundle endomorphism $\mathcal{E}$ such that

$$D_{gq\ell} = D_\nabla + \Gamma \otimes_A \mathcal{E}.$$

The corresponding connection 1-form ω^E is

$$\begin{bmatrix} \omega_a \otimes \begin{pmatrix} a & 0 \\ 0 & 0 \end{pmatrix} + \omega \otimes \begin{pmatrix} \frac{2}{3} & 0 \\ 0 & 2 \end{pmatrix} & 0 & 0 \\ 0 & \omega_a \otimes a - \omega \otimes \underline{\underline{\frac{4}{3}}} & 0 \\ 0 & 0 & \Omega \otimes \begin{pmatrix} \frac{1}{3} & 0 \\ 0 & 1 \end{pmatrix} + \underline{\omega_a} \otimes \begin{pmatrix} a & 0 \\ 0 & 0 \end{pmatrix} - \underline{\underline{\omega}} \otimes \begin{pmatrix} \frac{1}{3} & 0 \\ 0 & -1 \end{pmatrix} \end{bmatrix}$$
(10.11)

and the endomorphism $\mathcal{E}$, encoding the Higgs field φ and the mass matrices, is given by

$$\begin{bmatrix} 0 & 0 & \kappa^2(\varphi+\varepsilon)^* \otimes \begin{pmatrix} M_d^* & 0 \\ 0 & M_e^* \end{pmatrix} \\ 0 & 0 & \kappa^2(\underset{\sim}{\varphi}+\underset{\sim}{\varepsilon})^* \otimes (M_u^*, 0) \\ (\varphi+\varepsilon) \otimes \begin{pmatrix} M_d & 0 \\ 0 & M_e \end{pmatrix} & (\underset{\sim}{\varphi}+\underset{\sim}{\varepsilon}) \otimes \begin{pmatrix} M_u \\ 0 \end{pmatrix} & 0 \end{bmatrix}. \quad (10.12)$$

Proof. In Proposition 16, the differential operator $D_{gq\ell}$ on $\mathcal{C}^\infty(S \otimes_M E)$ is written in terms of the trivial bundle E, so that (10.9) must be specialized to "constant" sections $\xi = c \in \mathbb{G}_{q\ell}$. In this case

$$\begin{aligned} D_\nabla(\psi \otimes c) &= (D\psi) \otimes c + \gamma(dx^\mu)\psi \otimes_A \omega_\mu^E c \\ &= (D \otimes 1)(\psi \otimes c) + (\gamma(dx^\mu) \otimes \omega_\mu^E)(\psi \otimes c) \\ &= (D \otimes 1 + \gamma(dx^\mu) \otimes \omega_\mu^E)(\psi \otimes c). \end{aligned}$$

Decomposing c according to (9.4), the assertion follows from Proposition 16.

Theorem 2. *In terms of the tensor connection (10.10) we have*

$$-D_{gq\ell}^2 = \nabla^* \nabla + \mathcal{V}$$

where $\mathcal{V}$ is an endomorphism of $S \otimes_M E$ given by

$$-\mathcal{V} = \gamma^\mu \Gamma \otimes_A [\nabla_\mu^E, \mathcal{E}] + \tfrac{1}{2}\gamma^\mu\gamma^\nu \otimes_A R_{\mu\nu}^E + 1 \otimes_A (\mathcal{E}^2 - \tfrac{R}{4}1). \quad (10.13)$$

Here R^E is the curvature of ∇^E and $R = {}^i R_{ij}^j$ is the scalar curvature of g.

Proof. Since $D\Gamma + \Gamma D = 0$ by (7.72) and $\gamma^\mu \Gamma = -\Gamma \gamma^\mu$ by (7.5), it follows that

$$D_\nabla(\Gamma \otimes_A \mathcal{E}) + (\Gamma \otimes_A \mathcal{E})D_\nabla$$
$$= (D \otimes_A 1 + \gamma^\mu \otimes_A \nabla_\mu^E)(\Gamma \otimes_A \mathcal{E}) + (\Gamma \otimes_A \mathcal{E})(D \otimes_A 1 + \gamma^\mu \otimes_A \nabla_\mu^E)$$
$$= (D\Gamma + \Gamma D) \otimes_A \mathcal{E} + \gamma^\mu \Gamma \otimes_A \nabla_\mu^E \mathcal{E} + \Gamma \gamma^\mu \otimes_A \mathcal{E} \nabla_\mu^E = \gamma^\mu \Gamma \otimes_A [\nabla_\mu^E, \mathcal{E}].$$

Therefore the Lichnérowicz formula (Section 6)

$$D_\nabla^2 = -\nabla^* \nabla + \tfrac{1}{2} \gamma^\mu \gamma^\nu \otimes_A R_{\mu\nu}^E - \tfrac{1}{4} R\mathbf{1}$$

for E-valued Dirac operators implies

$$D_{gq\ell}^2 = (D_\nabla + \Gamma \otimes_A \mathcal{E})^2 = D_\nabla^2 + D_\nabla(\Gamma \otimes_A \mathcal{E}) + (\Gamma \otimes_A \mathcal{E})D_\nabla + \Gamma^2 \otimes_A \mathcal{E}^2$$
$$= -\nabla^* \nabla + \tfrac{1}{2} \gamma^\mu \gamma^\nu \otimes_A R_{\mu\nu}^E - \tfrac{1}{4} R\mathbf{1} + \gamma^\mu \Gamma \otimes_A [\nabla_\mu^E, \mathcal{E}] + 1 \otimes_A \mathcal{E}^2 = -\nabla^* \nabla - \mathcal{V}.$$

We now start calculating the various terms in the Lagrange functional.

Proposition 1. *We have $\mathcal{E}^2$ equal to*

$$\kappa^2 \begin{bmatrix} \|\varphi + \varepsilon\|^2 \otimes \begin{pmatrix} M_d^2 & 0 \\ 0 & M_e^2 \end{pmatrix} & 0 & 0 \\ 0 & \|\varphi + \varepsilon\|^2 \otimes \underline{M_u^2} & 0 \\ 0 & 0 & \begin{matrix} (\varphi + \varepsilon)(\varphi + \varepsilon)^* \otimes \begin{pmatrix} M_d^2 & 0 \\ 0 & M_e^2 \end{pmatrix} \\ + (\underset{\sim}{\varphi} + \underset{\sim}{\varepsilon})(\underset{\sim}{\varphi} + \underset{\sim}{\varepsilon})^* \otimes \begin{pmatrix} M_u^2 & 0 \\ 0 & 0 \end{pmatrix} \end{matrix} \end{bmatrix}$$

and $\mathcal{E}^4$ equal to

$$\|\varphi + \varepsilon\|^2 \begin{bmatrix} \|\varphi + \varepsilon\|^2 \otimes \begin{pmatrix} M_d^4 & 0 \\ 0 & M_e^4 \end{pmatrix} & 0 & 0 \\ 0 & \|\varphi + \varepsilon\|^2 \otimes \underline{M_u^4} & 0 \\ 0 & 0 & \begin{matrix} (\varphi + \varepsilon)(\varphi + \varepsilon)^* \otimes \begin{pmatrix} M_d^4 & 0 \\ 0 & M_e^4 \end{pmatrix} \\ + (\underset{\sim}{\varphi} + \underset{\sim}{\varepsilon})(\underset{\sim}{\varphi} + \underset{\sim}{\varepsilon})^* \otimes \begin{pmatrix} M_u^4 & 0 \\ 0 & 0 \end{pmatrix} \end{matrix} \end{bmatrix}.$$

Proof. This follows from (10.12) in view of the identities

$$\phi^* \underset{\sim}{\phi} = 0 = \underset{\sim}{\phi}^* \phi, \quad \phi^* \phi = \|\phi\|^2 = \underset{\sim}{\phi}^* \underset{\sim}{\phi},$$
$$\phi \phi^* \phi \phi^* = \|\phi\|^2 \phi \phi^*, \quad \underset{\sim}{\phi} \underset{\sim}{\phi}^* \underset{\sim}{\phi} \underset{\sim}{\phi}^* = \|\phi\|^2 \underset{\sim}{\phi} \underset{\sim}{\phi}^*, \quad \phi \phi^* \underset{\sim}{\phi} \underset{\sim}{\phi}^* = 0 = \underset{\sim}{\phi} \underset{\sim}{\phi}^* \phi \phi^*.$$

Since $\operatorname{tr} \phi \phi^* = \phi^* \phi = \operatorname{tr} \underset{\sim}{\phi} \overset{*}{\underset{\sim}{\phi}}$, Proposition 1 implies

$$\underline{\operatorname{tr}}_E \mathcal{E}^2 = 2\kappa^2 r \, \|\varphi + \varepsilon\|^2, \tag{10.14}$$

where

$$r := \underline{\operatorname{tr}}_G \, [3(M_d^2 + M_u^2) + M_e^2]$$

as in Section 3 (putting $I_{q\ell} = \operatorname{id}$). Similarly,

$$\underline{\operatorname{tr}}_E \mathcal{E}^4 = 2 \, \underline{\operatorname{tr}}_G \, (3(M_d^4 + M_u^4) + M_e^4) \cdot \|\varphi + \varepsilon\|^4. \tag{10.15}$$

As in (9.59), we consider the covariant derivative

$$\nabla_\mu^H \phi = \partial_\mu \phi + \Omega_\mu \phi - \phi \omega_\mu \tag{10.16}$$

acting on "Higgs" fields $\phi \in \mathcal{C}^\infty(M, \mathbb{C}^{2 \times 1})$.

Proposition 2. *The covariant derivative $[\nabla_\mu^E, \mathcal{E}]$ of $\mathcal{E}$ is given by*

$$\begin{bmatrix} 0 & 0 & \kappa^2 \left(\nabla_\mu^H (\varphi + \varepsilon) \right)^* \otimes \begin{pmatrix} M_d^* & 0 \\ 0 & M_e^* \end{pmatrix} \\ 0 & 0 & \kappa^2 \left(\nabla_\mu^H (\varphi + \varepsilon) \right)_{\sim}^* \otimes \left(M_u^* , 0 \right) \\ \nabla_\mu^H (\varphi + \varepsilon) \otimes \begin{pmatrix} M_d & 0 \\ 0 & M_e \end{pmatrix} & \nabla_\mu^H (\varphi + \varepsilon)_{\sim} \otimes \begin{pmatrix} M_u \\ 0 \end{pmatrix} & 0 \end{bmatrix}.$$

Proof. Since $\omega^E \mathcal{E}$ equals

$$\begin{bmatrix} 0 & 0 & \begin{matrix} \kappa^2 \underline{\omega}_a (\varphi + \varepsilon)^* \otimes \begin{pmatrix} a M_d^* & 0 \\ 0 & 0 \end{pmatrix} + \\ \kappa^2 \underline{\omega} (\varphi + \varepsilon)^* \otimes \begin{pmatrix} \frac{2}{3} M_d^* & 0 \\ 0 & 2 M_e^* \end{pmatrix} \end{matrix} \\ 0 & 0 & \begin{matrix} \kappa^2 \underline{\omega}_a (\varphi + \varepsilon)_{\sim}^* \otimes \left(a M_u^* , 0 \right) \\ -\kappa^2 \underline{\omega} (\varphi + \varepsilon)_{\sim}^* \otimes \left(\frac{4}{3} M_u^* , 0 \right) \end{matrix} \\ \begin{matrix} \Omega(\varphi + \varepsilon) \otimes \begin{pmatrix} M_d & 0 \\ 0 & M_e \end{pmatrix} + \\ \underline{\omega}_a (\varphi + \varepsilon) \otimes \begin{pmatrix} a M_d & 0 \\ 0 & M_e \end{pmatrix} \\ -\underline{\omega}(\varphi + \varepsilon) \otimes \begin{pmatrix} \frac{1}{3} M_d & 0 \\ 0 & -M_e \end{pmatrix} \end{matrix} & \begin{matrix} \Omega(\varphi + \varepsilon)_{\sim} \otimes \begin{pmatrix} M_u \\ 0 \end{pmatrix} + \\ \underline{\omega}_a (\varphi + \varepsilon)_{\sim} \otimes \begin{pmatrix} a M_u \\ 0 \end{pmatrix} \\ -\underline{\omega}(\varphi + \varepsilon)_{\sim} \otimes \begin{pmatrix} \frac{1}{3} M_u \\ 0 \end{pmatrix} \end{matrix} & 0 \end{bmatrix}$$

and $\mathcal{E}\omega^E$ equals

$$\begin{bmatrix} 0 & 0 & \begin{matrix} \kappa^2(\varphi+\varepsilon)^*\Omega\otimes\begin{pmatrix} M_d^* & 0 \\ 0 & M_e^* \end{pmatrix} -\kappa^2(\varphi+\varepsilon)^*\underline{\omega}\otimes\begin{pmatrix} \frac{1}{3}M_d^* & 0 \\ 0 & -M_e^* \end{pmatrix} \\ +\kappa^2(\varphi+\varepsilon)^*\underline{\underline{\omega_a}}\otimes\begin{pmatrix} M_d^*a & 0 \\ 0 & 0 \end{pmatrix} \end{matrix} \\ 0 & 0 & \begin{matrix} \kappa^2(\underline{\varphi+\varepsilon})^*\Omega\otimes\begin{pmatrix} M_u^* \\ 0 \end{pmatrix} -\kappa^2(\underline{\varphi+\varepsilon})^*\underline{\omega}\otimes\begin{pmatrix} \frac{1}{3}M_u^* \\ 0 \end{pmatrix} \\ +\kappa^2(\underline{\varphi+\varepsilon})^*\underline{\underline{\omega_a}}\otimes\begin{pmatrix} M_u^*a \\ 0 \end{pmatrix} \end{matrix} \\ \begin{matrix}(\varphi+\varepsilon)\omega_a\otimes\begin{pmatrix} M_d a & 0 \\ 0 & 0 \end{pmatrix} + \\ (\varphi+\varepsilon)\omega\otimes\begin{pmatrix} \frac{2}{3}M_d & 0 \\ 0 & 2M_e \end{pmatrix} \end{matrix} & \begin{matrix}(\underline{\varphi+\varepsilon})\omega_a\otimes\begin{pmatrix} M_u a \\ 0 \end{pmatrix} \\ -(\underline{\varphi+\varepsilon})\omega\otimes\begin{pmatrix} \frac{4}{3}M_u \\ 0 \end{pmatrix}\end{matrix} & 0 \end{bmatrix}$$

we obtain $[\omega^E, \mathcal{E}]$ as

$$\begin{bmatrix} 0 & 0 & \begin{matrix} \kappa^2(\omega(\varphi+\varepsilon)^* - (\varphi+\varepsilon)^*\Omega) \\ \otimes \\ \begin{pmatrix} M_d^* & 0 \\ 0 & M_e^* \end{pmatrix} \end{matrix} \\ 0 & 0 & \begin{matrix} -\kappa^2(\omega(\underline{\varphi+\varepsilon})^* + (\underline{\varphi+\varepsilon})^*\Omega) \\ \otimes \\ (\underline{M_u^*}, 0) \end{matrix} \\ \begin{matrix}(\Omega(\varphi+\varepsilon)-(\varphi+\varepsilon)\omega) \\ \otimes \\ \begin{pmatrix} M_d & 0 \\ 0 & M_e \end{pmatrix}\end{matrix} & \begin{matrix}(\Omega(\underline{\varphi+\varepsilon})+(\underline{\varphi+\varepsilon})\omega) \\ \otimes \\ \begin{pmatrix} M_u \\ 0 \end{pmatrix}\end{matrix} & 0 \end{bmatrix}.$$

Therefore $[\nabla^E_\mu, \mathcal{E}] = \partial_\mu \mathcal{E} + [\omega^E_\mu, \mathcal{E}]$ is given by

$$\begin{bmatrix} 0 & 0 & \begin{matrix} \kappa^2(\partial_\mu\varphi^*+\omega_\mu(\varphi+\varepsilon)^*-(\varphi+\varepsilon)^*\Omega_\mu) \\ \otimes \\ \begin{pmatrix} M_d^* & 0 \\ 0 & M_e^* \end{pmatrix} \end{matrix} \\ 0 & 0 & \begin{matrix} \kappa^2(\partial_\mu\underline{\varphi}^*-\omega_\mu(\underline{\varphi+\varepsilon})^*-(\underline{\varphi+\varepsilon})^*\Omega_\mu) \\ \otimes \\ (\underline{M_u^*}, 0) \end{matrix} \\ \begin{matrix}(\partial_\mu\varphi+\Omega_\mu(\varphi+\varepsilon)-(\varphi+\varepsilon)\omega_\mu) \\ \otimes \\ \begin{pmatrix} M_d & 0 \\ 0 & M_e \end{pmatrix}\end{matrix} & \begin{matrix}(\partial_\mu\underline{\varphi}+\Omega_\mu(\underline{\varphi+\varepsilon})+(\underline{\varphi+\varepsilon})\omega_\mu) \\ \otimes \\ \begin{pmatrix} M_u \\ 0 \end{pmatrix}\end{matrix} & 0 \end{bmatrix}.$$

In the following we use the normalization conventions for traces and inner products listed at the end of Section 7.

Corollary 1. We have $\kappa^{-2}[\nabla_\alpha^E, \mathcal{E}][\nabla_\mu^E, \mathcal{E}]$ equal to

$$\begin{bmatrix} (\nabla_\alpha^H(\varphi+\varepsilon))^*(\nabla_\mu^H(\varphi+\varepsilon)) & (\nabla_\alpha^H(\varphi+\varepsilon))^*(\nabla_\mu^H(\varphi+\varepsilon)) & 0 \\ \otimes & \otimes & \\ \begin{pmatrix} M_d^2 & 0 \\ 0 & M_e^2 \end{pmatrix} & \begin{pmatrix} M_d^* M_u \\ 0 \end{pmatrix} & \\ (\nabla_\alpha^H(\varphi+\varepsilon))^*(\nabla_\mu^H(\varphi+\varepsilon)) & (\nabla_\alpha^H(\varphi+\varepsilon))^*(\nabla_\mu^H(\varphi+\varepsilon)) & 0 \\ \otimes & \otimes & \\ (M_u^* M_d, 0) & M_u^2 & \\ 0 & 0 & (\nabla_\alpha^H(\varphi+\varepsilon))(\nabla_\mu^H(\varphi+\varepsilon))^* \otimes \begin{pmatrix} M_d^2 & 0 \\ 0 & M_e^2 \end{pmatrix} \\ & & + (\nabla_\alpha^H(\varphi+\varepsilon))(\nabla_\mu^H(\varphi+\varepsilon))^* \otimes \begin{pmatrix} M_u^2 & 0 \\ 0 & 0 \end{pmatrix} \end{bmatrix}$$

and hence

$$\underline{\mathrm{tr}}_E[\nabla_\alpha^E, \mathcal{E}][\nabla_\mu^E, \mathcal{E}] = 2\kappa^2 \, r \, \mathrm{Re}(\nabla_\alpha^H(\varphi+\varepsilon) | \nabla_\mu^H(\varphi+\varepsilon)) \, . \tag{10.17}$$

Lemma 1.
$$\underline{\mathrm{tr}}_{S \otimes E} \mathcal{V} = \frac{15}{4} R - 2\kappa^2 \, r \, \|\varphi + \varepsilon\|^2 \tag{10.18}$$

Proof. In view of the identities

$$\underline{\mathrm{tr}}_{S \otimes E} \gamma^\mu \Gamma \otimes [\nabla_\mu^E, \mathcal{E}] = \underbrace{\underline{\mathrm{tr}}_S \gamma^\mu \Gamma}_{=0} \underline{\mathrm{tr}}_E [\nabla_\mu^E, \mathcal{E}] = 0,$$

$$\underline{\mathrm{tr}}_{S \otimes E} \gamma^\mu \gamma^\nu \otimes R_{\mu\nu}^E = \underline{\mathrm{tr}}_S \gamma^\mu \gamma^\nu \underline{\mathrm{tr}}_E R_{\mu\nu}^E = -\underline{\mathrm{tr}}_S \gamma^\nu \gamma^\mu \underline{\mathrm{tr}}_E R_{\nu\mu}^E = 0,$$

it follows that

$$-\underline{\mathrm{tr}}_{S \otimes E} \mathcal{V} = \underline{\mathrm{tr}}_E (\mathcal{E}^2 - \frac{R}{4} 1).$$

Now the assertion follows with (10.14).

Lemma 2.
$$\underline{\mathrm{tr}}_{S \otimes E} \mathcal{V}^2 = 2\underline{\mathrm{tr}}_{\mathbb{G}}(3(M_d^4 + M_u^4) + M_e^4) \|\varphi + \varepsilon\|^4$$
$$- \kappa^2 r R \|\varphi + \varepsilon\|^2 + \frac{15}{16} R^2$$
$$- 2\kappa^2 r \|\nabla^H(\varphi + \varepsilon)\|^2 + \frac{1}{2} \|R^E\|^2.$$

Proof. In view of the identities

$$\underline{\operatorname{tr}}_{S \otimes E}\big(\gamma^\mu \Gamma \otimes [\nabla^E_\mu, \mathcal{E}]\big)\big(I \otimes (\mathcal{E}^2 - \frac{R}{4}\mathbf{1})\big) = \underbrace{\underline{\operatorname{tr}}_S \gamma^\mu \Gamma}_{=0} \underline{\operatorname{tr}}_E [\nabla^E_\mu, \mathcal{E}](\mathcal{E}^2 - \frac{R}{4}\mathbf{1}) = 0,$$

$$\underline{\operatorname{tr}}_{S \otimes E}\big(\gamma^\mu \gamma^\nu \otimes R^E_{\mu\nu}\big)\big(I \otimes (\mathcal{E}^2 - \frac{R}{4}\mathbf{1})\big) = \underline{\operatorname{tr}}_S \gamma^\mu \gamma^\nu \, \underline{\operatorname{tr}}_E R^E_{\mu\nu}(\mathcal{E}^2 - \frac{R}{4}\mathbf{1})$$
$$= -\underline{\operatorname{tr}}_S \gamma^\nu \gamma^\mu \, \underline{\operatorname{tr}}_E R^E_{\nu\mu}(\mathcal{E}^2 - \frac{R}{4}\mathbf{1}) = 0,$$

$$\underline{\operatorname{tr}}_{S \otimes E}\big(\gamma^\mu \Gamma \otimes [\nabla^E_\mu, \mathcal{E}]\big)\big(\gamma^\alpha \gamma^\beta \otimes R^E_{\alpha\beta}\big) = \underbrace{\underline{\operatorname{tr}}_S \gamma^\mu \Gamma \gamma^\alpha \gamma^\beta}_{=0} \underline{\operatorname{tr}}_E [\nabla^E_\mu, \mathcal{E}] R^E_{\alpha\beta} = 0$$

it follows from (10.13) that the mixed terms in $\mathcal{V}^2$ have trace 0. Therefore

$$\underline{\operatorname{tr}}_{S \otimes E} \mathcal{V}^2 = \underline{\operatorname{tr}}_{S \otimes E}\big(I \otimes (\mathcal{E}^2 - \frac{R}{4}\mathbf{1})^2 + (\gamma^\mu \Gamma \otimes [\nabla^E_\mu, \mathcal{E}])^2 + \frac{1}{4}(\gamma^\mu \gamma^\nu \otimes R^E_{\mu\nu})^2\big).$$

Moreover (7.52) implies

$$\underline{\operatorname{tr}}_{S \otimes E}(\gamma^\mu \gamma^\nu \otimes R^E_{\mu\nu})^2 = \underline{\operatorname{tr}}_S \gamma^\alpha \gamma^\beta \gamma^\mu \gamma^\nu \, \underline{\operatorname{tr}}_E R^E_{\mu\nu} R^E_{\alpha\beta}$$
$$= -(g^{\alpha\beta}g^{\mu\nu} - g^{\alpha\mu}g^{\beta\nu} + g^{\alpha\nu}g^{\beta\mu}) \underline{\operatorname{tr}}_E R^{E*}_{\mu\nu} R^E_{\alpha\beta}$$
$$= (g^{\alpha\mu}g^{\beta\nu} - g^{\alpha\nu}g^{\beta\mu})(R^E_{\mu\nu} \mid R^E_{\alpha\beta})_E$$
$$= g^{\alpha\mu}g^{\beta\nu}(R^E_{\mu\nu} \mid R^E_{\alpha\beta})_E + g^{\beta\mu}g^{\alpha\nu}(R^E_{\mu\nu} \mid R^E_{\beta\alpha})_E$$
$$= 2g^{\alpha\mu}g^{\beta\nu}(R^E_{\mu\nu} \mid R^E_{\alpha\beta})_E = 2(R^E_{\mu\nu} \mid R^{E\mu\nu})_E = 2\|R^E\|^2_E$$

and (10.17) yields

$$-\underline{\operatorname{tr}}_{S \otimes E}(\gamma^\mu \Gamma \otimes [\nabla^E_\mu, \mathcal{E}])^2 = -\underline{\operatorname{tr}}_S \gamma^\alpha \Gamma \gamma^\mu \Gamma \, \underline{\operatorname{tr}}_E [\nabla^E_\alpha, \mathcal{E}][\nabla^E_\mu, \mathcal{E}]$$
$$= \kappa^2 r g^{\alpha\mu} \, 2\operatorname{Re}\big(\nabla^H_\alpha(\varphi + \varepsilon) \mid \nabla^H_\mu(\varphi + \varepsilon)\big)$$
$$= 2\kappa^2 r\big(\nabla^{H\mu}(\varphi + \varepsilon) \mid \nabla^H_\mu(\varphi + \varepsilon)\big) = 2\kappa^2 r \|\nabla^H(\phi + \varepsilon)\|^2.$$

With (10.14) and (10.15), the assertion follows.

Lemma 3.

$$g^{\mu\alpha} g^{\nu\beta} \, \underline{\operatorname{tr}}_{S \otimes E} R^\nabla_{\alpha\beta} R^\nabla_{\mu\nu} = -\frac{15}{8}\|\dot{R}\|^2 - \|R^E\|^2_E. \tag{10.19}$$

Proof. We have
$$R^\nabla = R^S \otimes 1 + 1 \otimes R^E$$

since for any pair X, Y of vector fields

$$\begin{aligned}R^\nabla_{X,Y} &= \nabla_X \nabla_Y - \nabla_Y \nabla_X - \nabla_{[X,Y]} \\ &= (\nabla^S_X \otimes 1 + 1 \otimes \nabla^E_X)(\nabla^S_Y \otimes 1 + 1 \otimes \nabla^E_Y) \\ &\quad - (\nabla^S_Y \otimes 1 + 1 \otimes \nabla^E_Y)(\nabla^S_X \otimes 1 + 1 \otimes \nabla^E_X) - \nabla^S_{[X,Y]} \otimes 1 - 1 \otimes \nabla^E_{[X,Y]} \\ &= \nabla^S_X \nabla^S_Y \otimes 1 + \nabla^S_X \otimes \nabla^E_Y + \nabla^S_Y \otimes \nabla^E_X + 1 \otimes \nabla^E_X \nabla^E_Y \\ &\quad - \nabla^S_Y \nabla^S_X \otimes 1 - \nabla^S_Y \otimes \nabla^E_X - \nabla^S_X \otimes \nabla^E_Y - 1 \otimes \nabla^E_Y \nabla^E_X - \nabla^S_{[X,Y]} \otimes 1 - 1 \otimes \nabla^E_{[X,Y]} \\ &= R^S_{X,Y} \otimes 1 + 1 \otimes R^E_{X,Y}.\end{aligned}$$

Since $R^S_{\mu\nu} = \gamma(.\underset{\sim}{R}\dot{}_{\mu\nu})$ (cf. (7.60)) this implies

$$R^\nabla_{\mu\nu} = R^S_{\mu\nu} \otimes 1 + 1 \otimes R^E_{\mu\nu} = \gamma(.\underset{\sim}{R}\dot{}_{\mu\nu}) \otimes 1 + 1 \otimes R^E_{\mu\nu}.$$

Therefore

$$\begin{aligned}R^\nabla_{\alpha\beta} R^\nabla_{\mu\nu} &= (\gamma(.\underset{\sim}{R}\dot{}_{\alpha\beta}) \otimes 1 + 1 \otimes R^E_{\alpha\beta})(\gamma(.\underset{\sim}{R}\dot{}_{\mu\nu}) \otimes 1 + 1 \otimes R^E_{\mu\nu}) \\ &= \gamma(.\underset{\sim}{R}\dot{}_{\alpha\beta})\gamma(.\underset{\sim}{R}\dot{}_{\mu\nu}) \otimes 1 + \gamma(.\underset{\sim}{R}\dot{}_{\alpha\beta}) \otimes R^E_{\mu\nu} + \gamma(.\underset{\sim}{R}\dot{}_{\mu\nu}) \otimes R^E_{\alpha\beta} + 1 \otimes R^E_{\alpha\beta} R^E_{\mu\nu}\end{aligned}$$

and hence, by Proposition 6,

$$\begin{aligned}\underline{\mathrm{tr}}_{S \otimes E} R^\nabla_{\alpha\beta} R^\nabla_{\mu\nu} &= 15 \underline{\mathrm{tr}}_S \gamma(.\underset{\sim}{R}\dot{}_{\alpha\beta}) \gamma(.\underset{\sim}{R}\dot{}_{\mu\nu}) + \underline{\mathrm{tr}}_E R^E_{\alpha\beta} R^E_{\mu\nu} + \underbrace{\underline{\mathrm{tr}}_S \gamma(.\underset{\sim}{R}\dot{}_{\alpha\beta})}_{=0} \underline{\mathrm{tr}}_E R^E_{\mu\nu} \\ &\quad + \underbrace{\underline{\mathrm{tr}}_S \gamma(.\underset{\sim}{R}\dot{}_{\mu\nu})}_{=0} \underline{\mathrm{tr}}_E R^E_{\alpha\beta} \\ &\stackrel{*}{=} -\frac{15}{8}(.\underset{}{R}\dot{}_{\alpha\beta} \mid .\underset{}{R}\dot{}_{\mu\nu}) - \underline{\mathrm{tr}}_E R^E_{\alpha\beta} R^E_{\mu\nu} = -\frac{15}{8}(.\underset{}{R}\dot{}_{\alpha\beta} \mid .\underset{}{R}\dot{}_{\mu\nu}) - (R^E_{\alpha\beta} \mid R^E_{\mu\nu})_E.\end{aligned}$$

It follows that

$$\begin{aligned}-g^{\mu\alpha} g^{\nu\beta} \underline{\mathrm{tr}}_{S \otimes E} R^\nabla_{\alpha\beta} R^\nabla_{\mu\nu} &= g^{\mu\alpha} g^{\nu\beta} \left(\frac{15}{8}(.\underset{}{R}\dot{}_{\alpha\beta} \mid .\underset{}{R}\dot{}_{\mu\nu}) + (R^E_{\alpha\beta} \mid R^E_{\mu\nu})_E \right) \\ &= \frac{15}{8}(.\underset{}{R}\dot{}^{\mu\nu} \mid .\underset{}{R}\dot{}_{\mu\nu}) + (R^{E\mu\nu} \mid R^E_{\mu\nu})_E = \frac{15}{8} \|.\underset{}{R}\dot{}\|^2 + \|R^E\|^2_E.\end{aligned}$$

10.2 Spectral Action and Heat Kernel Invariants

We will now formulate the spectral action for the second order differential operator

$$P := -D^2_{g q \ell} \tag{10.20}$$

realized as a generalized Laplace operator according to Theorem 2 (the minus sign stems from the fact that $D_{g q \ell}$ is skew-hermitian).

For (compact) Riemannian manifolds M, the spectral properties of P are encoded in the well-known *heat kernel expansion* [91]

$$\{e^{-tP}\}(x,x) \sim (4\pi t)^{-n/2} \sum_{j\geq 0} t^j \, a_j(P,x) \tag{10.21}$$

of the (positive) operator P, regarded in the asymptotic sense as $t \to \infty$. Here $n = \dim M$ and $x \in M$. This series allows to express the trace

$$\operatorname{Tr} e^{-tP} \sim (4\pi t)^{-n/2} \sum_{j\geq 0} t^j \int_M d^n x \cdot \underline{\operatorname{tr}}_{S\otimes E} \, a_j(P,x)$$

of the trace-class operator e^{-tP}. Using the expansion (10.21) one can show [31] and Section 6 (Proposition 3.2) that for large m

$$\operatorname{Tr} \chi\left(\frac{P}{m^2}\right) \sim (4\pi)^{-n/2} \sum_{j\geq 0} m^{2j-n} \, \chi_j \, \operatorname{Tr} a_j(P) \tag{10.22}$$

holds for any positive "cut-off" function χ on $[0, \infty)$ which has an analytic extension to the right half-plane decaying sufficiently fast at ∞. Fixing such a function χ, we put

$$\chi_0 := \int_0^\infty du\, \chi(u)\, u\,, \qquad \chi_1 := \int_0^\infty du\, \chi(u) \tag{10.23}$$

and $\chi_{j+2} := (-1)^j \chi^{(j)}(0)$ for $j \geq 0$. Note that the coefficients χ_0, χ_1, χ_2 are positive. Putting $n = 4$ we obtain for $m \to \infty$

$$\operatorname{Tr} \chi\left(\frac{P}{m^2}\right) \sim \mathcal{L}^{(m)} + \mathcal{O}(m^{-2})\,,$$

where

$$\mathcal{L}^{(m)} = (4\pi)^{-2}\{m^4 \chi_0 \operatorname{Tr} a_0(P) + m^2 \chi_1 \operatorname{Tr} a_1(P) + \chi_2 \operatorname{Tr} a_2(P)\} \tag{10.24}$$

is considered as the "renormalized" action at mass scale m.

For manifolds of Minkowski signature $(+ - --)$ the analogue of the heat kernel expansion (10.21) is the so-called *Schwinger-de Witt expansion*

$$\{e^{-itP}\}(x,x) \sim i(4\pi it)^{-n/2} \sum_{j\geq 0} (it)^j \, a_j(P,x) \tag{10.25}$$

used extensively in quantum field theory on curved space-time [61][18][89]. While a mathematically rigorous study of this series and its spectral information is much more subtle (cf. [89] for discussion and references) it is nevertheless possible to derive an action analogous to (10.24) by using ideas from renormalization theory [18][61]. These ideas work best in the "adiabatic limit" obtained by considering $P + m^2$ at large mass-scale m. By (10.25) we have

$$\{e^{-it(P+m^2)}\}(x,x) \sim i(4\pi it)^{-n/2} \sum_{j \geq 0} (it)^j \, e^{-it\, m^2} \, a_j(P,x) \, . \tag{10.26}$$

Now consider (formally) the propagator

$$(P+m^2)^{-1} = \int_0^\infty i \, dt \, e^{-it(P+m^2)}$$

and its logarithm

$$ln\,(P+m^2)^{-1} = \int_0^\infty i\, dt\, (it)^{-1} e^{-it(P+m^2)} \, .$$

Inserting (10.26) we obtain the "effective Lagrangian density"

$$\mathcal{L}_{\text{eff}}(x) := \frac{i}{2} \text{tr}_{S \otimes E}\{ln(P+m^2)\}(x,x) = -\frac{i}{2} \text{tr}_{S \otimes E}\{ln(P+m^2)^{-1}\}(x,x)$$

$$\sim \frac{1}{2}(4\pi)^{-n/2} \sum_{j \geq 0} \text{tr}_{S \otimes E} a_j(P,x) \int_0^\infty i\, dt\, (it)^{j-1-n/2} e^{-it\, m^2}$$

$$= \frac{1}{2}(4\pi)^{-n/2} \sum_{j \geq 0} \text{tr}_{S \otimes E} a_j(P,x) \, \Gamma(j - \frac{n}{2}) m^{n-2j} \tag{10.27}$$

asymptotically as $m \to \infty$ [18, (6.41)]. The last equality follows by taking Laplace transforms. So far we have kept the "space-time" dimension n as a variable, but now we let $n \uparrow 4$. Since $\Gamma(j - \frac{n}{2})$ is regular at $n = 4$ for $j > 2$ we obtain

$$\mathcal{L}_{\text{eff}}(x) = \frac{1}{2}(4\pi)^{-n/2} \Big(\underline{\text{tr}}_{S \otimes E} a_0(P,x)\, \Gamma(-\frac{n}{2}) m^n +$$

$$\underline{\text{tr}}_{S \otimes E} a_1(P,x)\, \Gamma(1-\frac{n}{2}) m^{n-2} + \underline{\text{tr}}_{S \otimes E} a_2(P,x)\, \Gamma(2-\frac{n}{2}) m^{n-4} \Big) + O(m^{-2}) \, .$$

Put

$$c_n := \frac{2}{4-n} - \gamma$$

where γ is Euler's constant. Then

$$\begin{array}{rcl} \frac{1}{c_n} \Gamma\left(-\frac{n}{2}\right) & = & \frac{4}{n(n-2)} + o(4-n) \, , \\ \frac{1}{c_n} \Gamma\left(1-\frac{n}{2}\right) & = & \frac{-2}{n-2} + o(4-n) \, , \\ \frac{1}{c_n} \Gamma\left(2-\frac{n}{2}\right) & = & 1 + o(4-n) \, . \end{array}$$

Therefore (10.27) implies

$$\mathcal{L}_{\text{eff}}(x) \sim c_n \, \mathcal{L}_n^{(m)}(x) + \mathcal{O}(m^{-2})$$

as $n \to 4$, where

$$\mathcal{L}_n^{(m)}(x) = \tfrac{1}{2}(4\pi)^{-n/2}\left\{\tfrac{4m^n}{n(n-2)}\underline{\text{tr}}_{S\otimes E}a_0(P,x) - \tfrac{2m^{n-2}}{n-2}\underline{\text{tr}}_{S\otimes E}a_1(P,x) + m^{n-4}\underline{\text{tr}}_{S\otimes E}a_2(P,x)\right\}$$

is regular at $n = 4$, with value

$$\mathcal{L}^{(m)}(x) = \tfrac{1}{32\pi^2}\left\{\tfrac{m^4}{2}\underline{\text{tr}}_{S\otimes E}a_0(P,x) - m^2\underline{\text{tr}}_{S\otimes E}a_1(P,x) + \underline{\text{tr}}_{S\otimes E}a_2(P,x)\right\}. \tag{10.28}$$

This expression will be taken as the Lagrangian density in the non-euclidean case. Comparing (10.24) and (10.28) we see that for space-time of signature $\kappa \in \{1, i\}$ we may write the Lagrangian density as

$$\mathcal{L}^{(m)}(x) = \chi_0 m^4 \underline{\text{tr}}_{S\otimes E} a_0(P,x) - \kappa^2 \chi_1 m^2 \underline{\text{tr}}_{S\otimes E} a_1(P,x) + \chi_2 \underline{\text{tr}}_{S\otimes E} a_2(P,x) \tag{10.29}$$

where χ_0, χ_1, χ_2 are positive constants.

Theorem 3. *For $P = -D_{gq\ell}^2$ we have*

$$\underline{\text{tr}}_{S\otimes E} a_0(P) = 15 \;,$$

$$\underline{\text{tr}}_{S\otimes E} a_1(P) = 2\kappa^2 r \|\varphi + \varepsilon\|^2 - \tfrac{5}{4} R \;,$$

$$\underline{\text{tr}}_{S\otimes E} a_2(P) = \underline{\text{tr}}_G \left(3(M_d^4 + M_u^4) + M_e^4\right) \|\varphi + \varepsilon\|^4 \tag{10.30}$$

$$- \tfrac{1}{6} \kappa^2 r \, R \|\varphi + \varepsilon\|^2 + \kappa^2 r \, \|\nabla^H(\varphi + \varepsilon)\|^2$$

$$+ \tfrac{1}{6} \|R^E\|_E^2 - \tfrac{3}{16} C_{\mu\nu} C^{\mu\nu}$$

$$+ \text{surface terms} + \text{topological terms}.$$

Here $C_{\mu\nu}$ is the Weyl curvature tensor of g and

$$r := \underline{\text{tr}}_G \left(3(M_d^2 + M_u^2) + M_e^2\right).$$

Note that surface terms and topological terms do not contribute to the variation of the integrated Lagrangian.

Proof. Since $P = \nabla^*\nabla + V$ is a generalized Laplacian according to Theorem 2, we may apply [91, p. 336, Theorem 4.1.6] to obtain explicit geometric expressions for the low order terms $a_j(P,x)$ (in the non-euclidean case, with signature $+---$, cf. [18, p. 160]). Since $a_0(P) = \mathbf{1}$ we have

$$\underline{\text{tr}}_{S\otimes E} a_0(P) = \underline{\text{tr}}_E \mathbf{1} = 15 \;.$$

Since $a_1(P) = \frac{R}{6}\mathbf{1} - \mathcal{V}$, Lemma 1 implies

$$\underline{\mathrm{tr}}_{S\otimes E}\, a_1(P) = \frac{R}{6}\underline{\mathrm{tr}}_{S\otimes E}\,\mathbf{1} - \underline{\mathrm{tr}}_{S\otimes E}\,\mathcal{V}$$

$$= \frac{15}{6}R - \left(\frac{15}{4}R - 2\kappa^2 r\,\|\varphi+\varepsilon\|^2\right) = 2\kappa^2 r\,\|\varphi+\varepsilon\|^2 - \frac{5}{4}R\;.$$

In the formula

$$a_2(P) = \frac{1}{30}\nabla_\mu\nabla^\mu\,R - \frac{1}{6}\nabla_\mu\nabla^\mu\,\mathcal{V}$$

$$+ \frac{1}{180}(.R^{\cdot}{}_{\mu\nu}\,|.R^{\cdot\,\mu\nu})\,\mathbf{1} - \frac{1}{180}Ric_{\mu\nu}\,Ric^{\mu\nu}\,\mathbf{1} + \frac{R^2}{72}\mathbf{1}$$

$$- \frac{1}{6}R\mathcal{V} + \frac{\mathcal{V}^2}{2} + \frac{1}{12}R^\nabla_{\mu\nu}\,R^{\nabla\mu\nu}$$

we may neglect the first two terms which are total differentials (surface terms) and hence do not contribute to the integrated Lagrangian. For the remaining terms we obtain, using Lemma 1, Lemma 2 and Lemma 3

$$\underline{\mathrm{tr}}_{S\otimes E}\, a_2(P) = \frac{1}{180}((.R^{\cdot}{}_{\mu\nu}\,|.R^{\cdot\,\mu\nu}) - Ric_{\mu\nu}\,Ric^{\mu\nu})\,\underline{\mathrm{tr}}_{S\otimes E}\mathbf{1}$$

$$+ \frac{R^2}{72}\underline{\mathrm{tr}}_{S\otimes E}\mathbf{1} - \frac{1}{6}R\,\underline{\mathrm{tr}}_{S\otimes E}\mathcal{V} + \frac{1}{2}\underline{\mathrm{tr}}_{S\otimes E}\mathcal{V}^2 + \frac{1}{12}\underline{\mathrm{tr}}_{S\otimes E}\,R^\nabla_{\mu\nu}\,R^{\nabla\mu\nu} =$$

$$\frac{15}{180}((.R^{\cdot}{}_{\mu\nu}\,|.R^{\cdot\,\mu\nu}) - Ric_{\mu\nu}\,Ric^{\mu\nu}) + \frac{15}{72}R^2 - \frac{R}{6}\left(\frac{15}{4}R - 2\kappa^2 r\,\|\varphi+\varepsilon\|^2\right)$$

$$+ \frac{1}{2}\left\{2\underline{\mathrm{tr}}_{\mathbb{G}}\,(3(M_d^4 + M_u^4) + M_e^4)\,\|\varphi+\varepsilon\|^4 + \frac{15}{16}R^2 - \kappa^2 r\,R\,\|\varphi+\varepsilon\|^2\right.$$

$$\left. - 2\kappa^2 r\,\|\nabla^H(\varphi+\varepsilon)\|^2 + \frac{1}{2}\|R^E\|_E^2\right\}$$

$$+ \frac{1}{12}\left(\frac{15}{8}(.R^{\cdot}{}_{\mu\nu}|.R^{\cdot\,\mu\nu}) - \|R^E\|_E^2\right) = \frac{1}{96}\left(5R^2 - 8\,Ric_{\mu\nu}\,Ric^{\mu\nu} - 7(.R^{\cdot}{}_{\mu\nu}|.R^{\cdot\,\mu\nu})\right)$$

$$- \frac{1}{6}\kappa^2 rR\|\varphi+\varepsilon\|^2 + \underline{\mathrm{tr}}_{\mathbb{G}}(3(M_d^4 + M_u^4) + M_e^4)\|\varphi+\varepsilon\|^4 - \kappa^2 r\|\nabla^H(\varphi+\varepsilon)\|^2 + \frac{1}{6}\|R^E\|_E^2.$$

Now the assertion follows in view of the identity

$$5R^2 - 8\,Ric_{\mu\nu}\,Ric^{\mu\nu} - 7(.R^{\cdot}{}_{\mu\nu}\,|.R^{\cdot\,\mu\nu})$$

$$= -18((.R^{\cdot}{}_{\mu\nu}|.R^{\cdot\,\mu\nu}) - 2Ric_{\mu\nu}Ric^{\mu\nu} + \frac{R^2}{3}) + 11((.R^{\cdot}{}_{\mu\nu}|.R^{\cdot\,\mu\nu}) - 4Ric_{\mu\nu}\,Ric^{\mu\nu} + R^2)$$

$$= -18\,C_{\mu\nu}\,C^{\mu\nu} + 11\chi$$

for the Weyl tensor $C_{\mu\nu}$ and the Euler characteristic χ (a topological invariant which does not contribute to the variation of the Lagrangian).

Combining Theorem 3 with (10.29) it follows that the "spectral action" $\mathcal{L}^{(m)}$ at mass scale m has the density (up to surface terms and topological terms)

$$\mathcal{L}^{(m)}(x) = 15\, m^4 \chi_0 + m^2 \chi_1 (\frac{5}{4} \kappa^2 R - 2r\, \|\varphi + \varepsilon\|^2)$$
$$+ \chi_2 \{\mathrm{tr}_{\mathbb{G}}(3(M_d^4 + M_u^4) + M_e^4) \cdot \|\varphi + \varepsilon\|^4$$
$$- \frac{1}{6} \kappa^2 r\, R\, \|\varphi + \varepsilon\|^2 + \kappa^2 r\, \|\nabla^H(\varphi + \varepsilon)\|^2 + \frac{1}{6} \|R^E\|_E^2 - \frac{3}{16} C_{\mu\nu}\, C^{\mu\nu} \} \, .$$

Lemma 4. *The curvature R^E of ∇^E satisfies*

$$\frac{1}{4} \|R^E\|_E^2 = \frac{10}{3} \|F(\omega)\|^2 + \|F(\Omega)\|^2 + \|F(\omega_a \otimes a)\|^2 \, ,$$

where the connection 1-form ω^E is given by (10.11) and F denotes the respective curvature 2-forms (field strength).

Proof. According to (10.11), we have R^E equal to

$$\begin{bmatrix} \gamma\left(F\left(\omega_a \otimes \begin{pmatrix} a & 0 \\ 0 & 0 \end{pmatrix}\right)\right) \\ + \\ \gamma(F(\omega)) \otimes \begin{pmatrix} 2/3 & 0 \\ 0 & 2 \end{pmatrix} & 0 & 0 \\ \\ 0 & \begin{array}{c} \gamma(F(\omega_a \otimes a)) \\ \underline{} \\ \gamma(F(\omega)) \otimes 4/3 \end{array} & 0 \\ \\ 0 & 0 & \begin{array}{c} \gamma(F(\Omega)) \otimes \begin{pmatrix} 1 & 0 \\ 0 & 1 \end{pmatrix} \\ + \\ \gamma\left[F\left(\underline{\omega_a} \otimes \begin{pmatrix} a & 0 \\ 0 & 0 \end{pmatrix}\right)\right] \\ \underline{} \\ \gamma(F(\underline{\omega})) \otimes \begin{pmatrix} 1/3 & 0 \\ 0 & -1 \end{pmatrix} \end{array} \end{bmatrix}$$

Since $(\sigma_1 \otimes M_1 | \sigma_2 \otimes M_2) = (\sigma_1 | \sigma_2)\, \mathrm{tr}\, M_1^* M_2$ for 2-forms σ_1, σ_2 and matrices M_1, M_2 it follows that

$$\|R^E\|_E^2 = \|F(\omega_a \otimes a)\|^2 + \|F(\omega)\|^2 \left(\tfrac{4}{9} \cdot 3 + 4\right)$$
$$+ \|F(\omega_a \otimes a)\|^2 + \|F(\omega)\|^2\, \tfrac{16}{9} \cdot 3 + \|F(\Omega)\|^2 \cdot 4$$
$$+ \|F(\omega_a \otimes a)\|^2 \cdot 2 + 2\|F(\omega)\|^2 \left(\tfrac{1}{9} \cdot 3 + 1\right) \, .$$

11 The Higgs Mechanism and Spontaneous Symmetry Breaking

Florian Scheck

University of Mainz

11.1 Historical Note

As is well known all gauge bosons of a pure Yang-Mills theory are necessarily massless. This is so because any ad-hoc mass term such as

$$m_i^2 A_\mu^{(i)} A^{(i)\,\mu} \quad \text{or} \quad \sum_{ik} M_{ik} A_\mu^{(i)} A^{(k)\,\mu}$$

is incompatible with local gauge invariance. It is said that W. Pauli had developed nonabelian gauge theory for himself (or knew about it from the work of H. Weyl and O. Klein) before the work of C.N. Yang and R. Mills (1954) but dismissed it because he had realized that the gauge particles would all be massless. As there was only one massless spin-1 particle known at the time (the photon) nonabelian gauge theory was to be rejected on physical grounds. The few facts that were reported about this are collected in a recent paper by O'Raifeartaigh and Straumann [171] which also gives a more detailed account of the earlier work by H. Weyl and O. Klein.

The discovery of spontaneous symmetry breaking in gauge theories by Englert, Brout, Higgs, and Kibble in the mid-sixties was received as a genuine breakthrough: The principle of spontaneous symmetry breaking (SSB) had already appeared in Heisenberg's theory of ferromagnets (Landau). Also the theorem by J. Goldstone (1961) was known which says that in a manifestly Lorentz covariant theory which admits a continuous *global* symmetry, SSB implies the existence of massless *scalar* particles, the so-called *Goldstone bosons*. The new feature, realized by these authors in the mid-sixties, was that if this "covariant theory" is a local gauge theory with SSB then the Goldstone bosons appear in a disguise. Some of the formerly massless gauge bosons which, with regard to the Lorentz group, are classified in the two dimensional, helicity representation of the isotropy group of spatial 3-momentum, become genuine massive vector bosons classified in the triplet representation of the rotation group. The extra degree of freedom in the latter case is the remnant of a would-be Goldstone boson. (These matters are worked out, in all their filigree, in the reviews [15], [170] and in the book [169]).

However beautiful these ideas are, however subtle the interplay of Yang-Mills theory and spontaneous symmetry breaking, this theoretical framework still has (at least) two major weaknesses:
(i) The SSB mechanism is added in an ad hoc manner to the Yang-Mills

theory, post factum[1]. The specific choice of reduction of the original gauge group (obtained from the structure group G) to the residual gauge group (obtained from a subgroup H of G) is fixed, on purely empirical grounds, by the residual symmetry that the ground state of the theory is required to have.

(ii) The theory, in its spontaneously broken phase, at first describes the bosonic sector only, that is the kinetic energies and interactions of the spin-1 gauge bosons and the spin-0 scalar fields. In adding what we physicists call the *matter fields* to the theory, i.e. the quarks and the leptons, and in constructing the physically relevant Dirac operator much freedom is left. To witness, the choice of representations of G into which quarks and leptons are classified is guided by requiring anomalies to be absent. The presence of anomalies signals that symmetries of the classical theory, upon quantization "do not go through", i.e. no longer are symmetries of its quantized version. Also the Yukawa couplings of fermions to the scalar particles and the fermionic mass terms, to a large extent, remain undetermined and must be adjusted in accord with the empirical information.

Although this is still not fully recognized by the physics community noncommutative Geometry has brought about important progress in these matters. In all realistic versions of the standard model based on noncommutative geometry SSB is a consequence of the noncommutative structure and the ensuing differential calculus. The Higgs fields, in a rather natural way, are part of the connection, hence belong to the *gauge* rather than the *matter* sector of the model, and the electrically neutral Higgs appears from the start in the correct physical phase. No shift from the origin $\Phi = 0$ to an absolute minimum of the self-interaction is necessary. Quarks and leptons occur in representations of *algebras* and, therefore, the freedom of choice is reduced considerably. In the Mainz-Marseille model 13 for instance which is based on graded Lie algebras, the vanishing of the supertrace in every representation implies automatically the absence of chiral anomalies [190][2]. In all models the fermionic building blocks are the *chiral* fields

$$\Psi_R := \frac{1}{2}(\mathbb{1} + \gamma_5)\Psi, \qquad \Psi_L := \frac{1}{2}(\mathbb{1} - \gamma_5)\Psi, \qquad (11.1)$$

which are the ones that Nature seems to prefer in weak interactions. Conventional (commutative) Yang-Mills theories are indifferent to chirality. Indeed, it is possible to construct gauge theories where charged weak interactions involve left-handed and right-handed currents in arbitrary mixtures. Finally, in the model(s) based on Connes' Noncommutative Geometry 12 the Dirac operator is the driving agent that defines the relevant differential algebra $\Omega(\mathcal{A})$.

---

[1] There have been numerous attempts to generate SSB dynamically from the theory itself but none of these were conclusive or convincing.

[2] In the models of the Connes-Lott type the fixing of the weak hypercharges, responsible for anomaly cancellation, is somewhat artificial and remains to be studied further.

The choice of the algebra $\mathcal{A}$ and of the Dirac operator fixes the bosonic sector of the model as well as the fermionic couplings, thus establishing a novel link between the fermionic and bosonic sectors.

11.2 Spontaneous Symmetry Breaking and Goldstone Theorem

When introducing SSB into a nonabelian Yang-Mills theory the strategy is the following: The Lagrangian is constructed such as to be globally and locally invariant under the structure group G, a compact Lie group, and the gauge group $\mathcal{G}$ obtained from G, respectively. In the case of the standard model $G = SU(2)_L \times U(1)_Y$ or, including chromodynamic interactions $G = SU(2)_L \times U(1)_Y \times SU(3)_c$. One then identifies the residual symmetry $H \subset G$ (and the corresponding gauge group $\mathcal{H} \subset \mathcal{G}$, respectively) which should be obtained on grounds of what physics tells us. In the case of the minimal standard model, $H = U(1)$ or $H = U(1) \times SU(3)_c$, respectively. The symmetry must be broken in such a way that the gauge invariance of the theory is preserved – in view of its renormalizability after quantization – but that the original mass degeneracy of the multiplets of the gauge fields is lifted. In other words, while the Lagrangian still possesses the full symmetry, its physical ground state has less symmetry. It is for this reason that SSB is often called *hidden symmetry*, see e.g. [170]. So both G and H are inputs.

Let Φ be a set of scalar fields falling in a given, reducible or irreducible, representation of G. That is to say, in the case of the minimal electroweak model, Φ spans a representation of $SU(2)_L$ and is assigned a weak hypercharge y with respect to $U(1)_Y$. Now, as the photon eventually will be a mixture of a gauge field pertaining to the former and the gauge field pertaining to the latter factor, it is clear that the electric charges of the components ϕ_m of Φ will be fixed in terms of eigenvalues of the generators I_3 and Y of Lie G. The requirement is then that among the components of Φ there should be one which is electrically neutral. For a physicist this is an obvious requirement which is related to the absolute conservation of charge. Electric charges of single particle states can be defined only if the vacuum is an eigenstate of charge with eigenvalue zero. A component of the Higgs multiplet Φ can have a nonvanishing vacuum expectation value only if it is electrically neutral. Furthermore, in a theory without space-time supersymmetry the vacuum state is assigned spin zero. This implies that the Higgs fields must be scalars.

Let $V(\Phi)$ be a self-interaction ("potential") which satisfies the following conditions

(i) $V(\Phi)$ is (globally) invariant under G,

$$V(U_H(g)\Phi) = V(\Phi) \quad \text{for all } g \in G, \tag{11.2}$$

(ii) $V(\Phi)$ possesses an absolute minimum in Φ_0,
(iii) the minimum is *degenerate*, i.e. it is not invariant with respect to G.

Here and below $U_H(g)$ denotes the unitary representation which contains the Higgs multiplet. The degeneracy implies that there is at least one element $g \in G$ for which $U_H(g)\Phi_0 \neq \Phi_0$ or, equivalently, there is at least one generator T^k of Lie G for which $U_H(T^k)\Phi_0 \neq 0$. It is natural, therefore, to define the generators of the Lie algebra such that they fall in either of the two classes, (a) $\{T^1, \ldots, T^P\}$ which annihilate the "vacuum", $U_H(T^k)\Phi_0 = 0$, and (b) the remaining ones $\{T^{P+1}, \ldots, T^{\dim G}\}$ for which $U_H(T^l)\Phi_0 \neq 0$. Obviously class (a) generates a subgroup H of G, the *isotropy group* of Φ_0 for which $\dim H = P$.

Without having to specify the form of the self-interaction $V(\Phi)$ any further one shows that a model of the kind

$$\mathcal{L} = \frac{1}{2}(\partial_\mu \Phi^* \partial^\mu \Phi) - V(\Phi),$$

with $V(\Phi)$ satisfying the conditions (i) - (iii), predicts the existence of $N_{GB} = (\dim G - \dim H)$ massless scalar particles. This is the content of

Goldstone's Theorem 1. *A manifestly Lorentz invariant theory whose internal symmetry is spontaneously broken from G to the isotropy group $H \subset G$ of Φ_0, absolute minimum of $V(\Phi)$, contains massless scalar fields. The number of such fields is*

$$N_{GB} = \dim G - \dim H, \tag{11.3}$$

and is independent of the representation of G which classifies Φ. The Goldstone fields are tangent vectors to the group orbit of Φ_0.

The proof in this framework is easy. By assumption $V(\Phi)$ has an absolute minimum in Φ_0. Therefore, the matrix of second derivatives

$$\mathcal{M}_{mn} = \left.\frac{\partial^2 V}{\partial \phi_m \partial \phi_n}\right|_{\Phi_0}$$

is positive-semidefinite and takes the role of a mass matrix $\mathbf{M} = \{\mathcal{M}_{mn}\}$ for the tangent vectors $\boldsymbol{v}^{(k)} = U_H(T^k)\Phi_0$ of the orbit of Φ_0. From the minimum condition one concludes $\boldsymbol{v}^{(k)T}\mathbf{M}\boldsymbol{v}^{(l)} = 0$ and, from the positive semidefiniteness $\mathbf{M}\boldsymbol{v}^{(l)} = 0$. Now, $\boldsymbol{v}^{(l)}$ does not vanish whenever T^l is from class (b). As this happens for $(\dim G - \dim H)$ generators the mass matrix $\mathbf{M}$ must have N_{GB} vanishing eigenvalues.

Remark 1. (i) In physical terms this means that we are given a Lagrangian containing kinetic terms for the fields Φ, the self-interaction $V(\Phi)$ and, possibly, other fields, which possesses the global symmetry G. The equivalence class $[\Phi_0]$ where the potential has its absolute minimum, defines the ground state(s) of the model. In these ground states the original symmetry is hidden because the Lagrangian has the full symmetry G but the ground state, being invariant under $H \subset G$, has less symmetry.

(ii) Intuitively speaking the Goldstone fields are excitations *along* the orbit of Φ_0. As the minimum $V(\Phi_0)$ of the self-interaction is degenerate these excitations can have arbitrarily small frequency which is to say that they belong to massless particles.

(iii) For the mathematicians in the audience I may add that this realization of hidden symmetry (sometimes called *Goldstone* or *Goldstone-Nambu symmetry realization* – as opposed to *Wigner realization* where the symmetry results in mass-degenerate multiplets) is relevant in several branches of physics. It appears, for instance, in the context of massless excitations in solid state physics, but also in the effective, approximate, chiral phase of strong interactions at low energies.

(iv) Note that we did not need to specify the self-interaction $V(\Phi)$ any further. This is true in the classical version of the theory. In the process of quantizing the model a new requirement appears: A necessary condition for it to be renormalizable $V(\Phi)$ should contain no more than quartic terms. Thus, if the self-interaction is polynomial its generic form must be

$$V(\Phi) = -\frac{1}{2}\mu^2 \Phi^* \Phi + \frac{1}{4}\lambda (\Phi^*\Phi)^2 \quad \text{with } \mu^2 > 0, \, \lambda > 0. \tag{11.4}$$

Its minimum is attained at

$$\Phi_0^* \Phi_0 \equiv v^2 = \frac{\mu^2}{\lambda}. \tag{11.5}$$

11.3 Spontaneous Symmetry Breaking in Yang-Mills Theory

The realization of SSB in a Yang-Mills theory is dramatically different from the situation just described. Before even introducing fermions the gauge invariant Lagrangian

$$\mathcal{L} = -\frac{1}{4g^2}(F_{\mu\nu}, F^{\mu\nu}) + \frac{1}{2}((D_\mu \Phi)^*, D^\mu \Phi) - V(\Phi), \tag{11.6}$$

upon shifting the scalar fields to the absolute minimum of $V(\Phi)$, predicts that a number N_{GB}, eq. (11.3), of gauge bosons become massive. This is so because the *dynamical* scalar fields are

$$\widetilde{\Phi} = \Phi - \Phi_0.$$

Upon insertion into (11.6) this yields a term

$$\frac{1}{2}(U(W_\mu)\Phi_0^*, U(W^\mu)\Phi_0) = \frac{1}{2}\sum_{i=P+1}^{N}\sum_{k=P+1}^{N} M_{ik} W_\mu^{(i)} W^{(k)\,\mu}, \quad (N = \dim G),$$

where $M_{ik} = g^2(U(T^i)\Phi_0^*, U(T^k)\Phi_0)$ is a positive matrix. It has $N_{GB} = \dim G - \dim H$ positive eigenvalues, its eigenstates are obtained by an orthogonal transformation and yield the physical spin-1 particles. Thus, by a suitable choice of G and $H \subset G$ one fixes the number (11.3) of gauge bosons which should become massive. The remaining ones remain massless.

Before applying this general analysis to the minimal standard model let us comment on what happens when fermions are added to the Lagrangian (11.6). Fermions will have their own kinetic term, possibly a primordial mass term, and Yukawa couplings to the Φ, viz.

$$\mathcal{L}_f = i\left(\overline{\Psi}, \gamma^\mu D_\mu \Psi\right) - \left(\overline{\Psi}, \mathcal{M}\Psi\right) - h\left(\overline{\Psi}, \Phi\Psi\right) + \text{h.c.}. \tag{11.7}$$

The notation is symbolic but suggestive: as above D denotes the covariant derivative while the brackets mean a globally G-invariant coupling, $\mathcal{M}$ is a primordial fermionic mass matrix but is admissible only if the fermionic multiplets match to form a G-invariant. Again, upon shifting to the minimum of $V(\Phi)$ the Yukawa coupling produces a fermionic mass matrix, even when the primordial mass terms do not exist.

11.4 The Case of the Electroweak Model: Bosonic Sector

The minimal electroweak model is based on the structure group $U(2)$. Let me denote the generator of the $U(1)$ factor by T^0, and the generators of $SU(2)$ (in a more conventional notation) by $T^1 \equiv I_1$, $T^2 \equiv I_2$, $T^3 \equiv I_3$, and let g denote the coupling constant, so that the gauge potential reads

$$W_\mu(x) = ig \sum_{i=0}^{3} T^i W_\mu^{(i)}. \tag{11.8}$$

Furthermore, let us write the fields $W_\mu^{(3)}(x)$ and $W_\mu^{(0)}(x)$ as orthogonal linear combinations of the prospective, physical photon field $A_\mu(x)$ and Z^0-field $Z_\mu(x)$,

$$\begin{pmatrix} W_\mu^{(3)}(x) \\ W_\mu^{(0)}(x) \end{pmatrix} = \frac{1}{\sqrt{g^2 + g'^2}} \begin{pmatrix} g & g' \\ -g' & g \end{pmatrix} \begin{pmatrix} Z_\mu(x) \\ A_\mu(x) \end{pmatrix}. \tag{11.9}$$

Note that if $g' = 0$ the $U(1)$ field coincides with the photon field while the Z^0 is the isospin partner of the W-bosons. Inserting this into (11.8) one obtains

$$U(W_\mu)\Phi_0 = \ldots + \left\{ \frac{gg'}{\sqrt{g^2 + g'^2}} U(I_3) + \frac{g^2}{\sqrt{g^2 + g'^2}} U(T^0) \right\} A_\mu(x)\Phi_0 + \ldots, \tag{11.10}$$

where, for the moment, we have only written the photon term. Now, in analogy to a famous formula (due to Gell-Mann and Nishijima) of the old "eightfold way" in strong interactions one wishes to write the electric charge operator in the form

$$Q_{e.m.} = I_3 + \frac{1}{2} Y, \tag{11.11}$$

and – again by analogy to strong interactions – wishes to call I_3 the 3-component of *weak isospin*, and Y the *weak hypercharge*. This implies a redefinition of the abelian generator, viz.

$$T^0 = \frac{g'}{g}\frac{1}{2}Y; \qquad (11.12)$$

at the same time this allows to identify the elementary charge:

$$e = \frac{gg'}{\sqrt{g^2+g'^2}}. \qquad (11.13)$$

With this redefinition of T^0 the above term involving the photon field becomes simply

$$U(W_\mu)\Phi_0 = \ldots + eU(I_3 + Y/2)A_\mu \Phi_0 + \ldots,$$

while the corresponding term involving the Z^0 now reads

$$U(W_\mu)\Phi_0 = \ldots + \left\{\frac{g^2}{\sqrt{g^2+g'^2}}U(I_3) - \frac{1}{2}\frac{g'^2}{\sqrt{g^2+g'^2}}U(Y)\right\}Z_\mu(x)\Phi_0 + \ldots, \qquad (11.14)$$

Let y denote the eigenvalue of Y in the representation Φ and pick out the component $\phi^{(0)}$ of Φ that is electrically neutral. The eigenvalue t_3 of $\phi^{(0)}$, obviously, must be related to y by $y = -2t_3$ in order to obtain $Q_{e.m.}(\phi^{(0)}) = 0$. At the same time the coupling term (11.14) for the $\phi^{(0)}$ simplifies

$$\frac{g^2}{\sqrt{g^2+g'^2}}t_3 - \frac{1}{2}\frac{g'^2}{\sqrt{g^2+g'^2}}y = t_3\sqrt{g^2+g'^2}.$$

It remains to identify the charged $W^\pm$-fields. Taking, as usual, $I_\pm = I_1 \pm iI_2$ and $W_\mu^{(\pm)} = \pm(W_\mu^{(1)} \pm iW_\mu^{(2)})/\sqrt{2}$ one sees that the corresponding terms are

$$U(A_\mu)\Phi_0 = \ldots + \left\{\frac{g}{\sqrt{2}}U(I_+)W_\mu^{(-)} + \text{h.c.}\right\}\Phi_0 + \ldots. \qquad (11.15)$$

This shows that g is the coupling constant measured in charged current weak interactions. Comparison to the effective current-current interaction postulated by Fermi allows to relate this constant to the Fermi constant G_F, as measured in weak decays. The precise relation (that I do not work out here but that can be found, e.g. in [191]) is

$$\frac{G_F}{\sqrt{2}} = \frac{g^2}{8m_W^2}. \qquad (11.16)$$

The comparison of the electromagnetic coupling constant (11.13) and the weak interaction constant g (11.16) lead Weinberg to define what he calls the weak interaction angle,

$$\sin^2\vartheta_W := \frac{e^2}{g^2} = \frac{g'^2}{g^2+g'^2}, \quad \text{or} \quad \tan^2\vartheta_W = \frac{g'^2}{g^2}. \qquad (11.17)$$

(Note that the sign of ϑ_W is not physical.)

We are almost through with this somewhat tedious analysis. Inserting the results (11.15) and (11.14), with the neutral component $\phi^{(0)}$ in lieu of Φ_0, squaring these terms, and using $I_+I_- + I_-I_+ = 2(\mathbf{I}^2 - I_3^2)$, shows that m_W^2 and m_Z^2 are given by

$$m_W^2 = \frac{1}{2}g^2\left[t(t+1) - t_3^2\right]v^2,$$
$$m_Z^2 = (g^2 + g'^2)t_3^2 v^2.$$

Here, $t(t+1)$ is the eigenvalue of $\mathbf{I}^2$ in the Higgs multiplet, t_3 is the same as before, and v^2 is the squared modulus of the minimum position (11.5). Comparing these formulae leads to a famous relation

$$\rho := \frac{m_W^2}{m_Z^2 \cos^2 \vartheta_W} = \frac{m_W^2}{m_Z^2}\frac{g^2 + g'^2}{g^2} = \frac{t(t+1) - t_3^2}{2t_3^2}. \tag{11.18}$$

This formula shows clearly the arbitrariness of the choice in the Higgs sector: the only requirement that we have is that $t_3 \neq 0$, because otherwise the Z^0 would remain massless. There is nothing which tells us what the value of t should be. It is again experiment which allows us to measure the two masses and the Weinberg angle independently and which tells us that – up to radiative corrections – the Higgs must be in a doublet, $t = 1/2$. As you will learn in the contributions 13 and 12 the Noncommutative Geometry versions of the standard model cannot tolerate anything but doublets.

Finally, it is worthwhile to calculate the characteristic scale of SSB from these formulae and the Fermi constant: With $G_F = 1.16639 \times 10^{-5}$ GeV one finds

$$v^2 = \frac{1}{G_F\sqrt{2}} = (246 \text{ GeV})^2.$$

Summarizing up to this point we see that by the specific choices made $G = U(2)$ is spontaneously broken to $H = U_{e.m.}(1)$ which is generated by the generator (11.11) and which remains a gauge symmetry of the theory. The Higgs field is chosen to be a doublet,

$$\Phi = \begin{pmatrix} \phi^{(+)} \\ \phi^{(0)} \end{pmatrix}, \tag{11.19}$$

where the superscript denotes electric charge. It has weak hypercharge $y = 1$, in accord with (11.11). The two components being complex the Higgs doublet originally has 4 degrees of freedom only one of which is physical. By a gauge transformation the charged partner can be made to disappear completely, the neutral partner can be chosen real. The remaining real field $\tilde{\phi}^{(0)} = \phi^{(0)} - v$ is massive, its mass being $m_H^2 = 2\lambda v^2$ and remains undetermined. The other 3 degrees of freedom would have become Goldstone bosons in a model without local gauge invariance.

11.5 Electroweak Model: Adding Quarks and Leptons

Couplings to gauge bosons or massive spin-1 bosons always connect *like* chiralities, L to L and R to R, while couplings to scalars and, for that matter, fermionic mass terms of the Dirac type, connect *unlike* chiralities, L to R and R to L, so for instance[3],

$$\overline{\psi_L^{(n)}(x)}\,(\gamma_\mu \text{ or } \gamma_\mu\gamma_5)\,\psi_L^{(m)}(x)\,, \quad \left(\overline{\psi_L^{(n)}} M_{nm} \psi_R^{(m)} + \text{h.c.}\right).$$

In the minimal version of the standard model the left-chiral fermion fields are the ones that couple to the charged $W^\pm$-bosons. The coupling of the charged particles, i.e. e, μ, τ, and all *up*-type and *down*-type quarks, to the neutral Z^0 is a mixed left- and right-handed one. Neutrinos, on the other hand, seem to couple only via left-chiral fields (L type in (11.1))[4]. Accordingly, L-fields must be classified in doublets with respect to $SU(2)_L$, while R-fields must be singlets. Thus, a given lepton generation sits in a reducible triplet composed of a doublet and a singlet, a given quark generation is in a reducible quartet composed of a doublet and two singlets. As is customary in physics we write particle symbols in the place of the fields, e.g. $\nu_L \equiv \psi_L^{(\nu)}(x)$,

$$\Psi^{(l)}(x) = \left(\begin{pmatrix} \nu_L \\ e_L \end{pmatrix}\right), \quad \Psi^{(q)}(x) = \left(\begin{pmatrix} u_L \\ d_L \end{pmatrix}\right). \quad (11.20)$$

There is one more constraint: the couplings of the charged particles to the photon must be of pure vector type, i.e. via γ_μ, because electromagnetic interactions conserve parity. This is important because, even though the weak hypercharge assignment must be the same for the two members of a doublet, it could have any value for the singlets. This is so because weak hypercharge is a $U(1)$ factor. The only restriction there is comes from group theory: representations of the *group* $U(1)$ – not of the *algebra* – must be one-valued. This is Wigner's argument for one-parameter subgroups of the rotation group. In the present case this means that the ratio of the eigenvalues of the doublet and the singlet(s) must be rational. Calculating the couplings of, say, ν_L, e_L and e_R to the neutral bosons and redefining the $U(1)_Y$ generator as in (11.12), one has

$$e_L\,:\,-\frac{1}{2}\left(gW_\mu^{(3)} - y_d g' W_\mu^{(0)}\right)\overline{e_L}\gamma_\mu e_L\,, \quad e_R\,:\,\frac{1}{2}g' y_s W_\mu^{(0)}\overline{e_R}\gamma_\mu e_R\,,$$

$$\nu_L\,:\,\frac{1}{2}\left(gW_\mu^{(3)} + y_d g' W_\mu^{(0)}\right).$$

---

[3] Note, in particular, that *fermionic* mass matrices M_{nm}, due to the freedom in choosing phases of the fields, need not be hermitean.

[4] They also must have a right-chiral component if their masses are of Dirac type. However, these R-fields do not seem to couple to the gauge sector of the model.

Here y_d and y_s are the eigenvalues of weak hypercharge Y in the doublet and in the singlet, respectively.

Obviously, the linear combination multiplying ν_L must be proportional to the Z^0 field, for the neutrino to decouple from the photon. Comparison with the inverse of the orthogonal transformation (11.9) shows that this is true if $y_d = -1$. Parity conservation in the electron-photon interaction is equivalent to requesting e_L and e_R to couple to the photon field A_μ with the same strength. From our formulae it is easy to see that this is true if one chooses $y_s = -2$. An analogous reasoning for quark fields fixes the weak hypercharge of the doublet to be $1/3$, the weak hypercharges of the singlets to be $4/3$ and $-2/3$, respectively. In summary, for every lepton generation and every quark generation the assignments must be

$$Y^{(l)} = \operatorname{diag}(-1, -1, -2), \qquad Y^{(q)} = \operatorname{diag}\left(\frac{1}{3}, \frac{1}{3}, \frac{4}{3}, -\frac{2}{3}\right). \qquad (11.21)$$

Note that these assignments are in agreement with the formula (11.11). I stressed this construction because the $U(1)$ factor, in principle, allows any rational ratio of y_s and y_d (and likewise for quarks). The ratio is fixed, however, by an empirical constraint.

We also note a remarkable property of (11.21). If ones introduces a $\mathbb{Z}_2$-grading by defining the L - L sectors and the R - R sectors in the representation spaces to be *even*, the L - R and the R - L sectors to be *odd*, then the matrices (11.21) have the property

$$\operatorname{Str} Y = 0. \qquad (11.22)$$

This is precisely the condition for chiral anomalies to be absent, and there is also a tantalizing (but not fully understood) relation to charge quantization, s. [190] and references therein. As mentioned in the introduction the differential calculus on which the Mainz-Marseille model is built, has this grading. The assignment of the weak hypercharges, for the Higgs and for all fermions, is the correct one without having to fix it by hand.

To end this section let me write down the interaction Lagrangian in its full complexity. With $j_{e.m.}^\mu(x)$ denoting the standard *electromagnetic current*, $J^\mu(x)$ the *weak charged current*, (CC), and $K^\mu(x)$ the *weak neutral current*, (NC), the interaction reads

$$\mathcal{L}_{int} = -e\left\{j_{e.m.}^\mu(x)A_\mu(x) + \frac{1}{2\sqrt{2}\sin\vartheta_W}\left[J^\mu(x)W_\mu^{(-)}(x) + \text{h.c.}\right]\right.$$
$$\left. + \frac{1}{4\sin\vartheta_W \cos\vartheta_W}K^\mu(x)Z_\mu(x)\right\}. \qquad (11.23)$$

Essentially for historical reasons it is customary to write the currents in terms of the full Dirac fields, instead of the chiral fields, by writing out the projection $(\mathbb{1} - \gamma_5)/2$. The CC current then reads

$$J^\mu(x) = \sum \overline{\nu^{(i)}}\gamma^\mu(\mathbb{1}-\gamma_5)e^{(j)}V^{(mix)}_{ij} + \sum \overline{u^{(k)}}\gamma^\mu(\mathbb{1}-\gamma_5)d^{(l)}V^{(CKM)}_{kl}, \tag{11.24}$$

where $\mathbf{V}^{(mix)}$ denotes the mixing matrix in the leptonic sector, $\mathbf{V}^{(CKM)}$ the Cabibbo-Kobayashi-Maskawa mixing matrix in the quark sector. The neutral current, finally, is given by

$$\begin{aligned}K^\mu(x) = & \sum \overline{\nu^{(i)}}\gamma^\mu(\mathbb{1}-\gamma_5)\nu^{(i)} - \sum \overline{e^{(i)}}\gamma^\mu(\mathbb{1}-\gamma_5)e^{(i)} \\ & + \sum \overline{u^{(i)}}\gamma^\mu(\mathbb{1}-\gamma_5)u^{(i)} - \sum \overline{d^{(i)}}\gamma^\mu(\mathbb{1}-\gamma_5)d^{(i)} \\ & - 4\sin^2\vartheta_W j^\mu_{e.m.}(x). \end{aligned} \tag{11.25}$$

This is the interaction which has been tested over the last 25 years (including the radiative corrections that follow from it) and which appears to be in embarassingly perfect agreement with experiment.

11.6 Remarks About Fermionic Mass Generation

A pecularity of the minimal standard model is the fact that it does not admit primordial mass terms. This is so because mass terms involve L- and R-fields. As the former are doublets with respect to $SU(2)$ while the latter are singlets there is no way to form an invariant.

The Higgs field, on the other hand, is a weak isospin doublet. Therefore, when combined with SSB as described above, Yukawa couplings of fermions and Higgs,

$$h\left\{(\overline{\mathcal{L}}\Phi)\mathcal{R} + \overline{\mathcal{R}}(\Phi^{(c)}\mathcal{L})\right\},$$

will produce mass terms. In this expression I have used a short-hand notation $\mathcal{L}$ for an L-doublet, $\mathcal{R}$ for a R-singlet, and $\Phi^{(c)}$ for the charge conjugate of Φ. Although at first this looks fine, there are two problems with such mass terms. The first is that the coupling constant h in front of this term falls from heaven and must be adjusted for every generation. We note in passing that this, in turn, implies that the Yukawa coupling of the physical Higgs $\phi^{(0)}$ to a given quark or lepton is proportional to that particle's mass; this is one of the reasons why hunting the Higgs is so tedious and, hélas!, costly. The other problem is that the fermionic mass eigenstates do not coincide with the states participating in the weak CC interactions. This phenomenon, although parametrizable by means of the mixing matrices $\mathbf{V}^{(mix)}$ and $\mathbf{V}^{(CKM)}$, is completely outside the reach of the standard model.

Unfortunately embedding the standard model in Noncommutative Geometry does not help in that matter. In Connes' version(s) the fermionic mass matrices as well as the unitary mixing matrices are used as an *input*. In the Mainz-Marseille model generation mixing seems rather natural 13, but even though the model provides a certain pattern for the mass matrices, there is little predictive power.

Part III

New Directions in Noncommutative Geometry and Mathematical Physics

This part is devoted to more recent physical applications of current interest, presented by some of the leading researchers in the field. These individual contributions are an integral part of the overall program and are organized in a way compatible with the more expository first two chapters. Topics discussed include

- an investigation of the physical implications of the noncommutative geometry approach towards the Standard Model,
- alternative approaches, e.g. involving supergroups and superalgebras,
- general problems in quantum field theory involving noncommutative geometry,
- noncommutative spaces (NC torus etc.),
- Hopf algebras and renormalization,
- string duality from a noncommutative geometry point of view.

All of these are very active and rapidly developing research areas and involve a high degree of specialization. Nevertheless we have tried to stay close to the main topic of the conference by emphasizing throughout the connections to the established Standard Model and other realistic models of quantum field theory.

12 The Impact of NC Geometry in Particle Physics

Bruno Iochum

University of Marseille, Lumini

In this review, which could be entitled "Is noncommutative geometry the millenium bug of physics", we consider only the main impacts of noncommutative geometry in particle physics. Attention is paid to particle physics, to the standard model and renormalization of different models. So, most of the mathematical definitions are omitted since they can be found in Chapter 1. We look at ideas beyond technicalities.

12.1 Why Noncommutative Geometry?

Discreteness Versus Continuity

Noncommutative geometry was invented by Alain Connes [38] for improving the impact of classical geometry when there are no points. This is crucial in physics, since it addresses the old problem on the difference between a discrete and a continuous situation. Actually, it is tempting to say that, for a physicist, a point is just a mathematical representation which is convenient when one wants to describe the trajectory of a solid for instance. All the mass of the object is supposed to be concentrated in one point x in $\mathbb{R}^3$, the mass center, and one solves the equation of motion giving rise to a trajectory $x(t)$ and one concludes by adding the relative rotation of the solid around its center of mass at each time. So, for the physicist, this process appears just as an idealization of the physical situation. Nevertheless for a collection of solids, this construction is nothing but the replacement of a set of bodies by a collection of points. In other words, one has replaced a "continuous" 3-dimensional object just by a point. But it is wellknown that a discrete set and a continuous one are not in bijection, for instance the cardinalities of $\mathbb{N}$ and $\mathbb{R}$ are different. Consequently, in both mathematics and physics this situation is not satisfactory.

Let us give few other examples of the impassable wall between the discrete and continuous world:

- A surface cannot be considered only as a collection of points. More information is needed, like the distance between infinitesimally close points.

- The same problem occurs for the replacement of classical mechanics by quantum mechanics. Roughly speaking, a classical point is stretched in all $\mathbb{R}^3$ as a wave function, so a discrete object has been replaced by a continuous one.

- Similarly, the classical atoms, which are nothing but discrete objects like electrons, protons or neutrons are replaced by fields in quantum field theory. In particular many physical definitions turn around this confusion. Consider for instance, the magnetic moment of an electron. In most textbooks, it is introduced via an analogy with a current given by an electron moving on a circle. But again we face the same problem: either an electron is just a pointlike classical particle, or it is an "extended one" on the circle. This is very confusing since we have to explain how the electric charge is spread over the circle and so on.

It is very easy to continue this list proving that very elementary physical objects are not necessarily well defined, or better, are defined modulo an undefined background. One of the great advantages of noncommutative geometry is precisely to put on the same ground the discrete and the continuous cases. By this, we mean that the technical tools for solving a problem are based on the same objects. For us, this is probably the most important explanation of the successes of noncommutative geometry.

The impacts

Today, the two major impacts of this new geometry are in solid state physics and in particle physics:

- The explanation of the integer quantum Hall effect was given by Bellissard [11] using only the replacement of the classical Brillouin zone, which is a 2-torus, by its natural analogue, namely the noncommutative 2-torus. Of course, this is dressed with a lot of hard technicalities, but the main idea is precisely the exchange of a periodic situation essentially reflected by $\mathbb{N}$ or $\mathbb{Z}$, the multiple of a given period, and its generalization, the nonperiodic one, reflected by $\mathbb{R}$.

- The standard model constructed by Connes and Lott [46,47] and refined by Connes [39] is based on the product of two different parts: the spacetime which is continuous and the internal degrees of freedom which reveal the discrete part. Both ingredients are based on the notion of Dirac operators and spectral triples.

It is important to know that other interesting approaches exist. Let us mention [71,156].

12.2 Spectral Triples

A posteriori, the notion of spectral triple appears very natural, since, for instance, it mimics the notion of differentiable manifolds in an algebraic language . But this means that we face a problem of choice.

The choice
At this point, it is interesting to retrace history.
1) In the Connes-Lott model [46], there are :
- two finite algebras $\mathcal{A} = \mathbb{C} \oplus \mathbb{H}$, $\mathcal{B} = \mathbb{C} \oplus M_3(\mathbb{C})$,
- two representations $\rho = \rho_L \oplus \rho_R$ of $\mathcal{A}$ and $\mathcal{B}$ on $\mathcal{H} = \mathcal{H}_L \oplus \mathcal{H}_R$,
- a mass matrix $M : \mathcal{H}_L \longrightarrow \mathcal{H}_R$ and an operator $\mathcal{D} = \begin{pmatrix} 0 & M \\ M^* & 0 \end{pmatrix}$ from $\mathcal{H}$ to $\mathcal{H}$.

- some coupling constants due to the reducibility of ρ.

This yields an $(\mathcal{A}, \mathcal{B})$-module $(\mathcal{H}, \mathcal{D}, \chi)$ where χ is the chirality.

The interest of this construction is twofold:

- The Higgs potential is an output. The mass matrix M links the left and the right worlds and the "distance" between these worlds is related to the eigenvalues of $\mathcal{D}^{-1}$.

- The replacement of spacetime by noncommutative spacetime is motivated by the introduction of a possible cutoff.

Notice that the spirit of this geometry is that the only allowed groups are the unitary groups of *-algebras which will be used as gauge groups.

2) The notion of a real structure: It is based on an operator J corresponding in the commutative case to the charge conjugation and in the noncommutative one to the Tomita-Takesaki operator. Recall that if $\mathcal{M}$ is a von Neumann algebra equipped with a cyclic and separating vector ξ_0 in $\mathcal{H}$, the closure S of the operator applying $x\xi_0$ to $x^*\xi_0$ has a polar decomposition $S = J\Delta^{1/2}$ where Δ is a positive operator and J is an antilinear operator such that

$$J\mathcal{M}J^{-1} = \mathcal{M}'. \tag{12.1}$$

So we obtain a real spectral triple $(\mathcal{A}, \mathcal{D}, \mathcal{H}, \chi, J)$ satisfying some precise constraints. Let us emphasize two important axioms: χ is a Hochschild cycle and Poincaré duality is satisfied. There are two ideas: The electroweak sector is in Poincaré duality with the strong sector, and the standard model is, in some vague sense, minimal. Moreover, for all a and b in $\mathcal{A}$,

$$J^2 = \pm 1 \,,\ [\mathcal{D}, J] = 0, [\,[\mathcal{D}, \rho(a)], J\rho(b)J^{-1}] = 0 \text{ and}$$

$$[\rho(a),\ J\rho(b)J^{-1}] = 0 \tag{12.2}$$

<u>Consequences</u>: Connes has obtained an algebraic characterization of spin^c riemannian manifolds [39].

Problems

1) Classify the links between J_{Connes} and J_{Tomita}.

2) The commutative and noncommutative geometries [39] are based on different axioms. Note in particular that in (12.1), $J\mathcal{M}J^{-1} = J\rho(\mathcal{A})J^{-1}$ is the whole commutant of $\rho(\mathcal{A})$, while in (12.2), it is only imbedded in it. The constraints on χ are also different. This has important consequences on the computation of noncommutative distances.

3) Lattice gauge theory has been invented to remedy certain difficulties occuring in field theory. The idea was to discretize the fields, and we saw this idea in paragraph 1. So, it is natural to see what noncommutative geometry can do in this setting. Unfortunatly, a partial negative answer has been proposed [93]. It would be interesting to find a right notion of spectral triple which is compatible with lattice theory.

Present state of the art

Using all previous ingredients, Connes finally ended up with the replacement of the two algebras $\mathcal{A}$ and $\mathcal{B}$ by one algebra $\mathcal{A} = \mathbb{C} \oplus \mathbb{H} \oplus M_3(\mathbb{C})$ and $\mathcal{B} = J\mathcal{A}J^{-1}$.

At this point, we face different questions.

Problems:

1) $\mathbb{H}$ appears because its unitary group is $SU(2)$ while the unitary group of $M_2(\mathbb{C})$ is $U(2)$ and avoids an extra $U(1)$, so we must consider a fully satisfactory noncommutative geometry for real *-algebras and not only for C^*-algebras. Note that $\mathbb{H}$ has no order since the square of a quaternion is always a multiple of the identity, so $\mathbb{H}$ has only one trivial pure state. This creates problems when one computes distances using Connes' formula

$$d(\psi, \phi) = \operatorname{Sup} \{|\psi(a) - \phi(a)|,\ a \in \rho(\mathcal{A}),\ \|[\mathcal{D}, a]\| \leq 1\}$$

where ψ, ϕ are in the state space of $\rho(\mathcal{A})$.

2) In what sense is the noncommutative standard model minimal? Note that the algebra looks like $\mathcal{A} = M_1(\mathbb{C}) \oplus M_2(\mathbb{C}) \oplus M_3(\mathbb{C})$. What is the deeper reason for this simple form?

12.3 Technical Points

Scalar products

On the noncommutative differential n-forms $\Omega_\mathcal{D}^n(\mathcal{A})$, there is an easy way to construct a scalar product. For instance, in the finite dimensional case, take $< \omega, \omega' > = \operatorname{Re} \operatorname{Tr}(\omega^* \omega')$ and in the infinite case, $< \omega, \omega' > = \operatorname{Re} \operatorname{Tr}(\omega^* \omega' |\mathcal{D}|^{-dim})$.

It is well known that there are logarithmic divergences in this last expression, so we have to replace the usual trace Tr by the Dixmier trace $\operatorname{Tr}_{\text{Dix}}$.

Problem: Do there exist singular traces which are not of the Dixmier type? The consequence would be important: Either there are no other traces and from a physical point of view it is interesting to understand why the logarithmic divergences are privileged or, if they are others, do they get rid of other divergences?

Recall that in the commutative situation where M is a 4-dimensional manifold $(\mathcal{A} = C^\infty(M), \mathcal{H} = L^2(\text{Spinors}), \mathcal{D} = \not\partial)$, we get the identity spelled out for 2-forms ω, ω',

$$< \omega, \omega' > = \tfrac{1}{2} \int_M \omega_{\mu\nu}^* \omega'_{\mu'\nu'} g^{\mu\mu'} g^{\nu\nu'} (\det g_{..})^{1/2}\, d^4x.$$

The junk ideal

In the construction of $\Omega_\mathcal{D}^n(\mathcal{A})$, there occurs a natural quotient by the so-called junk ideal for getting a graded differential algebra. This junk modifies the scalar product. In particular it mixes the internal and external forms coming from the internal degrees of freedom and the external usual spacetime. So, from a physical point of view, it is also important.

Problems: A natural question is: what is the physical meaning of that junk ideal? It is important to quote that in noncommutative relativity [40,?,?], the junk ideal plays no role since only the one-forms are essential. What is behind this difference?

Moreover, there is the problem of the unimodularity condition. Do we have to perform it before or after the quotient by the junk ideal?

The choice of $\mathcal{D}$

For two given spectral triples $(\mathcal{A}_1, \mathcal{H}_1, \mathcal{D}_1, \chi_1, J_1)$ and $(\mathcal{A}_2, \mathcal{H}_2, \mathcal{D}_2, \chi_2, J_2)$, we have the choice of two candidates for the Dirac operator on their tensor product:

$$\mathcal{D}_t = \mathcal{D}_1 \otimes 1_2 + \chi_1 \otimes \mathcal{D}_2$$
$$\mathcal{D}'_t = \mathcal{D}_1 \otimes \chi_2 + 1_1 \otimes \mathcal{D}_2$$

Fortunately, the unitary operator $U = \frac{1}{2}(1_1 + \chi_1) \otimes 1_2 + \frac{1}{2}(1_1 - \chi_1) \otimes \chi_2$ intertwines these Dirac operators, $U \mathcal{D}'_t U^* = \mathcal{D}_t$, so their distinction is not relevant.

Importance of the scalar product

We saw before that there is a natural scalar product on $\Omega_\mathcal{D}^n(\mathcal{A})$. However, natural does not mean unique. So, we clearly have here some flexibility varying this theme. Playing around this notion, we conclude that the different scalar products must be in one-to-one correspondence with the physical gauge coupling constants g_1, g_2, g_3 which parameterize the invariant scalar products on the Lie algebra $u(1) \oplus su(2) \oplus su(3)$ of the standard model. These three physical quantities have been measured.

In order to get the coupling constants, we soften the scalar products to:

$$< \omega, \omega' >_z = \text{Re Tr}_{\text{Dix}} [z \omega^* \omega' |\mathcal{D}|^{-\dim}], \quad \omega, \omega' \in \Omega_\mathcal{D}^p \mathcal{A}. \quad (12.3)$$

z is a positive operator on Hilbert space that commutes with ρ, $J\rho J^{-1}$, $\mathcal{D}$ and χ. Whether or not z commutes with J will be a difficult choice. In the commutative case, we have anyhow that z is proportional to the identity.

Note that another scalar product is possible: For $\omega, \omega' \in \Omega_\mathcal{D}^p \mathcal{A}$,

$$< \omega, \omega' >_{z'} = \text{Re Tr}_{\text{Dix}} [z' (\omega + J\omega J^{-1})^* (\omega' + J\omega' J^{-1}) |\mathcal{D}|^{-\dim}]. (12.4)$$

12.4 The Noncommutative Highway

If one admits the fact that the real world is described by the classical spacetime plus the standard model, then everything is now fixed and we only have to follow the noncommutative highway:

Construction of a spectral triple

$$\mathcal{A}_t = C^\infty(M) \otimes (\mathbb{H} \oplus \mathbb{C} \oplus M_3(\mathbb{C})),$$
$$\mathcal{D}_t = \partial\!\!\!/ \otimes 1_{p \oplus \bar{p}} + \gamma_5 \otimes \mathcal{D},$$
$$\mathcal{H}_t = L^2(\text{Spinors}) \otimes \mathcal{H}_{p \oplus \bar{p}},$$
$$\chi_t = \gamma_5 \otimes \chi,$$
$$J_t = C \otimes J.$$

where $p, \bar{p}$ means particle and antiparticle.

Construction of the standard model
The internal part is defined as follows (cf Sections 2.2, 2.3, 2.5 for more details).

$$\mathcal{A} = \mathbb{H} \oplus \mathbb{C} \oplus M_3(\mathbb{C}) \ni (a,b,c),$$
$$\mathcal{H}_L = \left(\mathbb{C}^2 \otimes \mathbb{C}^N \otimes \mathbb{C}^3\right) \oplus \left(\mathbb{C}^2 \otimes \mathbb{C}^N \otimes \mathbb{C}\right),$$
$$\mathcal{H}_R = \left((\mathbb{C} \oplus \mathbb{C}) \otimes \mathbb{C}^N \otimes \mathbb{C}^3\right) \oplus \left(\mathbb{C} \otimes \mathbb{C}^N \otimes \mathbb{C}\right).$$

In each summand, the first factor denotes weak isospin doublets or singlets, the second N generations, $N = 3$, and the third denotes color triplets or singlets.

Let us choose the following basis of $\mathcal{H} = \mathbb{C}^{90}$:

$$\begin{pmatrix} u \\ d \end{pmatrix}_L, \begin{pmatrix} c \\ s \end{pmatrix}_L, \begin{pmatrix} t \\ b \end{pmatrix}_L, \begin{pmatrix} \nu_e \\ e \end{pmatrix}_L, \begin{pmatrix} \nu_\mu \\ \mu \end{pmatrix}_L, \begin{pmatrix} \nu_\tau \\ \tau \end{pmatrix}_L;$$

$$u_R, \quad c_R, \quad t_R,$$
$$d_R, \quad s_R, \quad b_R, \quad e_R, \quad \mu_R, \quad \tau_R;$$

$$\begin{pmatrix} u \\ d \end{pmatrix}_L^c, \begin{pmatrix} c \\ s \end{pmatrix}_L^c, \begin{pmatrix} t \\ b \end{pmatrix}_L^c, \begin{pmatrix} \nu_e \\ e \end{pmatrix}_L^c, \begin{pmatrix} \nu_\mu \\ \mu \end{pmatrix}_L^c, \begin{pmatrix} \nu_\tau \\ \tau \end{pmatrix}_L^c;$$

$$u_R^c, \quad c_R^c, \quad t_R^c,$$
$$d_R^c, \quad s_R^c, \quad b_R^c, \quad e_R^c, \quad \mu_R^c, \quad \tau_R^c.$$

The representation ρ acts on $\mathcal{H}$ by

$$\rho(a,b,c) := \begin{pmatrix} \rho_w(a,b) & 0 \\ 0 & \rho_s(b,c) \end{pmatrix} := \begin{pmatrix} \rho_{wL}(a) & 0 & 0 & 0 \\ 0 & \rho_{wR}(b) & 0 & 0 \\ 0 & 0 & \rho_{sL}(b,c) & 0 \\ 0 & 0 & 0 & \rho_{sR}(b,c) \end{pmatrix}$$

with

$$\rho_{wL}(a) := \begin{pmatrix} a \otimes 1_N \otimes 1_3 & 0 \\ 0 & a \otimes 1_N \end{pmatrix}, \quad \rho_{wR}(b) := \begin{pmatrix} B \otimes 1_N \otimes 1_3 & 0 \\ 0 & \bar{b} 1_N \end{pmatrix},$$
$$B := \begin{pmatrix} b & 0 \\ 0 & \bar{b} \end{pmatrix},$$

$$\rho_{sL}(b,c) := \begin{pmatrix} 1_2 \otimes 1_N \otimes c & 0 \\ 0 & \bar{b}1_2 \otimes 1_N \end{pmatrix}, \quad \rho_{sR}(b,c) := \begin{pmatrix} 1_2 \otimes 1_N \otimes c & 0 \\ 0 & \bar{b}1_N \end{pmatrix}.$$

Note the asymmetry between particles and antiparticles.

$$\chi = \begin{pmatrix} -1_{8N} & 0 & 0 & 0 \\ 0 & 1_{7N} & 0 & 0 \\ 0 & 0 & -1_{8N} & 0 \\ 0 & 0 & 0 & 1_{7N} \end{pmatrix}.$$

$$\mathcal{D} = \begin{pmatrix} 0 & \mathcal{M} & 0 & 0 \\ \mathcal{M}^* & 0 & 0 & 0 \\ 0 & 0 & 0 & \bar{\mathcal{M}} \\ 0 & 0 & \bar{\mathcal{M}}^* & 0 \end{pmatrix}.$$

The fermionic mass matrix of the standard model is

$$\mathcal{M} = \begin{pmatrix} \begin{pmatrix} M_u & 0 \\ 0 & M_d \end{pmatrix} \otimes 1_3 & 0 \\ 0 & \begin{pmatrix} 0 \\ M_e \end{pmatrix} \end{pmatrix},$$

with

$$M_u := \begin{pmatrix} m_u & 0 & 0 \\ 0 & m_c & 0 \\ 0 & 0 & m_t \end{pmatrix}, \quad M_d := C_{KM} \begin{pmatrix} m_d & 0 & 0 \\ 0 & m_s & 0 \\ 0 & 0 & m_b \end{pmatrix},$$

$$M_e := \begin{pmatrix} m_e & 0 & 0 \\ 0 & m_\mu & 0 \\ 0 & 0 & m_\tau \end{pmatrix}.$$

It follows that z of equation (12.3) involves $2(1+N) = 8$ strictly positive numbers x, y_1, y_2, y_N, $\tilde{x}$, $\tilde{y}_1$, $\tilde{y}_2$, $\tilde{y}_N$,

$$z := \begin{pmatrix} z_w & 0 \\ 0 & z_s \end{pmatrix},$$

$$z_w := \begin{pmatrix} x/3\, 1_2 \otimes 1_N \otimes 1_3 & 0 & 0 & 0 \\ 0 & 1_2 \otimes y & 0 & 0 \\ 0 & 0 & x/3\, 1_2 \otimes 1_N \otimes 1_3 & 0 \\ 0 & 0 & 0 & y \end{pmatrix},$$

$$z_s := \begin{pmatrix} \tilde{x}/3\, 1_2 \otimes 1_N \otimes 1_3 & 0 & 0 & 0 \\ 0 & 1_2 \otimes \tilde{y} & 0 & 0 \\ 0 & 0 & \tilde{x}/3\, 1_2 \otimes 1_N \otimes 1_3 & 0 \\ 0 & 0 & 0 & \tilde{y} \end{pmatrix},$$

$$y := \begin{pmatrix} y_1 & 0 & 0 \\ 0 & y_2 & 0 \\ 0 & 0 & y_N \end{pmatrix}, \quad \tilde{y} := \begin{pmatrix} \tilde{y}_1 & 0 & 0 \\ 0 & \tilde{y}_2 & 0 \\ 0 & 0 & \tilde{y}_N \end{pmatrix},$$

while z' only involves $1 + N = 4$ numbers x', y'_1, y'_2, y'_N:

$$z' := \begin{pmatrix} z'_w & 0 \\ 0 & z'_s \end{pmatrix},$$

$$z'_w = z'_s := \frac{1}{2} \begin{pmatrix} x'/3\, 1_2 \otimes 1_N \otimes 1_3 & 0 & 0 & 0 \\ 0 & 1_2 \otimes y' & 0 & 0 \\ 0 & 0 & x'/3\, 1_2 \otimes 1_N \otimes 1_3 & 0 \\ 0 & 0 & 0 & y' \end{pmatrix}.$$

Turning the crank
We give only the results.

$$\Omega^1_\mathcal{D}\mathcal{A} = \left\{ i \begin{pmatrix} 0 & \rho_{wL}(h)\mathcal{M} & 0 & 0 \\ \mathcal{M}^*\rho_{wL}(\tilde{h}^*) & 0 & 0 & 0 \\ 0 & 0 & 0 & 0 \\ 0 & 0 & 0 & 0 \end{pmatrix},\ h, \tilde{h} \in \mathbb{H} \right\}.$$

The following was noticed a long time ago in [51,52].

Lemma 1. *The map*

$$\omega : h \in \mathbb{H} \longrightarrow \omega(h) = \frac{1}{\sqrt{L}} \operatorname{diag}\left(\begin{pmatrix} 0 & \rho_L(h)\mathcal{M} \\ \mathcal{M}^*\rho_L(h) & 0 \end{pmatrix}; 0 \right) \in (\Omega^1_\mathcal{D}(\mathcal{A}))_{s.a.}$$

is an isometry with

$$< h, h' > := \operatorname{tr}(h^*h') \text{ and } < \omega(h), \omega(h') > := \operatorname{tr}(z\omega(h^*)\omega(h'))$$

where

$$L := \operatorname{tr}(z_{pL}\mathcal{M}\mathcal{M}^*) = \operatorname{tr}(z_{pR}\mathcal{M}^*\mathcal{M}) = x\,\operatorname{tr}(|M_u|^2 + |M_d|^2) + \operatorname{tr}(y|M_e|^2) \quad (2.5)$$

Since $\Omega^1_{\mathcal{D}_t}(C^\infty(M) \otimes \mathcal{A}) \approx (\Omega^1(M) \otimes \rho(\mathcal{A})) \oplus (C^\infty(M) \otimes \Omega^1_\mathcal{D}(\mathcal{A}))$, every antihermitian one-form H_t in $\Omega^1_{\mathcal{D}_t}(C^\infty(M) \otimes \mathcal{A})$ can be uniquely decomposed in $H_t = A \oplus H$, $A = f_0 df_1 \otimes \rho(x)$, $H = f\gamma_5 \otimes i\omega(h)$, $f, f_i \in C^\infty(M, \mathbb{R})$, $x = -x^* \in \mathcal{A}, h \in \mathbb{H}$.

The curvature C_{H_t} of H_t is decomposed (see [197]) into

$C_{H_t} = F + C_H - D\Phi$,
$F = (d \otimes 1)A + A^2$ is the field strength in $\Omega^2(M) \otimes \rho(\mathcal{A})$,
$d : f \in C^\infty(M) \to \gamma^\mu \partial_\mu f$,
$C_H = (\gamma_5 \otimes i[\mathcal{D}, .])H + H^2$ is the curvature of H,
$\Phi = if\gamma_5 \otimes \omega(h) - i\gamma_5 \otimes \omega(1_2) \in C^\infty(M) \otimes \Omega^1_\mathcal{D}(\mathcal{A})$ is a change of variable

corresponding to $f \to fh - 1_2$,
$D\Phi = (d \otimes 1)\Phi + [A, \Phi] \in \Omega^1(M) \otimes \Omega^1_\mathcal{D}(\mathcal{A})$ is the covariant derivative of Φ.

In this case, the Higgs potential is

$$V(H_t) = <C_{H_t}, C_{H_t}>$$
$$= \int_M \mathrm{Tr}(zF^* \wedge *F) + \int_M \mathrm{Tr}(zD\Phi^* \wedge *D\Phi) + V_h \int_M *|f|^2,$$
$$V_h := \mathrm{Re}\,\mathrm{Tr}(z(C_h - \alpha(C_h))^*(C_h - \alpha(C_h))) = \mathrm{Re}\,\mathrm{Tr}(z(C_h^2 - (\alpha(C_h))^2))$$
$$= (1 - \tfrac{1}{2}\mathrm{Tr}(\phi^*\phi))^2\, V_{1_2}$$

with

$$C_{H_h} = (1 - \tfrac{1}{2}\mathrm{Tr}(\phi^*\phi))\,C_{H_{1_2}}, \quad C_{H_{1_2}} = \mathrm{diag}(1_2 \otimes \Sigma, \mathcal{M}^*\mathcal{M}; 0), \quad (12.6)$$

where α is a linear map from $\Omega_\mathcal{D}^2(\mathcal{A})$ onto $\rho(\mathcal{A}) + J_\mathcal{D}^2$, $J_\mathcal{D}^2$ being the junk

$$J_\mathcal{D}^2 := \{\,[\mathcal{D}, \rho(x)]\,[\mathcal{D}, \rho(x')]\mid \rho(x)\,[\mathcal{D}, \rho(x')] = 0,\; x, x' \in \mathcal{A}\,\},$$

given by

$$\begin{aligned}\mathrm{Re}\,\mathrm{Tr}(\,z\,(C - \alpha(C))\,\rho(x)\,) &= 0, \quad x \in \mathcal{A},\\ \mathrm{Re}\,\mathrm{Tr}(z\alpha(C)j) &= 0, \quad j \in J_\mathcal{D}^2.\end{aligned} \quad (12.7)$$

This modified Higgs potential, compared with $V(C_h) = \mathrm{Re}\,\mathrm{Tr}(\,z\,C_h^*C_h\,)$, is due to the decomposition of the global two-forms:

$$\Omega_{\mathcal{D}_t}^2(C^\infty(M) \otimes \mathcal{A}) \approx (\Omega^2(M)1_4 \otimes \rho(\mathcal{A})$$
$$+ C^\infty(M) \otimes \Omega_\mathcal{D}^2(\mathcal{A})) \oplus (\Omega^1(M)\gamma_5 \otimes \Omega_\mathcal{D}^1(\mathcal{A})),$$
$$J_{\mathcal{D}_t}^2 = C^\infty(M)1_4 \otimes \rho(\mathcal{A}) + C^\infty(M) \otimes J_\mathcal{D}^2,$$

where the + are not direct sums.

Conclusions:
1) The Higgs scalar is nothing else but a natural connection in noncommutative geometry.
2) The Hilbert space representation of this scalar is $\mathcal{H}_S = \{H = -H^* \in \Omega_\mathcal{D}^1(\mathcal{A})\}$. This implies that all Connes-Lott models are Yang-Mill-Higgs models while the converse is far from being true.

12.5 Computation of Higgs and W Masses

We first use the scalar product (12.3). The normalization of the electroweak gauge boson which appears in the kinetic term of the Klein-Gordon action gives for the selfcoupling λ of the Higgs $\lambda = \frac{V_{1_2}}{16L^2}$ with

$$V_{1_2} := \mathrm{tr}(z\,(C_{1_2}^2 - \alpha(C_{1_2})^2))$$

$$= \frac{3}{2} x \, \text{tr}(|M_u|^4 + |M_d|^4) + x \, \text{tr}(|M_u|^2|M_d|^2) + \frac{3}{2}\text{tr}(y|M_e|^4)$$
$$- x \, \text{tr}(|M_u|^2 + |M_d|^2)$$
$$+ \text{tr}(y|M_e|^2)^2 \left((2Nx + 2\text{tr}(y))^{-1} + (2Nx + \text{tr}(y) + 3\text{tr}(\tilde{y}))^{-1}\right). \tag{12.8}$$

and the Higgs mass is $m_H = \sqrt{\frac{2V_{12}}{L}}$.

The normalization of gauge couplings is divided in three parts:

The isospin part : $g_2^{-2} = Nx + \text{tr}(y)$ implying $m_W := g_2 \frac{\mu}{4\sqrt{\lambda}} = \frac{L}{Nx+\text{tr}(y)}$.

The color part $g_3^{-2} = \frac{4}{3} N\tilde{x}$.

Since g_1 depends on $U(1)_Y$, recall that the Lie algebra of the unitary group of $\mathcal{A}$ is $\mathfrak{g} = \{x \in \mathcal{A} \mid x^* = -x\} = su(2) \oplus u(1) \oplus su(3) \oplus u(1)$. But here, a miracle occurs. The hypercharge generator Y is a linear combination of the two $u(1)$-generators: $Y = \frac{1}{i}\rho(0, \frac{i}{2}, \frac{i}{6}1_3)$, so $g_1^{-2} = Nx + \frac{2}{9} N\tilde{x} + \frac{1}{2} \text{tr } y + \frac{3}{2}\text{tr } \tilde{y}$.

Remark 1. The right Lie algebra $su(2) \oplus u(1)_Y \oplus su(3))$ is obtained by the *unimodularity condition*:

$$\text{tr}\left[J\rho(1_2, 0, 0)J^{-1} \, \rho(a, b, c)\right] = 0, \quad (a, b, c) \in \mathfrak{g}, \tag{12.9}$$

which is equivalent to
$$\text{tr}\left[P\left(\rho(a, b, c) + J\rho(a, b, c)J^{-1}\right)\right] = \text{tr}\left[\rho_p(a, b) + \rho_{\bar{p}}(b, c)\right] = 0,$$
where P is the projection on the space of particles, $\mathcal{H}_L \oplus \mathcal{H}_R$, and so appears more natural. This condition is also related to a condition of vanishing anomalies [2].

12.6 Parameter Counting

The standard model yields 18 parameters: 3 gauge couplings g_1, g_2, g_3, 2 bosons, W and Higgs, 9 fermions, the leptons - electron, muon and tau -, and the quarks, up, down, charm, strange, top and bottom, plus 4 mixings appearing in the Cabibbo-Kobayashi-Maskawa matrix

$$V := \begin{pmatrix} V_{ud} & V_{us} & V_{ub} \\ V_{cd} & V_{cs} & V_{cb} \\ V_{td} & V_{ts} & V_{tb} \end{pmatrix}.$$

In the noncommutative Yang-Mills model, there are 19 parameters: the 9 fermionic masses plus the 4 mixing angles encoded in $\mathcal{D}$, plus the 6 noncommutative gauge couplings related to z: x, y_1, y_2, y_3 $\tilde{x}$, $\text{Tr}(\tilde{y}_1, \tilde{y}_2, \tilde{y}_3)$. Consequently, there is a constraint which can be summarized:

We will use the following abbreviations: $f := m_f^2$ for all fermions f and $W := m_W^2, H := m_H^2$.

Here, we treat only the case given by the following hierarchies,

$$\begin{aligned} e < \quad \mu \quad < \quad \tau \quad < \quad W, \\ u+d < \min\{c,s\} < (1+\epsilon)^{-1}\max\{c,s\}, \\ c+s+\min\{c,s\} < \min\{t,b\} < (1+\epsilon)^{-1}\max\{t,b\}, \end{aligned} \right\} \quad (12.10)$$

where $\epsilon := 1 - \min\{|V_{tb}|^2, |V_{cs}|^2, |V_{ud}|^2\}$ measures the deviation of V from the identity.

With these hierarchies, the constant

$C := (3W^2)^{-1}(r^2 - q^2)$ is positive where
$q := t+b+c+s+u+d$,
$r^2 := 3(t^2 + b^2 + c^2 + s^2 + u^2 + d^2)$
$\quad +2ud|V_{ud}|^2 + 2us|V_{us}|^2 + 2ub|V_{ub}|^2 + 2cd|V_{cd}|^2 + 2cs|V_{cs}|^2 + 2cb|V_{cb}|^2$
$\quad +2td|V_{td}|^2 + 2ts|V_{ts}|^2 + 2tb|V_{tb}|^2.$

According to the definition of ϵ, $3\,CW^2/2 > u^2 + d^2 - \epsilon\,ud > 0$ because
$\frac{3}{2}CW^2 = [t^2 + b^2 + c^2 + s^2 + u^2 + d^2]$
$\quad + [tb\,(|V_{tb}|^2 - 1) + cs\,(|V_{cs}|^2 - 1) + ud\,(|V_{ud}|^2 - 1)]$
$\quad + [us\,|V_{us}|^2 + ub\,|V_{ub}|^2 + cd\,|V_{cd}|^2 + cb\,|V_{cb}|^2 + td\,|V_{td}|^2 + ts\,|V_{ts}|^2]$
$\quad - [(t+b)(c+s+u+d) + (c+s)(u+d)].$

Theorem 1 ([25]). *Assume (12.10): the heaviest lepton τ is lighter than the W and there is a hierarchy between quark masses and mixings. Then, the image, in the five dimensional space $(m_W, m_H, g_1, g_2, g_3)$, of the six strictly positive noncommutative gauge parameters $x, y_1, y_2, y_3, \tilde{x}, \tilde{y} = \mathrm{tr}(\mathrm{diag}(\tilde{y}_1, \tilde{y}_2, \tilde{y}_3))$, see below, is characterized by the following inequalities for $N = 3$ generations:*

$$\tau < W < \frac{q}{N}, \qquad (12.11)$$

$$H_{min}(W) < H(W) < H_{max}(W), \qquad (12.12)$$

$$0 < \sin^2\theta_w < \frac{2}{3}\left(1 + \frac{W-\tau}{q-3\tau} + \left(\frac{g_2}{3g_3}\right)^2\right)^{-1}. \qquad (12.13)$$

The saturated bounds for the Higgs are given by

$$H_{max}(W) := \frac{r^2 - 9e^2}{q-3e} - \frac{(r^2 - 3qe)e}{q-3e}\frac{1}{W} - \left(1 + \frac{g_2^{-2}}{g_1^{-2} - \frac{1}{6}g_3^{-2}}\right)W$$

$$H_{min}(W) := \frac{r^2 - 9\tau^2}{q-3\tau} - \frac{(r^2 - 3q\tau)\tau}{q-3\tau}\frac{1}{W} - \left(1 + \frac{g_2^{-2}}{g_1^{-2} - \frac{1}{6}g_3^{-2}}\right)W.$$

In particular, $H_{max}(W) - H_{min}(W) = (\tau - e)\left(1 + \frac{r^2-q^2}{(q-3e)(q-3\tau)}\right)\left(\frac{q}{W} - 3\right).$

This shows that the Higgs has a noncommutative fuzzy mass constraint.

As a side product, knowing $m_W = 80.41 \pm 0.1$ GeV and all quark masses, in particular the heaviest $m_t = 173.8 \pm 5.2$ GeV [28], we get that the number N of generations is less than 5.

It was noticed that this type of inequalities on masses are not invariant under renormalization [2,159].

Idea: The noncommutative Yang-Mills theory is an effective theory; there exists a cutoff Λ at which energy $\mathcal{A}_t = \mathcal{A}$ while the more the energy is decreased, the more the spacetime influence is increased. At our energy, for instance $m_Z = 91.187 \pm 0.007$ GeV, this influence is really significant.

Problem: Find an argument giving the energy ranking!

12.7 The Renormalization Machinery

According to the previous idea, the fuzzy constraints in theorem 1 are bare constraints. This means that we have to do renormalization to go beyond this tree level. This approach is standard in perturbative field theory, where, to include quantum perturbations, one computes the loop corrections. At this point, there are a lot of possibilities, since there exist different renormalization schemes which are not equivalent. We fix notations:

The kinetic term of the scalar field ϕ is normalized to $\frac{1}{2}$ in the Lagrangian which is written as

$$\mathcal{L} = \frac{1}{2}\partial_\mu \phi^* \partial_\mu \phi + \lambda(\phi^*\phi)^2 - \frac{\mu^2}{2}(\phi^*\phi) + \ldots$$

We adopt the mass independent $\overline{MS}$ renormalization scheme [82,194] in the approximation where all fermions masses are neglected except the top quark mass m_t. As running parameter associated to the energy E, we choose $r = \log_{10}(\frac{E}{m_Z})$. Since the flow of μ^2 is renormalization scheme dependent, we trade the running top mass for its Yukawa coupling, $m_t = g_t v$, $m_W = \frac{1}{2}g_2 v$, $m_H = 2\sqrt{2\lambda}\, v$, where the expectation value is $v = \frac{1}{2}\frac{\mu}{\sqrt{\lambda}}$. We interpret the five constraints on $g_1, g_2, g_3, \lambda, \mu^2$ in terms of running quantities at a noncommutative scale Λ

For the renormalization flow, the one-loop evolution equations of the above variables are the following first order differential equations

$$C\, g_1'(r) = \frac{41}{6} g_1(r)^3, \tag{12.14}$$

$$C\, g_2'(r) = -\frac{19}{6} g_2(r)^3, \tag{12.15}$$

$$C\, g_3'(r) = -7 g_3(r)^3, \tag{12.16}$$

$$C\, g_t'(r) = g_t(r)\left(-\frac{17}{12}g_1(r)^2 - \frac{9}{4}g_2(r)^2 - 8g_3(r)^2 + 9g_t(r)^2\right) \tag{12.17}$$

$$C\,\lambda'(r) = \lambda(r)(-3g_1(r)^2 - 9g_2(r)^2 + 24g_t(r)^2 + 96\lambda(r))$$
$$+ \frac{3}{32}g_1(r)^4 + \frac{9}{32}g_2(r)^4 - 6g_t(r)^4 + \frac{3}{16}g_1(r)^2 g_2(r)^2, \quad (12.18)$$

with $C = \frac{16\pi^2}{\ln(10)}$. The two-loops equations are more complicated, but are again first-order ordinary coupled differential equations. The initial conditions are taken from experiments: At $r = 0$, that is, at $m_Z = 91.187$ GeV, we have [28]

$$g_1(0) = 0.3575 \pm 0.0001, \quad (12.19)$$
$$g_2(0) = 0.6507 \pm 0.0007, \quad (12.20)$$
$$g_3(0) = 1.218 \pm 0.0026, \quad (12.21)$$
$$m_t(0) = 173.8 \pm 5.2 \text{ GeV}, \quad (12.22)$$
$$m_W(0) = 80.41 \pm 0.1 \text{ GeV}. \quad (12.23)$$

So we get for the central values

$$v(0) = 246.9 \text{ GeV},$$
$$g_t(0) = 0.004.$$

We neglect all threshold effects and identify pole masses with running masses at m_Z.

<u>Example</u>: If instead of using the scalar product (12.3), we use (12.4) for instance, then

$$g_3(\Lambda)^{-2} = \frac{1}{6\pi^2} N(\tilde{x} + x'), \quad (12.24)$$
$$g_2(\Lambda)^{-2} = \frac{1}{8\pi^2}[N(x + x') + y + y'], \quad (12.25)$$
$$g_1(\Lambda)^{-2} = \frac{1}{8\pi^2}[Nx + \frac{2}{9}N\tilde{x} + \frac{11}{9}Nx' + \frac{1}{2}y + \frac{3}{2}\tilde{y} + 3y'], \quad (12.26)$$
$$\lambda(\Lambda)^{-1} = \frac{2}{\pi^2}\frac{(x+x')^2}{k}, \quad (12.27)$$
$$g_t(\Lambda)^{-2} = \frac{1}{2\pi^2}(x + x'). \quad (12.28)$$

Here k is a number which can be computed in the top mass approximation (see [26]).
In terms of masses, we get in this case

$$\Lambda = 0.96 \cdot 10^{10} \text{ GeV}, \quad (12.29)$$
$$m_t = 214 \pm 0 \pm 4 \text{ GeV}, \quad (12.30)$$
$$m_H = 227 \pm 0 \pm 4 \text{ GeV} \quad (12.31)$$

or when $z' \in \rho(\text{center of } \mathcal{A})$:

$$m_t = 188 \pm 14 \pm 2 \pm 0 \text{ GeV}, \quad (12.32)$$
$$m_H = 198 \pm 8 \pm 2 \pm 0 \text{ GeV}. \quad (12.33)$$

The first error comes from the uncertainty in the noncommutative scale, $\Lambda \in [10^{13}, 10^{17}]$ GeV, the second from the present experimental uncertainty in the gauge couplings, $g_3 = 1.218 \pm 0.0026$, and the third from the uncertainty in the top mass, if needed as input, $m_t = 173.8 \pm 5.2$ GeV.

Problems:

1) Classify all possible scalar products preserving the important physical constraints and justify the choices. Different possibilities occur (see [174])

2) Do the same computations as in (12.24-12.33) for these scalar products.

12.8 Noncommutative Relativity

Using the approaches of Connes and Chamseddine [40,?], the effective action is chosen to be $S_{eff} = \text{tr} \, f(\mathcal{D}^2/\Lambda^2)$, where Λ is an energy cutoff and $f : \mathbb{R}_+ \to \mathbb{R}_+$ is a positive, smooth function with finite, strictly positive first 'momenta', $f_0 := \int_0^\infty u f(u) \, du$, $f_2 := \int_0^\infty f(u) \, du$, $f_4 := f(0)$. Using the Lichnérowicz formula and the heat kernel expansion, this action depends only on the three momenta f_0, f_2, f_4 and takes the form:

$$f(\mathcal{D}_t^2/\Lambda^2) \approx \int_M \left[-\tfrac{1}{16\pi} m_P(\Lambda)^2 R + \Lambda_C(\Lambda) + \tfrac{1}{2} g_3(\Lambda)^{-2} F^{(3)*}_{\mu\nu} F^{(3)\mu\nu} \right.$$
$$+ \tfrac{1}{2} g_2(\Lambda)^{-2} F^{(2)*}_{\mu\nu} F^{(2)\mu\nu} + \tfrac{1}{4} g_1(\Lambda)^{-2} F^{(1)*}_{\mu\nu} F^{(1)\mu\nu}$$
$$+ \tfrac{1}{2} (D_\mu \varphi)^* D^\mu \varphi + \lambda(\Lambda) |\varphi|^4 - \tfrac{1}{2} \mu(\Lambda)^2 |\varphi|^2$$
$$\left. - a(\Lambda) C_{\mu\nu\rho\sigma} C^{\mu\nu\rho\sigma} + \tfrac{1}{12} |R \varphi|^2 \right] (\det g_{..})^{1/2} \, d^4 x. \quad (12.34)$$

It follows that φ is an isospin doublet. After a normalization of the kinetic terms and a shift of the Higgs field by its vacuum expectation value, $|\varphi| = v(\Lambda)$, we can identify Newton's constant $G = \hbar c \, m_P^{-2}$, the cosmological constant Λ_C and the other coupling constants:

$$m_P(\Lambda)^2 = \tfrac{1}{3\pi} f_2 \left[15N - 2 \tfrac{L^2}{Q} \right] \Lambda^2,$$

$$\Lambda_C(\Lambda) = \tfrac{1}{4\pi^2} \left[15 N f_0 - \tfrac{f_2^2}{f_4} \tfrac{L^2}{Q} \right] \Lambda^4,$$

$$a(\Lambda) = \tfrac{3N}{64\pi^2} f_4$$

$$L(\Lambda) := 3q + \tau + \mu + e,$$

$$Q(\Lambda) := 3(t^2 + c^2 + u^2 + b^2 + s^2 + d^2) + \tau^2 + \mu^2 + e^2,$$

$$g_3(\Lambda)^{-2} = \tfrac{N}{3\pi^2} f_4, \quad (12.35)$$

$$g_2(\Lambda)^{-2} = \tfrac{N}{3\pi^2} f_4, \quad (12.36)$$

$$g_1(\Lambda)^{-2} = \tfrac{5}{3} \tfrac{N}{3\pi^2} f_4, \quad (12.37)$$

$$\lambda(\Lambda)^{-1} = \tfrac{1}{\pi^2} f_4 \tfrac{L(\Lambda)^2}{Q(\Lambda)} = \tfrac{3}{\pi^2} f_4 \left(1 + 2 \tfrac{m_b(\Lambda)^2}{m_t(\Lambda)^2} + O\left(\tfrac{m_\tau(\Lambda)^2}{m_t(\Lambda)^2}\right) \right), \quad (12.38)$$

$$\mu(\Lambda)^2 = 2 \tfrac{f_2}{f_4} \Lambda^2. \quad (12.39)$$

We ignore the gravitational part because we use the renormalization flow of the coupling constants. As before, we can soften the action (16.33) to $\mathrm{tr}\,[z'f(\mathcal{D}_t^2/\Lambda^2)]$. Then the constraints read [25]:

$$g_3(\Lambda)^{-2} = \tfrac{1}{9\pi^2} f_4 N x', \tag{12.40}$$

$$g_2(\Lambda)^{-2} = \tfrac{1}{12\pi^2} f_4 (N x' + y_1' + y_2' + y_N'), \tag{12.41}$$

$$g_1(\Lambda)^{-2} = \tfrac{1}{12\pi^2} f_4 (\tfrac{11}{9} N x' + 3(y_1' + y_2' + y_N')), \tag{12.42}$$

$$\lambda(\Lambda)^{-1} = \tfrac{1}{\pi^2} f_4 \frac{L(\Lambda)^2}{Q(\Lambda)}, \quad \mu(\Lambda)^2 = 2\frac{f_2}{f_4}\Lambda^2, \tag{12.43}$$

$$L(\Lambda) = x'(t + c + u + b + s + d) + y_3'\tau + y_2'\mu + y_1'e,$$

$$Q(\Lambda) = x'(t^2 + c^2 + u^2 + b^2 + s^2 + d^2) + y_3'\tau^2 + y_2'\mu^2 + y_1'e^2.$$

If $z' = 1_{90}$, then $x' = 3$, $y_1' = y_2' = y_N' = 1$ and we recover the stiff relations (12.37-12.39).

Consequences:

1) The constraints, $g_3(\Lambda) = g_2(\Lambda)$ and $\sin^2\theta_w(\Lambda) = \frac{g_2^{-2}}{g_1^{-2}+g_2^{-2}} = \tfrac{3}{8}$ (12.35-12.37), look as in grand unification. They disappear in (12.40-12.42), showing that, in noncommmutative geometry, slight variation on the hypotheses may generate important differences.

2) The noncommutative Yang-Mills and noncommutative relativity use completely different techniques. Nevertheless they intersect in the computation of the Higgs mass: $188 \leq m_H \leq 201$ GeV.

Problem: Is this intersection a pure coincidence or more profound?

12.9 Conclusions

1) Noncommutative geometry appears very successful in particle physics because a consistent set of axioms yields the following: the Higgs potential is an output, the gluons couple vectorially, are massless, the W couples axially, is massive, the weak sector is in Poincaré duality with the strong one. In particular, there are no right neutrinos. If such particles exist, then the model has to be changed.

It has not been proven that such an improvement is impossible.

See nevertheless [197]. See the different possibilities of finite spectral triples in [127], because we have the choice of the representation $\rho(\mathcal{A})$ and compatible $\mathcal{D}$, J and chirality.

The message is simple: the Higgs mechanism and the Higgs fields are just the magnetic fields of gravitational fields.

2) Noncommutative geometry is very rigid: Only few groups are allowed, all existing "super" versions do not satisfy the axioms. So the domain to understand is smaller than the classical one but covers nevertheless all important physical points (see [111] for this point).

3) Despite point 2, noncommutative geometry is very adaptive: a lot of phenomenology has to be done and some hypotheses to be clarified (cf the above problems).

A few words about the future: The main step to overcome is to give up the classical case and to get a true quantum noncommutative geometry. It is somehow funny to conclude that despite a natural affinity, noncommutative geometry does not intersect naturally with the Schrödinger quantum mechanics!

To go in this direction, we will have to consider the connection between Feynman diagrams, Hopf algebras and local computations of indices [129, 43, 128, 47, 48].

One can dream of the realization of Dirac's program: "Only one" Dirac operator is covariant under a Hopf algebra. If this is the case, this Dirac operator must somehow resemble the above one, at least its internal part!

So the answer to the question "Is noncommutative geometry the millenium bug of physics" is no. By the way, what about the opposite question?

Acknowledgments

I would like to thank F. Scheck and H. Upmeier for their kind invitation and the incredibly constructive attitude of all participants at this meeting in Hesselberg.

13 The $su(2|1)$ Model of Electroweak Interactions and Its Connection to NC Geometry

Rainer Häussling

University of Mainz

> I review the $su(2|1)$ model of electroweak interactions which is essentially based on the super Lie algebra $su(2|1)$, thus incorporating both usual gauge fields and Higgs fields in one generalized Yang-Mills field. Special emphasis is put on the natural appearance of spontaneous symmetry breaking and other appealing features of the model like generation mixing. Also the connection of the model to noncommutative geometry is briefly discussed.

13.1 Introduction and Motivation

Up to now, the minimal standard model (SM) of electroweak interactions is in almost perfect agreement with experiment, see e.g. [179]. But nevertheless, from a more theoretical point of view this model exhibits some very unpleasant features [196] which roughly can be divided into two classes:

- The SM contains too many free parameters in order to be a really fundamental and unifying theory.
- The SM incorporates a lot of apparent arbitrariness that is only justified by its experimental success, e.g.:
  - The higgs mechanism is an *ad-hoc* procedure rather than a nice and natural feature of gauge theories.
  - The Higgs field is taken to be a $SU(2)$-doublet, only justified by the experimental value of the ρ-parameter.
  - Generation mixing is put in by hand by adding suitable Yukawa terms to the action.
  - The assignment of quantum numbers for the fields needs further experimental input.

The aim of this talk is to review the $su(2|1)$ model, which was developed in a series of papers, see [52] – [54], and to explain how this model is able to give new and better insight into the second, more qualitative class of problems present in the ordinary formulation of the standard model.

13.2 The Bosonic Part of the Model

I want to begin the presentation of the $su(2|1)$ model by motivating the use and the rôle of the super Lie algebra $su(2|1)$ which in detail will be defined below. The fundamental starting point for all considerations to follow is the fact, very well established by experiment, that in nature quarks and leptons (which will be represented by some spinors $\Psi \in C$) appear in the form of purely chiral fields:

$$C \ni \Psi = \Psi_L + \Psi_R \in C^{(L)} \oplus C^{(R)} \tag{13.1}$$
$$\text{with} \quad \Psi_L = \tfrac{1}{2}(\mathbb{1} - \gamma_5)\Psi \;,\; \Psi_R = \tfrac{1}{2}(\mathbb{1} + \gamma_5)\Psi$$

Hence the space C of spinor fields carries a $\mathbb{Z}_2$-grading, the grading automorphism being given by γ_5 (i.e. $\gamma_5 \Psi_R = \Psi_R, \gamma_5 \Psi_L = -\Psi_L$). This $\mathbb{Z}_2$-grading is consistent with the $\mathbb{Z}_2$-grading of the Clifford algebra Γ acting on C,

$$\Gamma = \Gamma^{(0)} \oplus \Gamma^{(1)} \;, \tag{13.2}$$
$$\Gamma^{(0)} = \{\mathbb{1}, \gamma^\mu \gamma^\nu (\mu < \nu), i\gamma_5\} \;,\; \Gamma^{(1)} = \{\gamma^\mu, \gamma^\mu \gamma^\nu \gamma^\tau (\mu < \nu < \tau)\} \;,$$
$$\text{with} \quad [\Gamma^{(0)}, \gamma_5] = 0 \;,\; \{\Gamma^{(1)}, \gamma_5\} = 0 \;.$$

Furthermore, looking for a theory of electroweak interactions, experiment also tells us that the chiral fields Ψ_L and Ψ_R have to be classified into different representations of weak isospin (i.e. of $SU(2)_L$) and weak hypercharge (i.e. of $U(1)_Y$). As a consequence, the generators

$$T_k \;,\; k = 1, 2, 3, 4 \qquad (T_k = I_i \; (i = 1, 2, 3) \text{ or } \frac{1}{2}Y) \tag{13.3}$$

of $SU(2)_L \times U(1)_Y$ have a block structure of square matrices

$$T_k = \begin{pmatrix} [T_k]_L & 0 \\ 0 & [T_k]_R \end{pmatrix} \tag{13.4}$$

along the main diagonal. The generators T_k act on a space V,

$$V = V^{(0)} \oplus V^{(1)} \text{ with } V^{(0)} = C^{(L)} \otimes X^{(L)} \;,\; V^{(1)} = C^{(R)} \otimes X^{(R)} \;, \tag{13.5}$$

which inherits the $\mathbb{Z}_2$-grading from the space C (13.1). $X^{(L)}$ and $X^{(R)}$ are the different representation spaces for left- and right-chiral fields, respectively.

Now we can ask the following question: What happens if one embeds the $su(2) \times u(1)$-structure (13.4) in a minimal way into some larger structure by filling up the rectangular zeros in the off-diagonal of (13.4)? And, of course, what would one gain by following such a strategy?

The procedure of filling up the off-diagonal blocks naturally leads to the super Lie algebra $su(2|1)$ which, by definition, consists of all 3×3 antihermitean matrices

$$M = \left(\begin{array}{c|c} A_{2\times 2} & C_{2\times 1} \\ \hline D_{1\times 2} & B_{1\times 1} \end{array} \right)_{3\times 3} = M^{(0)} + M^{(1)} \;,\; M^\dagger = -M \;, \tag{13.6}$$

with vanishing supertrace:

$$\text{Str } M := \text{tr } A - B = 0 \tag{13.7}$$

In (13.6) $M^{(0)}$ comprises the blocks A and B along the main diagonal and is, by definition, *even*, $\partial M^{(0)} = 0$, whereas $M^{(1)}$ contains the rectangular blocks C and D in the off-diagonal and is *odd*, $\partial M^{(1)} = 1$. Thus, at this stage another $\mathbb{Z}_2$-grading, the $\mathbb{Z}_2$-grading of block matrices comes into the game. The Lie bracket in $su(2|1)$ is a graded one and is defined as a combination of the ordinary commutator $[\cdot,\cdot]$ and anticommutator $\{\cdot,\cdot\}$[1]:

$$[M,N]_g = [M^{(0)}, N^{(0)}] + [M^{(0)}, N^{(1)}] + [M^{(1)}, N^{(0)}] + i\{M^{(1)}, N^{(1)}\} \quad (13.8)$$

In direct analogy to ordinary Yang-Mills theory, the connection $\mathcal{A}$ of the $su(2|1)$ model will be a *super Lie algebra valued one-form*. Let me be more precise about this point: Call $\Omega_\pm$ and $\Omega'_\pm$ the four generators of the odd part of $su(2|1)$, viz.

$$\Omega_+ = \begin{pmatrix} 0 & 0 & 1 \\ 0 & 0 & 0 \\ \hline 0 & 0 & 0 \end{pmatrix}, \quad \Omega_- = \begin{pmatrix} 0 & 0 & 0 \\ 0 & 0 & 1 \\ \hline 0 & 0 & 0 \end{pmatrix}, \quad \Omega'_+ = (\Omega_-)^\dagger, \quad \Omega'_- = (\Omega_+)^\dagger. \quad (13.9)$$

(The generators of the even part are the generators T_k of $SU(2) \times U(1)$, see (13.3).)

Then the generalized connection one-form of the $su(2|1)$ model is given by:

$$\mathcal{A} = i\left(a\boldsymbol{I} \cdot \boldsymbol{W} + b\frac{Y}{2}W^{(8)}\right) + i\frac{c}{\mu}\left(\Phi^{(0)}\Omega'_- + \Phi^{(+)}\Omega'_+ + h.c.\right) \quad (13.10)$$

In (13.10) $\boldsymbol{W} = (W^{(1)}, W^{(2)}, W^{(3)})$ denotes the $SU(2)$ gauge fields, and $W^{(8)}$ is the $U(1)$ gauge field. Furthermore, a, b, c are dimensionless real parameters (which are irrelevant for the following) and μ is a mass scale, necessary to give $\mathcal{A}$ a unique dimension.

Because there is, roughly speaking, more space when enlarging the structure from $su(2) \times u(1)$ to $su(2|1)$, i.e. when filling up the rectangular blocks in the off-diagonal in (13.4), $\mathcal{A}$ (13.10) also contains some additional fields $\Phi^{(0)}, \Phi^{(+)}$ (and their conjugates). These fields have to be scalar fields due to the fact that $\mathcal{A}$ is a *one*-form, thus carrying total degree one. This total degree is defined to be the sum of the matrix degree and the exterior form degree (mod 2):

$$\hat{\partial}(M \otimes \omega) = \partial M + \partial \omega \quad \text{with} \quad M \in su(2|1), \; \omega \in \Lambda^\star(M) \quad (13.11)$$

Here we have used the fact that there is a vector space isomorphism between the Clifford algebra Γ and the algebra $\Lambda^\star(M)$ of exterior forms (the latter also being $\mathbb{Z}_2$-graded in an obvious way). Hence, the $\mathbb{Z}_2$-grading stemming

---

[1] One could, perhaps, wonder about the i-factor in front of the last term on the r.h.s. of (13.8). This factor is necessary in order to stay within the *real* super Lie algebra $su(2|1)$ which in turn is of some importance for controlling the physical degrees of freedom properly.

from the chirality of spinor fields, see (13.1), (13.5), and the $\mathbb{Z}_2$-grading of exterior forms can be identified [101].

As the forthgoing construction of the model will show, $\Phi^{(0)}$ and $\Phi^{(+)}$ are just the shifted (physical) Higgs fields. Thus, the $su(2|1)$ model combines both the usual gauge fields W_μ^k, $k = 1, 2, 3, 8$, and the Higgs fields $\Phi^{(0)}, \Phi^{(+)}$ (and their conjugates) into one single object, namely the generalized connection one-form $\mathcal{A}$ (13.10).

Please also note that due to the *minimal* embedding of $su(2) \times u(1)$ into a super Lie algebra, there is only space for a $SU(2)$-doublet of Higgs fields.[2]

In order to proceed we now have the task to construct a curvature two-form $\mathcal{F}$ out of the connection one-form $\mathcal{A}$, and, finally, to calculate the Yang-Mills Lagrangian $\mathcal{L}_{bos}$ by means of $\mathcal{F}$. In ordinary Yang-Mills theories (on flat spacetime) the construction of $\mathcal{F}$ relies on the (local) expression for the covariant derivative

$$\nabla = d_C + \mathcal{A} \qquad (13.12)$$

by taking the square of ∇:

$$\mathcal{F} = \nabla^2 = d_C \mathcal{A} + \frac{1}{2}[\mathcal{A}, \mathcal{A}] \qquad (13.13)$$

However, having merged all the $\mathbb{Z}_2$-gradings according to (13.11), it is natural to enlarge the exterior derivative d_C (which acts on differential forms, raising the exterior form degree by one) by an operation d_M acting on the $\mathbb{Z}_2$-graded matrices and raising the matrix degree by one (mod 2),

$$d_M : M^{(0),(1)} \longrightarrow M^{(1),(0)} \quad , \qquad (13.14)$$

such that the sum of d_C and d_M acts on both gradings. Without loss of generality[3], the action of d_M on $M \in su(2|1)$ can be given by the graded commutator of M with a special odd element $\eta \in su(2|1)$:

$$d_M M = [\eta, M]_g = [\eta, M^{(0)}] + i\{\eta, M^{(1)}\} \qquad (13.15)$$
$$\text{with} \quad \eta = i\rho \left(\Omega_+ + \Omega'_-\right) \qquad (13.16)$$

The parameter $\rho \in \mathbb{R}_+$ in (13.16), in fact, is not necessary and can be put equal to 1 without loss of generality. We keep it here in order to have the possibility of turning on and off the matrix derivative d_M later on.

Thus, with M a matrix in $su(2|1)$ and ω an exterior form, one has:

$$d(M \otimes \omega) = [\eta, M]_g \otimes \omega + (-1)^{\partial M} M \otimes (d_C \omega) \qquad (13.17)$$

---

[2] Furthermore, no other fields than the usual fields of the SM are introduced here. This is in contrast to other unifying theories like e.g. supersymmetric theories or GUTs in which additional fields occur.

[3] This is discussed in detail in [53].

The curvature two-form of the $su(2|1)$ model is then constructed by generalizing the standard structure equation (13.13) of Yang-Mills theory, i.e.[4]:

$$\mathcal{F} = d\mathcal{A} + \frac{1}{2}[\mathcal{A}, \mathcal{A}]_g \qquad (13.18)$$

Finally, the bosonic Lagrangian in the $su(2|1)$ model is obtained by taking the trace of the scalar product $\langle \mathcal{F}^\dagger \mathcal{F} \rangle$,

$$\mathcal{L}_{bos} \propto \mathrm{tr}\langle \mathcal{F}^\dagger \mathcal{F} \rangle \quad . \qquad (13.19)$$

($\langle \cdot \, \cdot \rangle$ denotes the scalar product in the space of exterior forms.)
Using the trace in (13.19) (instead of, e.g., the supertrace) explicitly breaks $su(2|1)$ invariance of the model, only offering the possibility for a $su(2) \times u(1)$ invariance of $\mathcal{L}_{bos}$ to survive eventually. This is equivalent to saying that $su(2|1)$ is not gauged in the $su(2|1)$ model. Instead, there is only the *action* of $su(2|1)$ on the fields of the theory, and, hence, $su(2|1)$ only serves as a classifying algebra for quarks and leptons. This *weak* [101] realization of the symmetry taken alone leads to nontrivial consequences, as will be seen below. Instead of going into the technical details of the explicit calculations, see e.g. [100], I now want to discuss the mechanism of spontaneous symmetry breaking in the $su(2|1)$ model in some depth.

The mechanism of spontaneous symmetry breaking

Formally, an infinitesimal gauge transformation is generated by an element $\mathcal{E}$ of $su(2|1) \otimes \Lambda^\star(M^4)$,

$$\mathcal{A}' = \mathcal{A} + d\mathcal{E} + [\mathcal{A}, \mathcal{E}]_g \quad , \qquad (13.20)$$

where $\mathcal{E} = \mathcal{E}^{(0)} + \mathcal{E}^{(1)}$ has total degree 0, $\hat{\partial}\mathcal{E} = 0$, i.e. $\mathcal{E}^{(0)}$ and $\mathcal{E}^{(1)}$ contain 0-forms and 1-forms, respectively. Later on, $\mathcal{E}$ will be further restricted to $[su(2|1)]^{(0)} \otimes \Lambda^0$, according to the fact mentioned above that $su(2|1)$ will not be gauged.
The key observation in the present context is the surprising fact that there is precisely one connection $\mathcal{A}_0$ which is invariant under all *constant* $su(2|1)$-transformations, viz.

$$\mathcal{A}_0 = -\eta = -i\rho\left(\Omega_+ + \Omega'_-\right) \quad . \qquad (13.21)$$

Indeed, one has:

$$\mathcal{A}'_0 = \mathcal{A}_0 + d\mathcal{E} + [\mathcal{A}_0, \mathcal{E}]_g = \mathcal{A}_0 + [\eta, \mathcal{E}]_g - [\eta, \mathcal{E}]_g = \mathcal{A}_0 \qquad (13.22)$$

Thus, $\mathcal{A}_0$ is a constant background connection that remains fixed under the action of the *whole* algebra $su(2|1)$. The corresponding curvature

$$\mathcal{F}_0 = d\mathcal{A}_0 + \frac{1}{2}[\mathcal{A}_0, \mathcal{A}_0]_g \qquad (13.23)$$

---

[4] The graded commutator of $su(2|1)$-valued exterior forms includes a slight and straightforward generalization of (13.8), see [53].

is given by (as a short calculation shows):

$$\mathcal{F}_0 = -i\eta^2 = i\rho^2(I_3 + \frac{1}{2}Y) \tag{13.24}$$

Therefore, $\mathcal{F}_0$ is proportional to the charge operator $Q = I_3 + \frac{1}{2}Y$.
With this in mind, it is suggestive to decompose the full superconnection $\mathcal{A}$ (13.10) into the constant part $\mathcal{A}_0$ and a remaining part $\mathcal{A}_\Theta$,

$$\mathcal{A} = \mathcal{A}_0 + \mathcal{A}_\Theta , \tag{13.25}$$

where $\mathcal{A}_\Theta$ is defined by this equation and easily seen to be:

$$\mathcal{A}_\Theta = i\left(a\mathbf{I}\cdot\mathbf{W} + b\frac{Y}{2}W^{(8)}\right) + i\frac{c}{\mu}\left(\Theta^{(0)}\Omega'_- + \Theta^{(+)}\Omega'_+ + h.c.\right) \tag{13.26}$$

with $\Theta^{(0)} = \Phi^{(0)} + \mu\rho$, $\Theta^{(+)} = \Phi^{(+)}$ \hfill (13.27)

So far, $\Theta^{(0)}$ and $\Theta^{(+)}$ are just symbols for some fields. However, it will turn out in the following that $\Theta^{(0)}$ and $\Theta^{(+)}$ are exactly the unshifted Higgs fields of the SM. Indeed, rewriting the gauge transformations (13.20) for the connection $\mathcal{A}_\Theta$ and restricting $\mathcal{E}$ to the even part of the algebra, i.e. $\mathcal{E}^{(1)} = 0$, these gauge transformations become the well-known gauge transformations for the gauge fields *and* Higgs fields of the SM:

$$\mathcal{A}'_\Theta = \mathcal{A}_\Theta + d_C\mathcal{E}^{(0)} + [\mathcal{A}_\Theta, \mathcal{E}^{(0)}] \tag{13.28}$$

Now, moving on to the curvature $\mathcal{F}$ of the $su(2|1)$ model by using the generalized structure equation (13.18), one finds that $\mathcal{F}$ also splits into two parts,

$$\mathcal{F} = \mathcal{F}_0 + \mathcal{F}_\Theta , \tag{13.29}$$

where the constant background curvature $\mathcal{F}_0$ was introduced in (13.23), (13.24) and where $\mathcal{F}_\Theta$ is given by:

$$\mathcal{F}_\Theta = d_C\mathcal{A}_\Theta + \frac{1}{2}[\mathcal{A}_\Theta, \mathcal{A}_\Theta]_g \tag{13.30}$$

The transformation behaviour of $\mathcal{F}$ under the generalized gauge transformations (13.20) is easily found to be:

$$\mathcal{F}' = (\mathcal{F}_\Theta + \mathcal{F}_0)' = \mathcal{F}_\Theta + \mathcal{F}_0 + [\mathcal{F}_\Theta, \mathcal{E}]_g \tag{13.31}$$

Thus, restricting as above the gauge transformations to the even part, i.e. $\mathcal{E}^{(1)} = 0$, one would recover the well-known transformation law for the curvature of the SM, *if $\mathcal{F}_0$ on the r.h.s. of (13.31) is absent*: This term evidently spoils the full $SU(2)_L \times U(1)_Y$ symmetry, see also below. Hence, in order to end up with the SM Lagrangian, at this stage an extra-term[5] has to be added to $\mathcal{F}$:

$$\hat{\mathcal{F}} := \mathcal{F} + \mathcal{C}_\mathcal{F} \tag{13.32}$$

The most general expression for $\mathcal{C}_\mathcal{F}$ (not spoiling $SU(2)_L \times U(1)_Y$ symmetry) is given by:

$$\mathcal{C}_\mathcal{F} = -\mathcal{F}_0 + i\rho^2 aY \quad \text{with } a \in \mathbb{R} \tag{13.33}$$

Finally, calculating the lagrangian, see (13.19), the supercurvature $\hat{\mathcal{F}}$ yields, among other terms[6], a Higgs potential $V(\Theta)$:

$$V(\Theta) = \frac{2}{g^2}\left[\overline{\Theta^{(0)}}\Theta^{(0)} + \overline{\Theta^{(+)}}\Theta^{(+)} - \frac{3}{2}a\mu^2\rho^2\right]^2 + const. \tag{13.34}$$

Let me come to the interpretation of the above formulae.

Clearly, in the $su(2|1)$ model spontaneous symmetry breaking is traced back to the introduction of the matrix derivative d_M and therefore to the "absolute element"

$$\eta = i\rho\left(\Omega_+ + \Omega'_-\right) \in su(2|1) \quad . \tag{13.35}$$

Hence, two cases can be distinguished:

(i) For $\rho = 0$ the matrix derivative d_M is trivial. At the same time, the constant background curvature $\mathcal{F}_0$ vanishes. The Higgs potential becomes

$$V(\Theta) = \frac{2}{g^2}\left[\overline{\Theta^{(0)}}\Theta^{(0)} + \overline{\Theta^{(+)}}\Theta^{(+)}\right]^2 \tag{13.36}$$

and has no local minimum away from the origin. The Lagrangian has the full $SU(2)_L \times U(1)_Y$ symmetry and there is no spontaneous symmetry breaking.

---

[5] A natural explanation for this extra-term can be given in an alternative, but equivalent derivation of the $su(2|1)$ model using 4×4 matrices instead of 3×3 matrices: In the 4×4 formalism $su(2|1)$ and, especially, the connection one-form $\mathcal{A}$ (13.10) are trivially embedded into 4×4 matrices. This is done in order to get a matrix *differential*, i.e. to achieve $d_M^2 = 0$ (which only holds in even dimensions). But because the action of d_M on $\mathcal{A}_{4 \times 4}$ also populates the former zeros in the fourth row and column one has to project back to 3×3 after the calculation of $\mathcal{F}$ has been done. This projection leads to an extra-term like above, see [53] for details.

[6] The other terms are the usual kinetic terms for the gauge fields and the Higgs fields, i.e.

$$\mathcal{L}_{bos} \propto -\frac{1}{4}\sum_a F_a^{\mu\nu}F_{\mu\nu}^a + 2(D_\mu\Theta^{(0)}, D^\mu\Theta^{(0)}) + 2(D_\mu\Theta^{(+)}, D^\mu\Theta^{(+)}) - V(\Theta) ,$$

see e.g. [53].

(ii) For $\rho \neq 0$ the matrix derivative is nontrivial and the model contains the constant background curvature $\mathcal{F}_0$ which is proportional to the charge operator $Q = I_3 + \frac{1}{2}Y$. Therefore, the background field $\mathcal{F}_0$ singles out a specific direction in the representation space of the algebra $su(2|1)$ and - no matter how the Lagrangian is constructed out of $\mathcal{F}$ - it can have no more symmetry than the isotropy group of $\mathcal{F}_0$, i.e. the $U(1)$ generated by the charge operator Q. This situation is very similar to the one of a constant magnetic field applied to a spherical atom; the magnetic field also singles out a specific direction in space and thus breaks the spherical symmetry of the atom.

In this sense, the mechanism of spontaneous symmetry breaking is a built-in feature of the $su(2|1)$ model which, in addition, here receives a new and geometrical interpretation[7].

13.3 The Fermionic Part of the Model

In principle, the fermionic Lagrangian $\mathcal{L}_{ferm}$ of the $su(2|1)$ model is given by a term like $\overline{\Psi} \not{D} \Psi$, where $\not{D}$ is the generalized gauged Dirac operator, $\not{D} = i\gamma^\mu \partial_\mu \otimes \mathbb{1} + \rho(\mathcal{A})$. ($\rho(\mathcal{A})$ denotes the representation of the connection one-form $\mathcal{A}$ on the fermionic multiplet Ψ in question.)

However, there are some pecularities which have to be handled carefully. Without going into the details [102], we just want to mention the most important one: The charge conjugated contribution, stemming from the multiplet $^c\Psi$ of antiparticles, has to be taken into account explicitly. Doing this in a democratic way the expression for the fermionic Lagrangian reads:

$$\mathcal{L}_{ferm} = \overline{\Psi}(i\gamma^\mu \partial_\mu \otimes \mathbb{1})\Psi + \frac{1}{4}(\mathcal{C} + {}^c\mathcal{C}) \tag{13.37}$$

$$\text{with} \quad \mathcal{C} = \overline{\Psi}\rho(\mathcal{A})\Psi + \text{h.c.} \tag{13.38}$$
$${}^c\mathcal{C} = \overline{{}^c\Psi}\rho({}^c\mathcal{A}){}^c\Psi + \text{h.c.}$$

In any case, because $su(2|1)$ serves as a classifying algebra for the fermions of the theory, see above, we will have to deal with the representation theory of $su(2|1)$ in order to specify the representations Ψ and $\rho(\mathcal{A})$ in (13.37), (13.38).

Representations of $su(2|1)$ and quantum numbers of fields

There exist different types of representations of $su(2|1)$, see [158], [192]. First, there are the so-called *typical* irreducible representations. $su(2|1)$ being a $\mathbb{Z}_2$-graded Lie algebra, the representation spaces consequently will carry a $\mathbb{Z}_2$-grading, too. A representation of $su(2|1)$ is called *typical* if the dimensions of the even and odd subspaces are equal. The simplest typical irreducible representation of $su(2|1)$ is given by $[y_0 = \frac{1}{3}, I_0 = \frac{1}{2}]$, where y_0 and

---
[7] Please note that also the correct shape of the Higgs potential is predicted, see (13.34).

I_0 are (functions of) the eigenvalues of the two Casimir operators K_2 and K_3 of $su(2|1)$. When looking for the subspaces which are invariant under the even part of $su(2|1)$, i.e. $su(2) \times u(1)$, this representation decomposes into an isospin doublet with hypercharge $y = \frac{1}{3}$ and two isospin singlets with hypercharges $y = \frac{4}{3}$ and $y = -\frac{2}{3}$, respectively:

$$[y_0 = \frac{1}{3}, I_0 = \frac{1}{2}] \xrightarrow{su(2) \times u(1)} (I = \frac{1}{2})_{y=\frac{1}{3}} \oplus (I = 0)_{y=\frac{4}{3}} \oplus (I = 0)_{y=-\frac{2}{3}} \tag{13.39}$$

Hence this supermultiplet describes just one family of quarks (u, d), and the correct assignment of quantum numbers for (u_L, d_L), u_R and d_R automatically follows from the choice of the representation (13.39) (without any further experimental input). The corresponding antiparticles $({}^c u, {}^c d)$ fit very well into the conjugated, but not equivalent, representation $[y_0 = -\frac{1}{3}, I_0 = \frac{1}{2}]$ of $su(2|1)$.

The second type of representations are the *nontypical* irreducible representations (the even and odd subspaces of the representation space not having the same dimension). In this type of representations, the smallest candidate is $[y_0 = -1, I_0 = \frac{1}{2}]$, whose decomposition with respect to $su(2) \times u(1)$ reads:

$$[y_0 = -1, I_0 = \frac{1}{2}] \xrightarrow{su(2) \times u(1)} (I = \frac{1}{2})_{y=-1} \oplus (I = 0)_{y=-2} \tag{13.40}$$

Therefore this representation describes one family of leptons, i.e. the left-handed doublet (e_L, ν_L) with hypercharge $y = -1$ and the right-handed singlet e_R with $y = -2$. The corresponding antiparticles again fall into the conjugated, but not equivalent, representation $[y_0 = 1, I_0 = \frac{1}{2}]$.

It should be remarked that the classification of one quark family or one lepton family according to representations of $su(2|1)$ was already discovered in [166]. (This is also true for the gauge fields and Higgs fields which fall into the adjoint representation of $su(2|1)$.) However, in these early attempts, it was tried to gauge $su(2|1)$ which led to serious difficulties with quantization and redundant or spurious degrees of freedom. For this reason $su(2|1)$ was discarded for a long time in elementary particle physics.

Remarks on Generation Mixing. For the classification of several families of quarks we can make use of another, third, type of representations, the so-called *reducible, but indecomposable* representations of $su(2|1)$. These representations can be written as the semi-direct sum of representations of, for instance, the first type. Hence, for three families of quarks we have:

$$[y_0 = \frac{1}{3}, I_0 = \frac{1}{2}] \not\in [y_0 = \frac{1}{3}, I_0 = \frac{1}{2}] \not\in [y_0 = \frac{1}{3}, I_0 = \frac{1}{2}] \tag{13.41}$$

In this case the generators of $su(2|1)$ have block triangular form due to the structure of (13.41) as a semi-direct sum:

$$\hat{\Omega} = \begin{pmatrix} \Omega_{11} & 0 & 0 \\ \Omega_{21} & \Omega_{22} & 0 \\ \Omega_{31} & \Omega_{32} & \Omega_{33} \end{pmatrix} \quad , \quad \hat{T} \text{ analogously} \tag{13.42}$$

Furthermore, it has to be remarked that the generators (13.42) of $su(2|1)$ contain a certain number of arbitrary parameters[8] which cannot be fixed by means of the super Lie algebra structure. Hence, when building the fermionic Lagrangian according to (13.37), it is clear that Yukawa terms will be produced automatically. In addition, the mass matrices originating from these Yukawa terms have triangular form as a short calculation shows. This triangular form of mass matrices has proven to be of relevance for the study of generation mixing, see [103], because of its maximal efficiency which in turn is due to a maximal number of zeros in the mass matrices.

Finally, the representation theory of $su(2|1)$ also offers the possibility to describe an extended leptonic family (e_L, ν_L), e_R and ν_R in case the neutrinos are massive Dirac particles. This is achieved by making use of a fourth type of representations which are reducible, but indecomposable, too, see [51] for details. The possible mixing of several of such extended leptonic families (via representations of the kind of (13.41)) was studied in [105].

13.4 The Connection to the Connes-Lott Model

In this section, I would like to briefly comment on the connection of the $su(2|1)$ model to Connes' noncommutative geometry [38], especially to the Connes-Lott (CL) model building [45]. There are, in principle, two possible lines to be followed when comparing the two models. On the one hand, the mathematical structures and ingredients can be compared, on the other hand, we can discuss the physical outputs of the two models. Of course, these two strategies are connected to each other.

Let me begin with some remarks concerning the mathematical structures: First of all, the mathematical apparatus underlying the $su(2|1)$ model evidently is much more elementary than the machinery of noncommutative geometry (NCG). On the other side, NCG has, of course, a higher character of generality: Beside its fruitful application in classical model building, it is a very interesting new branch in mathematics, possibly applicable to other problems of (elementary particle) physics.

The relevant mathematical objects for directly comparing the two models are the differential algebras generalizing the de Rham complex of ordinary differential geometry. In the CL model this differential algebra is given by $\Omega_D^\star(\mathcal{A})$ constructed out of the associative algebra $\mathcal{A} = (\mathbb{C} \oplus \mathbb{H}) \otimes \mathcal{C}^\infty(M)$ (for electroweak interactions) by means of a Dirac K-cycle (H, D, π). For the $su(2|1)$ model the underlying differential algebra $\Sigma^\star$ was discussed and explicitly constructed in [54]: Roughly speaking, $\Sigma^\star$, as a vector space, is given

---

[8] This is also the case for the representations (13.39), (13.40); the free parameters can be related to the quark and lepton masses, respectively, see [51].

by the set of matrix-valued exterior differential forms, this set carrying a $\mathbb{Z}$-grading in a natural way by making use of the exterior form grading and the matrix grading. The definition of the product $\odot$ and the differential d (only possible in even dimensions, see footnote [5]) is also straightforward. It can be shown that the spaces of generalized zero- and one-forms ($\Omega_D^0(\mathcal{A}), \Omega_D^1(\mathcal{A})$ and Σ^0, Σ^1, respectively) are isomorphic *as vector spaces*. However, there does not exist an algebra isomorphism between $\Sigma^\star$ and $\Omega_D^\star(\mathcal{A})$. This can be qualitatively seen from the fact that the construction of the product in the latter case involves the generalized Dirac operator D (and in particular the fermionic mass matrix) whereas the definition of the product of the former is totally independent of the Dirac operator[9].

Let me finally mention some of the main physical differences of the two models:

- In the CL model, spontaneous symmetry breaking only occurs if more than one generation is taken into account [173]. In the $su(2|1)$ model spontaneous symmetry breaking is always present.
- The CL model is formulated for Euclidean spacetimes. In order to arrive at the physical situation, a Wick rotation is necessary at the end. In the $su(2|1)$ model we can work in Minkowski space right from the beginning.

13.5 Conclusions

In this talk I reviewed the $su(2|1)$ model which is one of the noncommutative approaches to a more fundamental theory of electroweak interactions. Especially, I tried to show how some of the qualitative problems present in the ordinary formulation of the SM can be solved within this new approach. Of course, all the considerations presented here were at the classical level, and hence, there remains the task to find a noncommutative quantum theory, i.e. a quantum theory which takes into account the noncommutative structure seriously.

Acknowledgements

It is a pleasure for me to thank the organizers of the workshop for their kind hospitality at Hesselberg and for the excellent atmosphere there. Furthermore, my special thanks go to R. Coquereaux, G. Esposito-Farèse, F. Scheck and N.A. Papadopoulos for the fruitful collaboration and many enlightening discussions in the course of the development of the model presented here.

---

[9] Indeed, in the $su(2|1)$ model the Dirac operator is not an ingredient of the construction but rather a derived quantity.

14 Quantum Fields and Noncommutative Spacetime

Klaus Fredenhagen

University of Hamburg

14.1 Noncommutative Spacetime and Uncertainty Relations

A noncommutative structure of spacetime induced by quantum effects of gravity is proposed and the implementation of the locality principle on such a space is discussed. Heisenberg's uncertainty relation

$$\Delta x \geq \frac{\hbar}{\Delta p} \geq \frac{\hbar c}{E}$$

and Einstein's formula

$$E = mc^2$$

imply that measurements at short scales create strong gravitational fields which eventually shield the region of interest by a horizon whose size, in the spherical symmetric case, is given by the Schwarzschild radius $R = \frac{2Gm}{c^2}$. Hence, resolutions of distances which are smaller than the Planck length

$$\lambda_P := \sqrt{\frac{G\hbar}{c^3}} = 1.6 \times 10^{-33} \text{cm}$$

seem to be impossible. Therefore, the association of points of some smooth manifold with the localization of events has no operative meaning.

A more detailed investigation of the minimal uncertainties of coordinates of space-time events [69] leads to the result that there are no restrictions on the possible precision of a single coordinate. But a simultaneous determination of several coordinates might disturbe the spacetime structure heavily, so one finds the following uncertainty relations

$$\Delta t (\Delta x + \Delta y + \Delta z) \geq 1 \qquad (14.1)$$

$$\Delta x \Delta y + \Delta y \Delta z + \Delta z \Delta x \geq 1 \ . \qquad (14.2)$$

Here length and time are measured in so-called Planck units which are obtained when the fundamental constants of nature, c (velocity of light), $\hbar$ (Planck's constant divided by 2π) and Newton's gravitational constant G, are set equal to 1. So in this system of units lengths are multiples of the Planck length.

Similar relations have been found in Ashtekar's approach to quantum gravity and in string theory. The analysis in [69] is semiclassical and heuristic, but model independent.

We now make the following hypothesis: Space-time is to be replaced by a *noncommutative* space, i.e. the algebra of functions on spacetime is replaced by a noncommutative unital (complex) *-algebra $\mathcal{E}$, an idea which seems to have first been proposed by Snyder [203]. Probability measures on spacetime then correspond to states on $\mathcal{E}$, i.e. linear functionals ω which are positive, $\omega(a^*a) \geq 0$ and normalized, $\omega(1) = 1$.

The noncommutativity implies uncertainties for the coordinates. Therefore such spaces have been called fuzzy spaces in [155]. We want to find an algebra such that the uncertainty relations above hold true for all states where the uncertainty of a selfadjoint element $a = a^* \in \mathcal{E}$ in the state ω is defined as
$$\Delta(a) = \sqrt{\omega(a^2) - \omega(a)^2} \ .$$

We exploit the general quantum mechanical uncertainty relation which is an immediate consequence of the Cauchy-Schwartz inequality valid for the positive semidefinite sesquilinear form $\langle a, b \rangle = \omega(a^*b)$,
$$\Delta(a)\Delta(b) \geq \frac{1}{2}|\omega([a,b])|$$

and make the following Ansatz for $\mathcal{E}$:

The coordinates $q^\mu, \mu = 0, \ldots, 3$ of spacetime are selfadjoint generators of $\mathcal{E}$. The Poincaré group acts on $\mathcal{E}$ via
$$(a, \Lambda)q^\mu = \Lambda^{-1\mu}{}_\nu(q^\nu - a^\nu)$$

The algebra is then defined by the Poincaré invariant relations ($Q^{\mu\nu} := -i[q^\mu, q^\nu]$)
$$Q^{\mu\nu}Q_{\mu\nu} = 0$$
$$\frac{1}{8}Q^{\mu\nu}Q^{\rho\sigma}\epsilon_{\mu\nu\rho\sigma} = \pm 1$$
$$[Q^{\mu\nu}, q^\rho] = 0 \ .$$

The algebra above does not possess a C*-norm, but there is a C*-algebra to which the q^μ's and the $Q^{\mu\nu}$'s are affiliated as selfadjoint elements (i.e., bounded continuous functions of them are elements of the C*-algebra), and this C*-algebra has a dense set of states which can be extended to $\mathcal{E}$. The states of $\mathcal{E}$ obtained in this way are called regular.

One obtains the following theorem [69]:

Theorem 1. *Every regular state of $\mathcal{E}$ satisfies the uncertainty relations (14.1, 14.2) with $q^0 = t, q^1 = x, q^2 = y, q^3 = z$.*

The algebraic structure described above is not uniquely fixed by the uncertainty relations. In particular, the last condition that the commutators of coordinates lie in the center of the algeba can be weakened [67].

14.2 Noncommutative Spacetime and Quantum Field Theory

We now want to formulate physics on the noncommutative spacetime $\mathcal{E}$. Since the modification of spacetime by the noncommutativity of coordinates should be relevant only for very small distances we use the framework of quantum field theory.

Let us first look at the characterization of particles. According to Wigner, particles are mathematically modeled by unitary irreducible representations of the Poincaré group. This description of particles makes sense also on the noncommutative space $\mathcal{E}$ because of its Poincaré invariance. Moreover, asymptotically, at large distances, the space should look like the Minkowski space. Hence we expect that the Hilbert space of incoming scattering states, describing particles which are widely separated from each other at early times, can be identified with the Fock space of incoming free fields.

In Minkowski spacetime, quantum fields are operator valued distributions ("Wightman fields") on some Hilbert space, formally written as

$$\phi(f) = \int d^4x f(x) \phi(x) .$$

Physically, the smeared field $\phi(f)$ may be interpreted as a mean value of ϕ, provided fd^4x is a probability measure on Minkowski space.

It is therefore suggestive to define fields on a noncommutative space as affine mappings from the state space of $\mathcal{E}$ (which is a convex space) to operators on Fock space.

The physical interpretation is as follows: $\phi(\omega)$ describes a measurement in a "region" of spacetime which is characterized by the expectation values $\omega(a)$, $a \in \mathcal{E}$.

As an example let us look at the free scalar massless field

$$\phi(\omega) = \int d\mu(k) \left(\omega(e^{ikq}) a(k) + \omega(e^{-ikq}) a^*(k) \right)$$

where a, a^* are the usual Fock space annihilation and creation operators and μ is the Lorentz invariant measure on the boundary of the forward light cone. Here ω is a sufficiently "smooth" state such that the formula above makes mathematical sense. For details see [69].

The field defined above looks almost identical to the free field on Minkowski space, so one may ask where the noncommutativity of spacetime is encoded. But the physical content of a quantum field theory lies in the information where in spacetime a measurement corresponding to a selfadjoint element of the algebra of quantum fields is localized. In the noncommutative case described above the only possible localizations of observables are characterized by states on $\mathcal{E}$. In particular, there are no strictly localized observables, in accordance with the uncertainty relations.

The best localization near a point a of Minkowski space, in the sense that $\sum(q^\mu - a^\mu)^2$ is minimal, is obtained by states with

$$\omega_a(e^{ikq}) = e^{-\frac{1}{2}|k|^2} e^{ika}$$

(cf. ground state of harmonic oscillator). In terms of fields on Minkowski space, this means that fields must be smeared with a testfunction which is a Gaussian and which is concentrated in a region whose size is determined by the Planck length. Therefore, the commutator of two best localized field operators at spacelike separated points a and b will not vanish. One finds

$$[\phi(\omega_a), \phi(\omega_b)] =$$

$$\frac{-i}{4\pi|\mathbf{c}|}(8\pi)^{-\frac{1}{2}} \left(e^{-\frac{1}{8}(c^0 - |\mathbf{c}|)^2} - e^{-\frac{1}{8}(c^0 + |\mathbf{c}|)^2} \right)$$

($c = (c^0, \mathbf{c}) = a - b$). Hence, in any spacelike direction, the commutator decays like a Gaussian.

Let us study the behaviour of the commutator for large distances. We define the scaled field by $\phi_\lambda(a) := \frac{1}{\lambda}\phi(\omega_{\lambda a})$ and find in the limit $\lambda \to \infty$

$$[\phi_\lambda(a), \phi_\lambda(b)] \longrightarrow \frac{-i}{4\pi|\mathbf{c}|} \left(\delta(c^0 - |\mathbf{c}|) - \delta(c^0 + |\mathbf{c}|) \right) = iD(c)$$

(Pauli-Jordan-Function). This is the commutator function of the free massless field on Minkowski space. Hence the theory approaches in the long distance limit the theory of a massless free field on Minkowski space.

14.3 Interactions and Noncommutative Geometry

We now come to the crucial question of how interactions can be introduced [68]. In a naive approach, one replaces in perturbation theory the Feynman propagator by a linear functional on (a dense subspace of) the dual of $\mathcal{E} \otimes \mathcal{E}$,

$$\Delta_F(\omega) = (2\pi)^{-4} \int d^4k \frac{1}{k^2 + i\epsilon} \omega(e^{ikq} \otimes e^{-ikq})$$

This is equivalent to the introduction of a specific *nonlocal* interaction on Minkowski space. One might hope that the nonlocalities are harmless at large distances and that they regularize the theory at small distances. There is, however, the problem that no Lorentz invariant trace on the algebra exists which could represent the integral over spacetime, hence it is not clear how the integral over internal vertices of a Feynman integral should be performed. This problem is easier in an Euclidean version of our model, but it is not clear whether the so-called Wick rotation can be performed which makes the Euclidean and the Minkowskian theory equivalent on the basis of the Osterwalder-Schrader-Theorem [172].

It seems to be worthwhile to reconsider the problem from the general viewpoint of noncommutative geometry. Instead of using the differential calculus induced by the action of the translations on the algebra one may apply the universal differential calculus. Let us introduce the differential d as a map

$$d : \mathcal{E} \to \mathcal{E} \otimes \mathcal{E}$$

with

$$da = 1 \otimes a - a \otimes 1 .$$

We think of the differential of a function a of the noncommutative variable q as a function of two commuting variables q and q' and use the symbolic notation

$$da(q, q') = a(q') - a(q) .$$

A partial derivative may be obtained from the universal differential by "putting q' near to q". This means that we choose a positive linear functional ω on (some subspace of) $\mathcal{E} \otimes \mathcal{E}$ which describes a localization at neighbouring points. Then $\omega(da) =: \partial_\omega a$ may be interpreted as a partial derivative.

As an example we may look at the commutative case. Consider the algebra of smooth functions $\mathcal{C}^\infty(\mathbb{R}^n)$, and let $f \in \mathcal{C}^\infty(\mathbb{R}^n)$. Then the universal differential of f is given by $df(x, y) = f(y) - f(x)$. Now consider the functional

$$\omega(F) = \lim_{h \to 0, h > 0} \frac{1}{h} F(x_0, x_0 + ha)$$

with $F \in \mathcal{C}^\infty(\mathbb{R}^n \times \mathbb{R}^n)$. Then

$$\omega(df) = \partial_a f(x_0) \text{ (directional derivative) .}$$

The universal differential describes not only first derivatives. Let, e.g.,

$$\omega(F) = \lim_{r \to 0} \frac{2}{r^2 \Omega_n} \int_{S^{n-1}} F(x_0, x_0 + r\theta)$$

where Ω_n denotes the volume of the unit ball in $\mathbb{R}^n$. Then

$$\omega(df) = \Delta f(x_0) \text{ (Laplacian) .}$$

Now for nonabelian algebras, the limit $q' \to q$ is in general not possible because of the uncertainty relations. Instead one may try to put q' "as near as possible" to q. This may be made precize in the following way. Let E be the projection on the ground state of $|q' - q|^2$ where some Euclidean metric on Minkowski space was chosen. Let e be the positive mapping $\mathcal{E} \to \mathcal{E}$ defined by

$$e(a) = \varepsilon(a \otimes 1)$$

where ε is a unit preserving positive map from $\mathcal{E} \otimes \mathcal{E}$ into $\mathcal{E}$ with $\varepsilon(E) = 1$. Let ω be any state on $\mathcal{E}$ and consider the positive mapping $\Psi : \mathcal{E} \otimes \mathcal{E} \to \mathcal{E}$,

$$\Psi = \omega \otimes \text{id} + e \otimes \omega .$$

Then the "noncommutative Laplacian" may be defined by
$$\Delta a = \Psi(da) = e(a) - a .$$

In a similar way one may also define products of singular objects and obtain field equations, e.g.
$$\Delta a + \lambda \mathrm{Tr}_2(E(a \otimes a)E) = 0 .$$

Again, there is the problem that these concepts rely on the choice of a Euclidean metric. Unfortunately, it is by no means clear whether a Lorentz invariant notion of minimal distance exists.

14.4 Gauge Theories on Noncommutative Spacetime

We now want to introduce gauge theories on our spacetime. If one uses the recipe of spectral triples with a Dirac operator $D = \sum \gamma^\mu \partial_\mu$ with derivatives as infinitesimal translations, one finds a noncommutative electrodynamics. One obtains Yang-Mills equations and gets plane waves as special solutions. Due to the nonlinearity of the field equations, superpositions of plane waves are no longer solutions, and one obtains corrections by massive modes. It is tempting to use these formulas for an experimental test of our equations in the analysis of electromagnetic waves which travelled over astronomical distances. Unfortunately, there is an unsolved conceptual problem. As usual in nonabelian gauge theories, the field strength is not gauge invariant. But in our case, gauge transformations are unitary elements of the algebra $\mathcal{E}$, acting by the adjoint action on the field strength. Hence a gauge invariant function of the field strength must belong to the center of the algebra, and must in particular be translation invariant.

Because of these difficulties we work again in the general framework of noncommutative geometry and introduce gauge theories as follows. Let $\mathcal{H}$ be a right $\mathcal{E}$ module. A parallel transport on $\mathcal{H}$ may be defined as a module homomorphism
$$U : \mathcal{H} \to \mathcal{H} \otimes \mathcal{E}$$
with $m_0 U = 1$, where $m_0(\Phi \otimes a) = \Phi a$. A covariant differential is then given by
$$D\Phi := U\Phi - \Phi \otimes 1 ,$$
and it is easy to see that D satisfies the covariant Leibniz rule.

In the simplest case ("electrodynamics") we have $\mathcal{H} = \mathcal{E}$ and
$$U\Phi(q, q') = U(q, q')\Phi(q')$$
with some $U(q, q') \in \mathcal{E} \otimes \mathcal{E}$.

A gauge transformation is given in terms of an invertible $V \in \mathcal{E}$ and it transforms U according to
$$U(q, q') \to V(q)U(q, q')V(q')^{-1} .$$

The curvature $F = D^2$ can be described by left multiplication with an element of $\mathcal{E} \otimes \mathcal{E} \otimes \mathcal{E}$. In terms of the parallel transporters it is given by

$$F(q,q',q'') = U(q,q')U(q',q'') - U(q,q'') \ .$$

We now find gauge invariant quantities by a similar procedure as in lattice gauge theories.

Let $m_n : \mathcal{E}^{\otimes n} \to \mathcal{E}^{\otimes n-1}$, $m_n(a_1 \otimes \cdots \otimes a_n) = a_n a_1 \otimes \cdots \otimes a_{n-1}$, then the "Wilson loops"

$$W_n(U) := m_{n+1}(U(q_1,q_2) \cdots U(q_n,q_{n+1}))$$

are gauge invariant elements of $\mathcal{E}^{\otimes n}$.

Example: Let $u_i, v_i \in \mathcal{E}, i = 1, \ldots, d$ with $\sum u_i v_i = 1$ and $v_i u_j = \delta_{ij}$ (Cuntz algebra), then $U(q,q') = \sum u_i \otimes v_i$ is a parallel transport with curvature $F = 0$ and Wilson loops $W_n(U) = d1$.

We see that in the abstract approach described above many gauge invariant quantities exist. Unfortunately their physical interpretation is not evident. It is also not clear whether the principle of locality which is crucial for quantum field theory has an analogue in the noncommutative setting. Without such a principle there are far too many possibilities for writing down models, and without sufficient information from experiments a choice between these models is impossible.

15 NC Geometry and Quantum Fields: Simple Examples

Edwin Langmann

University of Stockholm

These notes contain four stories which are all related somehow to quantum field theory (QFT) and noncommutative geometry (NCG). (I) I sketch how the generalization of Yang-Mills theory to NCG motivates a useful and suggestive computation method to extract Chern-Simons terms from effective fermion actions. (II) Four nonconverging infinite series from a famous letter by Ramanujan are discussed as examples to illustrate a simple technique for regularization. (III) Regularized Hilbert space traces and their relation to the Wodzicki residue are discussed and illustrated for the simple example of matrix valued pseudodifferential operators on $\mathbb{R}^n$. (IV) To demonstrate the efficiency of the mathematical tools described in (III), the logarithmic divergence of the effective fermion action in four dimensions is computed. It is argued that the result of this computation provides a physical motivation for a particular form of the spectral action principle in NCG.

15.1 Introduction

In these lecture notes I will discuss examples where mathematical structures related to noncommutative geometry (NCG) are useful when treating traditional issues in quantum field theory (QFT). These examples all have to do with effective fermion actions which are fundamental in QFT, and what I can say I have learned in a few projects (partly together with Jouko Mickelsson) in which we have tried to confront QFT with NCG with the aim to make progress in both subjects: revisit traditional issues in QFT having in mind ideas from NCG to improve the 'tool kit' for QFT, but also to use problems in QFT as a motivation and guideline to study certain mathematical issues related to NCG. Rather than trying a review (a more systematic discussion of part of our work can be found in Ref. [141]) I will tell four different stories which, as I hope, will give the flavor of what we have been doing.

The idea to generalize geometry to situations without underlying manifold but rather algebras of Hilbert space operators is very powerful [38]. For example, it offers a natural way to understand the relation between the rich differential geometric structure of anomalies (anomalies as de Rham forms, characteristic classes, descendent equations relating anomalies in different dimensions etc.) and their explicit QFT derivation (see e.g. [115]). A general idea here is to interpret Feynman diagrams as regularized traces of certain operators on some Hilbert space, and to try to identify NCG structures based on the algebra of these operators. The regularization of the traces is necessary since one obtains operators which are not trace class. Anomalies can often be identified as regularized traces of commutators $[a, b] = ab - ba$ of certain operators a and b. Even though such an expression is always zero if $[a,b]$ is trace class, it can still be defined in more general cases and be non–zero then. Such regularized traces of commutators are also closely related to the Wodzicki residue playing a fundamental role in NCG.

The plan of these notes is as follows. In the next Section I summarize preliminaries from NCG and QFT (they can be skipped at first reading since I will refer to them in the other Sections when needed; Section 15.4 does not need any preliminaries). In Section 15.3 I sketch one example of a QFT computation of an anomaly, namely how to obtain the Chern-Simons terms from effective fermion actions. I show how some NCG generalization of Yang-Mills theory naturally appears there and how it can be exploited to make the computation efficient and suggestive. In Section 15.4 I discuss elementary examples for regularization in the context of nonconverging series. I included this Section as a 'warmup' for Section 15.5 where I give a general discussion of how to regularize the Hilbert space trace, something which is important in many QFT computations but also related to some basic notions in NCG. I first give a discussion on a general, abstract level and then illustrate this in more detail for one important example, namely pseudodifferential operators on $\mathbb{R}^n$. The discussion in Section 15.5 provides very efficient mathematical tools to compute effective fermion actions on $\mathbb{R}^n$. To demonstrate this I show in Section 15.6 how to compute the logarithmic divergent contribution to the effective fermion action on $\mathbb{R}^4$. I choose this example since the result of this computation is actually quite striking: In the final Section 15.7 I will give a physical interpretation of this result and explain why it has to be what it is. This then gives a simple physical interpretation of a specific form of the spectral action principle in NCG, see [38,31].

15.2 Preliminaries

In this Section I summarize some preliminaries. I first describe some basic notions from NCG, namely how to generalize de Rham forms to operator algebras where one does not have an underlying manifold. This allows to naturally extend the notion of Yang-Mills theory and gauge invariance in a manner which is very useful for QFT (e.g.). I then discuss the formal definition and some physics background to effective fermion actions. To be specific, I also give a down-to-earth definition of the mathematical ingredients for the computation of the effective fermion action on $\mathbb{R}^n$.

On noncommutative geometry

One important idea from NCG is to generalize the notion of de Rham forms. To motivate this I recall that it is possible to characterize de Rham forms on a manifold purely algebraically without reference to the underlying manifold. This algebraic characterization can be used as a definition of what is called a *graded differential algebra* (GDA). There are several interesting examples of

GDA's based on algebras of Hilbert space operators [38] (rather than algebras of functions on a manifold) some of which naturally occur in QFT. To be specific I will restrict my discussion to manifolds $\mathbb{R}^n$.

De Rham forms on $\mathbb{R}^n$. To construct de Rham forms one can start with an algebra of 'nice'[1] functions on $\mathbb{R}^n$, which one regards as zero forms, and define the exterior derivative[2] d such that it obeys $d^2 = 0$, linearity, and the Leibniz rule, $d(uv) = d(u)v + ud(v)$ for all zero forms u and v. Then one can define k-forms as the linear combination of elements

$$\omega_k = u_0 d(u_1) \cdots d(u_k)$$

with u_j zero forms, and the definition

$$d(u_0 d(u_1) \cdots d(u_k)) := d(u_0) d(u_1) \cdots d(u_k)$$

extends the linear map d to all k-forms such that d^2 holds. Moreover, the multiplication of forms is naturally defined by the Leibniz rule,

$$\begin{aligned} u_0 d(u_1) \cdots d(u_k) v_0 d(v_1) \cdots d(v_\ell) &= u_0 d(u_1) \cdots d(u_k v_0) d(v_1) \cdots d(v_\ell) \\ &\quad - u_0 d(u_1) \cdots d(u_{k-1}) u_k d(v_0) d(v_1) \cdots d(v_\ell), \end{aligned}$$

which by induction shows that this is a well-defined $(k + \ell)$-form. One can check that then the graded Leibniz rule holds,

$$d(\omega_k \omega'_\ell) = d(\omega_k)\omega'_\ell + (-)^k \omega_k d\omega'_\ell$$

for all k- and ℓ-forms ω_k and ω'_ℓ.

Generalized de Rham forms. One can mimic this construction of de Rham forms but instead of the algebra of functions of $\mathbb{R}^n$ take some algebra of Hilbert space operators which one regards as zero forms. An important example for a natural definition of the analog of exterior differentiation then is[3]

$$\hat{d}(\hat{u}) := [\varepsilon, \hat{u}] = \varepsilon \hat{u} - \hat{u}\varepsilon \qquad (15.1)$$

for all zero forms $\hat{u}$ where ε is a grading operator, i.e. ε is selfadjoint and its square is the identity. One then can define k-forms, multiplication of k- and ℓ-forms etc. as above, and the algebraic properties of de Rham forms carry over to these generalized de Rham forms. Note that these definitions imply

$$\hat{d}(\hat{\omega}_k) = \varepsilon \hat{\omega}_k - (-1)^k \hat{\omega}_k \varepsilon$$

for all k-forms $\hat{\omega}_k$.

---

[1] e.g. C_0^∞

[2] To compare with formulas in the other Sections, I note that I use the convention such that $d(u) = -i \sum_{\mu=1}^n \frac{\partial u}{\partial x_\mu} dx_\mu$

[3] Here and in the following I use hats in formulas for generalized de Rham forms.

Generalized Yang-Mills theory. It is often straightforward to generalize notions and results from de Rham forms on manifolds to general GDA's by formulating them in a way which only makes use of the defining relations of a GDA. An important example for us is the generalization of Yang-Mills theory to GDA's based on Hilbert space operators. As a motivation I recall that any Yang-Mills configuration A on $\mathbb{R}^n$ can be associated with a de Rham 1-form on $\mathbb{R}^n$,

$$A = \sum_{\mu=0}^{n} A_\mu(x)\mathrm{d}x^\mu \tag{15.2}$$

where the A_μ are functions on $\mathbb{R}^n$ with values in the algebra gl_N of $N \times N$ matrices. Zero forms u naturally act as (infinitesimal) gauge transformations on such 1-forms A,

$$\delta_u(A) = \mathrm{d}(u) + [A, u] = [\mathrm{d} + A, u], \tag{15.3}$$

and for any one form A we can define the corresponding Yang-Mills curvature,

$$F_A = \mathrm{d}(A) + A^2 = (\mathrm{d} + A)^2 \tag{15.4}$$

which is a 2-form (we have used $\mathrm{d}^2 = 0$ and the Leibniz rule i.e. $\mathrm{d}(u) = [\mathrm{d}, u]$ and $\mathrm{d}(A) = \mathrm{d}A + A\mathrm{d}$). The definitions then imply the following transformation rule for the Yang-Mills curvature,

$$\delta_u(F_A) = [F_A, u] \tag{15.5}$$

where I used that the definition of gauge transformations naturally extends to polynomials in A and $\mathrm{d}(A)$,

$$\delta_u(p[A]) = \frac{d}{dt}p[A + t\delta_u(A)]\bigg|_{t=0}. \tag{15.6}$$

One also can construct various other de Rham forms from A with interesting relations amongst them. One example which will be important for us in the next Section are forms which I denote as 'raw data' for the Chern-Simons terms: for all positive integers m, the $(2m-1)$-form[4]

$$Ch_{2m-1}[A] = \int_0^1 dt (F_{tA})^{m-1} A, \quad F_{tA} = t\mathrm{d}(A) + t^2 A^2 \tag{15.7}$$

obeys

$$(F_A)^m = \mathrm{d}(Ch_{2m-1}[A]) + (\cdots) \tag{15.8}$$

where '$(\cdots)$' vanishes when taking the matrix trace (i.e. it is of the form $A\omega_{2m-1} + \omega_{2m-1}A$). The last equation is one of the so-called descent equations

---

[4] Note that the following is not quite the Chern-Simons form since I do not assume that $2m-1$ equals the dimension of the underlying manifold and I do not take the matrix trace.

which play an important role for the understanding of anomalies in QFT. The reason for regarding $Ch_{2m-1}[A]$ as 'raw data' for the Chern-Simons term is as follows: in case of $2m - 1 = n$ we can define the integral of this form, $\int_{\mathbb{R}^{2m-1}} \operatorname{tr} Ch_{2m-1}[A]$ where tr is the matrix trace in gl_N, and this is precisely the Chern-Simons term on $\mathbb{R}^{2m-1}$.

All what I discussed here immediately carries over to other GDA's. For example, in the GDA based on the Hilbert space operator algebra I can regard

$$\hat{A} = \sum_j \hat{u}_j[\varepsilon, \hat{v}_j]$$

as generalized Yang-Mills fields, and all I said above for Yang-Mills theory on $\mathbb{R}^n$ immediately generalizes. Especially,

$$F_{\hat{A}} = \varepsilon \hat{A} + \hat{A}\varepsilon + \hat{A}^2, \qquad (15.9)$$

and we can also construct raw data for Chern-Simons terms related to $(F_{\hat{A}})^m$ as above. To obtain Chern-Simons terms we also need an analog of integration of de Rham forms. There is a natural definition for such a noncommutative generalization of integration of de Rham forms, see e.g. [140].

I finally note that it is possible to generalize, in a similar manner, all descent equations and thus obtain 'raw data' for anomalies in general GDA's [142].

On Effective Fermion Actions

Formal definition. Fermions on n-dimensional Euclidean spacetime in an external Yang–Mills field A can be described by an effective action which formally is given by the determinant of a Dirac operator $\slashed{D}_A$,

$$S_{\mathrm{eff}}(A) = - \text{``log det''}(\slashed{D}_A + iM) \qquad (15.10)$$

where M is a real constant which is interpreted as fermion mass and $\slashed{D}_A$ is self-adjoint. To be specific I will give a down-to-earth definition of $\slashed{D}_A$ for the simple (but important) special case where spacetime is $\mathbb{R}^n$. The quotation marks in (15.10) indicate the well-known fact that there are divergences present and the determinant has to be defined and computed using some regularization. A lot of interesting mathematics and physics has originated from studying such effective actions in various situations and from various points of view.

Chern-Simons terms. One interesting and important consequence of the regularizations is the occurrence of what is called *anomalies* by physicist; see e.g. [115]. One prominent example is the effective fermion action in odd dimensions: even though formally the effective action for massless fermions (i.e. for $M = 0$) is real, the imaginary part of the effective action $S_{\mathrm{eff}}(A)$

on $\mathbb{R}^{2m-1}$ does not vanish in the limit $M \to 0$ but is proportional to the Chern–Simons term discussed in the previous Section,

$$\lim_{M \to 0} \operatorname{Im} S_{\text{eff}}(A) \propto \int_{\mathbb{R}^{2m-1}} \operatorname{tr} CS_{2m-1}[A] \tag{15.11}$$

with $CS_{2m-1}[A]$ given in (15.7). In Section 15.3 I will outline a simple method to perform the computation proving (15.11), and in this computation a generalization of the CS term to NCG, as mentioned above, naturally appears. In this example we look out for *traces of NCG in a QFT computation*, and we will see that mathematical structures from NCG indeed appear rather naturally. Another such example is the scattering matrix of fermions in an external Yang-Mills field (closely related to the effective fermion action) as studied in Ref. [144].

Regularized traces and the logarithmic divergence. As mentioned above, one important part in the computation of $S_{\text{eff}}(A)$ is the regularization, and (in our approach) this amounts to making sense of the trace of Hilbert space operators which are not trace class: one can interpret "log det" in (15.10) as "Tr log". The logarithm of the Dirac operator can be defined in various ways (we will explain different possibilities for that in Sections 15.3 and 15.6), and one finally has to specify what is meant by the trace of the operator $\log(\slashed{D}_A + iM)$. I will explain a general method of how to define such a regularized trace "Tr" in Section 15.5, and I will mention also an interesting relation between "Tr" and the Wodzicki residue [224]: *regularized traces which are essential in QFT are closely related to traces playing an important role in NCG*.

In Section 15.6 I will discuss one interesting aspect of the effective fermion action resulting from one general feature of "Tr": As will be explained in Section 15.5, "Tr"(a) for many operators a is ambiguous: changing the regularization amounts to adding to "Tr"(a) a multiple of the operator residue of a. This ambiguity is what physicists refer to as *logarithmic divergence*. As is known by physicists since a long time, in four spacetime dimensions this logarithmic divergence actually is identical with the Yang-Mills action $S_{YM}(A)$ (see e.g. [114]). I will outline the computation proving this statement in order to illustrate the use of the mathematical tools (regularized traces, PSDO's etc.) in a effective action computation. In the final Section 15.7 I will give a physical interpretation of this result and argue that this suggests to *define* the Yang-Mills action also in more general cases as

$$S_{YM}(A) \propto -\operatorname{Res}(\log(\slashed{D}_A + iM) - \log(\slashed{D}_0 + iM)) \tag{15.12}$$

which can serve as an alternative formulation of the definition of the Yang-Mills action in NCG.

Dirac operators on $\mathbb{R}^n$. To be specific we restrict ourselves to space time described by the manifold $\mathbb{R}^n$, and we will consider gauge theories where the

Lie algebra of the gauge groups is represented by $N \times N$-matrices. In this case a Yang–Mills field configurations A is represented by gl_N-valued functions A_μ on $\mathbb{R}^n$, $\mu = 1, 2, \ldots, N$, which we assume to be 'nice'.[5] To define $\slashed{D}_A$ we also need a spin structure, which in the present case is given by self-adjoint matrices γ_μ, $\mu = 1, 2, \ldots, n$, obeying

$$\gamma_\mu \gamma_\nu + \gamma_\nu \gamma_\mu = 2\delta_{\mu\nu} 1 \tag{15.13}$$

with 1 the identity matrix. One can construct such matrices in $\mathrm{gl}_{\nu(n)}$ where $\nu(n) = 2^{(n-1)/2}$ for n odd and $\nu(n) = 2^{n/2}$ for n even (explicit formulas can be found e.g. in [140].) Then $\slashed{D}_A$ is defined by the following differential operator acting on the space of differentiable and compactly supported functions on $\mathbb{R}^n$ with values in $\mathbb{C}^{\nu(n)} \otimes \mathbb{C}^N$,

$$\slashed{D}_A = \sum_{\mu=1}^n \gamma_\mu \left(-i \frac{\partial}{\partial x_\mu} + A_\mu(x) \right) \tag{15.14}$$

(by abuse of notation I write γ_μ short for $\gamma_\mu \otimes 1$ and $A_\mu(x)$ short for $1 \otimes A_\mu(x)$). This operator extends to a self-adjoint operator acting on the Hilbert space

$$\mathcal{H} = L^2(\mathbb{R}^n, d^n x) \otimes \mathbb{C}^{\nu(n)} \otimes \mathbb{C}^N \tag{15.15}$$

and which I usually denote by the same symbol $\slashed{D}_A$ (except in Section 15.6). Especially $\slashed{D}_0 = -i \sum_{\mu=1}^n \gamma_\mu \frac{\partial}{\partial x_\mu}$. We shall also say that $\slashed{D}_A$ is the *Dirac operator coupled to the Yang–Mills field* $A = \sum_{\mu=1}^n A_\mu(x) dx_\mu$, and $\slashed{D}_0$ is the *free Dirac operator*.

15.3 Story I: Chern–Simons Terms from Effective Actions

(In this Section I will use notation explained in Section 15.2.)

As discussed above, the effective action for massless fermions in the external Yang–Mills field A formally is (minus) the logarithm of the determinant of the Dirac operator $\slashed{D}_0 + \slashed{A}$, or equivalently, the trace of its logarithm. Due to divergences some regularization of this trace is necessary (one has to add a small massterm and a large momentum cutoff, e.g.) but we will ignore this in this Section for simplicity.

I write this log of the determinant as "Tr" $\int_0^1 dt \frac{d}{dt} \log(\slashed{D}_0 + t\slashed{A})$, which formally is equivalent to[6]

$$S_{\mathrm{eff}}(A) = -\text{``Tr''} \int_0^1 dt (\slashed{D}_0 + t\slashed{A})^{-1} \slashed{A}; \tag{15.16}$$

---

[5] C_0^∞, e.g., but this conditions could be easily relaxed.
[6] In Section 15.6 I will give a refined definition, but for our purposes here the following simplified definition is sufficient.

I take this as definition of the effective fermion action in this Section. I now define
$$\hat{A} = |\not{D}_0|^{-1}\not{A} \quad \text{and} \quad \varepsilon = |\not{D}_0|^{-1}\not{D}_0 \,. \tag{15.17}$$

At this point this can be regarded as a useful notation which is motivated by what we discussed in Section 15.2 above (note that ε is a grading operator). Then we can write
$$(\not{D}_0 + t\not{A})^{-1}\not{A} = (\varepsilon + t\hat{A})^{-1}\hat{A} = (\varepsilon + t\hat{A})(1 + F_{t\hat{A}})^{-1}\hat{A}$$

where we introduced the suggestive notation
$$F_{t\hat{A}} = t(\varepsilon\hat{A} + \hat{A}\varepsilon) + t^2\hat{A} \tag{15.18}$$

which is very natural after our discussion in Section 15.2 above (I used $(\varepsilon + t\hat{A})^2 = 1 + F_{t\hat{A}}$). The imaginary part of the action thus becomes
$$\text{Im}S_{\text{eff}}(A) \sim \sum_{k=1}^{\infty}(-)^{k-1}\text{Im"Tr"}\int_0^1 dt(\varepsilon + t\hat{A})(F_{t\hat{A}})^{k-1}\hat{A}. \tag{15.19}$$

This should be equal, up to a constant, to the Chern–Simons term
$$\int_{\mathbb{R}^{2m-1}} \text{tr} \int_0^1 dt\, (F_{tA})^{m-1}A \tag{15.20}$$

where $F_{tA} = td(A) + t^2 A^2$ (de Rham 2–form on $\mathbb{R}^n$) and $n = 2m - 1$. This result is now very plausible by notation. To prove it requires a nontrivial calculation: namely to show that for all k
$$\text{Im"Tr"}\hat{A}(F_{t\hat{A}})^{k-1}\hat{A} = 0$$
$$\text{Im"Tr"}\varepsilon(F_{t\hat{A}})^{k-1}\hat{A} \propto \delta_{k,m}\int_{\mathbb{R}^{2m-1}} \text{tr}\,(F_{tA})^{m-1}A. \tag{15.21}$$

This can be proven in a computation using the mathematical tools explained in Section 15.5 [145]. The second result in (15.21) can be interpreted as follows: Im"Tr"$(\varepsilon\cdot)$ is an integration of generalized de Rham forms, as mentioned in Section 15.2. Note also that all manipulations from (15.16) to (15.19) did not use any property of the operators $\not{D}_0$ and $\not{A}$ except that they are self-adjoint operators on some Hilbert space, and thus everything we said above is valid also for much more general situations. The place where specific properties of these operators and the Hilbert space we are using enters is the step from (15.19) to (15.20). Especially only here the dimension of the underlying spacetime manifold enters: it is only the (regularized) trace that distinguishes the different dimensions, and it picks up exactly one term *viz.* the anomaly.

15.4 Story II: Regularization: Elementary Examples

(This Section is self-contained.)

A general feature in quantum field theory is the appearance of infinities, so-called *divergences*. Over the years physicists have learned how to deal with them, to do computations and extract meaningful, i.e. finite, answers from mathematical expressions involving diverging series and integrals. Physicists are not always very careful in their notation, and in this context this can result in rather strange looking equations, for example

$$1 - 2 + 3 - 4 + \ldots = \frac{1}{4}, \tag{15.22}$$

$$1 - 1! + 2! - 3! + \ldots = 0.596\ldots, \tag{15.23}$$

$$1 + 2 + 3 + 4 + \cdots = -\frac{1}{12}, \tag{15.24}$$

$$1^3 + 2^3 + 3^3 + 4^3 + \cdots = \frac{1}{120}. \tag{15.25}$$

In fact, I have taken these equations not from a physics text but from a letter a prominent mathematician wrote to another, quite some time before quantum field theory: these equations are stated in the first letter which S. Ramanujan wrote to G.H. Hardy on January 16, 1913 [14].[7] However, such formulas play indeed an important role in QFT and string theory.[8]

To explain the basic idea of regularization I will now discuss an elementary method allowing to give precise meaning to these and similar equations. We start with (15.22). Consider the series

$$f(\varepsilon) = \sum_{n=1}^{\infty} (-1)^{n-1} e^{-\varepsilon n} \quad (\varepsilon > 0)$$

which is absolutely convergent and can be summed to

$$f(\varepsilon) = 1 - \frac{1}{1 + e^{-\varepsilon}}.$$

Formally, the (ill-defined) series $\sum_{n=1}^{\infty}(-1)^{n-1}n^s$ for a positive integer s is equal to $(-1)^s d^s f(\varepsilon)/d\varepsilon^s|_{\varepsilon=0}$. We can expand $f(\varepsilon)$ in a series and obtain,

$$f(\varepsilon) = \frac{1}{2} - \frac{1}{4}\varepsilon + \mathcal{O}(\varepsilon^3)$$

---

[7] I should cite here the complete paragraph in this letter containing these equations (see [14], p.29 ff): "XI. I have got theorems on divergent series, theorems to calculate the convergent values corresponding divergent series, viz." (here come Eqs. (15.22)–(15.25)). "Theorems to calculate such values for any given series ..., and the meaning of such values."

[8] see e.g. Ref. [176], (2.9.19) on p. 73.

which shows that we can *define*

$$\sum_{n=1}^{\infty}{}' (-1)^{n-1} n^s := \lim_{\varepsilon \downarrow 0} (-1)^s \frac{d^s}{d\varepsilon^s} \sum_{n=1}^{\infty} (-1)^{n-1} e^{-\varepsilon n} \quad (s \in \mathbb{N}) \quad (15.26)$$

where the prime is to indicate that this sum is defined with a specific summation prescription. For $s = 1$ we obtain an equation giving precise meaning to (15.22).

To give a precise meaning to (15.23) we use $n! = \int_0^\infty dt\, e^{-t} t^n$ and *define*

$$\sum_{n=0}^{\infty}{}' (-1)^n n! := \int_0^\infty dt \sum_{n=0}^{\infty} e^{-t}(-t)^n \quad (15.27)$$

equal to $\int_0^\infty dt\, e^{-t}/(1+t) = 0.59635\ldots$. (I find it quite remarkable that this 'dangerous looking' series can be 'tamed' simply by changing the order of a summation and an integration.)

We thus see that the series in (15.22) and (15.23) are actually conditionally convergent, and what we did is to specify summation prescriptions giving well-defined (finite!) results. We now turn to the series in (15.24) and (15.25) which are different (they are obviously not conditionally convergent). To give meaning to them we consider

$$g(\varepsilon) = \sum_{n=0}^{\infty} e^{-\varepsilon n} \quad (\varepsilon > 0)$$

which is absolutely convergent and equals

$$g(\varepsilon) = \frac{1}{1 - e^{-\varepsilon}}.$$

Formally, the (diverging) sum $\sum_{n=0}^{\infty} n^s$ for $s \in \mathbb{N}$ equals $(-1)^s\, d^s g(\varepsilon)/d\varepsilon^s|_{\varepsilon=0}$. Expanding $g(\varepsilon)$ in a series we obtain

$$g(\varepsilon) = \frac{1}{\varepsilon} + \frac{1}{2} + \frac{1}{12}\varepsilon - \frac{1}{720}\varepsilon^3 + \mathcal{O}(\varepsilon^5)$$

which shows that it is the term $\frac{1}{\varepsilon}$ which is responsible for $(-1)^s\, \partial^s g(\varepsilon)/\partial \varepsilon^s|_{\varepsilon=0}$ to be infinite. We can remove this divergence by subtracting this term. We therefore *define* the *regularized sums*

$$\sum_{n=0}^{\infty}{}' n^s := \lim_{\varepsilon \downarrow 0} (-1)^s \frac{d^s}{d\varepsilon^s} \left(\sum_{n=0}^{\infty} e^{-\varepsilon n} - \frac{1}{\varepsilon} \right) \quad (s \in \mathbb{N}) \quad (15.28)$$

where the prime here indicates the regularization. For $s = 1$ and 3 we thus obtain equations which give a precise meaning to (15.24) and (15.25).

15.5 Story III: Regularized Traces of Operators

(This Section is essentially self-contained except for some definitions and motivation given in Section 15.2.)

In QFT computation one is often led to compute the trace of a Hilbert space operator which is not trace class. To give a precise meaning to that one has to do a regularization. In this section I discuss a rather general method for this. I will first give a discussion on a general abstract level which is in the spirit of NCG: given some unbounded operator D_0 and some algebra $\mathcal{A}$ of Hilbert space operators with certain properties (depending on D_0) I show how a regularized trace Tr_C and an operator residue Res on $\mathcal{A}$ can be defined, and I will discuss a few interesting properties of Tr_C and Res. To simplify the presentation I will be somewhat vague in the precise definition of $\mathcal{A}$. Instead I will consider a specific example, namely the pseudodifferential operators (PSDO) on $\mathbb{R}^n$, where all statements concerning Tr_C and Res can be proven by rather elementary computations.

Conditional traces of commutators and the operator residue

Let D_0 be an unbounded, self-adjoint operator and f some real valued, smooth, function on $\mathbb{R}^+$ obeying

$$f(t) \geq 0 \quad \forall t \geq 0, \quad f(0) = 1, \quad f(t) = \mathcal{O}(t^{-\infty}). \tag{15.29}$$

Specific examples which we have in mind are

$$f(t) = e^{-t} \quad \text{or} \quad f(t) = \theta_\varepsilon(1-t) \tag{15.30}$$

(θ_ε is an approximate Heaviside step function), but it is useful to leave f general in the following discussion: Choosing f is regarded as a particular choice for a regularization. We are mainly interested in results independent of the regularization, i.e. results which to not change under $f \to \tilde{f}$.[9]

Given D_0 and f we can construct a family of regularization operators

$$P_\Lambda := f(\tfrac{|D_0|}{\Lambda}), \quad 0 < \Lambda < \infty \tag{15.31}$$

which are defined by the spectral theorem (note that P_Λ converges strongly to the identity operator for $\Lambda \to \infty$). For all operators a such that aP_Λ is trace class for $\Lambda < \infty$, we then can define a trace with a cut-off as follows,

$$\text{Tr}_\Lambda(a) := \text{Tr}(aP_\Lambda), \quad 0 < \Lambda < \infty \tag{15.32}$$

where Tr is the usual Hilbert space trace. The *conditional trace class* $B_{1,C}$ then can be defined as the set of all such operators for which $\text{Tr}_\Lambda(a)$ has a

---

[9] More generally, one could also change the regularization by changing $D_0 \to \tilde{D}_0$ such that $\tilde{D}_0 - D_0$ is a bounded operator. The discussion of this is beyond the scope of these notes.

finite limit $\Lambda \to \infty$, and for those operators we can define the *conditional trace*

$$\mathrm{Tr}_C(a) = \lim_{\Lambda \to \infty} \mathrm{Tr}_\Lambda(a). \qquad (15.33)$$

Obviously $B_{1,C}$ contains all trace class operators, and if a is trace class then $\mathrm{Tr}_C(a) = \mathrm{Tr}(a)$. However, Tr_C is not cyclic: There are operators a and b such that the commutator $[a,b] = ab - ba$ is in $B_{1,C}$ but $\mathrm{Tr}_C([a,b]) \neq 0$.[10] In fact, the conditional trace of commutators is a very interesting object, as will be further discussed below.

In many cases (e.g. applications in QFT) one is interested in operators such that aP_Λ is trace class for all $\Lambda > 0$ and which allow an expansion of $\mathrm{Tr}_\Lambda(a)$ as follows,

$$\mathrm{Tr}_\Lambda(a) = \Lambda^N c_N(a) + \Lambda^{N-1} c_{N-1}(a) + \ldots + \Lambda c_1(a) + \\ + \log(\Lambda) c_{\log}(a) + c_0(\Lambda) + \mathcal{O}(\Lambda^{-1}) \qquad (15.34)$$

where N is some non-negative integer (depending on a, of course). Below we will discuss an important example of such operators, namely pseudodifferential operators (PSDO) on $\mathbb{R}^n$. This will give a specific example of what we now discuss.

We will see below that in general the terms c_k in the expansion above depend on the function f chosen to define Tr_Λ i.e. on the regularization. These terms therefore are not particularly interesting. There is, however, one term which is regularization independent, namely $c_{\log}$, and this indicates that this term is interesting. In fact, one can prove that for a large class of PSDO a (and D_0 a Dirac operator), $c_{\log}(a)$ is proportional to the Wodzicki residue [224] (we will discuss this in more detail below). We therefore introduce the notation[11]

$$\mathrm{Res}(a) = c_{\log}(a) \qquad (15.35)$$

also in more general cases.

There is one more term in the expansion of Tr_Λ which is interesting, namely c_0: in case of operators $a \in B_{1,C}$,

$$\mathrm{Tr}_C(a) := c_0(a), \qquad (15.36)$$

and it is natural to use this equation to extend the definition of the conditional trace to all operators allowing for an expansion (15.34). It is important to note, however, that the conditional trace of an operator is not quite regularization independent in general: For example, if we change $\Lambda \to \Lambda s$ with $s > 0$ (which corresponds to a legitimate change of the regularization function) then

$$\mathrm{Tr}_C(a) \to \mathrm{Tr}_C(a) + \log(s) \mathrm{Res}(a), \qquad (15.37)$$

---

[10] One can therefore say, somewhat paradoxically: the conditional trace is not a trace.

[11] Our normalization is different from the one in [224]

and more generally one can prove that a regularization change $f \to \tilde{f}$ amounts also to a change (15.37) with s depending on f and $\tilde{f}$. However, operators a with $\mathrm{Res}(a) = 0$ have a regularization independent conditional trace. Especially, conditional traces of commutators are regularization independent, another indication that these are interesting objects. In fact, one can prove that the trace of a commutator can be computed as a residue, namely

$$\mathrm{Tr}_C([a,b]) = \mathrm{Res}(a[\log|D_0|, b]) . \qquad (15.38)$$

We will sketch a proof of these relations for PSDO on $\mathbb{R}^n$ below. We note, however, that the equations (15.29)–(15.38) can be extended to large classes of operators.

The objects discussed above have several important applications in QFT. For examples, many anomalies in QFT can be computed as conditional traces of commutators of certain PSDO [141], and this is one explanation why anomalies are interesting geometric objects. We will also discuss one interesting QFT application of the residue in Section 15.6 below.

Example: Pseudodifferential operators

We consider the Hilbert space $\mathcal{H}$ defined in (15.15) and PSDO which are operators on $\mathcal{H}$.

Let $\mathcal{D}$ be the dense subset of smooth functions $f(x)$ in $\mathcal{H}$ vanishing exponentially for $|x| \to \infty$. We consider linear operators $\mathcal{H} \to \mathcal{H}$ which can be defined as follows,

$$(af)(x) = \int_{\mathbb{R}^n} \frac{d^n p}{(2\pi)^n} e^{-ip \cdot x} \sigma[a](p,x) \int_{\mathbb{R}^n} d^n x\, e^{ip \cdot y} f(y) \qquad (15.39)$$

for all $f \in \mathcal{D}$, where $\sigma[a](p,x)$ is a matrix valued[12] function on $\mathbb{R}^n \times \mathbb{R}^n$ obeying certain conditions[13] such that all I write down is well-defined ($p \cdot x = \sum_{\mu=1}^n p_\mu x_\mu$). An example important in this Section are the operators $f(\frac{|D_0|}{\Lambda})$ which obviously can be represented as in (15.39) with

$$\sigma[f(\tfrac{|D_0|}{\Lambda})](p,x) = f(\tfrac{|p|}{\Lambda})$$

independent of x ($|p| = \sqrt{p \cdot p}$). Another example are the Dirac operators defined in (15.40) above with

$$\sigma[\slashed{D}_A](p,x) = \sum_{\mu=1}^n \gamma_\mu [p_\mu + A_\mu(x)] := \gamma \cdot [p + A(x)] . \qquad (15.40)$$

(We will use the notation introduced here in the next Section.)

---

[12] i.e. $\mathrm{gl}_{\nu(n)} \otimes \mathrm{gl}_N$-valued

[13] to keep our presentation short I do not spell out these conditions; see e.g. in [108]

Note that (15.39) implies

$$\sigma[ab](p,x) = \int_{\mathbb{R}^n} \frac{d^n q}{(2\pi)^n} \int_{\mathbb{R}^n} d^n y \, e^{i(x-y)\cdot(p-q)} \sigma[a](q,x)\sigma[b](p,y) \quad (15.41)$$

(the order of terms is important since the matrix product is understood here, of course) and one can show that if a is trace–class,

$$\text{Tr}(a) = \int_{\mathbb{R}^n} \frac{d^n p}{(2\pi)^n} \int_{\mathbb{R}^n} d^n x \, \text{tr}\, \sigma[a](p,x)$$

where tr is the usual matrix trace (i.e. sum of diagonal elements).

We now use $D_0 = \slashed{D}_0$ and some function f obeying (15.29) for a regularization, as discussed in Section 15.5. From our discussions above we obtain, $\sigma[a f(\frac{|\slashed{D}_0|}{\Lambda})](p,x) = \sigma[a](p,x) f(\frac{|p|}{\Lambda})$ and

$$\text{Tr}_\Lambda(a) = \int_{\mathbb{R}^n} \frac{d^n p}{(2\pi)^n} \int_{\mathbb{R}^n} d^n x \, \text{tr}\, \sigma[a](p,x) f(\frac{|p|}{\Lambda}) =$$
$$\int_0^\infty d|p| |p|^{n-1} f(\frac{|p|}{\Lambda}) \int_{\mathbb{R}^n} \frac{d^n \xi}{(2\pi)^n} \delta(|\xi|-1) \int_{\mathbb{R}^n} d^n x \, \text{tr}\, \sigma[a](|p|\xi, x) \quad (15.42)$$

where we introduced spherical coordinates, $p = |p|\xi$ (note that $\int_{\mathbb{R}^n} d^n \xi \, \delta(|\xi|-1)$ is just the integral over the unit sphere in $\mathbb{R}^n$). We now define $\mathcal{A}$ as the class of all operators which can be represented as in (15.39) with a smooth function $\sigma[a](p,x)$ on $\mathbb{R}^n \times \mathbb{R}^n$ compactly supported[14] in x and such that that there is an *asymptotic expansion*

$$\sigma[a](p,x) \sim \sum_{j=0}^\infty \sigma[a]_{N_a-j}(p,x) \quad (15.43)$$

for some finite integer N_a, where $\sigma_k[a](p,x)$ is homogeneous of degree k in p (i.e. $\sigma_k[a](sp,x) = s^k \sigma_k[a](p,x)$ for all $s > 0$ and $p \neq 0$) and goes like $|p|^k$ for $|p| \to \infty$. We will also use the notation

$$\sigma[a](p,x) = \sum_{j=0}^\ell \sigma_{N_a-j}[a](p,x) + \mathcal{O}(|p|^{N_a-\ell-1})$$

for all $\ell = 0, 1, 2 \ldots$. We note also that (15.41) implies

$$\sigma[ab](p,x) \sim \sum_{k=0}^\infty \sum_{\mu_1 \ldots \mu_k = 1}^n \frac{(-i)^k}{k!} \frac{\partial^k \sigma[a](p,x)}{\partial p_{\mu_1} \cdots \partial p_{\mu_k}} \frac{\partial^k \sigma[b](p,x)}{\partial x_{\mu_1} \cdots \partial x_{\mu_k}} \quad (15.44)$$

This allows to determine the asymptotic expansion of $\sigma[ab]$ from the ones of $\sigma[a]$ and $\sigma[b]$. Especially if $\sigma[a]$ is $\mathcal{O}(|p|^{N_a})$ and $\sigma[b]$ is $\mathcal{O}(|p|^{N_b})$ then $\sigma[ab]$ is $\mathcal{O}(|p|^{(N_a+N_b)})$.

---
[14] This condition is not essential and could be easily relaxed.

We now show that all operators in $\mathcal{A}$ allow for an expansion (15.34), and the asymptotic expansion (15.43) actually allows a simple computation of the terms $c_k(a)$. For that we compute the contribution of $\sigma_k[a](p,x)$ to $\text{Tr}_\Lambda(a)$. Using homogeneity of $\sigma_k[a]$ we obtain,

$$\int_0^\infty d|p| |p|^{k+n-1} f(\tfrac{|p|}{\Lambda}) \int_{\mathbb{R}^n} \frac{d^n\xi}{(2\pi)^n} \delta(|\xi|-1) \int_{\mathbb{R}^n} d^n x \ \text{tr}\, \sigma_k[a](\xi,x) \,.$$

Changing variables, $|p| \to u = \tfrac{|p|}{\Lambda}$, and comparing with (15.34) we see that for all $k = -n+1, -n+2, \ldots$,

$$c_{k+n}(a) = N_{n,k} \int_{\mathbb{R}^n} \frac{d^n\xi}{(2\pi)^n} \delta(|\xi|-1) \text{tr}\, \sigma_k[a](\xi,x) \tag{15.45}$$

with $N_{n,k} = \int_0^\infty du\, u^{k+n-1} f(u)$ constants depending on f. For $k = -n$ the computation above does not make sense (the constant $N_{n,-n}$ diverges). However, we can compute $c_{\log}(a)$ as follows. We first subtract from the symbol of a the part which we already accounted for and define,

$$\sigma^\perp_{-n+1}[a](p,x) := \sigma[a](p,x) - \sum_{j=0}^{N_a+n-1} \sigma_{N_a-j}[a](p,x) = \sigma_{-n}[a](p,x) + \mathcal{O}(|p|^{-n-1}).$$

Then (15.34) suggests that

$$c_{\log}(a) = \lim_{\Lambda \to \infty} \frac{1}{\log(\Lambda)} \int_{\mathbb{R}^n} \frac{d^n p}{(2\pi)^n} f(\tfrac{|p|}{\Lambda}) \int_{\mathbb{R}^n} d^n x \ \text{tr}\, \sigma[a]^\perp_{-n+1}(p,x) \,.$$

The computation of this is somewhat delicate, but a formal argument giving the correct result is as follows: Using L'Hospital's rule we get

$$c_{\log}(a) = \lim_{\Lambda \to \infty} \Lambda \int_{\mathbb{R}^n} \frac{d^n p}{(2\pi)^n} f'(\tfrac{|p|}{\Lambda})(-\tfrac{|p|}{\Lambda^2}) \int_{\mathbb{R}^n} d^n x \ \text{tr}\, \sigma[a]^\perp_{-n+1}(p,x)$$

$$= \lim_{\Lambda \to \infty} \left(\int_{\mathbb{R}^n} \frac{d^n p}{(2\pi)^n} (-f'(\tfrac{|p|}{\Lambda})\tfrac{|p|}{\Lambda}) \int_{\mathbb{R}^n} d^n x \ \text{tr}\, \sigma[a]_{-n}(p,x) + \mathcal{O}(\Lambda^{-1}) \right) \,.$$

Changing variables etc. as above and using $\int_0^\infty du(-f'(u)) = f(0) = 1$ (independent of f!) we obtain

$$c_{\log}(a) = \int_{\mathbb{R}^n} \frac{d^n\xi}{(2\pi)^n} \delta(|\xi|-1) \int_{\mathbb{R}^n} d^n x \ \text{tr}\, \sigma_{-n}(a)(\xi,x). \tag{15.46}$$

We now see that this is independent of the regularizatiation function f used, and it coincides (up to a constant $1/n$) with the Wodzicki residue on $\mathbb{R}^n$ [224].

We now discuss conditional traces of commutators of PSDO on $\mathbb{R}^n$. For simplicity we restrict ourselves to PSDO a and b such that $[a,b]$ is conditionally trace class (the extension of our argument to the general case is quite straightforward but tedious; see e.g. [29]). We assume that $\sigma[a]$ is $\mathcal{O}(|p|^{N_a})$ and similarly for b. We first determine the conditions on N_a and N_b where we expect $[a,b]$ to be conditionally trace class. From (15.44) we see that the leading term in the asymptotic expansion of the symbol of $[a,b]$ is the commutator of the leading symbols of a and b which vanishes when taking the matrix trace. Thus $\operatorname{tr}\sigma[[a,b]]$ is $\mathcal{O}(|p|^{N_a+N_b-1})$. From our discussion above it thus follows that $\operatorname{Tr}_\Lambda([a,b])$ has a finite limit $\Lambda \to \infty$ if $N_a + N_b < 1-n$ and (potentially) a log-divergence if

$$N_a + N_b = 1 - n. \tag{15.47}$$

Actually, if $N_a+N_b < 1-n$, $[a,b]$ is trace class, and $\operatorname{Tr}_C([a,b]) = \operatorname{Tr}([a,b]) = 0$ follows from the cyclicity of the Hilbert space trace. In case $N_a + N_b = 1-n$ one can show that only the leading term in the asymptotic expansion of $\operatorname{tr}\sigma[[a,b]]$ can contribute to $\operatorname{Tr}_\Lambda([a,b])$ in the limit $\Lambda \to \infty$. We therefore restrict our attention to this term. Using (15.44) and the cyclicity of the matrix trace we obtain that this leading term equals $(-i)$ times

$$\operatorname{tr}\sum_{j=1}^n \left(\frac{\partial \sigma_{N_a}[a](p,x)}{\partial p_j}\frac{\partial \sigma_{N_b}[b](p,x)}{\partial x_j} - \frac{\partial \sigma_{N_a}[a](p,x)}{\partial x_j}\frac{\partial \sigma_{N_b}[b](p,x)}{\partial p_j}\right)$$

$$= \operatorname{tr}\sum_{j=1}^n \left(\frac{\partial}{\partial p_j}\sigma_{N_a}[a](p,x)\frac{\partial \sigma_{N_b}[b](p,x)}{\partial x_j} - \frac{\partial}{\partial x_j}\sigma_{N_a}[a](p,x)\frac{\partial \sigma_{N_b}[b](p,x)}{\partial p_j}\right)$$

which is a sum of total derivatives.[15] The contribution of this term to $\operatorname{Tr}_\Lambda([a,b])$ therefore is $(-i)$ times

$$\int_{\mathbb{R}^n}\frac{d^n p}{(2\pi)^n}\int_{\mathbb{R}^n} d^n x\, f(\tfrac{|p|}{\Lambda})\sum_{j=1}^n \frac{\partial}{\partial p_j}\operatorname{tr}\sigma_{N_a}[a](p,x)\frac{\partial \sigma_{N_b}[b](p,x)}{\partial x_j} =$$

$$\int_{\mathbb{R}^n}\frac{d^n p}{(2\pi)^n}\int_{\mathbb{R}^n} d^n x\, [-f'(\tfrac{|p|}{\Lambda})]\sum_{j=1}^n \frac{1}{\Lambda}\frac{p_j}{|p|}\operatorname{tr}|p|^{N_{ab}}\sigma_{N_a}[a](\tfrac{p}{|p|},x)\frac{\partial \sigma_{N_b}[b](\tfrac{p}{|p|},x)}{\partial x_j},$$

$N_{ab} = N_a + N_b$, where we performed a partial integration. Introducing new variables $\xi = \frac{p}{|p|}$ and $u = \frac{|p|}{\Lambda}$ etc. as above we see that this term is proportional to $\Lambda^{N_a+N_b+n-1}$ and thus finite in the limit $\Lambda \to \infty$ precisely if $N_a+N_b \leq 1-n$, as expected. In the interesting case (15.47) this term is independent of Λ and equal to

$$\operatorname{Tr}_C([a,b]) = (-i)\int_{\mathbb{R}^n}\frac{d^n \xi}{(2\pi)^n}\delta(|\xi|-1)$$

$$\times \int_{\mathbb{R}^n} d^n x\operatorname{tr}\sigma_{N_a}[a](\xi,x)\sum_{j=1}^n \xi_j \frac{\partial \sigma_{N_b}[b](\xi,x)}{\partial x_j}. \tag{15.48}$$

---

[15] One should expect and can actually prove that all of $\sigma([a,b])$ is a sum of total derivatives.

On the other hand, using (15.44) it is easy to see that if (15.47) holds then

$$\sigma_{-n}(a[\log|D_0|,b])(p,x) = (-i)\sum_{j=1}^{n}\sigma_{N_a}[a](p,x)\frac{p_j}{|p|}\frac{\partial\sigma_{N_b}[b](p,x)}{\partial x_j}. \quad (15.49)$$

Inserting this in (15.46) and comparing with (15.48) and (15.35) we confirm (15.38).

15.6 Story IV: Yang-Mills Actions from Dirac Operators

(Here we apply the calculus of PSDO and regularized Hilbert space traces, as explained in Section 15.5. Some background from QFT and the definition of the Dirac operator $\slashed{D}_A$ can be found in Section 15.2. A more detailed version of this appeared in [143].)

In this Section I demonstrate that the mathematics discussed in the previous Section provides powerful tools to compute effective fermion actions on $\mathbb{R}^n$. We first indicate the general method and in the end specialize and extract the logarithmic divergence for $n = 4$.

According to the discussion in Sections 15.2 and 15.5 we compute

$$S_\Lambda(A) = -\text{Tr}_\Lambda\left(\log(\slashed{D}_A + iM) - \log(\slashed{D}_0 + iM)\right) \quad (15.50)$$

where we also need to specify what we mean by $\log(\slashed{D}_A + iM)$. We will then compute $S_\Lambda(A)$ as a series in Λ; see (15.34). In our computation we will finally specialize to four dimensions and the term $\propto \log(\Lambda)$ i.e. we compute

$$S_{\log}(A) = -\text{Res}\left(\log(\slashed{D}_A + iM) - \log(\slashed{D}_0 + iM)\right) \quad (15.51)$$

for $n = 4$. As discussed in the next Section, this term corresponds to an ambiguity in the definition of of the effective action on $\mathbb{R}^4$ and therefore is of special interest.

The definition we use in this Section is[16]

$$\log(\slashed{D}_A + iM)) := \int_0^1 \frac{ds}{s}\left(1 - (1 + s[\slashed{D}_A + iM - 1])^{-1}\right) \quad (15.52)$$

and is motivated by the following computation

$$\log(1+a) = \sum_{k=1}^{\infty}\frac{(-1)^{k-1}}{k}a^k = -\sum_{k=1}^{\infty}\int_0^1\frac{ds}{s}(-sa)^k = \int_0^1\frac{ds}{s}\left(1 - (1+sa)^{-1}\right)$$

---

[16] Our presentation here is sketchy, especially we will ignore the following detail [143]: in our computation the following formula should be used for $(\slashed{D}_A + iM)/\Lambda_0$ instead of $(\slashed{D}_A + iM)$, where Λ_0 is an arbitrary constant. Of course, the results will be independent of $|\Lambda_0|$, but one should have a constant Λ_0 with an infinitesimal imaginary part since this will specify the branch of log (i.e. the integration prescription in otherwise ill-defined integrals below).

which is valid for bounded operators with norm less than one but which we boldly extend to Dirac operators.

Next we use that there is a simple formula allowing to compactly write the asymptotic expansion for the symbol of resolvents of the Dirac operators, namely

$$\sigma[(c_1 + c_2 \slashed{D}_A)^{-1} a](p, x) = (c_1 + c_2 \gamma \cdot [p - i\partial + A(x)])^{-1} \sigma[a](x, p) \quad (15.53)$$

for all complex numbers $c_{1,2}$ (we use the notation $\gamma \cdot [p - i\partial + A(x)] = \sum_{\mu=1}^{n} \gamma_\mu [p_\mu - i\partial_\mu + A_\mu(x)] = \gamma \cdot p - i\gamma \cdot \partial + \gamma \cdot A(x)$ etc. where $\partial_\mu = \frac{\partial}{\partial x_\mu}$). The interpretation of this formula is as follows: expand in powers of $|p|^{-1}$ and apply the differentiations (which come from $-i\gamma \cdot \partial$) to the right using the Leibniz rule of differentiation [143].

A simple argument to see that this is true is as follows. We note that obviously (cf. (15.40))

$$\sigma[c_1 1 + c_2 \slashed{D}_A](p, x) = c_1 + c_2 \gamma \cdot [p + A(x)],$$

and since $(\slashed{D}_A f)(x) = \gamma \cdot [-i\partial + A(x)] f(x)$ for all $f \in \mathcal{D}$,

$$((c_1 + c_2 \slashed{D}_A) af)(x) = (c_1 + c_2 \gamma \cdot [-i\partial + A(x)])(af)(x) =$$
$$\int_{\mathbb{R}^n} \frac{d^n p}{(2\pi)^n} \int_{\mathbb{R}^n} d^n y \, e^{ip \cdot (x-y)} (c_1 + c_2 \gamma \cdot [p - i\partial + A(x)]) \sigma[a](p, x) f(y)$$

where we used (15.39) and the Leibniz rule. Replacing a by $(c_1 + c_2 \slashed{D}_A)^{-1} a$ and using (15.39) again we obtain (15.53).

We now use these equations to compute $S_\Lambda(A)$ and obtain

$$S_\Lambda(A) = \int_{\mathbb{R}^n} \frac{d^n p}{(2\pi)^n} f(\tfrac{|p|}{\Lambda}) \int_{\mathbb{R}^n} d^n x \operatorname{tr} \mathcal{S}(x, p) 1$$

where

$$\mathcal{S}(x, p) = \int_0^\infty \frac{dv}{v} \left([1 + v(\gamma \cdot (p - i\partial + A(x)) + iM)]^{-1} - [1 + v(\gamma \cdot p + iM)]^{-1} \right)$$

(we used (15.52) and (15.53) and changed the integration variable $s \to v$ with $s = v/(1+v)$). As already mentioned, $\mathcal{S}$ here is to be regarded as differential operators which acts on 1. It is straightforward to expand the integrand in this equation in powers of $\gamma \cdot [-i\partial + A(x)]$,

$$\mathcal{S} = \sum_{k=1}^{N} \mathcal{S}_k + \mathcal{R}_{N+1} \quad (15.54)$$

where

$$\mathcal{S}_k = (-1)^k \int_0^\infty \frac{dv}{v} \frac{1}{1 + v(\gamma \cdot p + iM)} \left(v\gamma \cdot [-i\partial + A(x)] \frac{1}{1 + v(\gamma \cdot p + iM)} \right)^k \quad (15.55)$$

and $\mathcal{R}_{N+1}$ is a remainder term which we can ignore for N large enough (since then it does not contribute to the logarithmic divergence, as will become obvious further below).

We thus obtain an expansion $S_\Lambda \sim \sum_{k=1}^\infty S_k$ where

$$S_k = \int_{\mathbb{R}^n} \frac{d^n p}{(2\pi)^n} f(\tfrac{|p|}{\Lambda}) \int_{\mathbb{R}^n} d^n x \operatorname{tr} \mathcal{S}_k(x,p) = \qquad (15.56)$$

$$= \sum_{\mu_1,\ldots,\mu_k=1}^n \mathcal{M}^{(k)}_{\mu_1\ldots\mu_k} \int_{\mathbb{R}^n} d^n x \operatorname{tr} D_{\mu_1} \cdots D_{\mu_k} 1 \qquad (15.57)$$

where we introduced the notation $D_\mu = -i\partial_\mu + A_\mu(x)$ and

$$\mathcal{M}^{(k)}_{\mu_1\ldots\mu_k} := (-1)^k \int_{\mathbb{R}^n} \frac{d^n p}{(2\pi)^n} f(\tfrac{|p|}{\Lambda}) \int_0^\infty dv\, v^{k-1}$$
$$\times \operatorname{tr} \frac{1}{1+v[\gamma\cdot p+iM]} \gamma_{\mu_1} \frac{1}{1+v[\gamma\cdot p+iM]} \cdots \gamma_{\mu_k} \frac{1}{1+v[\gamma\cdot p+iM]}$$

(we split the matrix trace in traces over gl_N and $\operatorname{gl}_{\nu(n)}$ and denote both traces by the same symbol). We thus have obtained an expansion of the effective action in polynomials of the covariant derivatives D_μ. Since these covariant derivatives have simple transformation properties under gauge transformations we denote our expansion as *quasi-gauge covariant* (see [143] for a more detailed discussion). It is now easy to see how to obtain an expansion in powers of Λ: Introducing $u = |p|/\Lambda$ and $v' = v\Lambda$ one finds that $\mathcal{M}^{(k)} \propto \Lambda^{n-k}$ (times a function depending on M/Λ). Especially, the only terms contributing to $\log(\Lambda)$ are for $k = n$. The remaining part of the computation is to evaluate the constants $\mathcal{M}^{(k)}$ which amounts to computing traces of products of γ-matrices and computing integrals which symbolic computation programs like MAPLE actually can handle. In the following we concentrate on the case $n = k = 4$ and extract the logarithmic divergence of the effective fermion action on $\mathbb{R}^4$.

After straightforward computations we obtain

$$\mathcal{M}^{(4)}_{\mu_1\mu_2\mu_3\mu_4} = \log(\Lambda) \frac{1}{24\pi^2} (2\delta_{\mu_1\mu_4}\delta_{\mu_2\mu_3} - 2\delta_{\mu_1\mu_3}\delta_{\mu_2\mu_4}) + \mathcal{O}(\Lambda^0)\,,$$

and $S_4 = \log(\Lambda) S_{\log} + \mathcal{O}(\Lambda^0)$ with

$$S_{\log} = \sum_{\mu,\nu=1}^4 \frac{1}{24\pi^2} \int_{\mathbb{R}^4} d^4 x \operatorname{tr}(F_{\mu\nu}^2) \qquad (15.58)$$

where

$$F_{\mu\nu} = i[D_\mu, D_\nu] = \frac{\partial}{\partial x_\mu}(A_\nu) - \frac{\partial}{\partial x_\nu}(A_\mu) + i[A_\mu, A_\nu] \qquad (15.59)$$

are the components of the Yang-Mills curvature(!). This result is very remarkable (as I will explain in more detail in the next Section): *The logarithmic divergent part of the effective fermion action in four dimensions is proportional to the Yang-Mills action.*

15.7 Final Remarks

One important theory in physics is Yang-Mills theory coupled to fermions on $\mathbb{R}^n$, especially for $n = 4$. It formally corresponds to computing what physicist refer to as a path integral and symbolically write as

$$Z = \text{``}\int \mathcal{D}[A] \int \mathcal{D}[\psi]\text{''} e^{-S(A,\psi)}$$

where A and ψ stand for the 'quantum fields' which occur in this theory: A is short for matrix valued functions A_μ, $\mu = 1,\ldots,n$, on $\mathbb{R}^n$ and represent the Yang-Mills fields, and ψ is short for certain functions on $\mathbb{R}^4$ representing the fermion fields which we do not need to further specify for what we want to say here. The theory is defined by the so-called *action* $S(A, \psi)$ which is a functional about which we will say more further below.

It is difficult to give a precise mathematical meaning to this path integral Z (even though physicists have developed remarkably successful computation methods to extract meaningful answers from such path integrals), and doing that is a very interesting and challenging problem at the borderline between mathematics and physics. However, one part of this path integral is comparably simple, namely the integral over the fermion fields ψ: The action S is a sum of two terms, the Yang-Mills action S_{YM} depending only on A and another term S_F depending on ψ and A. The path integral above thus can be (formally) computed in two steps: first by performing the fermion path integral, "$\int \mathcal{D}[\psi]$" $\exp(-S_F(A, \psi))$, and then the Yang-Mills path integral

$$Z = \text{``}\int \mathcal{D}[A]\text{''} e^{-S_{\text{tot}}(A)}$$

where

$$S_{\text{tot}}(A) = S_{YM}(A) + S_{\text{eff}}(A) \tag{15.60}$$

with S_{eff} the effective fermion action which is the contribution from the fermion path integral,

$$e^{-S_{\text{eff}}(A)} = \text{``}\int \mathcal{D}[\psi]\text{''} e^{-S_F(A,\psi)}.$$

According to the rules of fermion path integrals this equals

$$S_{\text{eff}}(A) = -\text{``log det''}(\slashed{D}_A + iM)$$

as discussed in Section 15.2 ($\slashed{D}_A$ is the Dirac operator of fermions coupled to the Yang-Mills field A).

Our discussion in Section 15.5 suggests that we can give a precise definition of $S_{\text{eff}}(A)$ as follows,

$$S_{\text{eff}}(A) = -\text{Tr}_C \left(\log(\slashed{D}_A + iM) - \log(\slashed{D}_0 + iM)\right) \qquad (15.61)$$

which we can compute as outlined in the last Section. However, as discussed in Section 15.5, the definition of Tr_C is ambiguous, and changing the regularization, e.g. $\Lambda \to s\Lambda$, amounts to changing

$$S_{\text{eff}}(A) \to S_{\text{eff}}(A) + \log(s) S_{\log}(A) \qquad (15.62)$$

with $S_{\log}$ defined in (15.51). *The* log-*divergence is (potentially) 'dangerous' since it can make the theory ambiguous (i.e. regularization dependent).* However, the Yang-Mills action is

$$S_{YM}(A) = \frac{1}{2g^2} \sum_{\mu,\nu=1}^{4} \int_{\mathbb{R}^4} d^4x \, \text{tr}\,(F_{\mu\nu}^2) \qquad (15.63)$$

with $F_{\mu\nu}$ given in (15.59) and g the Yang-Mills coupling constant, and it is proportional to $S_{\log}$ as discussed in the last Section. This is why the result obtained in the last Section is so important: Since $S_{\log}$ is proportional to the Yang-Mills action S_{YM}, a change in the regularization can be compensated by a change in the Yang-Mills coupling constant,

$$\frac{1}{2g^2} \to \frac{1}{2g^2} - \frac{1}{24\pi^2} \log(s). \qquad (15.64)$$

To have an unambiguous theory one thus needs to assume that the Yang-Mills coupling depends on the regulations, $g = g(\Lambda)$, and one has to adjust the Λ-dependence of g such that one obtains regularization independent results from the theory. As is well-known amongst physicists, this is possible.

One important contribution of NCG to physics is the spectral action principle [38,31]: a generalized Dirac operator determines a generalized Yang-Mills-fermion theory, and especially there is a formula allowing to compute the generalized Yang-Mills action, up to a constant, from the generalized Dirac operator. I believe that my discussion above gives a physical interpretation of a particular form of this spectral action principle: Assume one has a generalized Dirac operator $\slashed{D}_A$ which is given by a matrix valued differential operator. Then the corresponding Yang-Mills action has to be

$$S_{YM}(A) \propto -\text{Res}((\slashed{D}_A + iM) - (\slashed{D}_0 + iM)) \qquad (15.65)$$

since this guarantees a regularization independent Yang-Mills-fermion theory. However, as far as I can see, for all practical purposes this definition should coincide with the standard one [31].

Acknowledgments

I would like to thank Ari Laptev, Jouko Mickelsson and Florian Scheck for interesting discussions related to what is described in these notes.

16 Dirac Eigenvalues as Dynamical Variables

Giovanny Landi

University of Trieste

We review some work done with Carlo Rovelli on the use of the eigenvalues of the Dirac operator on a curved spacetime as dynamical variables, the main motivation coming from their invariance under the action of diffeomorphisms. The eigenvalues constitute an infinite set of 'observables' for general relativity and can be taken as variables for an invariant description of the gravitational field dynamics.

16.1 Introduction

A (generalized) Dirac operator D is the main ingredient in Alain Connes program of noncommutative geometry [38]. This operator encodes the full information about spacetime geometry in a way usable for describing the dynamics of the latter. Not only the geometry is reconstructed from the (normed) algebra generated by D and the smooth functions on spacetime, but the Einstein-Hilbert action of the Standard Model coupled to gravity is approximated by the trace of a simple function of D [40,30,31]. One should stress that the model obtained is both classical and Euclidean. But there is a new emphasis and a new conceptual interpretation of particle physics. The latter is used to unravel the fine geometric structure of spacetime pointing to a noncommutative structure at short distance scales and to an intrinsic coupling between gravity and other fundamental interactions. Recently [43] there has been a step in the direction of quantum field theories and it has been suggested in which sense the spacetime itself and its geometrical structure should be regarded as a concept which is derived from properties of quantum field theory.

The previous attitude also suggests the possibility of taking the eigenvalues λ_n of D as 'dynamical variables' for general relativity. They form an infinite family of diffeomorphism[1] invariant quantities and are therefore, truly *observables* for general relativity. It is a central point of the latter theory that fundamental physics is invariant under diffeomorphisms: there is no fixed non-dynamical structure with respect to which location or motion could be defined. Consequently, a fully diffeomorphism invariant description of the geometry has long been sought [13] and would be extremely useful also for quantum gravity [113]. Although this noncommutative approach has limitations, notably its Euclidean character, it definitely opens new paths in the study of the dynamics of spacetime.

As a first step for the use of these ideas in classical and/or quantum theories, an expression for the Poisson brackets of the Dirac eigenvalues has

---

[1] In fact, the eigenvalues of the Dirac operator are invariant only under diffeomorphisms which preserve the spin structure [20].

been derived [138,139,59]. Surprisingly, the brackets can be expressed in terms of the energy-momentum tensors of the Dirac eigenspinors. These tensors form the Jacobian matrix of the change of coordinates between metric and eigenvalues. The brackets are quadratic with a kernel given by the propagator of the linearized Einstein equations. The energy-momentum tensors of the Dirac eigenspinors provide the key tool for analyzing the representation of spacetime geometry in terms of Dirac eigenvalues.

In [138,139] we also study the Chamseddine-Connes spectral action. As given in [40,30,31] it is rather unrealistic as a pure gravity action, because of a huge cosmological term implying that geometries for which the action approximates the Einstein-Hilbert action are *not* solutions of the theory. We introduce a minor modification which eliminates the cosmological term. The equations of motion, derived directly from the (modified) spectral action, are solved if the energy momenta of the high mass eigenspinors scale linearly with the mass. This scaling requirement approximates the vacuum Einstein equations. These results suggest that the Chamseddine-Connes gravitational theory can be viewed as a manageable theory possibly with powerful applications to classical and quantum gravity.

16.2 Noncommutative Geometry and Gravity

We refer to [136,156,96] for friendly introductions to noncommutative geometry. In Connes' program [38], noncommutative C^*-algebras are the dual arena for noncommutative topology. We recall that a C^*-algebra $\mathcal{A}$ is an algebra over the complex numbers $\mathbb{C}$, which is complete with respect to a norm $||\cdot|| : \mathcal{A} \to \mathbb{C}$. Furthermore, there is an involution $* : \mathcal{A} \to \mathcal{A}$ and these two structures are related by suitable compatibility conditions. The (commutative) Gel'fand-Naimark theorem provides a geometric interpretation for commutative C^*-algebras and concludes that there is a complete equivalence between the category of (locally) compact Hausdorff spaces and the dual category of commutative C^*-algebras (not necessarily with a unit). Any commutative C^*-algebra is realized as the C^*-algebra of complex valued continuous functions over a (locally) compact Hausdorff space, endowed with the *sup* norm. And the points of the space are seen as the maximal ideals (or equivalently, the irreducible representations or the pure states) of the algebra. A noncommutative C^*-algebra will now be thought of as an algebra of operator valued, continuous functions on some 'virtual noncommutative space'. The attention will be switched from spaces, which in general do not even exist 'concretely', to algebras of functions. This fact allows one, for instance, to treat on the same footing 'continuum' and discrete spaces. It also permits one to address problems associated with spaces of orbits or foliation spaces or even fractal sets for which the usual notion of space is inadequate.

A metric structure is constructed out of a *real spectral triple* $(\mathcal{A}, \mathcal{H}, D)$[2]. Now $\mathcal{A}$ is a noncommutative *-algebra (indeed, in general not necessarily a C^*-algebra); $\mathcal{H}$ is a Hilbert space on which $\mathcal{A}$ is realized as an algebra of bounded operators; and D is a self-adjoint unbounded operator on $\mathcal{H}$ with suitable additional properties and which contains all (relevant) 'geometric' information. With any compact n-dimensional Riemannian spin manifold (M, g) there is associated a canonical spectral triple. The algebra is $\mathcal{A} = C^\infty(M)$, the algebra of complex valued smooth functions on M. The Hilbert space is $\mathcal{H} = L^2(M, S)$, the Hilbert space of square integrable sections of the irreducible spinor bundle over M, its rank being $2^{[n/2]}$[3]. The scalar product in $L^2(M, S)$ is the usual one of the measure $d\mu(g)$ associated with the metric g,

$$(\psi, \phi) = \int d\mu(g) \overline{\psi(x)} \phi(x), \tag{16.1}$$

with bar indicating complex conjugation and scalar product in the spinor space being the natural one in $\mathbb{C}^{2^{[n/2]}}$. Finally, D is the Dirac operator associated with the Levi-Civita connection $\omega = dx^\mu \omega_\mu$ of the metric g. If $(e_a, a = 1, \ldots, n)$ is an orthonormal basis of vector fields which is related to the natural basis $(\partial_\mu, \mu = 1, \ldots, n)$ via the n-beins, with components e_a^μ, the components $\{g^{\mu\nu}\}$ and $\{\eta^{ab}\}$ of the curved and the flat metrics respectively, are related by[4]

$$g^{\mu\nu} = e_a^\mu e_b^\nu \eta^{ab}, \quad \eta_{ab} = e_a^\mu e_b^\nu g_{\mu\nu}. \tag{16.2}$$

The coefficients $(\omega_{\mu a}{}^b)$ of the Levi-Civita (metric and torsion-free) connection of the metric g, defined by $\nabla_\mu e_a = \omega_{\mu a}{}^b e_b$, are the solutions of the equations

$$\partial_\mu e_\nu^a - \partial_\nu e_\mu^a - \omega_{\mu b}{}^a e_\nu^b + \omega_{\nu b}{}^a e_\mu^b = 0. \tag{16.3}$$

Also, let $C(M)$ be the Clifford bundle over M whose fiber at $x \in M$ is the complexified Clifford algebra $Cliff_\mathbb{C}(T_x^* M)$ and $\Gamma(M, C(M))$ be the module of corresponding sections. We have an algebra morphism into bounded operators $\mathcal{B}(\mathcal{H})$ on $\mathcal{H}$,

$$\gamma : \Gamma(M, C(M)) \to \mathcal{B}(\mathcal{H}), \tag{16.4}$$

defined by

$$\gamma(dx^\mu) =: \gamma^\mu(x) = \gamma^a e_a^\mu, \quad \mu = 1, \ldots, n, \tag{16.5}$$

and extended as an algebra map and by requiring $\mathcal{A}$-linearity. The curved and flat gamma matrices $\{\gamma^\mu(x)\}$ and $\{\gamma^a\}$, which we take to be Hermitian,

---

[2] In fact, when constructing gauge theories one needs a 'quintuple' $(\mathcal{A}, \mathcal{H}, D, \Gamma, J)$, with Γ a grading operator on $\mathcal{H}$ and J a antilinear isometry on $\mathcal{H}$ [39,40]. We shall not dwell upon these in this paper.

[3] The symbol $[k]$ indicates the integer part in k.

[4] Curved indices $\{\mu\}$ and flat ones $\{a\}$ run from 1 to n and as usual we sum over repeated indices. Curved indices are lowered and raised by the curved metric g, while flat indices are lowered and raised by the flat metric η.

obey the relations

$$\gamma^\mu(x)\gamma^\nu(x) + \gamma^\nu(x)\gamma^\mu(x) = -2g(dx^\mu, dx^\nu) = -2g^{\mu\nu} , \quad \mu,\nu = 1,\ldots,n ;$$
$$\gamma^a\gamma^b + \gamma^b\gamma^a = -2\eta^{ab} , \quad a,b = 1,\ldots,n . \tag{16.6}$$

The lift ∇^S of the Levi-Civita connection to the bundle of spinors is then

$$\nabla^S_\mu = \partial_\mu + \omega^S_\mu = \partial_\mu + \frac{1}{4}\omega_{\mu ab}\gamma^a\gamma^b , \tag{16.7}$$

while the Dirac operator, defined by

$$D = \gamma \circ \nabla^S , \tag{16.8}$$

can be written locally as

$$D = \gamma(dx^\mu)\nabla^S_\mu = \gamma^\mu(x)(\partial_\mu + \omega^S_\mu) = \gamma^a e^\mu_a(\partial_\mu + \frac{1}{4}\omega_{\mu ab}\gamma^a\gamma^b) . \tag{16.9}$$

For this canonical triple Connes' construction gives back the usual differential calculus on M together with a metric structure. First of all, exterior forms on M are represented as bounded operators on $L^2(M,S)$. Elements of $C^\infty(M)$ act as multiplicative operators on $\mathcal{H}$ and for any function f it makes sense to consider the commutator $[D,f] = \gamma^\mu\partial_\mu f$, which results into a multiplicative and a fortiori bounded operator, and which realizes the exterior derivative df. From this Connes proceeds to obtain forms of higher degree. In this algebraic framework, the usual geodesic distance between any two points p and q of M is expressed as

$$d(p,q) = \sup_{f\in\mathcal{A}}\{|f(p) - f(q)| \; : \; ||[D,f]|| \leq 1\} , \tag{16.10}$$

where the norm $||[D,f]||$ is the operator norm. The formula (16.10) does not make use of curves on the manifold M. As it stands, for a general triple, it will provide a distance on the state space of the C^*-algebra $\bar{\mathcal{A}}$, the norm closure of the algebra $\mathcal{A}$, once any point $p \in M$ is thought of as a state on the algebra of functions and one writes $p(f)$ for $f(p)$ (remember that a point is the same as a representation of the algebra of functions). In a sense, (16.10) identifies the infinitesimal unit of length as the *bare* Dirac propagator,

$$ds = D^{-1} , \tag{16.11}$$

the ambiguity coming from possible zero modes being inconsequential (one can always add a mass term)[5].

---

[5] In fact, (16.11) shows all its classical character since quantum effects will necessarily dress the bare propagator. That the dressed propagator will produce quantum effects on the geometry is a challenging and fascinating suggestion [43].

What is more, the Einstein-Hilbert action of general relativity is obtained as the *noncommutative integral* (also known as the Wodzicki residue) of the infinitesimal unit of 'area' $ds^{n-2} = D^{2-n}$ [40,119,125],

$$Res_W(D^{2-n}) =: \frac{1}{n(2\pi)^n} \int_{S^*M} tr(\sigma_{-n}(x,\xi))dxd\xi = c_n \int_M R dx ,$$

$$c_n = \frac{(2-n)}{12} \frac{2^{[n/2]-n/2}}{(2\pi)^{n/2}} \Gamma(\frac{n}{2}+1)^{-1} . \quad (16.12)$$

Here,

$$\sigma_{-n}(x,\xi) = \text{part of order } -n \text{ of the total symbol of } D^{2-n} , \quad (16.13)$$

R is the scalar curvature of the metric of M and tr is a normalized Clifford trace. This result follows from the realization that $Res_W(D^{2-n})$ is (proportional) to the integral of the second coefficient of the heat kernel expansion of D^2. Furthermore, the result does not depend upon extra contributions coming from couplings to gauge potentials like $U(1)$ which are present, for instance, in a spin^c structure.

It may be worth noticing that the dimension n itself can be extracted from the operator D as well, the Weyl formula giving $\lambda_k(|D|) \sim k^{1/n}$ for large values of k.

16.3 From the Metric to the Eigenvalues

The idea that the phase space of a physical theory should be identified with the space of solutions of the equations of motion (modulo gauge transformations) can be traced back to Lagrange [135] and has been given a new emphasis in more recent work [33,58,228]. In the case of general relativity, gauge transformations are diffeomorphisms of the space(-time) which are connected to the identity. Thus, the phase space Γ of general relativity is the space of the metric fields that solve Einstein equations, modulo diffeomorphisms (Ricci flat geometries). Corresponding observables are functions on Γ [112,186,184].

The Dirac operator allows one to define an infinite family of observables. The operator D is a self-adjoint operator on $\mathcal{H}$ admitting a complete set of real eigenvalues λ_n and eigenspinors ψ_n. The manifold M being compact, the spectrum is discrete

$$D\psi_n = \lambda_n \psi_n , \quad (16.14)$$

and the eigenvalues are labeled so that $\lambda_n \leq \lambda_{n+1}$, with repeated multiplicity. Here n is integer (positive and negative) and we choose λ_0 to be the positive eigenvalue closest to zero. As already mentioned, for simplicity we assume that there are no zero modes. The eigenvalues have dimension of an inverse length.

We shall denote the space of smooth metric fields as $\mathcal{M}$ and the space of the orbits of the gauge group in $\mathcal{M}$ as $\mathcal{G}$ (geometries). To stress the dependence upon a metric g of the Dirac operator and of its eigenvalues, we shall also write $D[g]$ and $\lambda_n[g]$. The latter then, define a discrete family of real-valued functions on $\mathcal{M}$, $\lambda_n : g \longmapsto \lambda_n[g]$. Equivalently, we have a function λ from $\mathcal{M}$ into the space of infinite sequences $\mathbb{R}^\infty$

$$\lambda : \mathcal{M} \longrightarrow \mathbb{R}^\infty, \quad g \longmapsto \{\lambda_n[g]\}, \qquad (16.15)$$

the image $\lambda(\mathcal{M})$ of $\mathcal{M}$ under this map being contained in the cone $\lambda_n \leq \lambda_{n+1}$ of $\mathbb{R}^\infty$. As we shall also see explicitly later on, the functions λ_n are invariant under diffeomorphisms (in fact, the invariance is only under diffeomorphisms which preserve the spin structure; however, only large diffeomorphisms can change the spin structure). Therefore they are well defined functions on $\mathcal{G}$. In particular, they are well defined on the phase space Γ. Thus, they are *observables* of general relativity.

Unfortunately, life is not easy: *we cannot (completely) hear the shape of a drum*, even if it is spinorial, namely the eigenvalues λ_n's need not be a set of coordinates for $\mathcal{G}$ and/or the phase space Γ. Two metric fields with the same collection of eigenvalues $\{\lambda_n\}$ are called *isospectral*. Isometric g fields are isospectral, but the converse need not be true. There exist Dirac isospectral deformations; continuous 1-parameter families of mutually non-isometric metrics with the same Dirac spectrum have been constructed [5,4]. They are of the form $M_s = G/F_s$, $s \in \mathbb{C}$, with G a nilpotent group (e.g. the Heisenberg group) and F_s a nilpotent subgroup. Also, there exist known examples of Laplace-isospectral 4-dimensional flat tori [50] which are also Dirac-isospectral (at least for the trivial spin structure). Not even the topology is determined [8][6]. Let us recall that a spherical space form is a manifold of the form S^n/F where S^n is the n-dimensional sphere and F is a finite fixed point free subgroup of $SO(n+1)$ (the group of orientation preserving isometries of S^n). Then, it has been proven in [8] that there exist two non-isometric spherical space forms of dimension $4d - 1$ with d an odd integer greater that 5, having the same Dirac spectrum and the same fundamental group. The smallest example would be in dimension 19! However, from what we understand, all the (counter)-examples constructed so far are very particular and by no means generic. The question of whether in the generic situation, the spectrum of the Dirac operator characterizes the metric is still open.

Before we proceed, let us mention another problem, namely the possibility of spectral flows (communicated by A. P. Balachandran and C. Bär): the map λ in (16.15) is only defined up to index shift: there may exist non-contractible loops in Γ such that by following the eigenvalues along the loop they come back with index shifted by some number. A possible way out could be to substitute the target space $\mathbb{R}^\infty$ by $\mathbb{R}^\infty/(\text{index shift})$; however, the map λ

---

[6] Indeed, it is rather the interplay between the Dirac operator D and the algebra $\mathcal{A}$ that determines topological/geometric properties.

would not be globally (continuously) defined. Locally, in a neighborhood of some geometry, things are fine.

Let us then proceed locally by working out the Jacobian of the transformation from metric to eigenvalues. The variation of λ_n for a variation of g can be computed using standard time independent quantum mechanics perturbation theory. For a self-adjoint operator $D(v)$ depending on a parameter v and whose eigenvalues $\lambda_n(v)$ are non-degenerate, we have

$$\frac{d\lambda_n(v)}{dv} = (\psi_n(v)| \left(\frac{d}{dv}D(v)\right) |\psi_n(v)). \tag{16.16}$$

This equation is well known for its application in elementary quantum mechanics. It can be obtained by varying v in the eigenvalue equation for $D(v)$, taking the scalar product with one of the eigenvectors, and noticing that the terms with the variation of the eigenvectors cancel. We now apply this equation to our situation, assuming generic metrics with non-degenerate eigenvalues (we refer to [20] for the general situation). We wish to compute the variation of $\lambda_n[g]$ for a small variation of the metric field g. Let $k(x) = (k_{\mu\nu}(x))$ be an arbitrarily chosen metric field and v a real parameter, and consider a 1-parameter family of metric fields g_v

$$g_v = g + vk. \tag{16.17}$$

Then, the variation $\delta\lambda_n[g]/\delta g_{\mu\nu}(x)$ of the eigenvalues under a variation of the metric, is the distribution defined by

$$\int d\mu(g) \frac{\delta\lambda_n[g]}{\delta g_{\mu\nu}(x)} k_{\mu\nu}(x) = \frac{d\lambda_n[g_v]}{dv}\bigg|_{v=0} \tag{16.18}$$

Using (16.16), we have

$$\frac{d\lambda_n[g_v]}{dv} = (\psi_n[g_v]| \frac{dD[g_v]}{dv} |\psi_n[g_v]). \tag{16.19}$$

Explicitly

$$\frac{d\lambda_n[g_v]}{dv} = \int d\mu(g_v) \, \bar\psi_n[g_v] \frac{dD[g_v]}{dv} \psi_n[g_v]. \tag{16.20}$$

In $v = 0$ we have

$$\frac{d\lambda_n[g_v]}{dv}\bigg|_{v=0} = \int d\mu(g) \, \bar\psi_n[g] \frac{dD[g_v]}{dv}\bigg|_{v=0} \psi_n[g]. \tag{16.21}$$

We can rewrite this equation as

$$\begin{aligned}\frac{d\lambda_n[g_v]}{dv}\bigg|_{v=0} &= \frac{d}{dv}\bigg|_{v=0} \int d\mu(g_v) \, \bar{\psi}_n[g] \, D[g_v] \, \psi_n[g] \\ &\quad - \int \frac{d}{dv}(d\mu(g_v))\bigg|_{v=0} \bar{\psi}_n[g] \, D[g] \, \psi_n[g] \\ &= \frac{d}{dv}\bigg|_{v=0} \int d\mu(g_v) \, \bar{\psi}_n[g] \, D[g_v] \, \psi_n[g] \\ &\quad - \int \frac{d}{dv}(d\mu(g_v))\bigg|_{v=0} \bar{\psi}_n[g] \, \lambda_n[g] \psi_n[g] \\ &= \frac{d}{dv}\bigg|_{v=0} \int d\mu(g_v) \, (\bar{\psi}_n D[g_v]\psi_n - \lambda_n \bar{\psi}_n \psi_n). \end{aligned} \quad (16.22)$$

The last formula gives the variation of the action of a spinor field with 'mass' λ_n under a variation of the metric (computed for the n-th eigenspinor of the operator $D[g]$). But the variation of the action under a variation of the metric is a well known quantity: it provides the general definition of the energy momentum tensor $T^{\mu\nu}(x)$. Indeed, the Dirac energy-momentum tensor is defined in general by

$$T^{\mu\nu}(x) := \frac{\delta}{\delta g_{\mu\nu}(x)} S_{\text{Dirac}}, \qquad (16.23)$$

where $S_{\text{Dirac}} = \int d\mu(g) \, (\bar{\psi} D\psi - \lambda \bar{\psi}\psi)$ is the Dirac action of a spinor with 'mass' λ. (Since there is no Planck constant in the Dirac action, λ has dimensions of an inverse length, rather than of a mass.) See for instance [60] where the explicit form of this tensor is also given. By denoting the energy momentum tensor of the eigenspinor ψ_n as $T_n{}^{\mu\nu}(x)$, we obtain, from (16.18), (16.22) and (16.23), that

$$\frac{\delta \lambda_n[g]}{\delta g_{\mu\nu}(x)} = T_n{}^{\mu\nu}(x). \qquad (16.24)$$

This equation gives the variation of the eigenvalues λ_n under a variation of the metric $g_{\mu\nu}(x)$, namely the Jacobian matrix of the map λ in (16.15). The matrix elements of this Jacobian are given by the energy momentum tensor of the Dirac eigenspinors. This fact suggests that we can study the map λ locally in the space of the metrics, by studying the space of the eigenspinor's energy-momenta. As far as we know, little is known on the topology of the space of solutions of Euclidean Einstein's equations on a compact manifold. A local analysis on Γ would of course miss information on disconnected components of Γ.

It is now easy to prove that the eigenvalues λ_n are invariant under the action of diffeomorphisms in the connected component of the identity in the sense that their variation vanishes when we vary the metric g by the action of any such diffeomorphism. If ξ is a vector field on M, the variation of the metric under the action of the infinitesimal diffeomorphism generated by ξ is given by

$$(\delta_\xi g)_{\mu\nu} = (\mathcal{L}_\xi g)_{\mu\nu} = 2\xi_{(\mu;\nu)} . \qquad (16.25)$$

Here $\mathcal{L}$ denotes Lie derivative, the semicolon denotes covariant derivative with respect to the Levi-Civita connection and the round brackets denote symmetrization. Then, by using (16.24) and integrating by parts, we get

$$\begin{aligned}\delta_\xi \lambda_n &= \int d\mu(g)\, \frac{\delta \lambda_n[g]}{\delta g_{\mu\nu}} (\delta_\xi g)_{\mu\nu} = 2\int d\mu(g)\, T_n{}^{\mu\nu} \xi_{(\mu;\nu)}\\ &= -2 \int d\mu(g)\, T_n{}^{\mu\nu}{}_{;\nu}\xi_\mu\end{aligned} \qquad (16.26)$$

and this expression vanishes by the 'equation of motion' for the spinor field ψ_n, that is $D\psi_n - \lambda_n \psi_n = 0$, which just states that ψ_n is an eigenspinor with eigenvalue λ_n.

It is worth stressing that the quantities λ_n are not invariant under *arbitrary* changes of the metric fields, i.e. the left hand side of (16.24) does not vanish in general.

Finally, we mention that the above derivations would go through for several other operators, beside the Dirac operator. In [161] a formula similar to (16.24) has been derived for any second order elliptic self-adjoint operator.

16.4 Action and Field Equations

We now turn to the gravitational sector of the spectral action introduced in [40,30,31]. This action contains a cutoff parameter l_0 with units of a length, which determines the scale at which the defined gravitational theory departs from general relativity. We may assume that l_0 is the Planck length $l_0 \sim 10^{-33} cm$ (although we make no reference to quantum phenomena in the present context). We use also $m_0 = 1/l_0$, which has the same dimension as D and the eigenvalues λ_n. The action depends also on a dimensionless cutoff function $\chi(u)$, which vanishes for large u. The spectral action is then defined as

$$S_G[D] = \kappa\, Tr\left[\chi(l_0^2 D^2)\right]. \qquad (16.27)$$

Here κ is a multiplicative constant to be chosen to recover the right dimensions of the action and the multiplicative overall factor.

To be definite, we shall work in dimension 4, although much of what follows can be easily generalized. The action (16.27) approximates the Einstein-Hilbert action with a large cosmological term for 'slowly varying' metrics with small curvature (with respect to the scale l_0). Indeed, the heat kernel expansion [30,31,91] allows to write

$$S_G(D) = (l_0)^{-4} f_0 \kappa \int_M \sqrt{g}\, dx \;+\; (l_0)^{-2} f_2 \kappa \int_M R\, \sqrt{g}\, dx \;+\; \ldots \; . \qquad (16.28)$$

The momenta f_0 and f_2 of the function χ are defined by

$$f_0 = \frac{1}{4\pi^2} \int_0^\infty \chi(u)u\,du\;, \qquad f_2 = \frac{1}{48\pi^2} \int_0^\infty \chi(u)\,du\;. \qquad (16.29)$$

The other terms in the expansion (16.28) are of higher order in l_0.

The expansion (16.28) shows that the action (16.27) is dominated by the Einstein-Hilbert action with a Planck-scale cosmological term. The presence of this term is a problem for the physical interpretation of the theory because the solutions of the equations of motions would have Planck-scale Ricci scalar, and therefore they would *all* be out of the regime for which the approximation taken is valid! However, the cosmological term can be cancelled by replacing the function χ with $\widetilde{\chi}$ defined by

$$\widetilde{\chi}(u) = \chi(u) - \epsilon^2 \chi(\epsilon u)\;, \qquad (16.30)$$

with $\epsilon \ll 1$. Indeed, one finds for the new momenta $\widetilde{f_0} = 0$, $\widetilde{f_2} = (1-\epsilon)f_2$. The modified action becomes

$$\widetilde{S}_G(D) = \frac{\widetilde{f_2}\kappa}{l_0^2} \int_M R\sqrt{g}\,dx + \cdots\;. \qquad (16.31)$$

We obtain the Einstein-Hilbert action in dimension four by fixing

$$\kappa = \frac{l_0^2}{16\pi G \widetilde{f_2}}\;. \qquad (16.32)$$

If l_0 is the Planck length $\sqrt{\hbar G}$, then $\kappa = \frac{3}{2}h$, where h is the Planck constant, up to terms of order ϵ. Low curvature geometries, for which the expansion (16.28) holds, *are now* solutions of the theory. Thus we obtain a theory that genuinely approximates pure general relativity at scales which are large compared to l_0.

Next, let us consider the equations of motion derived from the previous action when we regard the λ_n's as the gravitational variables. The action can easily be expressed in terms of these variables:

$$\widetilde{S}_G[\lambda] = \kappa \sum_n \widetilde{\chi}(l_0^2 \lambda_n^2). \qquad (16.33)$$

However, we cannot obtain (approximate) Einstein equations by simply varying (16.33) with respect to the λ_n's. We must minimize (16.33) on the surface $\lambda(\mathcal{M})$, not on the entire $\mathbb{R}^\infty$. In other words, the λ_n's are not independent variables, there are relations among them and these relations encode the complexity of general relativity. We can still obtain the equations of motion by varying $\widetilde{S}_G$ with respect to the metric field:

$$0 = \frac{\delta \widetilde{S}_G}{\delta g_{\mu\nu}} = \sum_n \frac{\partial \widetilde{S}_G}{\partial \lambda_n} \frac{\delta \lambda_n}{\delta g_{\mu\nu}} = \sum_n \frac{d\widetilde{\chi}(l_0^2 \lambda_n^2)}{d\lambda_n} T_n{}_I^\mu. \qquad (16.34)$$

By defining $f(u) =: \frac{d}{du}\tilde{\chi}(u)$, (16.34) becomes

$$\sum_n f(l_0^2 \lambda_n^2) \, \lambda_n \, T_{nI}^\mu = 0. \tag{16.35}$$

These are the Einstein equations in the Dirac eigenvalues formalism.

Up to now, the cutoff function $\chi(u)$ is arbitrary. The simplest choice is to take it to be smooth and monotonic on $\mathbb{R}^+$ with

$$\chi(u) = \begin{cases} 1 \text{ if } u < 1 - \delta \\ 0 \text{ if } u > 1 + \delta \end{cases} \tag{16.36}$$

where $\delta \ll 1$. Namely $\chi(u)$ is the smoothed-out characteristic function of the interval $[0, 1]$. With this choice, the action (16.27) is essentially (κ times) the *number* of eigenvalues λ_n with absolute value smaller that m_0 (up to corrections of order δ). Then the function $f(u)$ vanishes everywhere except on two narrow peaks. A negative one (width 2δ and height $1/2\delta$) centered at one; and a positive one (width $2\delta/\epsilon$ and height $\epsilon^3/2\delta$) around the arbitrary large number $1/\epsilon =: s \gg 1$. The first of these peaks gets contributions from λ_n's such that $\lambda_n \sim m_0$, namely from Planck scale eigenvalues. The second from ones such that $\lambda_n \sim sm_0$. Equations (16.35) are solved if the contributions of the two peaks cancel. This happens if below the Planck scale the energy momentum tensor scales as

$$\lambda_{n(m_0)} \rho(1) \, T_{n(m_0)I}^\mu(x) = s^{-2} \lambda_{n(sm_0)} \rho(s) \, T_{n(sm_0)I}^\mu(x). \tag{16.37}$$

Here $\rho(1)$ and $\rho(s)$ are the densities of eigenvalues of $l_0^2 D^2$ at the two peaks and the index $n(t)$ is defined by

$$l_0 \lambda_{n(t)}^2 = t. \tag{16.38}$$

For large n the growth of the eigenvalues of the Dirac operator is given by the Weyl formula $\lambda_n \sim \sqrt{2\pi} V^{-1/4} n^{1/4}$, where V is the volume. Using this one derives immediately the eigenvalue densities and simple algebra yields

$$T_{nI}^\mu(x) = \lambda_n \, l_0 \, T_{0I}^\mu(x) \,, \tag{16.39}$$

for $n \gg n(m_P)$, where $T_{0I}^\mu(x) = T_{n(m_0)I}^\mu(x)$ is the energy momentum at the Planck scale. We have shown that *the dynamical equations for the geometry are solved if below the Planck length the energy-momentum of the eigenspinors scales as the eigenspinor's mass*. In other words, we have expressed the Einstein equations as a scaling requirement on the energy-momenta of the very-high-frequency Dirac eigenspinors. This scaling requirement yields vacuum Einstein equations at low energy scale [138,139].

16.5 Poisson Brackets for the Eigenvalues

A symplectic structure on the phase space Γ can be constructed in covariant form [6]. First of all, we recall that a vector field X on the space $\mathcal{S}$ of solutions of Einstein field equations can be written as a differential operator

$$X = \int d^4x \, X_{\mu\nu}(x)[g] \, \frac{\delta}{\delta g_{\mu\nu}(x)}, \qquad (16.40)$$

where $X_{\mu\nu}(x)[g]$ is any solution of the Einstein equations for the metric field, *linearized* over the background g. A vector field $[X]$ on Γ is given by an equivalence class of such vector fields X, modulo linearized gauge transformations of $X_{\mu\nu}(x)$. A linearized gauge transformation is given by

$$g \mapsto g + \delta_\xi g = \mathcal{L}_\xi g, \qquad (16.41)$$

where ξ is a vector field on the spacetime M (generating an infinitesimal diffeomorphism). Two linearized fields X and Y (around the metric g) are gauge equivalent if

$$Y = X + \delta_\xi g \qquad (16.42)$$

for some vector field ξ. The symplectic two-form Ω of general relativity is given by [6]

$$\Omega(X,Y) = \int_\Sigma d^3\sigma \, n_\rho \, (X_{\mu\alpha} \overset{\leftrightarrow}{\nabla}_\tau Y_{\nu\beta}) \, \epsilon^{\tau\alpha\beta}{}_\upsilon \, \epsilon^{\upsilon\rho\mu\nu} \qquad (16.43)$$

where

$$(X_{\mu\alpha} \overset{\leftrightarrow}{\nabla}_\tau Y_{\nu\beta}) =: (X_{\mu\alpha} \nabla_\tau Y_{\nu\beta} - Y_{\mu\alpha} \nabla_\tau X_{\nu\beta}). \qquad (16.44)$$

Moreover, $\Sigma \ni \sigma \mapsto x(\sigma) \subset M$ is chosen to be a (compact non-contractible) three-dimensional surface, such that, topologically, $M = \Sigma \times S^1$ (so that it gives the single non trivial 3-cycle of M), but otherwise arbitrary, and n_ρ is its normal one-form.

Both sides of (16.43) are functions of the metric g, namely scalar functions on $\mathcal{S}$. The form Ω is degenerate precisely in the gauge directions,

$$\Omega(X,Y)[g] = 0 \quad \text{iff} \quad Y = \delta_X g, \qquad (16.45)$$

thus it defines a non-degenerate *symplectic* two form on the space of orbits of the diffeomorphism group, namely on Γ. The coefficients of Ω can be written as

$$\Omega^{\mu\nu;\alpha\beta}(x,y) = \int_\Sigma d^3\sigma \, n_\rho \, [\delta(x, x(\sigma)) \overset{\leftrightarrow}{\nabla}_\tau \delta(y, x(\sigma))] \, \epsilon^{\tau\alpha\beta}{}_\upsilon \, \epsilon^{\upsilon\rho\mu\nu}. \qquad (16.46)$$

Because of the degeneracy, Ω has no inverse on $\mathcal{S}$. However, let us (arbitrarily) fix a gauge (choose a representative field g for any four geometry, and, consequently, choose a field X in any equivalence class $[X]$). On the space of the gauge fixed fields, Ω is non degenerate and we can invert it. Let

$P_{\mu\nu;\alpha\beta}(x,y)$ be the inverse of the symplectic form matrix on this subspace, namely

$$\int d^4y \int d^4z \, P_{\mu\nu;\alpha\beta}(x,y) \, \Omega^{\nu\rho;\beta\gamma}(y,z) \, F_{\rho\gamma}(z) = \int d^4z \, \delta(x,z) \, \delta^\rho_\mu \, \delta^\gamma_\alpha \, F_{\rho\gamma}(z) \quad (16.47)$$

for all solutions F of the linearized Einstein equations, satisfying the gauge condition chosen. Integrating over the delta functions, and using (16.46), we have

$$\int_\Sigma d^3\sigma \, n_\rho \, [P_{\mu\nu;\alpha\beta}(x,x(\sigma))\overleftarrow{\nabla}_\rho F_{\tau\gamma}(x(\sigma))] \, \epsilon^{\rho\beta\gamma}{}_\upsilon \, \epsilon^{\upsilon\nu\tau\sigma} = F_{\mu\alpha}(x). \quad (16.48)$$

This equation, where F is any solution of the linearized equations, defines P in the chosen gauge. Then, we can write the Poisson bracket between two functions f, g on $\mathcal{S}$ as

$$\{f,g\} = \int d^4x \int d^4y \, P_{\mu\nu;\sigma\tau}(x,y) \, \frac{\delta f}{\delta g_{\mu\sigma}(x)} \, \frac{\delta g}{\delta g_{\nu\tau}(y)}. \quad (16.49)$$

If the functions f and g are gauge invariant, i.e. are well defined on Γ, the r.h.s of (16.49) is independent of the gauge chosen. But (16.48) is precisely the definition of the propagator of the linearized Einstein equations over the background g in the chosen gauge. By combining (16.48),(16.49) and (16.24) we obtain the Poisson bracket for any two eigenvalues of the Dirac operator as

$$\{\lambda_n, \lambda_m\} = \int d^4x \int d^4y \, T_{[n}{}^{\mu\alpha}(x) \, P_{\mu\nu;\alpha\beta}(x,y) \, T_{m]}{}^{\nu\beta}(y) . \quad (16.50)$$

This equation gives the Poisson bracket of two eigenvalues in terms of the energy-momentum tensor of the two corresponding eigenspinors and of the propagator of the linearized Einstein equations. The right hand side does not depend on the gauge chosen.

16.6 Final Remarks

Recent work of Connes and Chamseddine on a spectral description of fundamental interactions and in particular of gravity, has motivated our attempt to describe gravity by means of the eigenvalues of the Dirac operator. This approach could open new paths in the exploration of the physics of spacetime and find applications in classical and quantum gravitation. The main obstacle for a full development of this approach is its natural euclidean character since, at the moment, there does not exist a satisfactory 'Lorentzian' version of Connes' program. Some interesting steps in the direction of a 'quantum spectral approach' have also been recently presented [43,185]. An extension of these ideas to supergravity has been considered in [213,214,36,175].

We have analyzed some aspects of the dynamical structure of the theory in the λ_n variables by computing their Poisson algebra (16.50). In the way it is presented, the Poisson algebra is not in closed form, since the right hand side of (16.50) is not expressed in terms of the λ_n themselves, and it is unclear if this can be done in general. Still, representations of this algebra could give information on a diffeomorphism invariant quantum theory.

The central and important feature of the approach that we have presented is that the theory is formulated in terms of diffeomorphism invariant quantities. The λ_n's are a family of diffeomorphism invariant observables in euclidean general relativity, which is presumably complete or 'almost complete': it would fail to distinguish possible isospectral and not isometric geometries, although at the moment it is not clear what is the generic situation. Another remarkable aspect of the spectral approach is that there is a physical cutoff and an elementary physical length in the action that does not break diffeomorphism invariance. All high frequency modes are cut off without introducing background structures, thus in a diffeomorphic invariant manner. Since the number of the remaining modes is determined by the ratio of the spacetime volume to the Planck scale, one may expect that such a theory would have infrared divergences but not ultraviolet ones in the quantum regime.

The key open problem is, of course, a better (complete) understanding of the map λ given in (16.15) and its range. Namely a characterization of the constraints that a sequence of real numbers λ_n must satisfy in order to represent the spectrum of the Dirac operator of some geometry. We have partially addressed this problem locally in the phase space of the theory by studying the *tangent* map to λ. This tangent map is given explicitly in terms of the eigenspinor's energy-momenta and of the propagator of the linearized Einstein equation. The constraints on the λ_n's are the core of the formulation of the gravitational theory that we have begun to explore here. They should be contained in Connes' axioms for D in its axiomatic definition of a spectral triple [38]. The equations in these axioms capture the notion of Riemannian manifold algebraically and they should encode the constraints satisfied by the eigenvalues λ_n's of the operator D.

Finally, we mention that it would be very interesting to extend the approach presented here to the noncommutative manifolds recently constructed in [44,41].

Acknowledgments

I thank F. Scheck and H. Upmeier for they kind invitation to Hesselberg. Together with W. Werner they provided a very friendly and stimulating 'atmosphere'. I am grateful to Christian Bär for several suggestions and conversations and for making me aware of his papers. And I have no words to thank Carlo for his collaboration. This work is supported by the Italian MURST.

17 Hopf Algebras in Renormalization and NC Geometry

Raimar Wulkenhaar

University of Vienna

We review the appearance of Hopf algebras in the renormalization of quantum field theories and in the study of diffeomorphisms of the frame bundle important for index computations in noncommutative geometry.

17.1 Introductory Remarks

This contribution focuses on two applications of the Hopf algebra of rooted trees discovered during the last two years. They suggest an amazing link between mathematics and physics. There exists an excellent review [43] of these topics, written by the authors of these ideas. In what follows I am going to explain parts of this development.

In mathematics, foliations provide a large class of examples of noncommutative spaces and lead to an index problem for the transverse hypoelliptic operator [47]. The computation of the cocycles in the local index formula turned out to be extremely lengthy even in dimension one. Alain Connes and Henri Moscovici [48] were looking for an organizing principle for that calculation, which they found in the cyclic cohomology of a Hopf algebra $\mathcal{H}_T$ obtained by the action of vector fields on a crossed product of functions by diffeomorphisms.

Concerning physics, Dirk Kreimer [129] discovered that a perturbative quantum field theory carries in a natural way a Hopf algebra structure $\mathcal{H}_R$ given by operations on Feynman graphs. The antipode reproduces precisely the combinatorics of renormalization, i.e. it produces the local counterterms to make the divergent integral corresponding to the Feynman graph finite.

Noticing that both Hopf algebras have formally a very similar structure, Connes and Kreimer gave the precise relation [42] between $\mathcal{H}_T$ and $\mathcal{H}_R$. This is very transparent in the language of rooted trees they used. The commutative Hopf subalgebra $\mathcal{H}^1$ of Connes–Moscovici is (in dimension 1) a Hopf subalgebra of Kreimer's Hopf algebra for a quantum field theory with a single primitively divergent graph.

Recently it was pointed out [23] that the same algebra of rooted trees plays a role in Runge–Kutta methods of numerical analysis.

17.2 The Hopf Algebra of Connes–Moscovici

In principle, the Hopf algebra of Connes and Moscovici can be understood from classical differential geometry [34]. We give here a somewhat shortened version of the derivation and refer to [225] for more details. We recommend [157] for a useful introduction to Hopf algebras and related topics.

We consider the oriented frame bundle F^+ of a manifold M and in particular the vector fields on F^+. There is a natural notion of vertical vector fields, these are the tangent vectors to curves in F^+ obtained by the right action of the group $Gl^+(n)$ of $n \times n$ matrices with positive determinant. The horizontal vector fields are not canonically given, they are determined once a connection is specified. For our purpose we can work in local coordinates.

Let $\{x^\mu\}_{\mu=1\ldots,n}$ be the coordinates of $x \in M$ within a local chart of M and $\{y_i^\mu\}_{\mu,i=1,\ldots n}$ be the coordinates of n linearly independent vectors of the tangent space T_xM with respect to the basis ∂_μ. On F^+ there exist the following geometrical objects, written in terms of the local coordinates (x^μ, y_i^μ) of $p \in F^+$:

1) an R^n-valued (soldering) 1-form α with $\alpha^i = (y^{-1})_\mu^i dx^\mu$,
2) a $\mathfrak{gl}(n)$-valued (connection) 1-form ω with $\omega_j^i = (y^{-1})_\mu^i (dy_j^\mu + \Gamma_{\alpha\beta}^\mu y_j^\alpha dx^\beta)$, where $\Gamma_{\alpha\beta}^\mu$ depends only on x^ν,
3) n^2 vertical vector fields $Y_j^i = y_j^\mu \partial_\mu^i$,
4) n horizontal (with respect to ω) vector fields $X_i = y_i^\mu (\partial_\mu - \Gamma_{\alpha\mu}^\nu y_j^\alpha \partial_\nu^j)$.

A local diffeomorphism ψ of M has a lift $\tilde\psi : (x^\mu, y_i^\mu) \mapsto (\psi(x)^\mu, \partial_\nu \psi(x)^\mu y_i^\nu)$ to the frame bundle and induces the following transformations of the previous geometrical objects:

1') $(\tilde\psi^*\alpha)\big|_p = \alpha\big|_p$.
2') $(\tilde\psi^*\omega)\big|_p = (y^{-1})_\mu^i (dy_j^\mu + \tilde\Gamma_{\alpha\beta}^\mu y_j^\alpha dx^\beta)$ is again a connection form, with
$\tilde\Gamma_{\alpha\beta}^\mu\big|_x = ((\partial\psi(x))^{-1})_\gamma^\mu \Gamma_{\delta\epsilon}^\gamma\big|_{\psi(x)} \partial_\alpha\psi(x)^\delta \partial_\beta\psi(x)^\epsilon + ((\partial\psi(x))^{-1})_\gamma^\mu \partial_\beta\partial_\alpha\psi(x)^\gamma$,
3') $(\tilde\psi_* Y_i^j)\big|_p = Y_i^j\big|_p$,
4') $(\tilde\psi_*^{-1} X_i)\big|_p = y_i^\mu(\partial_\mu - \tilde\Gamma_{\alpha\mu}^\nu y_j^\alpha \partial_\nu^j)$ is horizontal to $\tilde\psi^*\omega$.

We refer to [225] for the proof.

Given these tools of classical differential geometry, the new idea is to apply the vector fields X, Y to a crossed product $\mathcal{A} = C_c^\infty(F^+) \rtimes \Gamma$ of the algebra of smooth functions on F^+ with compact support by the action of the pseudogroup Γ of local diffeomorphisms of M. As a set, $\mathcal{A}$ can be regarded as the tensor product of $C_c^\infty(F^+)$ with Γ. It is generated by the monomials

$$fU_\psi^*, \qquad f \in C_c^\infty(\mathrm{Dom}(\tilde\psi)), \quad \psi \in \Gamma, \qquad (17.1)$$

where $\tilde\psi$ is the diffeomorphism of F^+ obtained as the lift of $\psi \in \Gamma$. As an algebra, the multiplication rule in $\mathcal{A}$ is defined by

$$f_1 U_{\psi_1}^* \, f_2 U_{\psi_2}^* := f_1 (f_2 \circ \tilde\psi_1) U_{\psi_2 \psi_1}^*. \qquad (17.2)$$

The function $f_1(f_2 \circ \tilde\psi_1)$ evaluated at p (in the domain of definition) gives $f_1(p) f_2(\tilde\psi_1(p))$, i.e. we have a non-local product on the function algebra.

The action of vector fields on $\mathcal{A}$ is defined as the action on the function part. Interesting is the application to the product (17.2), because the non-locality in the function part leads to a deviation from the Leibniz rule. For V being a vector field on F^+ one computes

$$V(f_1 U^*_{\psi_1} f_2 U^*_{\psi_2}) = V(f_1 U^*_{\psi_1}) f_2 U^*_{\psi_2} + f_1 U^*_{\psi_1} \left(\tilde{\psi}_{1*}(V)\right)(f_2 U^*_{\psi_2}) \ . \quad (17.3)$$

Since diffeomorphisms and right group action commute, we get the unchanged Leibniz rule for the vertical vector fields,

$$Y^j_i(ab) = Y^j_i(a) b + a Y^j_i(b) \ , \qquad a, b \in \mathcal{A} \ . \quad (17.4)$$

For the horizontal vector fields, however, there will be an additional term $a(\psi_{1*} X_i - X_i)(b)$. Comparing 4), 4') and 3) above we have $\psi_{1*} X_i - X_i = \tilde{\delta}^k_{ji} Y^j_k$ for some function $\tilde{\delta}^k_{ji}$. Using (17.2) we commute this function in front of a and obtain

$$X_i(ab) = X_i(a) b + a X_i(b) + \delta^k_{ji}(a) Y^j_k(b) \ , \qquad a, b \in \mathcal{A} \ . \quad (17.5)$$

The operator δ^k_{ji} on $\mathcal{A}$ is computed to

$$\delta^k_{ji}(f U^*_{\psi}) = (\tilde{\Gamma}^\nu_{\alpha\mu} - \Gamma^\nu_{\alpha\mu}) y^\alpha_j y^\mu_i (y^{-1})^k_\nu f U^*_{\psi} \ , \quad (17.6)$$

where $\tilde{\Gamma}^\nu_{\alpha\mu}$ are the connection coefficients belonging to $\tilde{\psi}^*\omega$. It turns out that δ^k_{ji} is a derivation:

$$\delta^k_{ji}(ab) = \delta^k_{ji}(a) b + a \delta^k_{ji}(b) \ . \quad (17.7)$$

These formulae can now be immediately interpreted in the dual sense, for instance $X_i(ab) = \Delta(X_i)(a \otimes b)$, which leads to a structure of a coalgebra on the linear space $R(1, X_i, Y^j_k, \delta^k_{ji})$,

$$\begin{aligned}
\Delta(Y^j_k) &= Y^k_j \otimes 1 + 1 \otimes Y^j_k \ , \\
\Delta(X_i) &= X_i \otimes 1 + 1 \otimes X_i + \delta^k_{ji} \otimes Y^j_k \ , \\
\Delta(\delta^k_{ji}) &= \delta^k_{ji} \otimes 1 + 1 \otimes \delta^k_{ji} \ , \\
\Delta(1) &= 1 \otimes 1 \ ,
\end{aligned} \quad (17.8)$$

with 1 being the identity on $\mathcal{A}$. Coassociativity $(\Delta \otimes \mathrm{id}) \circ \Delta = (\mathrm{id} \otimes \Delta) \circ \Delta$ is easy to check.

Vector fields form a Lie algebra, so the next step is to ask whether $R(1, X_i, Y^j_k, \delta^k_{ji})$ is closed under the Lie bracket. The first commutators are OK,

$$\begin{aligned}
{[Y^i_j, Y^k_l]}(fU^*_\psi) &= (\delta^i_l Y^k_j - \delta^k_j Y^i_l)(fU^*_\psi) \ , \\
{[Y^k_j, X_i]}(fU^*_\psi) &= \delta^k_i X_j(fU^*_\psi) \ , \\
{[Y^i_j, \delta^k_{lm}]}(fU^*_\psi) &= (\delta^i_l \delta^k_{jm} + \delta^i_m \delta^k_{lj} - \delta^k_j \delta^i_{lm})(fU^*_\psi) \ .
\end{aligned} \quad (17.9)$$

The next one between horizontal fields

$$[X_i, X_j] = R^k_{lij} Y^l_k + \Theta^k_{ij} X_k \qquad (17.10)$$

leads to new generators, because curvature R and torsion Θ are no structure 'constants'. Therefore, one uses a different strategy and considers instead of $\mathcal{A}$ a *Morita equivalent* algebra $\mathcal{A}'$ based on a *flat* manifold $N = \coprod U_\alpha$ – the disjoint union of the charts U_α of M. Now, there is neither curvature nor torsion, and horizontal vector fields commute. There remain the commutators of X with δ, which lead indeed to new generators of the Lie algebra:

$$\delta^k_{ji,\ell_1\ldots\ell_n}(fU^*_\psi) := [X_{\ell_n},\ldots,[X_{\ell_1},\delta^k_{ji}]\ldots](fU^*_\psi) \qquad (17.11)$$
$$= \partial_{\lambda_n}\ldots\partial_{\lambda_1}\Big(((\partial\psi(x))^{-1})^\nu_\beta \partial_\mu \partial_\alpha \psi(x)^\beta\Big) y^\mu_j y^\alpha_i (y^{-1})^k_\nu y^{\lambda_1}_{\ell_1}\ldots y^{\lambda_n}_{\ell_n} fU^*_\psi.$$

All these generators $\delta^k_{ji,\ell_1\ldots\ell_n}$ commute with each other.

Now having established a Lie algebra, we call $\mathcal{H}$ its enveloping algebra, i.e. the algebra of polynomials in $\{1, X_i, Y^k_j, \delta^k_{ji}, \delta^k_{ji,\ell_1\ldots\ell_n\ldots}\}$, with the commutation relations inherited from the Lie algebra. With the coproduct Δ on the Lie algebra, $\mathcal{H}$ becomes automatically a bialgebra, where the coproduct is defined via the algebra homomorphism axiom:

$$\Delta(h^1 h^2) = \Delta(h^1)\,\Delta(h^2) := \sum h^1_1 h^1_2 \otimes h^2_1 h^2_2 \,, \qquad \Delta(h_i) = \sum h^1_i \otimes h^2_i \,, \qquad (17.12)$$

for $h_1, h_2 \in \mathcal{H}$. The counit $\varepsilon : \mathcal{H} \to \mathbb{C}$ is defined by

$$\varepsilon(1) = 1_\mathbb{C} \,, \qquad \varepsilon(h) = 0 \quad \forall h \neq 1 \,. \qquad (17.13)$$

The counit axiom $(\varepsilon \otimes \mathrm{id}) \circ \Delta(h) = (\mathrm{id} \otimes \varepsilon) \circ \Delta(h) = h$ is straightforward to check.

There also exists an antipode on $\mathcal{H}$ which makes it into a Hopf algebra. The antipode is the unique antiautomorphism of $\mathcal{H}$ satisfying

$$S(h_1 h_2) = S(h_2) S(h_1) \,,$$
$$m \circ (S \otimes \mathrm{id}) \circ \Delta(h) = 1\varepsilon(h) = m \circ (\mathrm{id} \otimes S) \circ \Delta(h) \,, \qquad (17.14)$$

for $h, h_1, h_2 \in \mathcal{H}$, and where m denotes the multiplication. From the second line and (17.8) one easily obtains

$$\begin{aligned} S(1) &= 1 \,, \\ S(Y^j_k) &= -Y^j_k \,, \\ S(\delta^k_{ji}) &= -\delta^k_{ji} \,, \\ S(X_i) &= -X_i + \delta^k_{ji} Y^j_k \,. \end{aligned} \qquad (17.15)$$

The action of S on the other generators of $\mathcal{H}$ can be derived from (17.14).

The purpose of this Hopf algebra $\mathcal{H}$ is to facilitate the computation [48] of cocycles in the local index formula [47] of Connes and Moscovici. A good way to learn it would be to consult [56].

17.3 Rooted Trees

Coproduct and antipode for the generators $\delta^k_{ji,\ell_1...\ell_n...}$ are only recursively defined via the axioms of coproduct and antipode. Now we are going to present an explicit solution – via the concept of rooted trees. This was introduced by Connes and Kreimer [42] to clarify the relation between the two Hopf algebras in the theory of foliations and in perturbative quantum field theory. In [225] we generalize their construction from dimension 1 to arbitrary dimension of the manifold M. To the first three classes of δ's we associate the following trees:

$$\delta^k_{ji} = \bullet^{\,k}_{ji},$$

$$\delta^k_{ji,l} = \begin{array}{c} \bullet^{\,k}_{ji} \\ | \\ \bullet \, l \end{array},$$

$$\delta^k_{ji,lm} = \begin{array}{c} \bullet^{\,k}_{ji} \\ | \\ \bullet \, l \\ | \\ \bullet \, m \end{array} + \begin{array}{c} \bullet^{\,k}_{ji} \\ \diagup \, \diagdown \\ \bullet\, l \quad \bullet\, m \end{array}. \qquad (17.16)$$

The rule is obvious. A symbol $\delta^k_{ji,A\ell}$, for A a string of $|A|$ indices, is obtained from $\delta^k_{ji,A} = \sum_{a=1}^{|A|!} t_a^{|A|}$ by attaching to each of its trees $t_a^{|A|}$ a new vertex with label ℓ successively to the right of each vertex. The root (with three indices) remains the same and order is important.

Coproduct and antipode require the definition of cuts of a tree. An elementary cut along a chosen edge splits a tree into two – the trees above (trunk) and below (cut branch) the cut. It is clear that we have to add 2 indices to complete the root of the cut branch. This will be a pair of summation indices. We define the action of a cut as the shift of one index of the vertex above the cut to the first position of the new root of the cut branch. The remaining position to complete the root of the cut branch is filled with a summation index and the same summation index is put into the vacant position of the trunk. In the case of cutting immediately below the root, we have to sum over the three possibilities of picking up indices of the root, adding a minus sign if we pick up the unique upper index. The following examples illustrate the definition of a cut, where we write the trunk as the rhs of the tensor product and the cut branch as the lhs:

$$\begin{array}{c} \bullet^{\,k}_{ji} \\ | \\ \bullet\, l \end{array} = \bullet^{\,a}_{jl} \otimes \bullet^{\,k}_{ai} + \bullet^{\,a}_{il} \otimes \bullet^{\,k}_{ja} - \bullet^{\,k}_{al} \otimes \bullet^{\,a}_{ij},$$

$$\begin{array}{c} \bullet^{\,k}_{ji} \\ \diagup \, \diagdown \\ \bullet\, l \quad \bullet\, m \end{array} = \bullet^{\,a}_{jm} \otimes \bullet^{\,k}_{ai,l} + \bullet^{\,a}_{im} \otimes \bullet^{\,k}_{ja,l} - \bullet^{\,k}_{am} \otimes \bullet^{\,a}_{ji,l},$$

$$\begin{array}{c}\bullet\,{}^{k}_{ji}\\ \Big|\,l\\ \bullet\,m\end{array} = \bullet\,{}^{a}_{lm} \otimes \begin{array}{c}\bullet\,{}^{k}_{ji}\\ \Big|\\ \bullet\,a\end{array}. \qquad (17.17)$$

A multiple cut consists of several elementary cuts, where *the order of cuts is from top to bottom and from left to right*. An *admissible* cut is a multiple cut such that on the path from any vertex to the root there is at most one elementary cut. The product of all cut branches forms the lhs of the tensor product, whereas the trunk alone containing the old root serves as the rhs.

The purpose of these definitions is to give an explicit formula for coproduct and antipode. Indeed, by induction one can prove the following:

Proposition 1. *The coproduct of* $\delta^k_{ji,A} = \sum_{a=1}^{|A|!} t_a^{|A|}$ *is given by*

$$\Delta(\delta^k_{ji,A}) = \delta^k_{ji,A} \otimes 1 + 1 \otimes \delta^k_{ji,A} + \sum_{a=1}^{|A|!} \sum_{\mathcal{C}} P^{\mathcal{C}}(t_a^{|A|}) \otimes R^{\mathcal{C}}(t_a^{|A|}) , \qquad (17.18)$$

where for each $t_a^{|A|}$ the sum is over all admissible cuts $\mathcal{C}$ of $t_a^{|A|}$. In (17.18), $R^{\mathcal{C}}(t_a^{|A|})$ is the trunk and $P^{\mathcal{C}}(t_a^{|A|})$ the product of cut branches obtained by cutting $t_a^{|A|}$ via the multiple cut $\mathcal{C}$.

Proof. We start from

$$\Delta(\delta^k_{ji,A\ell}) = [\Delta(\delta^k_{ji,A}), \Delta(X_\ell)] = \delta^k_{ji,A\ell} \otimes 1 + 1 \otimes \delta^k_{ji,A\ell} + R^k_{ji,A\ell} ,$$
$$R^k_{ji,A\ell} = [X_\ell \otimes 1 + 1 \otimes X_\ell, R^k_{ji,A}] + [\delta^m_{n\ell} \otimes Y^n_m, R^k_{ji,A} + (1 \otimes \delta^k_{ji,A})] \in \mathcal{H} \otimes \mathcal{H} .$$

By definition of the tree, the commutator with X_ℓ attaches a vertex ℓ successively to all previous vertices, where $X_\ell \otimes 1$ attaches to the cut branches and $1 \otimes X_\ell$ attaches to the trunk. Next, the commutator with $\delta^m_{n\ell} \otimes Y^n_m$ puts for each vertex of the trunk (due to the commutator with Y) a cut branch consisting of a single vertex to the lhs of the tensor product. Both contributions together yield precisely all admissible cuts of the trees corresponding to $\delta^k_{ji,A\ell}$. □

The antipode is obtained by applying the antipode axiom $m \circ (S \otimes \mathrm{id}) \circ \Delta = 0$ to (17.18). By recursion one proves

Proposition 2. *The antipode S of* $\delta^k_{ji,A} = \sum_{a=1}^{|A|!} t_a^{|A|}$ *is given by*

$$S(\delta^k_{ji,A}) = -\delta^k_{ji,A} - \sum_{a=1}^{|A|!} \sum_{\mathcal{C}_a} (-1)^{|\mathcal{C}_a|} P^{\mathcal{C}_a}(t_a^{|A|}) R^{\mathcal{C}_a}(t_a^{|A|}) , \qquad (17.19)$$

where the sum is over the set of all non-empty multiple cuts $\mathcal{C}_a$ of $t_a^{|A|}$ (multiple cuts on paths from bottom to the root are allowed) consisting of $|\mathcal{C}_a|$ individual cuts. □

17.4 Feynman Graphs and Rooted Trees

In a perturbative quantum field theory it is convenient to symbolize contributions to Green's functions by Feynman graphs. These Feynman graphs stand for analytic expressions of momentum variables. Internal momentum variables have to be integrated out. Very often some of these integrations formally yield infinity. The art of obtaining meaningful results out of these integrals is called renormalization. A central problem is the existence of subdivergences which cannot be regularized by a simple subtraction of the divergent part. Bogoliubov [19] found a recursion formula for the regularization of Feynman graphs with subdivergences and Zimmermann gave an explicit solution – the forest formula [227].

In 1997 Dirk Kreimer discovered [129] that there is the structure of a Hopf algebra behind this art of renormalization, with the combinatorics of the forest formula produced by the antipode. Kreimer's idea was to visualize the divergence structure of Feynman graphs in terms of parenthesized words, which are in 1:1 correspondence to rooted trees [42]. Let us exemplify this idea by a Feynman graph from QED:

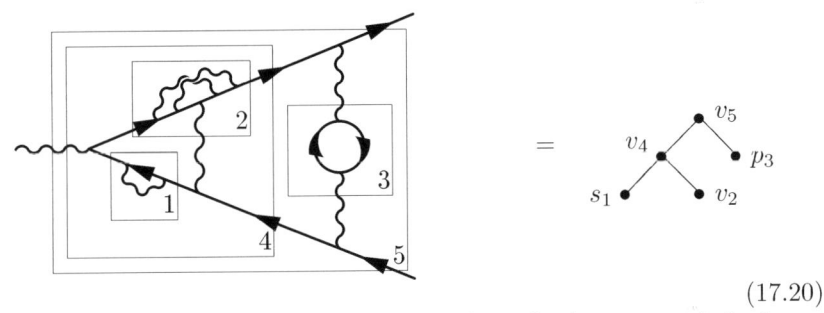

$$(17.20)$$

Straight lines stand for fermions and wavy lines for bosons, and the boxes contain divergent sectors. A criterion for superficial divergence of a region confined in a box is power counting. If a box has n_B bosonic and n_F fermionic outgoing legs, the power counting degree of divergence d is (in four dimensions) defined by $d := 4 - n_B - \frac{3}{2}n_F \geq 0$. Owing to symmetries the actual degree of divergence of one graph or a sum of graphs can be lower than d, see [114]. The construction of the rooted tree from the Feynman graphs with identified divergent sectors is clear: The outermost (superficial) divergence (5) is the root v_5. The box (5) contains the boxes (3) and (4) as immediate subdivergences, hence we connect two vertices p_3 and v_4 directly to the root v_5. The box (4) contains the subdivergences (1) and (2), so we attach the vertices s_1 and v_2 to v_4. This works as long as there are no overlapping divergences, which must be resolved before in terms of disjoint and nested ones and lead to a sum of rooted trees [128,130].

Having identified the trees to Feynman graphs, apply the same cutting operations on trees as before which give us coproduct and antipode. Here, a cut splits a Feynman graph into several subgraphs – a standard operation in renormalization. It is very remarkable that the antipode obtained in this way reproduces the combinatorics of renormalization [129]. These surprising facts have been extended to a complete renormalization of a toy model [131], which we review in the next section.

Before, let us ask an interesting question: What is the role of the operators $\delta^k_{ji,\ell_1...\ell_n}$ in quantum field theory, and what is the meaning of the individual trees for diffeomorphisms? I am not aware of an answer, but there is an interesting observation [225] concerning the relation of the *decorated* rooted trees (17.16) to Feynman graphs. The trees emerging from the Connes–Moscovici Hopf algebra are decorated by spacetime indices (three for the root) whereas in QFT the decoration is a label for divergent Feynman graphs without subdivergences. Although the operators δ are invariant under permutation of the indices after the comma, for instance $\delta^k_{ji,lm} = \delta^k_{ji,ml}$, see (17.11), this symmetry is lost on the level of individual trees. That leads us to speculate that *the sum of Feynman graphs according to the collection of rooted trees to δ's has more symmetry than the individual Feynman graphs*. This should be checked in QFT calculations. Another interpretation would be the observation from (17.16)

$$\begin{array}{c} \overset{k}{\underset{ji}{\bullet}} \\ \bullet \, l \\ \bullet \, m \end{array} + \overset{k}{\underset{\bullet l \quad \bullet m}{\overset{ji}{\bigwedge}}} - \begin{array}{c} \overset{k}{\underset{ji}{\bullet}} \\ \bullet \, m \\ \bullet \, l \end{array} - \overset{k}{\underset{\bullet m \quad \bullet l}{\overset{ji}{\bigwedge}}} = 0 \,, \qquad (17.21)$$

which could possibly be regarded as a relation between Feynman graphs similar to those derived in [132]. According to a private communication by Kreimer, (17.21) is satisfied in QFT for the leading divergences, as it can be derived from Section V.C in [133]. For non-leading singularities there will be (probably systematic) modifications.

In mathematics, Connes and Kreimer extended the investigation of the commutative Hopf subalgebra $\mathcal{H}^1$ in [48] to the level of individual trees [42]. They showed that the Hopf algebra of rooted trees $\mathcal{H}_R$ is the solution of a universal problem in Hochschild cohomology. We recall [48] that $\mathcal{H}^1$ is the dual of the enveloping algebra of the Lie algebra $\mathcal{L}^1$ of formal vector fields on R vanishing to order 2 at the origin, and that $\mathcal{H}^1$ itself is isomorphic to the Hopf algebra of coordinates on the group of diffeomorphisms of R of the form $\psi(x) = x + o(x)$. By analogy, Connes and Kreimer regard $\mathcal{H}_R$ as the Hopf algebra of coordinates on a nilpotent formal group $\mathcal{G}$ whose Lie algebra $\mathcal{L}^1$ they succeed to compute. This group was recently found to be related to the Butcher group in numerical analysis [23]. It will certainly contain precious information for quantum field theory because the antipode in $\mathcal{H}_R$ governing renormalization is the dual of the inversion operation in $\mathcal{G}$. Renormalization seems to provide a new mathematical calculus which generalizes differential calculi.

17.5 A Toy Model: Iterated Integrals

In the spirit of Kreimer [131] we are going to give the reader a feeling for renormalization by considering a toy model. The toy model is given by iterated divergent integrals, in close analogy to QFT. The only difference is that the integrals are very simple to compute.

Let us take the integral

$$\Gamma^1(t) = \int_t^\infty \frac{dp_1}{p_1^{1+\epsilon}}, \qquad (17.22)$$

which diverges logarithmically for $\epsilon \to 0$. We can regard it as the analytic expression to the Feynman graph

To a Feynman graph with subdivergence there corresponds an iterated integral:

$$\Gamma^2(t) = \int_t^\infty \frac{dp_1}{p_1^{1+\epsilon}} \int_{p_1}^\infty \frac{dp_2}{p_2^{1+\epsilon}},$$

$$\Gamma^3(t) = \int_t^\infty \frac{dp_1}{p_1^{1+\epsilon}} \int_{p_1}^\infty \frac{dp_2}{p_2^{1+\epsilon}} \int_{p_2}^\infty \frac{dp_3}{p_3^{1+\epsilon}}. \qquad (17.23)$$

Clearly, these iterated integrals form a Hopf algebra of rooted trees without side branches, and the coproduct is given by the admissible cuts of the trees. The renormalization of these integrals requires an algebra homomorphisms ϕ_a on iterated integrals, which represents a certain way of evaluation under "a set of conditions a". For our purpose we take

$$\phi_a\left(\prod_{i \in I} \Gamma^i(t)\right) := \prod_{i \in I} \Gamma^i(a), \qquad (17.24)$$

the evaluation of the integrals at $t = a$. In QFT, a should be regarded as an energy scale, and ϕ_a evaluates the Feynman graphs at this scale.

The essential idea [131] is now to consider the *convolution product* of these homomorphisms, defined via the Hopf algebra structure

$$(\phi \star \psi)(h) := m \circ (\phi \otimes \psi) \circ \Delta(h), \qquad h \in \mathcal{H}. \qquad (17.25)$$

The antipode axiom can be written in the compact form $S \star \mathrm{id} = 1\varepsilon$. It is however more interesting to consider the following modification:

$$\varepsilon_{a,b} = S_a \star \mathrm{id}_b := (\phi_a \circ S) \star \phi_b. \qquad (17.26)$$

Due to the Hopf algebra properties, the $\varepsilon_{a,b}$ satisfy a groupoid law. We give the derivation in full detail, using 1) associativity of m and coassociativity of Δ, 2) the antipode axiom, 3) homomorphism property of ϕ, 4) $\phi \circ 1\epsilon = 1\epsilon$, 5) the counit axiom:

$$\begin{aligned}
\varepsilon_{a,b} \star \varepsilon_{b,c} &= m \circ \Big(\big(m \circ (S_a \otimes \phi_b) \circ \Delta \big) \big) \otimes \big(m \circ (S_b \otimes \phi_c) \circ \Delta \big) \big) \Big) \circ \Delta \\
&= m \circ (m \otimes m) \circ \Big(S_a \otimes \phi_b \otimes S_b \otimes \phi_c \Big) \circ (\Delta \otimes \Delta) \circ \Delta \\
&= m \circ (\mathrm{id} \otimes m) \circ (m \otimes \mathrm{id} \otimes \mathrm{id}) \circ \Big(S_a \otimes \phi_b \otimes S_b \otimes \phi_c \Big) \circ \\
&\qquad \circ (\Delta \otimes \mathrm{id} \otimes \mathrm{id}) \circ (\mathrm{id} \otimes \Delta) \circ \Delta \\
&=^1 m \circ (m \otimes \mathrm{id}) \circ (m \otimes \mathrm{id} \otimes \mathrm{id}) \circ \Big(S_a \otimes \phi_b \otimes S_b \otimes \phi_c \Big) \circ \\
&\qquad \circ (\Delta \otimes \mathrm{id} \otimes \mathrm{id}) \circ (\Delta \otimes \mathrm{id}) \circ \Delta \\
&= m \circ \Big(\big(m \circ (m \otimes \mathrm{id}) \circ (S_a \otimes \phi_b \otimes S_b) \circ (\Delta \otimes \mathrm{id}) \circ \Delta \big) \otimes \phi_c \Big) \circ \Delta \\
&=^1 m \circ \Big(\big(m \circ (\mathrm{id} \otimes m) \circ (S_a \otimes \phi_b \otimes S_b) \circ (\mathrm{id} \otimes \Delta) \circ \Delta \big) \otimes \phi_c \Big) \circ \Delta \\
&= m \circ \Big(\big(m \circ \{ S_a \otimes \big(m \circ (\phi_b \otimes \phi_b) \circ (\mathrm{id} \otimes S) \circ \Delta \big) \} \circ \Delta \big) \otimes \phi_c \Big) \circ \Delta \\
&=^{2,3} m \circ \Big(\big(m \circ \{ S_a \otimes \big(\phi_b \circ 1\epsilon \big) \} \circ \Delta \big) \otimes \phi_c \Big) \circ \Delta \\
&=^4 m \circ \Big(\big(m \circ (S_a \otimes \mathrm{id}) \circ (\mathrm{id} \otimes 1\epsilon) \circ \Delta \big) \otimes \phi_c \Big) \circ \Delta \\
&= m \circ (m \otimes \mathrm{id}) \circ (S_a \otimes \mathrm{id} \otimes \phi_c) \circ (\mathrm{id} \otimes 1\epsilon \otimes \mathrm{id}) \circ (\Delta \otimes \mathrm{id}) \circ \Delta \\
&=^{1,4} m \circ (\mathrm{id} \otimes m) \circ (S_a \otimes \phi_c \otimes \phi_c) \circ (\mathrm{id} \otimes 1\epsilon \otimes \mathrm{id}) \circ (\mathrm{id} \otimes \Delta) \circ \Delta \\
&=^3 m \circ (S_a \otimes \phi_c) \circ (\mathrm{id} \otimes (m \circ (1\epsilon \otimes \mathrm{id}) \circ \Delta)) \circ \Delta \\
&=^5 m \circ (S_a \otimes \phi_c) \circ \Delta \ = \ \varepsilon_{a,c} \ .
\end{aligned}$$

We apply now the $\varepsilon_{a,b}$ operation to the divergent integrals to compute $\varepsilon_{a,b}(\Gamma^i(t)) = \Gamma^i_{a,b}$:

$$\begin{aligned}
\Gamma^1_{a,b} &= m \circ (\phi_a \otimes \phi_b) \circ (S \otimes \mathrm{id}) \circ \Delta(\bullet) \\
&= m \circ (\phi_a \otimes \phi_b) \circ \big(- \bullet \otimes 1 + 1 \otimes \bullet \big) \\
&= -\Gamma^1(a) + \Gamma^1(b) = \int_b^a \frac{dp}{p^{1+\epsilon}} \ .
\end{aligned}$$

The result $\Gamma^1_{a,b}$ is finite for $\epsilon \to 0$ and vanishes for $a = b$. We proceed with the next integral, using the definition of Δ as given by the admissible cuts and S as given by all cuts (with sign from the number of elementary cuts) of the graphs:

$$\Gamma^1_{a,b} = m \circ (\phi_a \otimes \phi_b) \circ (S \otimes \mathrm{id}) \circ \Delta\left(\begin{array}{c}\bullet\\|\\\bullet\end{array}\right)$$

$$= m \circ (\phi_a \otimes \phi_b) \circ \left(S\left(\begin{array}{c}\bullet\\|\\\bullet\end{array}\right) \otimes 1 + S(\bullet) \otimes \bullet + 1 \otimes \begin{array}{c}\bullet\\|\\\bullet\end{array}\right)$$

$$= m \circ (\phi_a \otimes \phi_b) \circ \left(-\begin{array}{c}\bullet\\|\\\bullet\end{array} \otimes 1 + \bullet\bullet \otimes 1 - \bullet \otimes \bullet + 1 \otimes \begin{array}{c}\bullet\\|\\\bullet\end{array}\right)$$

$$= -\Gamma^2(a) + \Gamma^1(a)\Gamma^1(a) - \Gamma^1(b)\Gamma^1(a) + \Gamma^2(b)$$

$$= \left(-\int_a^\infty \int_{p_1}^\infty + \int_a^\infty \int_a^\infty - \int_b^\infty \int_a^\infty + \int_b^\infty \int_{p_1}^\infty\right) \frac{dp_1}{p_1^{1+\epsilon}} \frac{dp_2}{p_2^{1+\epsilon}}$$

$$= \int_b^a \frac{dp_1}{p_1^{1+\epsilon}} \int_{p_1}^a \frac{dp_2}{p_2^{1+\epsilon}} .$$

Again, the result is finite. Note that in $\bullet \otimes \bullet$ the root which stands for the p_1 integration is the right vertex and hence is evaluated at $t = b$. The computation for Γ^3_{ab} is left as an exercise.

From the identity $\varepsilon_{a,b} \star \varepsilon_{b,c} = \varepsilon_{a,c}$ and the coproduct rule given by admissible cuts of a tree without side branches we get Chen's Lemma [32]:

$$\Gamma^i_{a,c} = \Gamma^i_{a,b} + \Gamma^i_{b,c} + \sum_{j=1}^{i-1} \Gamma^j_{a,b}\Gamma^{i-j}_{b,c} . \tag{17.27}$$

For $i = 2$ it reads

$$\int_c^a \frac{dp_1}{p_1} \int_{p_1}^a \frac{dp_2}{p_2} = \int_b^a \frac{dp_1}{p_1} \int_{p_1}^a \frac{dp_2}{p_2} + \int_c^b \frac{dp_1}{p_1} \int_{p_1}^b \frac{dp_2}{p_2} + \int_c^b \frac{dp_1}{p_1} \int_b^a \frac{dp_2}{p_2} .$$

The purpose of these considerations was the renormalization of a QFT. Let us assume a theory where all contributions to the coupling constant come from the following ladder diagrams:

$$\Gamma \;=\; \Gamma^0 \;+\; \Gamma^1 \;+\; \Gamma^2 \;+\; \Gamma^3 \;+\; \cdots$$

Formally, this series evaluates to infinity, but this infinity can be renormalized to a finite but *undetermined* value. That value has to be adapted to experiment and yields a *normalization condition*. At some energy scale a we are allowed to fix the coupling constant $\Gamma_a = \Gamma^0(a)$. But suppose we measure now the value of the coupling constant at another energy scale b. The normalization condition is fixed so that in the diagrams we have to use in all vertices the normalized coupling constant $\!\!-\!\!\!<\, = \varGamma_a$. Since the renormalization removing the infinities was scale dependent, the loop diagrams $\varGamma^i$ now give a contribution, and this contribution is precisely $\varGamma^i_{a,b}$. Hence,

$$\varGamma_b = \varGamma_a + \varGamma^1_{a,b} + \varGamma^2_{a,b} + \varGamma^3_{a,b} + \ldots \qquad (17.28)$$

Assuming the series converges, we get a finite shift of the coupling constant. In realistic quantum field theories, the agreement of this value with experiment is overwhelming. In particular, in first order we recover the familiar logarithmic energy dependence of the coupling constant. We also learn from (17.28) that one can completely avoid talking about infinities.

As it is clear from our model, the running coupling constants resulting from renormalization are governed by the Hopf algebra structure together with the convolution product. The Hopf algebra structure not only produces the combinatorics of the forest formula, it also allows to compare different renormalization schemes, which arise from each other by a finite renormalization. The theory is consistent without a preferred scale or preferred renormalization scheme. They are always related by the convolution identity $\varepsilon_{a,c} = \varepsilon_{ab} \star \varepsilon_{bc}$, where a, b, c stand for parameterizations of different renormalization schemes. Applications of these ideas to QFT calculations are starting [133].

Acknowledgements

I am grateful to the organizers of the Hesselberg'99 conference, Florian Scheck, Harald Upmeier and Wend Werner, for the invitation and the opportunity to present these ideas. It is a pleasure to thank my colleagues Bruno Iochum, Thomas Krajewski, Serge Lazzarini, Thomas Schücker and Daniel Testard for collaboration and numerous discussions. Finally, I would like to thank Alain Connes and Dirk Kreimer for important advice at various stages of my study of Hopf algebras.

18 NC Geometry of Strings and Duality Symmetry

Fedele Lizzi

Università di Napoli

The aim of this talk is to introduce the spectral triple of strings, together with some of the main string symmetries, like duality in the target space etc. It is a report of work done in collaboration with G. Landi and R.J Szabo.

Classical geometry has been an excellent tool for the description of the physical world, the whole of classical mechanics and general relativity are based on it. Nevertheless there is a growing belief that the onset of quantum mechanics in the gravitational theory will require a change in the very structure of space time, which may not be describable with the tools of *classical geometry*. A suitable candidate for this new mathematics goes under the name of *Noncommutative Geometry*, [38]. In noncommutative geometry *classical* notions such as point, lines etc. are substituted by algebraic entities, using the duality between Hausdorff topological spaces and commutative C^*-algebras, seen as the algebra of complex valued functions on the space, with the obvious generalization given by noncommutative algebras.

18.1 String Theory and T-duality

Recently a strong candidate for a theory of quantum gravity has emerged in *String Theory* [97]. There is no space for the introduction of string theory, so we will just point out that string theory is a field theory defined on a two dimensional surface, the world sheets of the theory, and space time appears as a *field* of this theory. Interacting strings are described by higher genus surfaces, and in the nonperturbative regime we have the appearance of nonstandard phenomena, such as ∞-genus surfaces, duality, branes whose coordinates are (noncommuting) matrices...

Crucial to string theory is the conformal invariance of the world-sheet, the two dimensional generalization of the world line, swept out by the string. The richness of string theory stems from the fact that in two dimensions the conformal group is infinite dimensional, giving rise to a wealth of interesting phenomena. Spacetime is a derived concept, and it appears as the fields of a field theory defined on the world surface. In this talk I will assume spacetime to be toroidally compact, d dimensional and the bosonic sector of the string theory. Spacetime is thus $\mathbb{R}^d$ divided by an abelian infinite group (a lattice) Γ generated by d generators e_i. On the generators of Γ we define an inner product which provides a metric (of Euclidean signature) on the torus $\mathcal{T}_d$:

$$\langle e_i, e_j \rangle \equiv g_{ij} \ . \tag{18.1}$$

Classically the string is described by a two dimensional nonlinear σ model, whose fundamental objects are the Fubini–Veneziano fields, which, for the case of a closed string are:

$$X^i(\tau,\sigma) = x^i + g^{ij}p_j\tau + g^{ij}w_j\sigma + \sum_{k\neq 0} \frac{1}{ik} \alpha_k^{(\pm)i} e^{ik(\tau\pm\sigma)}, \qquad (18.2)$$

where x represents the centre of mass of the string, p its momentum and w is the winding number, the number of times the string wraps around the direction defined by the e_i.

Notice that, since the space is compact, the momentum is quantized, and in fact it must be $p \in \tilde{\Gamma}$ with $\tilde{\Gamma}$ the dual lattice spanned by the basis e^i with

$$\langle e^i, e_j \rangle = \delta^i_j . \qquad (18.3)$$

(we implicitly complexify Γ and extend the product). The inner products of the e^i's define a metric which is the inverse of g_{ij}, that is:

$$\langle e^i, e^j \rangle \equiv g^{ij} . \qquad (18.4)$$

Notice that, if all of the e_i are quantities of order R (we take Planck's length to be unity unless otherwise stated), with $\det g$ of order R^d, then the 'size' of the dual lattice is a quantity of order $1/R$. In this sense, if to a given lattice corresponds a large universe, to its dual there will correspond a small one, the dual torus $\tilde{\mathcal{T}}_d^*$.

The winding number, on the other side must belong to the original lattice $w \in \Gamma$. For $R \to \infty$ the momentum spectrum becomes continuous, while the values of the windings different from 0 drop out. On the other side $R \to 0$ inverts the situation, in fact the role of p and w in (18.2) is symmetric. In the following I will mainly deal with the zero modes of the string, mostly ignoring the oscillator modes. These are internal excitations, not sensible to the target space in which the strings live, and will therefore not play a central role in considerations dealing with the structure of spacetime at least at low energy, as the oscillators describe excitations at the Planck mass. Although the excitations are the hallmarks of the "stringy" character of the theory we will see that already the zero mode describe behaviour different from particles.

Consider the following action for a nonlinear σ model:

$$S = \frac{1}{4\pi} \int d\sigma d\tau \sqrt{\eta}\eta^{\alpha\beta}\partial_\alpha X^i g_{ij}\partial_\beta x^j + \varepsilon^{\alpha\beta}b_{ij}\partial_\alpha X^i \partial_\beta X^j, \qquad (18.5)$$

where η is the world sheet two dimensional metric, ε is the antisymmetric tensor with $\varepsilon_{12} = 1$, G is the metric defined in (18.1), and b is an antisymmetric tensor which represents the 'torsion' of the string.

---

* Strictly speaking however this conclusion is only valid only in the absence of torsion (introduced below) in the action [70].

Perform a chiral decomposition of the X's defining:

$$X_\pm^i(\tau \pm \sigma) = x_\pm^i + g^{ij} p_j^\pm(\tau \pm \sigma) + \sum_{k \neq 0} \frac{1}{ik} \alpha_k^{(\pm)i} e^{ik(\tau \pm \sigma)} . \qquad (18.6)$$

The zero modes $x_\pm^i$ (the centre of mass coordinates of the string) and the (centre of mass) momenta $p_i^\pm = 2\pi p_i \pm (g \mp b)_{ij} w^j$ are canonically conjugate variables,

$$[x_\pm^i, p_i^\pm] = -i\delta_i^j \qquad (18.7)$$

with all other commutators vanishing. The left-right momenta are

$$p_i^\pm = \frac{1}{\sqrt{2}} (p_i \pm \langle e_i, w \rangle) \qquad (18.8)$$

The $p^\pm$'s belong to the lattice:

$$\Lambda = \tilde{\Gamma} \oplus \Gamma \qquad (18.9)$$

The fields $X = X_+ + X_-$ are defined in a natural way, but equally natural is the definition of the fields $\tilde{X} \equiv X_+ - X_-$, whose zero mode we will indicate as $\tilde{x}$.

Exchange of a lattice with its dual is a symmetry called T-duality[126,92,163]. It corresponds to an exchange of the momentum quantum number with the winding, and of the zero mode corresponding to x, the position of the centre of mass of the string with its dual $\tilde{x}$. This is a symmetry of the Hamiltonian:

$$H = \frac{1}{2} \left((2\pi)^2 p_i g^{ij} p_j + w^i (g - bg^{-1}b)_{ij} w^j + 4\pi w^i b_{ik} g^{kj} p_j \right)$$

$$+ \sum_{k>0} g_{ij} \alpha_{-k}^{(+)i} \alpha_k^{(+)j} + \sum_{k>0} g_{ij} \alpha_{-k}^{(-)i} \alpha_k^{(-)j} - \frac{d}{12} \qquad (18.10)$$

$$= \frac{1}{2}(p_+^2 + p_-^2) + \text{Oscillators} - \frac{d}{12} \qquad (18.11)$$

the term $-\frac{d}{12}$ due to normal ordering. Since momenta and windings belong to a lattice, the spectrum is discrete.

The physics of T-dual spaces is the same, they are indistinguishable if one exchanges windings with momenta. This can be seen heuristically [21] if one considers the fact that in ordinary quantum mechanics position is just a derived concept, it is the Fourier transform of momentum spaces. In a string theory it is possible however to consider winding (and its eigenstates) rather than momentum. If the compactification radius is of the order one, the two choices are equivalent, but for a very large radius the eigenvalues of momentum are nearly continuous, while the ones of winding are far apart, the first one above zero being at a very large energy. It is therefore difficult to make "localized wave packets" with the Fourier transform of winding. Conversely, with a small radius of compactification, it is the winding which gives the possibility to create localized wave packets.

The case $b=0$ (called the torsionless case) is simpler and in it T-duality corresponds to an exchange of g with its inverse g^{-1}, and the change of size of the target space in which the radius $R \to 1/R$. In the presence of torsion the exchange is $g^{-1} \leftrightarrow g - bg^{-1}b$ and $bg^{-1} \leftrightarrow -g^{-1}b$, and it depends crucially on the values of the b_{ij}. In the toroidal case it is possible to exchange only some of the generators of the lattice with their duals, giving rise to a group of factorized T-dualities.

The theory has two continuous symmetries common basically to all string theories:

- Target space reparametrization:

$$X_\pm(z_\pm) \to X_\pm(z_\pm) + \delta X_\pm(z_\pm) \tag{18.12}$$

- World sheet conformal invariance, represented by two Virasoro algebras:

$$\begin{aligned}[L_k^\pm, L_m^\pm] &= (k-m)L_{k+m}^\pm + \frac{c}{12}\left(k^3 - k\right)\delta_{k+m,0} \\ [L_k^-, L_m^+] &= 0 ,\end{aligned} \tag{18.13}$$

both these symmetries play a crucial role in the theory.

The full group of symmetry is even larger: it is in fact $O(d,d,\mathbb{Z})$[164,165,92], generated by three kinds of transformations:

- We can change the basis of the lattice without changing the physics, this is done by a matrix in
- The factorized dualities we have already discussed: $G(d, \mathbb{Z})$,
- We can effectuate the factorized dualities we mentioned earlier.
- We can transform $b_{ij} \to b_{ij} + c_{ij}$, this is a symmetry as long as c is antisymmetric with integer entries.

A further $\mathbb{Z}_2$ symmetry is obtained exchanging σ and τ on the world sheet, this last symmetry does not affect the target space.

In the following I will try to construct the *Noncommutative Geometry of String Theory* following ideas originated by Fröhlich and Gawędzki [87], and describe some later developments in [149,?,151,137,204,148,152].

18.2 Interacting Strings and Spectral Triples

Given this scenario I will now proceed to construct a *noncommutative geometry* of interacting strings. We will therefore construct, in the spirit of Connes, a spectral triple, the Fröhlich-Gawędzki Spectral Triple [87]

The first ingredient is the Hilbert space of string states. Upon first quantization the oscillator modes become creation and annihilation operators:

$$\left[\alpha_k^{(\pm)i}, \alpha_m^{(\pm)j}\right] = kg^{ij}\delta_{k+m,0} \tag{18.14}$$

while the zero modes have the usual commutation relations (18.7). The Hilbert space of (excited) string states therefore is:

$$\mathcal{H} = L_2(sp(T^d))^\Gamma \otimes \mathcal{F}^+ \otimes \mathcal{F}^- \qquad (18.15)$$

where $L_2(sp(T^d))^\Gamma$ (spinors on T^d) is a set of spinors for each winding sector, labelled by the lattice Γ. These are the so called 'tachyon states', although, depending on the actual string theory at hand, they may not be tachyons (and hopefully they are not, as in superstring theory). The spaces $\mathcal{F}^\pm$ are the Fock spaces of higher excitations (graviton, dilaton etc.) acted upon by the oscillator creation and annihilation modes. They represent the internal excitations of the strings and have an indirect effect on spacetime, which is described by the zero modes.

The description of interacting strings is done via the insertion on the world–sheet of *vertex operators*. The fundamental operator is the "tachyon vertex operator"

$$V_{q^\pm}(z_\pm) = \; :e^{-iq_i^\pm X_\pm^i(\tau\pm\sigma)}: \qquad (18.16)$$

where $:\cdots:$ represents normal ordering obtained by putting creators to the left of annihilators:

$$:\alpha_k^{(\pm)i}\alpha_m^{(\pm)j}: = \alpha_k^{(\pm)i}\alpha_m^{(\pm)j} \quad \text{for} \;\; k < m \qquad (18.17)$$

$$= \alpha_m^{(\pm)j}\alpha_k^{(\pm)i} \quad \text{for} \;\; k > m \qquad (18.18)$$

and $x_\pm^i$ to the left of $p_i^\pm$. The tachyon vertex operator represents the insertion on the world sheet of a ground state (tachyon) of a given momentum. Higher states (the dilaton, graviton etc.) are obtained acting with the appropriate combination of creation operators.

One of the aspects of vertex operator algebras which is important in this context is the *operator–state correspondence*. We can put the generic vertex operator:

$$V(z_+, z_-)_\psi =: i\, V_{q+q^-}(z_+, z_-) \prod_j \frac{r_i^{(j)+}}{(n_j-1)!} \partial_{z_+}^{n_j} X_+^i \prod_k \frac{r_j^{(k)-}}{(m_k-1)!} \partial_{z_-}^{m_k} X_-^j : \qquad (18.19)$$

in correspondence with the state:

$$|\psi\rangle = |q^+; q^-\rangle \otimes \prod_j r_i^{(j)+} \alpha_{-n_j}^{(+)i}|0\rangle_+ \otimes \prod_k r_j^{(k)-}\alpha_{-m_k}^{(-)j}|0\rangle_- \qquad (18.20)$$

of $\mathcal{H}$, where $(q^+, q^-), (r^+, r^-) \in \Gamma \oplus \Gamma^*$.

We thus have the second element of the spectral triple, an algebra of vertex operators. A warning however: a vertex operator algebra (in the common use of the term) is *not a C^*-algebra*. In general vertex operators are not even bounded operators! The problem stems from the fact that vertex operators are not defined at coinciding points giving rise to nontrivial *operator product* expansions [97]. One can do two things to regularize the theory: *smear* the vertex operators [87,88]:

$$V(\psi, f) = \int dz V_\psi(z) f(z) \tag{18.21}$$

but this not always cures the problem, as discussed for example in [49]. An alternative is to consider *truncated vertex operators*:

$$V_{q\pm}^N(z) = \mathcal{N}_N \prod_{n=0}^N W_n \tag{18.22}$$

where W_0 contains the zero modes x and p, while the W_n's ($n \neq 0$) involve only the nth oscillator modes $\alpha_n^{(\pm)}$ and $\alpha_{-n}^{(\pm)}$. This is equivalent to an ultraviolet cutoff on the world sheet, a standard practice in string theory to avoid the infinities arising from the product of operators at coincident points. At the end one considers $N \to \infty$.

It is however fair to say that, at present, the rigorous definition of a C^*-algebra of operators representing interacting strings is (at least to my knowledge) still an open problem. We have the tachyon operators and the higher spin states (18.19), and one should regularize them, and create an algebra with the appropriate completion. It is in a sense like attempting to construct $C(\mathbb{R})$ from the knowledge of plane waves e^{ipx}. The general idea is present but many (crucial) details have to be filled. This is an area in which the collaboration of mathematicians would be of paramount importance. In the following we will indicate with the generic term vertex operator algebra a proper completion of the regularized operators.

It is easy to recognize the two fundamental continuous symmetries of the theory in the vertex operator algebra. As we said the tachyon operators are in a sense a "Fourier", or plane wave, basis on the space of conformal field configurations. The tachyon states are highest weight states of the level a pair of $u(1)_+^d \oplus u(1)_-^d$ current algebra (18.14), so that the entire Hilbert space can be built up from the actions of the α_k's for $k < 0$ on these states. This current algebra represents the target space reparametrization symmetry of the string theory. On the other side, the two Virasoro algebras which represent the world sheet conformal invariance have irreducible representations whose highest weights grade the Hilbert space $\mathcal{H}$. The Virasoro operators in the present case are $L_k^\pm = \frac{1}{2} \sum_{m \in \mathbb{Z}} g_{ij} : \alpha_m^{(\pm)i} \alpha_{k-m}^{(\pm)j} :$, with $\alpha_0^{(\pm)i} \equiv g^{ij} p_j^\pm$. They generate a representation of the Virasoro algebra (18.13) of central charge d. The grading is defined on the subspaces $\mathcal{H}_{\Delta_q} \subset \mathcal{H}$ of states (18.20) which are highest weight vectors,

$$L_0^\pm |\psi\rangle = \Delta_q^\pm |\psi\rangle \quad , \quad L_k^\pm |\psi\rangle = 0 \ \forall k > 0 , \tag{18.23}$$

where $\Delta_q^+ = \frac{1}{2} g^{ij} q_i^+ q_j^+ + \sum_j n_j$ and $\Delta_q^- = \frac{1}{2} g^{ij} q_i^- q_j^- + \sum_k m_k$. The corresponding operator-valued distributions (18.19) are called *primary fields*.

Vertex operator algebras are very important in several branches of mathematics: modular functions, Monster group, Lie algebras and there are many good reviews, including [85,90,116]. In this talk however the relevant aspect is that they are operators on $\mathcal{H}$ describing (via the operator-state correspondence) the interaction of strings.

The last element to complete the spectral triple is the Dirac operator. We have not one but *two* natural Dirac operators:

$$D^{\pm} = \gamma_i^{\pm} \alpha_{\pm}^i \quad \alpha_{\pm}^i = -i\partial_{\pm} X_{\pm}^i \tag{18.24}$$

These two operators generate target space reparametrization of $X_{\pm}$. Moreover it can be seen that they are square roots of the Laplace–Beltrami operator. They are also naturally related to the other symmetry of string theory, in fact worldsheet conformal symmetry has the conserved stress energy:

$$T^{\pm}(z_{\pm}) = -\frac{1}{2} : D^{\pm}(z_{\pm})^2 : \ = \sum_k L_k^{\pm} z_{\pm}^{-k-2} \tag{18.25}$$

In analogy with the X and $\tilde{X}$ we can define:

$$D = D^+ + D^- \ ; \ \tilde{D} = D^+ - D^- \tag{18.26}$$

The spectral triple $\mathcal{T}$ of string geometry therefore is:

$$\mathcal{A} \quad \mathcal{H} \quad D \tag{18.27}$$

One can ask now what happened to ordinary spacetime? Spacetime emerges as a "subtriple" $\mathcal{T}_0$, that is, a spectral triple with a subalgebra, a subspace of Hilbert space, and an operator which is the reduction of the Dirac operator on the subspace:

$$\mathcal{A}_0 \quad \mathcal{H}_0 \quad \partial\!\!\!/ \tag{18.28}$$

In order to construct the low energy subtriple we first have to project out all of the oscillator modes to obtain $\mathcal{A}_0$ and $\mathcal{H}_0$. The rationale behind this is that, since the excited oscillators start at the Planck mass, and this is much larger than 'ordinary' space time energies, we have to isolate the modes of the string which will be accessible at low energies. This is still not sufficient however, as, in the case of large uncompactified directions, the modes associated to the winding are also highly energetic. We therefore choose:

$$\begin{array}{l} C(\mathbb{R}^d): \quad f \in \mathcal{A}_0 : [\tilde{D}, f] = 0 \ \text{commutant of } \tilde{D} \\ L_2(T^d, sp): \psi \in \mathcal{H}_0: \ \tilde{D}\psi = 0 \quad \text{kernel of } \tilde{D} \end{array} \tag{18.29}$$

It is easy to relate $\mathcal{A}_0$ with the algebra of complex valued function on spacetime, it is sufficient to notice that it is constructed from the (commutative) 'vertex operator' e^{ipx_0}. Here we encounter the already mentioned problems of the appropriate completion in order to obtain a well defined C^*-algebra. The essence of T-duality lies in the relatively simple observation that instead of $\tilde{D}$ we could have chosen D as well. In this case we would have obtained the triple pertaining to the torus whose coordinates are the $\tilde{x}$, that is the T-dual torus $\tilde{\mathcal{T}}_0$, with all the radii of compactifications inverted:

$$D^{\pm} \to D^{\mp} \quad D \leftrightarrow \tilde{D} \tag{18.30}$$

That the full theory is invariant under this change is ensured by the observation that this transformation is a *gauge* transformation. In fact there are (many) $u \in \mathcal{A}$ unitary such that:

$$uDu^{-1} = \tilde{D} \tag{18.31}$$

For example:

$$u = e^{i\mathcal{G}_{\pm}} \tag{18.32}$$

with

$$\mathcal{G}_{\chi} = \int \frac{dz_+ \, dz_-}{4\pi z_+ z_-} \left(\chi^a_{+,i}[X] \, J_a^{+(i)}(z_+) + \chi^a_{-,i}[X] \, J_a^{-(i)}(z_-) \right) f_S(z_+, z_-) \tag{18.33}$$

where $a = \pm$, $J_a^{\pm(i)} =: e^{aik_j^{(i)} X_{\pm}^j} :$, $\chi_{\pm}$ are sections of the spin bundle and f_S is a smearing function.

This T-duality is however a gauge transformation only in the full FG-triple. When this is projected to the subtriple so to give a *commutative* spacetime, in general the process will give rise to very different spacetimes. We can in fact consider T-duality to be the commutativity of the following diagram:

$$\begin{array}{ccc} \mathcal{T}_D & \xrightarrow{u} & \mathcal{T}_{\tilde{D}} \cong \mathcal{T}_D \\ \mathcal{P}_0 \downarrow & & \downarrow \tilde{\mathcal{P}}_0 \\ \mathcal{T}_0 & \xrightarrow{T_0} & \tilde{\mathcal{T}}_0 \end{array} \tag{18.34}$$

The operation T_0 is what we call T-duality, and from the previous discussion it is clear that it is just the low energy projection of a gauge transformation. All of the remaining $O(d,d,\mathbb{Z})$ dualities can be obtained in the same way, as gauge transformations [150].

There are many more inner automorphisms, *gauge transformations*, which project down to non trivial transformations. Defining the currents:

$$J^i_{\pm}(\tau \pm \sigma) = \partial_{\pm} X^i_{\pm}(\tau \pm \sigma) = \sum_{k=-\infty}^{\infty} \alpha_k^{(\pm)i} \, e^{ik(\tau \pm \sigma)} \tag{18.35}$$

a general spacetime coordinate transformation $X \to \xi(X)$, with $\xi(X)$ a local section of $\mathrm{spin}(T^d)$, is generated by $\mathcal{G}_{\chi} = \mathcal{G}_{\xi}$ with

$$\mathcal{G}_{\xi} = \int \frac{dz_+ \, dz_-}{4\pi z_+ z_-} \xi_i(X) \left(J^i_+(z_+) + J^i_-(z_-) \right) f_S(z_+, z_-) \tag{18.36}$$

This means that also the diffeomorphisms of the (low energy) target space are gauge transformations of the full spectral triple. The inner automorphisms project down to outer automorphisms of spacetime. This is, in my opinion, one of the best justifications of the often heard statement that "general relativity is a gauge theory". We can see also a glimpse of a huge group of symmetries, which when projected down connects different low energy theories.

18.3 Compactification and Noncommutative Torus

If we try to uncover the structure of spacetime at higher energies we would have to consider momentum and winding modes on a par. This will be relevant when the radius of compactification is comparable with Planck's length, as in this case it is not possible to ignore the former over the latter. We will however limit ourselves for the time being to the tachyonic case. Nevertheless the oscillators (at least the lower ones) do play an important role. Consider therefore tachyon vertex operators, for which we only excite the first N oscillators (for the basis $e_\pm^i$ of $\Gamma \oplus \Gamma^*$). The commutation relation among the elementary operators are:

$$V^N_{e^i_\pm}(z_{\pm i})\, V^N_{e^j_\mp}(z_{\mp j}) = V^N_{e^j_\mp}(z_{\mp j})\, V^N_{e^i_\pm}(z_{\pm i})$$

$$V^N_{e^i_\pm}(z_{\pm i})\, V^N_{e^i_\pm}(z_{\pm i})^\dagger = V^N_{e^i_\pm}(z_{\pm i})^\dagger\, V^N_{e^i_\pm}(z_{\pm i}) = \mathbb{I}$$

$$V^N_{e^i_\pm}(z_{\pm i})\, V^N_{e^j_\pm}(z_{\pm j}) = e^{2\pi i \omega^{ij}_{N\pm}}\, V^N_{e^j_\pm}(z_{\pm j})\, V^N_{e^i_\pm}(z_{\pm i}),\quad i\neq j \quad (18.37)$$

where the $z_\pm{}^i$ are distinct points, and

$$\omega^{ij}_{N\pm} = \pm g^{ij}\left(\log\left(\frac{z_i^\pm}{z_j^\pm}\right) - \sum_{n=1}^N \frac{1}{n}\left(\left(\frac{z_i^\pm}{z_j^\pm}\right)^n - \left(\frac{z_j^\pm}{z_i^\pm}\right)^n\right)\right). \quad (18.38)$$

One can easily recognize in (18.37) a *noncommutative torus* structure [181]. If we enclose more and more oscillators:

$$\lim_{N\to\infty} \omega^{ij}_{N\pm} = \omega^{ij}_\pm = \pm g^{ij}\,\mathrm{sgn}(\arg z_i^\pm - \arg z_j^\pm)\ i\neq j \quad (18.39)$$

The symmetries of the theory are still present, even in this truncated version, in fact theories related by $O(d,d,\mathbb{Z})$ transformations give rise to Morita equivalent tori [181,180,182,183,137]. The commutative case is recovered in the uncompactified/large compactification radius because when $R\to\infty$, the off diagonal elements of $g^{ij}\to 0$ and we recover the commutative torus.

"Turning the NCG crank" it is also possible to write a *low energy dual symmetric action* [137]:

$$\mathcal{L} = (F + {}^\star F)_{ij}\,(F + {}^\star F)^{ij}$$
$$-i\overline{\psi}\gamma^i\left(\partial_i + i\overset{\leftrightarrow}{A_i}\right)\psi - i\overline{\tilde\psi}\tilde\gamma_i\left(\tilde\partial^i + i\overset{\leftrightarrow}{\tilde A^i}\right)\tilde\psi \quad (18.40)$$

where the dual field strength is defined:

$$F_{ij} = \partial_i A_j - \partial_j A_i + i\left[A_i, A_j\right]$$
$$-g_{ik}g_{jl}\left(\tilde{\partial}^k \tilde{A}^l - \tilde{\partial}^l \tilde{A}^k + i\left[\tilde{A}^k, \tilde{A}^l\right]\right) \quad (18.41)$$

All of the $O(d,d,\mathbb{Z})$ transformations, being unitary transformations, do not change the *spectrum* of D. Let us analyse in some details the transformation which changes the components of the antisymmetric second rank tensor b_{ij} by the addition of an arbitrary, integer valued, constant matrix. Although this transformation does not change the lattice Γ, it will change the momenta conjugated to the zero modes of X and $\tilde{X}$. In particular, in the spectrum (18.11), the relative contribution of the momenta (represented by the first term) with respect to the windings and the mixed term will change. Choosing the components of the antisymmetric tensor b arbitrarily large, we can make the contribution of the second and third term arbitrarily large. We have therefore concentrated the lowest eigenvalues of the Hamiltonian in the momentum part. The low energy spectrum is made only of the momentum eigenvalues. The lattice is still the same, but the strings are extremely twisted, and we have transferred the lowest eigenvalues of the energy from winding to momentum. Roughly speaking, low energy strings, which in the original (small radius) lattice had a combination of momentum and winding, will now be twisted in such a way that it will appear to have just momentum, it is like the lattice "repeats itself over and over".

Again, as in the case of the $R \leftrightarrow 1/R$ symmetry, we have to ask ourselves 'what is position'? 'How is it measured'? And using the same heuristic arguments of [21], we can think of making wave packets using superpositions of the eigenvalues of the momentum. In the case of large torsion the eigenvalues of momentum are continuous for all practical purposes, therefore the superposition will have the character of an uncompactified space, rather than a string moving on a lattice. And this will be the situation until energies in which the new eigenvalues (coming from windings or the oscillatory modes) start to play a role.

18.4 Noncommutative Configuration Space and Spectral Geometry

Let us briefly discuss the role of the classical configuration space in ordinary quantum mechanics in the language and formalism of noncommutative geometry. We will be very brief and refer to [148] for further details and references. Consider a purely quantum observer, that is a set of operators which form an algebra. For example bounded operators constructed from p and x. The information on the topology of M, the manifold on which the motion is living, can be recovered in the programme of noncommutative geometry by considering the algebra of *position operators*, that is, the algebra of continuous**, complex valued, functions on M, seen as operators on $L^2(M)$, with a norm given by the maximum of the modulus of the function. This is a simple application of the Gel'fand–Naimark theorem.

We will consider the configuration space of a quantum mechanical space therefore not as a set of points, but rather as an abelian C^*-algebra. The Hilbert space could also be easily constructed a posteriori by giving a sesquilinear form (a scalar product) on the algebra, and completing it under the norm given by this product. Other choices for the Hilbert space are possible, a relevant one for instance is the space of spinors. A quantum observer will have at his disposal, among the bounded operators on the Hilbert space, an abelian subalgebra $\mathcal{A}_0$ which he will identify with the continuous function on his space.

The "size" of this configuration space is given by the Dirac operator via Connes' distance formula [38,81,136,156,216]:

$$d(x,y) = \sup_{\|[D,a]\| \leq 1} |a(x) - a(y)| \quad a \in \mathcal{A}_0 \ . \tag{18.42}$$

Noncommutative geometry equips our quantum observer with a series of tools: algebras of operators, traces etc. In the commutative case these tools reconstruct the usual differential geometry, but they can be used in the noncommutative case as well. If we are in a commutative case, the quantum observer has therefore at his disposal an algebra of observables, in this algebra he recognizes an abelian subalgebra, that he calls the space on which he lives, and with formula (18.42) he calculates distances, metric etc.

In string theory, spacetime, as described for example by (18.37), the quantum observer finds himself on a *noncommutative space*. That is, among his set of quantum observables he does not identify an abelian algebra giving him the configuration space, he can however define some sort of "noncommutative" space, to which corresponds a noncommutative algebra.

To specify the meaning of "low energy" we will resort to the spectral action principle [30,31], and argue that low energy means a theory in which only the lowest eigenvalues of D are accessible to experiment. In fact the spectrum of D (and its fluctuations) hold the information on the physics, in the form of metric properties of the noncommutative space.

Let us briefly describe the main aspects of the spectral action principle. The main idea is, as we said, that the covariant Dirac operator holds all of the information on the metric properties of the space, and that the action principle is based on the variations of the covariant form of the operator, which of course depends on the algebra, remembering that 1-forms are obtained by combining the elements of the algebra with their commutators with D. The action must be read in a Wilson renormalization scheme sense, and

---

** In the following we will consider M compact, therefore continuous functions are bounded as well.

it depends on an ultraviolet cutoff m_0:

$$S_{m_0} = \operatorname{Tr} \chi \left(\frac{D_A^2}{m_0^2} \right) \qquad (18.43)$$

where D_A is the covariant Dirac operator and $\chi(x)$ is a function which is 1 for $x \leq 1$ and then goes rapidly to zero (some smoothened characteristic function). The action (18.43) effectively counts the eigenvalues of the covariant Dirac operator up to the cutoff. The trace in the action can be calculated using known heat kernel techniques [30], and the resulting theory contains a cosmological constant, the Einstein–Hilbert and Yang–Mills actions, plus some terms quadratic in the Riemann tensor.

It is suggestive [138,139] to consider the sequences of real numbers which form the spectrum of D_A as dynamical variables. In this theory the spectral action is then nothing but the general relativity action of this space.

Let us stress the importance of the *spectral principle*, which considers as starting point the spectrum of an operator, and its variations as the backgrounds fields (the 1–form A in this case) change. One can ask, in fact, what is the role of the algebra in the spectral action, as the latter depends just on the trace of the Dirac operator. Of course the role of the algebra lies in the fact that in (18.43) appears the *covariant* Dirac operator. And the form $A = \sum a_i [D, b_i]$ depends on the algebra chosen. Let us now apply these considerations to the Fröhlich-Gawędzki spectral triple.

The spectrum of D and $\tilde{D}$, or of any operator obtained from them with an $O(d,d,\mathbb{Z})$ unitary transformation, are the same. Let us call D for convenience the one for which the lowest eigenvalues are the one relative to momentum. Here by lowest we mean the ones which are lower than the energy of the oscillatory modes (of the order of the Planck mass m_p). If the cutoff m_0 is lower than m_p, the cutoff function χ effects the projection of the operator on the Hilbert space $\mathcal{H}_0$. Elements of the algebra which commute with D (such as the elements of $\tilde{\mathcal{A}}$) will not contribute to the variations of the action, and will therefore be unobservable. This algebra can be constructed as the commutant of the T-dual operator $\tilde{D}$. This means that the winding mode degrees of freedom are unobservable. Since the Dirac operator has a near continuous spectrum, the tachyonic, low energy, algebra is spanned by operators of the kind

$$V_p = e^{ipx} , \qquad (18.44)$$

considered as the Fourier modes describing an uncompactified space.

In fact, a quantum observer with a spectral action will be able to measure (in the form of fields, potentials etc.) only the elements of the algebra which give low energy perturbations of the lowest eigenvalues of D, always with the assumption of the cutoff $m_0 < m_p$ so that oscillatory modes do not play a role. This is the abelian algebra of functions on some space time. If, as we have seen, there are many low eigenvalues, the observer will experience an effectively decompactified space time. The algebra which he will measure will be composed of the operators which will create low energy perturbation to D. At this point we have to make the sole assumption that D has a spectrum with several small eigenvalues. In this way the quantum observer will experience a (nearly) continuous spectrum of the momentum, the sign of an uncompactified space.

The strings could still be seen as compactified on a "small" lattice, but the presence of a very large torsion term b has drastically changed the operator content of the theory, and this has rendered space effectively uncompactified.

18.5 Conclusions

String Theory and Noncommutative Geometry are good examples of the fruitfulness of the interaction between physics and mathematics. But while the mathematics used in String theory has been essentially "classical" differential geometry, noncommutative geometry requires the indroduction of new tools, essentially the ones introduced by quantum mechanics. Here I tried to give an impressionistic way on how the mathematics well suited to describe strings in the high energy regime (which is proper to them) should be some sort of noncommutative geometry. While from the physical point of view some (initial) results are already to be seen: *duality, gauge transformations ...*, from the mathematical point of view the structures to use are still in need of proper definition.

A proper mathematical sharpening of the tools is necessary not so much for abstract mathematical rigour, but to help uncover the beauty which lies behind such a rich structure.

Acknowledgements

I would like to thank G. Landi and R.J. Szabo for a very fruitful collaboration, F. Scheck, H. Upmeier and W. Werner for organizing the Arbeitstagung and inviting me to participate in it, and principally *all* of the participants for making the workshop a most interesting and enjoyable one! Some part of these notes have appeared in the proceedings Euroconference *Noncommutative Geometry and Hopf Algebras, Torino 1999*, Int. J. Mod. Phys. B14 (2000) 2287-2509.

References

1. Th. Ackermann: *A note on the Wodzicki residue.* J. Geom. Phys. 20 **4** (1996) 404–406.
2. E. Alvarez, J. M. Gracia-Bondía, C. P. Martín: *Anomaly cancellation and the gauge group of the Standard model in non-commutative geometry.* Phys. Lett. B **364** (1995), 33
3. E. Alvarez, J. M. Gracia-Bondía, C. P. Martín: *Parameter restrictions in non-commutative geometry model do not survive standard quantum corrections.* Phys. Lett. B **306** (1993), 55
4. B. Ammann: *The Dirac operator on collapsing S^1-bundles.* Sém. Th. Spec. Géom. Inst. Fourier Grenoble 16 (1998) 33–42
5. B. Ammann, C. Bär: *The Dirac operator on nilmanifolds and collapsing circle bundles.* Ann. Global Anal. Geom. **16** (1998) 181–200
6. A. Ashtekar, L. Bombelli, O. Reula: *The covariant phase space of asymptotically flat gravitational fields.* In: *Mechanics, Analysis and Geometry: 200 Years after Lagrange*, edited by M. Francaviglia, Elsevier, 1991
7. M. F. Atiyah: *K-theory and reality.* Quart. J. Math. Oxford **17** (1966), 367–386
8. C. Bär: *The Dirac operator on space forms of positive curvature.* J. Math. Soc. Japan **48** (1998), 69–83
9. G. Barenboim, F. Scheck, Phys. Lett. B **440** (1998), 332
10. H. Baum: *Spin-Strukturen und Dirac-Operatoren über Pseudoriemannschen Mannigfaltigkeiten.* Teubner, 1981
11. J. Bellissard: *Ordinary quantum Hall effect and non-commutative cohomology.* Proc. Bad Schandau 1986, Teubner, 1988
12. N. Berline, E. Getzler, M. Vergne: *Heat kernels and Dirac operators*, Springer, 1991
13. P. G. Bergmann: *Observables in general relativity.* Rev. Mod. Phys. **33** (1961), 510–514
14. B. C. Berndt, R. A. Rankin: *Ramanujan. Letters and commentary.* History of Math. **9**, Amer. Math. Soc. (1995)
15. J. Bernstein, Rev. Mod. Phys. **46** (1974), 7
16. R. A. Bertlmann: *Anomalies in Quantum Field Theory.* Oxford Univ. Press, 1996
17. A. Besse: *Einstein Manifolds.* Springer, 1987
18. N. D. Birrell, P. C.W. Davies: *Quantum Fields in Curved Space-Time.* Cambridge Univ. Press, 1982
19. N. N. Bogoliubov, D. V. Shirkov: *Introduction to the Theory of Quantized Fields.* Wiley, 1980
20. J.-P. Bourguignon, P. Gauduchon: *Spineurs, opérateurs de Dirac et variations de métrique.* Comm. Math. Phys. **144** (1992), 581–599
21. R. Brandenberger, C. Vafa: *Superstrings in the early universe.* Nucl. Phys. B **316** (1989), 391
22. O. Bratteli, D. W. Robinson: *Operator Algebras and Quantum Statistical Mechanics.* Springer, 1979
23. C. Brouder: *Runge-Kutta methods and renormalization.* hep-th/ 9904014
24. J. W. Calkin: *Two-sided ideals and congruences in the ring of bounded operators in Hilbert space.* Ann. Math. **42** (1941), 839–873

25. L. Carminati, B. Iochum, D. Kastler, T. Schücker: *On Connes' new principle of general relativity: can spinors hear the forces of spacetime?*. In: *Operator algebras and quantum field theory*, eds.: S. Doplicher et al., Proc. Roma 1996. International Press (1997), 359–373
26. L. Carminati, B. Iochum, T. Schücker: *The noncommutative constraints on the standard model à la Connes*. J. Math. Phys. **38** (1997), 1269; hep-th/9604169
27. L. Carminati, B. Iochum, T. Schücker: *Noncommutative Yang-Mills and non-commutative relativity: A bridge over troubled water*. Europ. Phys. J. C (1999), 697
28. C. Caso et al.: *Review of Particle Physics*. Europ. Phys. J. C3 (1998) 1 and off-year partial update available on http://pdg.lbl.gov/
29. M. Cederwall, G. Ferretti, B. E.W. Nilsson, A. Westerberg: *Schwinger terms and cohomology of pseudodifferential operators*. Comm. Math. Phys. **175** (1996), 203; hep-th/9410016
30. A. H. Chamseddine, A. Connes: *Universal formula for noncommutative geometry actions: Unification of gravity and the standard model*. Phys. Rev. Lett. **77** (1996), 4868
31. A. H. Chamseddine, A. Connes: *The spectral action principle*. Com. Math. Phys. **186** (1997), 731–750
32. T. K. Chen, Bull. Amer. Math. Soc. **83** (1977), 831–879
33. P. Chernoff, J. Marsden: *Infinite Dimensional Hamiltonian Systems*. Springer Lect. Notes Math. **452** (1974)
34. Y. Choquet-Bruhat, C. De Witt-Morette, M. Dillard-Bleick: *Analysis, Manifolds and Physics*. North Holland, 1982
35. F. Cipriani, D. Guido, S. Scarlatti: *A remark on trace properties of K-cycles*. J. Operator Theory, **35** (1996), 179–189
36. C. Ciuhu, I.V. Vancea: *Constraints on spacetime manifold in euclidean supergravity in terms of Dirac eigenvalues*. J. Mod. Phys. A15 (2000), 2093–2104
37. A. Connes: *The action functional in non-commutative geometry*. Comm. Math. Phys. **117** (1988), 673–683
38. A. Connes: *Noncommutative Geometry*. Academic Press, 1994
39. A. Connes: *Noncommutative geometry and reality*. J. Math. Phys. **36** (1995), 6194–6231
40. A. Connes: *Gravity coupled with matter and the foundation of noncommutative geometry*. Comm. Math. Phys. **182** (1996), 155–176
41. A. Connes, M. Dubois-Violette: *Noncommutative finite-dimensional manifolds. I. Spherical manifolds and related examples*. math.QA/0107070
42. A. Connes, D. Kreimer: *Hopf algebras, renormalization and noncommutative geometry*. Comm. Math. Phys. **199** (1998), 203–242; hep-th/9808042
43. A. Connes, D. Kreimer: *Lessons from quantum field theory. Hopf algebras and spacetime geometries*. Lett. Math. Phys. 48 (1999), 85–96
44. A. Connes, G. Landi: *Noncommutative manifolds, the instanton algebra and isospectral deformations*. Comm. Math. Phys. **221** (2001) 141–159
45. A. Connes, J. Lott, Nucl. Phys. B **18** (1990), 28
46. A. Connes, J. Lott: *The metric aspect of noncommutative geometry*. Proc. Cargèse 1991, eds.: J. Fröhlich et al., Plenum Press (1992)
47. A. Connes, H. Moscovici: *The local index formula in noncommutative geometry*. Geom. Funct. Anal. **5** (1995), 174–243
48. A. Connes, H. Moscovici: *Hopf algebras, cyclic cohomology and the transverse index theorem*. Comm. Math. Phys. **198** (1998), 198–246; math.DG 9806109

49. F. Constantinescu, G. Scharf: *Smeared and unsmeared chiral vertex operators.* hep-th/9712174
50. J. H. Conway, N. J.A. Sloane: *Four dimensional lattices with the same theta series.* Intern. Math. Res. Notes **4** (1992), 93
51. R. Coquereaux, G. Esposito-Farèse, F. Scheck: *Noncommutative geometry and graded algebras in electroweak interactions.* Int. J. Mod. Phys. A **7** (1992), 6555
52. R. Coquereaux, G. Esposito-Farèse, G. Vaillant: *Higgs fields as Yang-Mills fields and discrete symmetries.* Nucl. Phys. B **353** (1991), 689
53. R. Coquereaux, R. Häußling, N. A. Papadopoulos, F. Scheck, Int. J. Mod. Phys. A **7** (1992), 2809
54. R. Coquereaux, R. Häußling, F. Scheck, Int. J. Mod. Phys. A **10** (1995), 89
55. J. F. Cornwell: *Group Theory in Physics I,II.* Academic Press, 1984
56. M. Cranic: *Cyclic cohomology of Hopf algebras, and non-commutative Chern-Weil theory.* math.QA/9812113
57. B. Crell: *Zum Beweis des Connes'schen Spurtheorems.* Unpublished notes, 1995
58. Č Crnković, E. Witten: *Covariant description of canonical formalism in geometrical theories.* In: *Newton's tercentenary volume*, edited by S. W. Hawking and W. Israel, Cambridge Univ. Press, 1987
59. R. De Pietri, C. Rovelli: *Eigenvalues of the Weyl operator as observables of general relativity.* Class. Quant. Grav. **12** (1995), 1279–1285
60. S. Deser, P. vanNieuwenhuizen: *Nonrenormalizability of the quantized Dirac-Einstein system.* Phys. Rev. D **10** (1974), 411–419
61. B. S. DeWitt: *Dynamical Theory of Groups and Fields.* Gordon-Breach, 1965
62. J. Dieudonné: *Treatise on Analysis.* Vol. III, Academic Press, 1972
63. J. Dieudonné: *Treatise on Analysis.* Vol. VII, Academic Press, 1988
64. J. Dixmier: *Les C^*-algèbres et leurs Représentations.* Gauthier-Villars, Paris, 1969
65. J. Dixmier: *Existence de traces non normales.* C. R. Acad. Sci. Paris **262** (1966), 1107–1108
66. P. H. Dondi, P. D. Jarvis, Phys. Lett. B **84** (1979) 75
67. S. Doplicher: *Quantum spacetime.* Ann. IHP Phys. Theor. **64** (1996), 543–553
68. S. Doplicher, K. Fredenhagen, work in preparation
69. S. Doplicher, K. Fredenhagen, J. E. Roberts: *The quantum structure of space-time at the Planck scale and quantum fields.* Comm. Math. Phys. **172** (1995), 187–220
70. M. Douglas, C. Hull: *D-branes and the noncommutative torus.* JHEP **02** (1998) 008
71. M. Dubois-Violette, R. Kerner, J. Madore: *Gauge bosons in a noncommutative geometry.* Phys. Lett. B **217** (1989), 485
72. M. J. Dupré, R. M. Gilette: *Banach Bundles, Banach Modules and Automorphisms of C^*-Algebras.* Pitman, 1983
73. A.L. Durán and R. Estrada. *Strong moment problems for rapidly decreasing smooth functions.* Proc. Amer. Math. Soc. *120*, 529–534, 1994.
74. Y. Egorov: *Linear differential equations of principal type.* Izd. Nauka, Moscow, 1984 (Russian)
75. Y. Egorov: *Lectures on partial differential equations. Supplementary chapters.* Izd. Mosk. Univ., Moscow, 1985 (Russian)

76. Y. Egorov, M. Shubin: *Linear partial differential equations. Elements of the modern theory*. Encycl. Math. Sciences **31** (Partial Differential Equations II), Springer, 1985
77. R. Estrada: *The Cesàro behaviour of distributions*. Proc. Royal Soc. London **454** (1998), 2425–2443
78. R. Estrada, J. M. Gracia-Bondia, J. C. Várilly: *On summability of distributions and spectral geometry*. Comm. Math. Phys. **191** (1998), 219–248
79. D. B. Fairlie, J. Phys. G**5** (1979) L55
80. H. D. Fegan, P. Gilkey: *Invariants of the heat equation*. Pac. J. Math.
81. H. Figueroa, J. M. Gracia-Bondía, J. C. Várilly: *Elements of Noncommutative Geometry*. Birkhäuser, 1999
82. C. Ford, D. R.T. Jones, P. W. Stephenson, M. B. Einhorn: *The effective potential and the renormalisation group*. Nucl. Phys. B **395** (1993), 17
83. M. Frank, D. R. Larson: *Frames in Hilbert C^*-modules and C^*-algebras*. preprint, Univ. of Houston, 1998
84. M. Frank, D. R. Larson: *A module frame concept for Hilbert C^*-modules*. In: Functional and Harmonic Analysis of Wavelets (San Antonio 1999), ed.: D. R. Larson, L. W. Baggett, *Contemp. Math.* **247** (1999), 207–233.
85. I. B. Frenkel, J. Lepowsky, A. Meurman: *Vertex Operator Algebras and the Monster*. Academic Press, 1988
86. T. Friedrich: *Dirac-Operatoren in der Riemannschen Geometrie – mit einem Ausblick auf die Seiberg-Witten-Theorie*. Vieweg, 1997
87. J. Fröhlich, K. Gawędzki: *Conformal field theory and geometry of strings*. CRM Proc. Lect. Notes **7** (1994), 57; hep-th/9310187
88. J. Fröhlich, O. Grandjean, A. Recknagel: *Supersymmetric quantum theory, noncommutative geometry and gravitation*. Quantum Symmetries, Proc. LXIV Les Houches Session, eds. A. Connes and K. Gawędzki, to appear; hep-th/9706132
89. S. A. Fulling: *Aspects of Quantum Field Theory in Curved Space-Time*. Cambridge Univ. Press, 1989
90. R. W. Gebert: *Introduction to vertex algebras, Borcherds algebras and the monster Lie algebra*. Int. J. Mod. Phys. A **8** (1993), 5441
91. P. Gilkey: *Invariance Theory, the Heat Equation, and the Atiyah-Singer Index Theorem*. CRC Press, 1995; http://www.emis.de/monographs/gilkey/index.html
92. A. Giveon, M. Porrati, E. Rabinovici: *Target space duality in string theory*. Phys. Rep. **244** (1994), 77; hep-th/9401139
93. M. Göckeler, T. Schücker: *Noncommutative geometry and reality*. Phys. Lett. B **434** (1998), 80
94. J. Goldstone, Nuovo Cim. **19** (1961), 154
95. J. Gracia-Bondia, J. Varilly: *Connes' noncommutative differential geometry and the Standard model*. J. Geom. Phys. **12** (1993), 223–301
96. J. M. Gracia-Bondiá, J. C. Várilly, H. Figueroa: *Elements of Noncommutative Geometry*. Birkhäuser, 2001
97. M. B. Green, J. H. Schwarz, E. Witten: *Superstring Theory*. Cambridge Univ. Press, 1987
98. G. Grosche, V. Ziegler, D. Ziegler, E. Zeidler, editors: *Teubner-Taschenbuch der Mathematik, Teil II*. Teubner 1995.
99. V. Guillemin: *A new proof of Weyl's formula on the asymptotic distribution of eigenvalues*. Adv. Math. **55** (1985), 131–160

100. R. Häußling, N. A. Papadopoulos, F. Scheck, Phys. Lett B **260** (1991), 125
101. R. Häußling, N. A. Papadopoulos, F. Scheck, Phys. Lett. B **303** (1993), 265
102. R. Häußling, Ph.D. thesis, Mainz 1994
103. R. Häußling, F. Scheck, Phys. Lett. B **336** (1994), 477
104. R. Häußling, F. Scheck, Phys. Rev. D **57** (1998), 6656
105. R. Häußling, M. Paschke, F. Scheck, Phys. Lett. B **417** (1998), 312
106. G. Hardy: *Divergent Series*. Oxford Univ. Press, 1949
107. G. Hardy, J. Littlewood: *The Riemann zeta function and the theory of the distribution of primes.* Acta Math. **41** (1918), 119–196
108. L. Hörmander: *The Analysis of Linear Partial Differential Operators III.* Springer, 1985
109. S. Ikehara: *An extension of Landau's theorem in the analytic theory of numbers* **10** (1931), 1–12
110. B. Iochum, D. Kastler, T. Schücker: *On the universal Chamseddine-Connes action. I. Details of the action computation.* J. Math. Phys. **38** (1997), 4929; hep-th/9607015
111. B. Iochum, T. Schücker: *Yang-Mills-Higgs versus Connes-Lott.* Comm. Math. Phys. **178** (1996), 1
112. C. J. Isham: *Topological and global aspects of quantum theory.* Relativity, Groups and Topology: Les Houches, edited by B.S. DeWitt and R. Stora, North-Holland 1984
113. C. J. Isham: *Structural issues in quantum gravity.* gr-qc/9510063, Proceedings of GR14 meeting
114. C. Itzykson, J.-B. Zuber: *Quantum Field Theory.* McGraw-Hill, 1985
115. R. Jackiw: *Topological investigations of quantized gauge theories.* Relativity, Groups and Topology II, Les Houches 1983, B. S. DeWitt and R. Stora (eds.), North Holland, 1984
116. V. Kac: *Vertex Algebras for Beginners.* AMS Univ. Lect. Series **10** (1997)
117. R. Kadison, J. Ringrose: *Fundamentals of the Theory of Operator Algebras* I, Academic Press, 1983
118. R. Kadison, J. Ringrose: *Fundamentals of the Theory of Operator Algebras* II, Academic Press, 1986
119. W. Kalau, M. Walze: *Gravity, non-commutative geometry and the Wodzicki residue.* J. Geom. Phys. **16** (1995), 327–344
120. W. Kalau, N. A. Papadopoulos, J. Plass, J.-M. Warzecha, J. Geom. Phys. **16** (1995), 149
121. H. Karcher: *Riemannian comparison constructions*, Global Differential Geometry (S. S. Chern, ed.), Math. Assoc. America, 1989, 170–222
122. M. Karoubi: *K-Theory.* Springer, 1978
123. G. G. Kasparov: *The operator K-functor and extensions of C^*-algebras.* Math. USSR Izv. **16** (1981), 513–572
124. C. Kassel: *Le résidu non commutatif (d'après M. Wodzicki).* Sem. Bourbaki **708**, 1988-89
125. D. Kastler: *The Dirac operator and gravitation.* Comm. Math. Phys. **166** (1995), 633–643
126. K. Kikkawa, M. Yamasaki: *Casimir effects in superstring theories.* Phys. Lett. B **149** (1997)
127. T. Krajewski: *Classification of finite spectral triples.* hep-th/9701081
128. T. Krajewski, R. Wulkenhaar: *On Kreimer's Hopf algebra structure of Feynman graphs.* Europ. Phys. J. C **7** (1999), 697–708; hep-th/9805098

129. D. Kreimer: *On the Hopf algebra structure of perturbative quantum field theories.* Adv. Theor. Math. Phys. **2** (1998), 303–334; q-alg/9707029
130. D. Kreimer: *On overlapping divergences.* Comm. Math. Phys.; hep-th/9810022
131. D. Kreimer: *Chen's iterated integral represents the operator product expansion.* hep-th/9901099
132. D. Kreimer: *Weight systems from Feynman diagrams.* J. Knot Theor. Ramifications **7** (1998), 61–85; hep-th/9612011
133. D. Kreimer, R. Delbourgo: *Using the Hopf algebra structure of QFT in calculations.* hep-th/9903249
134. Ky-Fan: *Maximum properties and inequalities for the eigenvalues of completely continuous operators.* Proc. Amer. Math. Soc. **37** (1951), 760–766
135. J. L. Lagrange, Mem. Cl. Sci. Math. Phys. Inst. France (1808)
136. G. Landi: *An Introduction to Noncommutative Spaces and their Geometries.* Springer Lect. Notes Phys. **51** (1997)
137. G. Landi, F. Lizzi, R. J. Szabo: *String geometry and the noncommutative torus.* Comm. Math. Phys.; hep-th/9806099
138. G. Landi, C. Rovelli: *General relativity in terms of Dirac eigenvalues.* Phys. Rev. Lett. **78** (1997), 3051
139. G. Landi, C. Rovelli: *Gravity from Dirac eigenvalues.* Mod. Phys. Lett. A **13** (1998), 479
140. E. Langmann: *Noncommutative integration calculus.* J. Math. Phys. **36** (1995), 3822; hep-th/9501092
141. E. Langmann: *Quantum gauge theories and noncommutative geometry.* Proc. Zakopane 1995; Acta Phys. Pol. B **27** (1996), 2477; hep-th/9608003
142. E. Langmann: *Descent equations for Yang-Mills anomalies in noncommutative geometry.* J. Geom. Phys. **22** (1997), 259; hep-th/9508003
143. E. Langmann: *Generalized Yang-Mills actions from Dirac operator determinants.* J. Math. Phys. (to appear); math-ph/0104011
144. E. Langmann, J. Mickelsson: *Scattering matrix in external field problems.* J. Math. Phys. **37** (1996), 3933; hep-th/9509034
145. E. Langmann, J. Mickelsson: *Elementary derivation of Chern-Simons terms from effective fermion actions.* "GROUP21 Physical Applications and Mathematical Aspects of Geometry, Groups, and Algebras", Vol. 2, Editors: H.-D. Doebner, W. Scherer, C. Schulte, World Scientific, (1997) 617
146. H. B. Lawson, M.-L. Michelson: *Spin Geometry*, Princeton Univ. Press, 1989
147. A. Lichnérowicz: *Spineurs harmoniques*, C. R. Acad. Sci. Paris. Sér. A **257** (1963), 7–9
148. F. Lizzi: *Strings, noncommutative geometry and the size of the target space*, hep-th/9902187
149. F. Lizzi, R. J. Szabo: *Target space duality in noncommutative geometry.* Phys. Rev. Lett. **79** (1997), 3581; hep-th/9706107
150. F. Lizzi, R. J. Szabo: *Duality symmetries and noncommutative geometry of string spacetimes.* Comm. Math. Phys. **197** (1998), 667; hep-th/9707202
151. F. Lizzi, R. J. Szabo: *Noncommutative geometry and spacetime gauge symmetries of string theory.* Chaos, Solitons and Fractals **10** (1999) 445
152. F. Lizzi, R. J. Szabo: *Noncommutative geometry and string duality.* Proc. Corfu 1998; PRHEP-corfu98/073
153. G. Luke, A. S. Mishchenko: *Vector Bundles and Their Applications.* Kluwer, 1998

154. V. Macaev: *A class of completely continuous operators.* Sov. Math. Dokl. **2** (1961), 972–975
155. J. Madore: *Fuzzy physics.* Ann. Phys. **219** (1992), 187–198
156. J. Madore: *An Introduction to Noncommutative Differential Geometry and its Physical Applications.* LMS Lect. Notes **206**, Cambridge Univ. Press, 1995
157. S. Majid: *Quasitriangular Hopf algebras and Yang-Baxter equations.* Int. J. Mod. Phys. A **5** (1990), 1–91
158. M. Marcu, J. Math. Phys. **21** (1980), 1277 and 1284
159. C. P. Martin, J. M. Gracia-Bondia, J. C. Várilly: *The standard model as a noncommutative geometry: the low energy regime.* Phys. Rep. **294** (1998), 363–406; hep-th/9605001
160. R. Matthes, G. Rudolph, R. Wulkenhaar: *On the structure of a differential algebra used by Connes and Lott.* Rep. Math. Phys. **38** (1996), 45–66
161. V. Moretti: *Direct ζ-function approach and renormalization of one-loop stress tensors in curved spacetimes.* Phys. Rev. D **56** (1997), 7797–7819
162. G. J. Murphy: C^*-*Algebras and Operator Theory.* Academic Press, 1991
163. V. P. Nair, A. Shapere, A. Strominger, F. Wilczek: *Compactification of the twisted heterotic string.* Nucl. Phys. B **289** (1987), 414
164. K. S. Narain: *New heterotic string theories in uncompactified dimensions < 10.* Phys. Lett. B **169** (1986), 41
165. K. S. Narain, M. H. Sarmadi, E. Witten: *A note on toroidal compactification of heterotic string theory.* Nucl. Phys. B **279** (1987), 369
166. Y. Ne'eman, Phys. Lett. B **81** (1979), 190
167. Y. Ne'eman, J. Thierry-Mieg, Phys. Lett. B **108** (1982), 399
168. M. A. Neumark: *Normierte Algebren.* VEB Deutscher Verlag der Wissenschaften, Berlin, 1959.
169. L. O'Raifeartaigh: *Group Structure of Gauge Theories.* Cambridge Univ. Press, 1986
170. L. O'Raifeartaigh, Rep. Progr. Phys. **42** (1979), 159
171. L. O'Raifeartaigh, N. Straumann: *Early history of gauge theories and Kaluza-Klein theories.* hep-ph/9810524
172. K. Osterwalder, R. Schrader: *Axioms for euclidean Green's functions.* Comm. Math. Phys. **31** (1973), 83 and **41** (1975), 281
173. N. A. Papadopoulos, J. Plass, F. Scheck, Phys. Lett. B **324** (1994), 380
174. M. Paschke: *Mass relations in noncommutative geometry revisited.* hep-th/9703043
175. N. Pauna, I. V. Vancea: *On local constraints of $D=4$ supergravity in terms of Dirac eigenvalues.* Mod. Phys. Lett. A **13** (1998), 3091–3098
176. J. Polchinski: *String Theory* I, Cambridge Univ. Press, 1998
177. C. Portenier: *Calcul symbolique.* Seminar Notes, 1999, Univ. Marburg
178. R. J. Plymen: *Strong Morita equivalence, spinors and symplectic spinors.* J. Operator Theory **16** (1986), 305–324
179. *Reports of the Working Group on Precision Calculations for the Z-resonance,* CERN Yellow Report, 95–03, eds. D. Bardin, W. Hollik, G. Passarino
180. M. A. Rieffel: *Morita equivalence for operator algebras.* Proc. Symp. Pure Math. **38** (1982), 285–298
181. M. A. Rieffel: C^*-*algebras associated with irrational rotations.* Pac. J. Math. **93** (1981), 415
182. M. A. Rieffel: *Induced representations of C^*-algebras.* Bull. Amer. Math. Soc. **78** (1972) 606–609; Adv. Math. **13** (1974) 176–257

183. M. A. Rieffel, A. Schwarz: *Morita equivalence of multidimensional noncommutative tori.* math.QA/9803057, to appear in: Int. J. Math.
184. C. Rovelli: *Quantum reference systems.* Class. Quant. Grav. 8 (1991) 317–332
185. C. Rovelli: *Spectral noncommutative geometry and quantization: a simple example.* Phys. Rev. Lett. 83 (1999), 1079–1083
186. C. Rovelli: *What is observable in classical and quantum gravity.* Class. Quant. Grav. **8** (1991), 297–316
187. S. Sakai: C^*-*Algebras and* W^*-*Algebras.* Springer, 1971
188. N. Sakai, I. Senda: *Vacuum energies of string compactified on torus.* Progr. Theor. Phys. **75** (1986), 692
189. R. Schatten: *Norm Ideals of Completely Continuous Operators.* Springer, 1970
190. F. Scheck, Phys. Lett. B **284** (1992), 303
191. F. Scheck: *Electroweak and Strong Interactions - An Introduction to Theoretical Particle Physics.* Springer, 1996
192. M. Scheunert, Springer Lect. Notes Math. **716** (1979)
193. K. Schmüdgen: *Die Dixmier-Spur.* Lecture held at 'Oberseminar Mathematische Physik', Leipzig, 1994. (notes by R. Matthes).
194. B. Schrempp, M. Wimmer: *Top quark and Higgs boson masses: Interplay between infrared and ultraviolet physics.* Progr. Particle Nucl. Phys. **37** (1996)
195. E. Schrödinger: *Diracsches Elektron im Schwerefeld I.*, Sitzungsber. Preuss. Akad. Wiss., Phys.-Math. Kl. (1932), 105–128
196. T. Schücker: *Geometries and forces.* Proc. Summer School on Noncommutative Geometry and Applications, Portugal, 1997, ed.: Paulo Almeida
197. T. Schücker, J.-M. Zylinski: *Connes's model building kit.* J. Geom. Phys. **16** (1994), 207
198. L. Schwartz: *Transformations de Laplace. Les cours de Sorbonne. Méthodes mathématiques de la physique.* VI. Paris, 1955
199. A. Schwarz: *Morita equivalence and duality.* Nucl. Phys. B **534** (1998) 720; hep-th/9805034
200. R. Seeley: *Complex powers of an elliptic operator.* Proc. Symp. Pure Math. **10** (1966), 288–307
201. J.-P. Serre: *Modules projectifs et espaces fibrés à fibre vectorielle.* Sém. Dubreil-Pisot **11** (1957/58), 531–543
202. M. A. Shubin: *Pseudodifferential Operators and Spectral Theory.* Izd. Nauka, Moscow, 1978 (Russian) and Springer, 2001 (English)
203. H. S. Snyder: *Quantized space-time.* Phys. Rev. **71** (1948), 38
204. D. Song, R. J. Szabo: *Spectral geometry of heterotic compactification.* hep-th/9812235
205. S. Sternberg: *Group Theory and Physics.* Cambridge Univ. Press, 1994
206. A. Sudbery: *Quantum Mechanics and the Particles of Nature.* Cambridge Univ. Press, 1986
207. R. G. Swan: *Vector bundles and projective modules.* Trans. Amer. Math. Soc. **105** (1962), 264–277
208. M. Takesaki: *On the conjugate space of an operator algebra.* Tôhoku Math. J. **10** (1958), 194–203
209. M. Takesaki: *Theory of Operator Algebras.* Springer, 1979
210. M. Taylor: *Pseudodifferential Operators.* Princeton Univ. Press, 1981
211. J. G. Taylor, Phys. Lett. B **84** (1979), 79
212. F. Treves: *Introduction to Pseudodifferential and Fourier Integral Operators.* Plenum Press, 1980

213. I. V. Vancea: *Observables of the euclidean supergravity.* Phys. Rev. Lett. **79** (1997), 3121
214. I. V. Vancea: *Euclidean supergravity in terms of Dirac eigenvalues.* Phys. Rev. D **58** (1998), 045005
215. J. C. Várilly, J. M. Gracia-Bondía: *Connes' noncommutative differential geometry and the Standard Model.* J. Geom. Phys. **12** (1993), 223–301
216. J. C. Várilly: *An introduction to noncommutative geometry.* EMS Summer School on Noncommutative Geometry and Applications, 1997, Portugal; (physics/9709045)
217. A. Walfisz: *Gitterpunkte in mehrdimensionalen Kugeln.* PWN, Warszawa, 1957
218. J.-M. Warzecha: *Von Fredholmmoduln zu Spektralen Tripeln in Nichtkommutativer Geometrie,* Ph. D. thesis, Universität Mainz 1997, Shaker Verlag, Aachen, 1998
219. N. E. Wegge-Olsen: *K-Theory and C^*-Algebras: a friendly approach.* Oxford Univ. Press, 1993
220. H. Weyl: *Das asymptotische Verteilungsgesetz der Eigenwerte linearer partieller Differentialgleichungen.* Math. Ann. **71** (1912), 441–479
221. D. V. Widder: *The Laplace Transform.* Princeton Univ. Press, 1972
222. N. Wiener: *The Fourier Integral and certain of its Applications.* Dover Publications, 1960
223. M. Wodzicki: *Local invariants of spectral asymmetry,* Invent. Math. **75** (1984), 143–178
224. M. Wodzicki: *Noncommutative residue.* In: *K-theory, arithmetic and geometry,* Yu. I. Manin (ed.), Springer Lect. Notes Math. **1289** (1987), 320–399
225. R. Wulkenhaar: *On the Connes-Moscovici Hopf algebra associated to the diffeomorphisms of a manifold.* math-ph/9904009
226. E. Zeidler: *Pseudodifferentialoperatoren,* Lectures held at 'Oberseminar Mathematische Physik', Leipzig, 1994 (notes by R. Matthes)
227. W. Zimmermann: *Convergence of Bogoliubov's method of renormalization in momentum space.* Comm. Math. Phys. **15** (1969), 208–234
228. G. J. Zuckerman: *Action principles and global geometry.* In: *Mathematical Aspects of String Theory,* edited by S. T. Yau, World Scientific 1997

Lecture Notes in Physics

For information about Vols. 1–556
please contact your bookseller or Springer-Verlag

Vol. 557: J. A. Freund, T. Pöschel (Eds.), Stochastic Processes in Physics, Chemistry, and Biology. X, 330 pages. 2000.

Vol. 558: P. Breitenlohner, D. Maison (Eds.), Quantum Field Theory. Proceedings, 1998. VIII, 323 pages. 2000

Vol. 559: H.-P. Breuer, F. Petruccione (Eds.), Relativistic Quantum Measurement and Decoherence. Proceedings, 1999. X, 140 pages. 2000.

Vol. 560: S. Abe, Y. Okamoto (Eds.), Nonextensive Statistical Mechanics and Its Applications. IX, 272 pages. 2001.

Vol. 561: H. J. Carmichael, R. J. Glauber, M. O. Scully (Eds.), Directions in Quantum Optics. XVII, 369 pages. 2001.

Vol. 562: C. Lämmerzahl, C. W. F. Everitt, F. W. Hehl (Eds.), Gyros, Clocks, Interferometers...: Testing Relativistic Gravity in Space. XVII, 507 pages. 2001.

Vol. 563: F. C. Lázaro, M. J. Arévalo (Eds.), Binary Stars. Selected Topics on Observations and Physical Processes. 1999.IX, 327 pages. 2001.

Vol. 564: T. Pöschel, S. Luding (Eds.), Granular Gases. VIII, 457 pages. 2001.

Vol. 565: E. Beaurepaire, F. Scheurer, G. Krill, J.-P. Kappler (Eds.), Magnetism and Synchrotron Radiation. XIV, 388 pages. 2001.

Vol. 566: J. L. Lumley (Ed.), Fluid Mechanics and the Environment: Dynamical Approaches. VIII, 412 pages. 2001.

Vol. 567: D. Reguera, L. L. Bonilla, J. M. Rubí (Eds.), Coherent Structures in Complex Systems. IX, 465 pages. 2001.

Vol. 568: P. A. Vermeer, S. Diebels, W. Ehlers, H. J. Herrmann, S. Luding, E. Ramm (Eds.), Continuous and Discontinuous Modelling of Cohesive-Frictional Materials. XIV, 307 pages. 2001.

Vol. 569: M. Ziese, M. J. Thornton (Eds.), Spin Electronics. XVII, 493 pages. 2001.

Vol. 570: S. G. Karshenboim, F. S. Pavone, F. Bassani, M. Inguscio, T. W. Hänsch (Eds.), The Hydrogen Atom: Precision Physics of Simple Atomic Systems. XXIII, 293 pages. 2001.

Vol. 571: C. F. Barenghi, R. J. Donnelly, W. F. Vinen (Eds.), Quantized Vortex Dynamics and Superfluid Turbulence. XXII, 455 pages. 2001.

Vol. 572: H. Latal, W. Schweiger (Eds.), Methods of Quantization. XI, 224 pages. 2001.

Vol. 573: H. M. J. Boffin, D. Steeghs, J. Cuypers (Eds.), Astrotomography. XX, 434 pages. 2001.

Vol. 574: J. Bricmont, D. Dürr, M. C. Galavotti, G. Ghirardi, F. Petruccione, N. Zanghi (Eds.), Chance in Physics. XI, 288 pages. 2001.

Vol. 575: M. Orszag, J. C. Retamal (Eds.), Modern Challenges in Quantum Optics. XXIII, 405 pages. 2001.

Vol. 576: M. Lemoine, G. Sigl (Eds.), Physics and Astrophysics of Ultra-High-Energy Cosmic Rays. X, 327 pages. 2001.

Vol. 577: I. P. Williams, N. Thomas (Eds.), Solar and Extra-Solar Planetary Systems. XVIII, 255 pages. 2001.

Vol. 578: D. Blaschke, N. K. Glendenning, A. Sedrakian (Eds.), Physics of Neutron Star Interiors. XI, 509 pages. 2001.

Vol. 579: R. Haug, H. Schoeller (Eds.), Interacting Electrons in Nanostructures. X, 227 pages. 2001.

Vol. 580: K. Baberschke, M. Donath, W. Nolting (Eds.), Band-Ferromagnetism: Ground-State and Finite-Temperature Phenomena. IX, 394 pages. 2001.

Vol.581: J. M. Arias, M. Lozano (Eds.), An Advanced Course in Modern Nuclear Physics. XI, 346 pages. 2001.

Vol.582: N. J. Balmforth, A. Provenzale (Eds.), Geomorphological Fluid Mechanics. X, 579 pages. 2001.

Vol.583: W. Plessas, L. Mathelitsch (Eds.), Lectures on Quark Matter, XIII, 334 pages. 2002.

Vol.584: W. Köhler, S. Wiegand (Eds.), Thermal Nonequilibrium Phenomena in Fluid Mixtures. XVII, 470 pages. 2002.

Vol.585: M. Lässig, A. Valleriani (Eds.), Biological Evolution and Statistical Physics. XI, 337 pages. 2002.

Vol.586: Y. Auregan, A. Maurel, V. Pagneux, J.-F. Pinton (Eds.), Sound-Flow Interactions. XVI, 286 pages. 2002

Vol.587: D. Heiss (Ed.), Fundamentals of Quantum Information. Quantum Computation, Communication, Decoherence and All That. XIII, 265 pages. 2002.

Vol.588: Y. Watanabe, S. Heun, G. Salviati, N. Yamamoto (Eds.), Nanoscale Spectroscopy and Its Applications to Semiconductor Research. XV, 306 pages. 2002.

Vol.589: A. W. Guthmann, M. Georganopoulos, A. Marcowith, K. Manolakou (Eds.), Relativistic Flows in Astrophysics. XII, 241 pages. 2002

Vol.590: D. Benest, C. Froeschlé (Eds.), Singularities in Gravitational Systems. Applications to Chaotic Transport in the Solar System. XI, 215 pages. 2002

Vol.591: M. Beyer (Ed.), CP Violation in Particle, Nuclear and Astrophysics. XI, 333 pages. 2002

Vol.592: S. Cotsakis, L. Papantonopoulos (Eds.), Cosmological Crossroads. An Advanced Course in Mathematical, Physical and String Cosmology. XVI, 477 pages. 2002

Vol.593: D. Shi, B. Aktaş, L. Pust, F. Mikhailov (Eds.), Nanostructured Magnetic Materials and Their Applications. XII, 289 pages. 2002

Vol.594: S. Odenbach (Ed.),Ferrofluids. Magnetical Controllable Fluids and Their Applications. XI, 255 pages. 2002

Vol.595: C. Berthier, L. P. Lévy, G. Martinez (Eds.), High Magnetic Fields. Applications in Condensed Matter Physics and Spectroscopy. X, 493 pages. 2002

Vol.596: F. Scheck, H. Upmeier, W. Werner (Eds.), Noncommutative Geometry and the Standard Model of Elememtary Particle Physics. XII, 346 pages. 2002

Monographs

For information about Vols. 1–29
please contact your bookseller or Springer-Verlag

Vol. m 30: A. J. Greer, W. J. Kossler, Low Magnetic Fields in Anisotropic Superconductors. VII, 161 pages. 1995.

Vol. m 31 (Corr. Second Printing): P. Busch, M. Grabowski, P.J. Lahti, Operational Quantum Physics. XII, 230 pages. 1997.

Vol. m 32: L. de Broglie, Diverses questions de mécanique et de thermodynamique classiques et relativistes. XII, 198 pages. 1995.

Vol. m 33: R. Alkofer, H. Reinhardt, Chiral Quark Dynamics. VIII, 115 pages. 1995.

Vol. m 34: R. Jost, Das Märchen vom Elfenbeinernen Turm. VIII, 286 pages. 1995.

Vol. m 35: E. Elizalde, Ten Physical Applications of Spectral Zeta Functions. XIV, 224 pages. 1995.

Vol. m 36: G. Dunne, Self-Dual Chern-Simons Theories. X, 217 pages. 1995.

Vol. m 37: S. Childress, A.D. Gilbert, Stretch, Twist, Fold: The Fast Dynamo. XI, 406 pages. 1995.

Vol. m 38: J. González, M. A. Martín-Delgado, G. Sierra, A. H. Vozmediano, Quantum Electron Liquids and High-Tc Superconductivity. X, 299 pages. 1995.

Vol. m 39: L. Pittner, Algebraic Foundations of Non-Com-mutative Differential Geometry and Quantum Groups. XII, 469 pages. 1996.

Vol. m 40: H.-J. Borchers, Translation Group and Particle Representations in Quantum Field Theory. VII, 131 pages. 1996.

Vol. m 41: B. K. Chakrabarti, A. Dutta, P. Sen, Quantum Ising Phases and Transitions in Transverse Ising Models. X, 204 pages. 1996.

Vol. m 42: P. Bouwknegt, J. McCarthy, K. Pilch, The W3 Algebra. Modules, Semi-infinite Cohomology and BV Algebras. XI, 204 pages. 1996.

Vol. m 43: M. Schottenloher, A Mathematical Introduction to Conformal Field Theory. VIII, 142 pages. 1997.

Vol. m 44: A. Bach, Indistinguishable Classical Particles. VIII, 157 pages. 1997.

Vol. m 45: M. Ferrari, V. T. Granik, A. Imam, J. C. Nadeau (Eds.), Advances in Doublet Mechanics. XVI, 214 pages. 1997.

Vol. m 46: M. Camenzind, Les noyaux actifs de galaxies. XVIII, 218 pages. 1997.

Vol. m 47: L. M. Zubov, Nonlinear Theory of Dislocations and Disclinations in Elastic Body. VI, 205 pages. 1997.

Vol. m 48: P. Kopietz, Bosonization of Interacting Fermions in Arbitrary Dimensions. XII, 259 pages. 1997.

Vol. m 49: M. Zak, J. B. Zbilut, R. E. Meyers, From Instability to Intelligence. Complexity and Predictability in Nonlinear Dynamics. XIV, 552 pages. 1997.

Vol. m 50: J. Ambjørn, M. Carfora, A. Marzuoli, The Geometry of Dynamical Triangulations. VI, 197 pages. 1997.

Vol. m 51: G. Landi, An Introduction to Noncommutative Spaces and Their Geometries. XI, 200 pages. 1997.

Vol. m 52: M. Hénon, Generating Families in the Restricted Three-Body Problem. XI, 278 pages. 1997.

Vol. m 53: M. Gad-el-Hak, A. Pollard, J.-P. Bonnet (Eds.), Flow Control. Fundamentals and Practices. XII, 527 pages. 1998.

Vol. m 54: Y. Suzuki, K. Varga, Stochastic Variational Approach to Quantum-Mechanical Few-Body Problems. XIV, 324 pages. 1998.

Vol. m 55: F. Busse, S. C. Müller, Evolution of Spontaneous Structures in Dissipative Continuous Systems. X, 559 pages. 1998.

Vol. m 56: R. Haussmann, Self-consistent Quantum Field Theory and Bosonization for Strongly Correlated Electron Systems. VIII, 173 pages. 1999.

Vol. m 57: G. Cicogna, G. Gaeta, Symmetry and Perturbation Theory in Nonlinear Dynamics. XI, 208 pages. 1999.

Vol. m 58: J. Daillant, A. Gibaud (Eds.), X-Ray and Neutron Reflectivity: Principles and Applications. XVIII, 331 pages. 1999.

Vol. m 59: M. Kriele, Spacetime. Foundations of General Relativity and Differential Geometry. XV, 432 pages. 1999.

Vol. m 60: J. T. Londergan, J. P. Carini, D. P. Murdock, Binding and Scattering in Two-Dimensional Systems. Applications to Quantum Wires, Waveguides and Photonic Crystals. X, 222 pages. 1999.

Vol. m 61: V. Perlick, Ray Optics, Fermat's Principle, and Applications to General Relativity. X, 220 pages. 2000.

Vol. m 62: J. Berger, J. Rubinstein, Connectivity and Superconductivity. XI, 246 pages. 2000.

Vol. m 63: R. J. Szabo, Ray Optics, Equivariant Cohomology and Localization of Path Integrals. XII, 315 pages. 2000.

Vol. m 64: I. G. Avramidi, Heat Kernel and Quantum Gravity. X, 143 pages. 2000.

Vol. m 65: M. Hénon, Generating Families in the Restricted Three-Body Problem. Quantitative Study of Bifurcations. XII, 301 pages. 2001.

Vol. m 66: F. Calogero, Classical Many-Body Problems Amenable to Exact Treatments. XIX, 749 pages. 2001.

Vol. m 67: A. S. Holevo, Statistical Structure of Quantum Theory. IX, 159 pages. 2001.

Vol. m 68: N. Polonsky, Supersymmetry: Structure and Phenomena. Extensions of the Standard Model. XV, 169 pages. 2001.

Vol. m 69: W. Staude, Laser-Strophometry. High-Resolution Techniques for Velocity Gradient Measurements in Fluid Flows. XV, 178 pages. 2001.

Vol. m 70: P. T. Chruściel, J. Jezierski, J. Kijowski, Hamiltonian Field Theory in the Radiating Regime. VI, 172 pages. 2002.

Vol. m 71: S. Odenbach, Magnetoviscous Effects in Ferrofluids. X, 151 pages. 2002.

Vol. m 72: J. G. Muga, R. Sala Mayato, I. L. Egusquiza (Eds.), Time in Quantum Mechanics. XII, 419 pages. 2002.

Printed in Poland
by Amazon Fulfillment
Poland Sp. z o.o., Wrocław